Algorithms and Computation in Mathematics

Volume 33

Series Editors

William J. Cook, Faculty of Mathematics, University of Waterloo, Waterloo, Canada

David Eisenbud, Department of Mathematics, University of California, Berkeley, Berkeley, CA, USA

Bernd Sturmfels, Department of Mathematics, University of California, Berkeley, Berkeley, CA, USA

Bianca Viray, Department of Mathematics, University of Washington, Seattle, USA

With this forward-thinking series Springer recognizes that the prevailing trend in mathematical research towards algorithmic and constructive processes is one of long-term importance. This series is intended to further the development of computational and algorithmic mathematics. In particular, Algorithms and Computation in Mathematics emphasizes the computational aspects of algebraic geometry, number theory, combinatorics, commutative, non-commutative and differential algebra, geometric and algebraic topology, group theory, optimization, dynamical systems and Lie theory. Proposals or manuscripts that center on content in non-computational aspects of one of these fields will also be regarded if the presentation gives consideration to the contents' usefulness in algorithmic processes.

Stephen S.-T. Yau • Xiuqiong Chen • Xiaopei Jiao •
Jiayi Kang • Zeju Sun • Yangtianze Tao

Principles of Nonlinear Filtering Theory

Stephen S.-T. Yau
Beijing Institute of Mathematical Sciences
and Applications (BIMSA)
Beijing, China

Department of Mathematical Sciences
Tsinghua University
Beijing, China

Xiaopei Jiao
Beijing Institute of Mathematical Sciences
and Applications (BIMSA)
Beijing, China

Zeju Sun
Department of Mathematical Sciences
Tsinghua University
Beijing, China

Xiuqiong Chen
School of Mathematics
Renmin University of China
Beijing, China

Jiayi Kang
Beijing Institute of Mathematical Sciences
and Applications (BIMSA)
Beijing, China

Yangtianze Tao
Department of Mathematical Sciences
Tsinghua University
Beijing, China

ISSN 1431-1550
Algorithms and Computation in Mathematics
ISBN 978-3-031-77683-0   ISBN 978-3-031-77684-7   (eBook)
https://doi.org/10.1007/978-3-031-77684-7

Mathematics Subject Classification: 35K15, 68T07, 62M20, 93E11, 60G35, 17B30

© The Editor(s) (if applicable) and The Author(s), under exclusive license to Springer Nature Switzerland AG 2024, corrected publication 2025

This work is subject to copyright. All rights are solely and exclusively licensed by the Publisher, whether the whole or part of the material is concerned, specifically the rights of reprinting, reuse of illustrations, recitation, broadcasting, reproduction on microfilms or in any other physical way, and transmission or information storage and retrieval, electronic adaptation, computer software, or by similar or dissimilar methodology now known or hereafter developed.
The use of general descriptive names, registered names, trademarks, service marks, etc. in this publication does not imply, even in the absence of a specific statement, that such names are exempt from the relevant protective laws and regulations and therefore free for general use.
The publisher, the authors and the editors are safe to assume that the advice and information in this book are believed to be true and accurate at the date of publication. Neither the publisher nor the authors or the editors give a warranty, expressed or implied, with respect to the material contained herein or for any errors or omissions that may have been made. The publisher remains neutral with regard to jurisdictional claims in published maps and institutional affiliations.

This Springer imprint is published by the registered company Springer Nature Switzerland AG
The registered company address is: Gewerbestrasse 11, 6330 Cham, Switzerland

If disposing of this product, please recycle the paper.

I would like to dedicate this book to my father Chun Ying Yau and mother Yeuk Lam Yau-Leung. Although our family was quite poor, my parents were determined to give me the opportunity to receive a high-level education. I am also very grateful to my wife Rong He for her continuous support on my research.

–Stephen S.-T. Yau

Preface

Nonlinear filtering theory explores how to estimate the value of a stochastic process based on generative models of the process, noisy observations, and initial probability distributions. Estimation is of importance in any applied area involving the processing of noisy signals. In fact, it is a fundamental tool in modern industry addressing a wide range of challenges from aircraft navigation and guidance, satellite orbit determination, radar tracking, and solar mapping. Historically it goes back to Gauss on astronomical observations. In the first half of the last century, Wiener and Kolmogorov initiated the subject by estimating the trajectories of aircraft from the ground.

Basically, we are given a vector valued stochastic process $x(t)$ generated by the output of system excited by white noise. Suppose we observe a noisy version of a function of the signal corrupted by the addition of a second white noise. The goal of nonlinear filtering theory is to find the probability density of $x(t)$, conditioned on the available past observations, in a real time manner.

In the early 1960s, the Kalman-Bucy filter gave a fundamental solution to the problem of the finite dimensional recursive estimation of a signal observed in noise. This basic result requires the signal to be generated by a finite dimensional linear system with Gaussian initial condition. However, the effort to provide a complete theory and efficient computational methodology for filters in the nonlinear case has been a challenging area of systems and control for more than six decades.

In the late 1970s, Brockett, Clark, and Mitter proposed the classification problem for finite dimensional filters from the Lie algebraic point of view. This classification of the so-called estimation algebras is quite important since the filters are finite dimensional whenever the algebras are.

For filtering system with no finite dimensional filter, it becomes essential to study the evolution equation for the probability density of $x(t)$, conditioned on the available past observations. In 1959, Stratonovich in the USSR and subsequently Kushner in the USA derived a stochastic partial differential equation to describe the evolution of this conditional density function. The general Stratonovich-Kushner equation is a nonlinear equation. In the mid-sixties, Duncan, Mortensen, and Zakai derived a linear stochastic partial differential equation for unnormalized conditional density.

Recent years have witnessed substantial growth in filtering theory, prompting the need for a contemporary resource. While existing classical textbooks like Bain and Crisan's "Fundamentals of Stochastic Filtering" (2009) and Jazwinski's "Stochastic Processes and Filtering Theory" (1970) offer valuable insights, they lack the updated methodologies and applications crucial in today's landscape. This book aims to fill this gap by providing a comprehensive yet accessible treatment of filtering theory, balancing rigorous mathematical foundations with practical methods, including novel algorithms based on deep learning techniques.

This book is structured into three parts. In the initial part, we lay the groundwork by reviewing the basic concepts in probability theory (Chap. 1), stochastic processes (Chap. 2), stochastic differential equations (Chap. 3), and optimization (Chap. 4). While these topics serve as a foundation for the subsequent discussions, experienced readers familiar with this theoretical framework may wish to skip to the following sections.

The second part of the book rigorously explores filtering theory, offering a comprehensive examination of filtering equations (Chap. 5), which are satisfied by the conditional probability density of state conditioned on the available past observations. Additionally, we also introduce the estimation algebra (Chap. 6), which are used in construction and classification of finite-dimensional filters from geometric and algebraic aspects.

In the final part, we delve into the practical realm by introducing a spectrum of numerical filtering algorithms. These include the Yau-Yau algorithm (Chap. 7) and direct methods (Chap. 8) grounded in filtering equations. Furthermore, we delve into classical filtering methods (Chap. 9) such as the extended Kalman filter, unscented Kalman filter, and particle filter. Embracing the forefront of innovation, we also present algorithms using newly developed deep learning techniques (Chap. 10), thus providing readers with a holistic view of contemporary filtering methodologies.

I taught the Nonlinear Filtering Theory based on the material of this book at University of Illinois at Chicago and Tsinghua University in various years. The final version grows out from the course which I taught in the Spring of 2020 at Tsinghua University. It was coordinated by my student Dr. Xiuqiong Chen and written by my students Xiaopei Jiao, Jiayi Kang, Zeju Sun, Yangtianze Tao, and Xiuqiong Chen.

The first principle guiding the preparation of the book was to include those important theories not previously available in the book format. Indeed, the finite dimensional filter theory is the first time appeared in the book format. We also incorporate our recently developed theory, which utilizes neural networks to address filtering problems in this book. The second guiding principle was to ensure that the book is self-contained.

Acknowledgments We gratefully acknowledge the financial support from the National Natural Science Foundation of China (Grants Nos. 42450242, 12201631, 123B2020, 11471184, 11961141005) and the Tsinghua University Education Foundation. We extend our sincere gratitude to the anonymous reviewers for their insightful comments and constructive suggestions, which significantly improved the quality of this work. Special thanks to our colleagues, collaborators, and family members for their support and encouragement throughout this project.

Beijing, China
Spring, 2024

Stephen S.-T. Yau

The original version of the book has been revised. A correction to this book can be found at https://doi.org/10.1007/978-3-031-77684-7_12

Contents

Part I   Preliminary Knowledge

1   Probability Theory 3
   1.1   Probability Space 3
   1.2   Random Variables 6
        1.2.1   Definition and Properties 6
        1.2.2   Jointly Distributed Random Variables 14
   1.3   Numerical Characteristics 19
        1.3.1   Characteristic Function 19
        1.3.2   Examples 21
   1.4   Conditional Probability 22
        1.4.1   Definition and Independence 22
        1.4.2   Conditional Probability Density 24
        1.4.3   Conditional Expectation 26
   1.5   Limit Theorem 30
        1.5.1   Strong Law of Large Numbers 30
        1.5.2   Central Limit Theorem 31
   1.6   Exercises 32
   References 33

2   Stochastic Processes 35
   2.1   Introduction to Stochastic Processes and Filtrations 35
        2.1.1   Discrete and Continuous Stochastic Processes 36
        2.1.2   Sameness Between Two Processes 36
        2.1.3   Filtrations and Measurement of Processes 37
        2.1.4   Stopping Time and Associated σ-Fields 40
   2.2   Discrete Martingales 44
   2.3   Continuous Martingales 49
        2.3.1   Introduction of Continuous Martingales 49
        2.3.2   The Important Inequalities for Martingales 51
        2.3.3   The Doob-Meyer Decomposition 57
   2.4   Continuous Local Martingale 61

	2.5	Square-Integrable Martingale	64
		2.5.1 The Quadratic Variation of a Continuous Martingale	64
		2.5.2 The Quadratic Variation of a Continuous Local Martingale	66
	2.6	Exercises	69
	References	71
3	**Stochastic Differential Equations**	73
	3.1	Stochastic Integral	73
		3.1.1 Construction of Stochastic Integral	74
		3.1.2 Itô's Formula	78
		3.1.3 Girsanov's Theorem and Novikov Condition	81
		3.1.4 Burkholder-Davis-Gundy Inequality	89
		3.1.5 Stratonovich's Integral	93
	3.2	Formulations of Stochastic Differential Equations	95
		3.2.1 Strong Solutions	96
		3.2.2 Weak Solutions	99
		3.2.3 The Martingale Problem of Stroock and Varadhan	101
	3.3	Connections Between Stochastic Differential Equations and Partial Differential Equations	105
		3.3.1 Feynman-Kac Representation	105
		3.3.2 Kolmogorov Equation	107
	3.4	Exercises	110
	References	111
4	**Optimization**	113
	4.1	Background	113
		4.1.1 Basic Concepts	114
		4.1.2 Examples	116
	4.2	Optimal Condition and Duality Theory	118
		4.2.1 Optimal Condition	118
		4.2.2 Constraint Qualification	122
		4.2.3 Duality Theory	123
	4.3	Convex Optimization	124
		4.3.1 Basic Properties	124
		4.3.2 Optimal Condition	127
		4.3.3 Examples	128
		4.3.4 Quadratic Convex Optimization	129
	4.4	Non-convex Optimization	132
		4.4.1 Newton-Lagrange Method	133
		4.4.2 SQP Constrained by Equalities	133
	4.5	Exercises	137
	References	138

Part II Filtering Theory

5 Filtering Equations 141
  5.1 Introduction to Filtering Equations 141
  5.2 The Change of Probability Measure Method 143
  5.3 The DMZ Equation 148
  5.4 The Innovation Process Method 152
  5.5 The Density Function of Conditional Distribution 157
  5.6 Robust DMZ Equation 161
  5.7 Exercises 165
  References 167

6 Estimation Algebra 169
  6.1 Introduction 169
  6.2 Basic Concepts and Preliminaries 172
  6.3 Algebraic Classification of Finite-Dimensional Filter 178
    6.3.1 Maximal Rank Classification: Structures of Quadratic Forms 178
    6.3.2 Maximal Rank Classification: Hessian Matrix Nondecomposition Theorem 191
    6.3.3 Maximal Rank Classification: Complete Classification Theorem 195
  6.4 Wei-Norman Approach 198
  6.5 Classification with Nonmaximal Rank 201
    6.5.1 State Dimension 2 201
    6.5.2 State Dimension 3 222
  6.6 Novel Finite-Dimensional Filter 241
    6.6.1 State Dimension 3 241
    6.6.2 Arbitrary State Dimension 247
  6.7 Exercises 257
  References 258

Part III Numerical Algorithms

7 Yau-Yau Algorithm 263
  7.1 Introduction 263
  7.2 The Formulation of Yau-Yau Algorithm 264
  7.3 L^1-Convergence 266
  7.4 Lower Bound Estimation of Density Function 276
  7.5 Algorithm in Time-Variant Systems 279
  7.6 Numerical Methods for Solving Parabolic Differential Equations . 283
  7.7 Numerical Results 293
  7.8 Exercises 295
  References 296

8 Direct Methods 297
  8.1 Introduction 297

8.2 Explicit Solution of DMZ Equation for Finite-Dimensional Filters 300
8.3 Direct Method for Yau Filtering System with Nonlinear Observations 310
8.4 Nonlinear Filtering and Time-Varying Schrödinger Equation I 314
8.5 Nonlinear Filtering and Time-Varying Schrödinger Equation II 325
8.6 Nonlinear Filtering and Time-Varying Schrödinger Equation III 332
8.7 Exercises 336
References 337

9 Classical Filtering Methods 339
9.1 Introduction 339
9.2 Filtering Algorithm Based on Bayesian Framework 341
 9.2.1 Linear System and KF 343
 9.2.2 Discrete KF 343
 9.2.3 From KF to EKF 347
 9.2.4 UKF 348
 9.2.5 Discrete Particle Methods for Filtering 351
9.3 Filtering Algorithm of DMZ Framework 356
 9.3.1 DMZ Equation After Applying the Hopf-Cole Transformation 356
 9.3.2 Linear Filtering System and KBF 357
 9.3.3 From KBF to Continuous EKF 358
 9.3.4 Control-Oriented Particle Filtering for Multidimensional System 360
9.4 Robust Filtering 363
 9.4.1 Nonlinear Regression Form and Robust Optimization Framework 364
 9.4.2 Iterative Outlier-Robust Extended Kalman Filtering 366
9.5 Numerical Results 368
 9.5.1 Linear Filtering Problem 368
 9.5.2 Nonlinear Filtering Problem 374
9.6 Exercises 379
References 381

10 Estimation Algorithms Based on Deep Learning 385
10.1 Overview 385
10.2 Estimation Problems 385
 10.2.1 State Estimation 386
 10.2.2 Parameter Estimation 386
 10.2.3 Dual Estimation 387
10.3 Feedforward Neural Networks 387
 10.3.1 Mathematical Forms for FNNs 387

		10.3.2	Universal Approximation Theorem for FNNs	389
	10.4	Optimization and Backpropagation		393
		10.4.1	Optimization Algorithms for Neural Networks	393
		10.4.2	Backpropagation	395
	10.5	Recurrent Neural Networks		397
		10.5.1	Mathematical Forms for Networks	397
		10.5.2	Universal Approximation Theorem for RNNs	398
	10.6	The Application of Deep Learning in Nonlinear Filtering		400
		10.6.1	Preliminaries	401
		10.6.2	Universal Approximation Theorem for RNN with Stochastic Inputs	404
		10.6.3	RNN-Based Filtering for Discrete-Time Systems	409
		10.6.4	RNN-Based Yau-Yau Algorithm for Continuous-Time Systems	412
		10.6.5	Numerical Results	422
	10.7	Exercises		423
	References			425
11	**Solutions**			429
	Problems of Chap. 1			429
	Problems of Chap. 2			432
	Problems of Chap. 3			438
	Problems of Chap. 4			440
	Problems of Chap. 5			442
	Problems of Chap. 6			447
	Problems of Chap. 7			454
	Problems of Chap. 8			457
	Problems of Chap. 9			460
	Problems of Chap. 10			470
	Reference			470
	Correction to: Principles of Nonlinear Filtering Theory			C1

Acronyms

$\mathcal{D}(f)$	The domain of the function f
$C^\infty(U)$	The set of smooth functions defined on U
$C^p(U)$	The set of p-times differentiable functions defined on U
$C_0(U)$	The sets of compactly supported continuous functions defined on U
$C_b(U)$	The set of bounded continuous functions defined on U
$\mathcal{B}$	Borel σ-algebra
det	Determinant
$L^p(U)$	p-times Lebesgue integrable functions on U
P	Probability measure
exp	Exponential function
$E[\,\cdot\,]$	The expectation of a random variable
R^n	Euclidean space with dimension n
R_+	The positive real numbers
$\mathcal{S}^n$	The set of $n \times n$ symmetric matrix defined on R
$\mathcal{S}^n_+$	The set of $n \times n$ positive semi-definite matrix defined on R
$\mathcal{S}^n_{++}$	The set of $n \times n$ positive definite matrix defined on R
$\mathcal{F}$	The $\sigma-$ algebra
$\mathcal{F}_t$	The filtrations
$\mathcal{Y}_t$	The filtrations generated by the process $\{Y_t\}$
v	And other lower Latin letters denote the vectors of corresponding dimensions
$v^\top$	The transpose of a vector v
X_t, ξ_t	And other capital Latin letters or lower Greek letters $(\cdot)_t$ denotes the stochastic process and the corresponding lower Latin letters denotes a particular trajectory of the stochastic process
M_t, N_t	Especially, the continuous martingales
W_t, V_t	Especially, the Brownian motions
$\mathcal{M}$	The set of all finite measures

$\mathcal{M}_2^c$	The set of continuous, square-integrable martingales
$\mathcal{M}^{c,loc}$	The set of continuous local martingales
$\langle M \rangle$	The quadratic variation process of a martingale M
$\langle M, N \rangle$	The cross variation process of two martingale M and N
$\mathscr{L}(M)$	The set of all progressively measurable processes which are square-integrable with respect to the quadratic variation process of M
$RCLL$	The abbreviation of 'right-continuous with left limits'
$P-a.s.$	The abbreviation of 'almost surely' under the probability measure P
$a.e.$	The abbreviation of 'almost everywhere'
$\|\cdot\|_p$	(or $\|\cdot\|_{L^p(\cdot)}$) the p-norm of a function, with $1 \leq p \leq \infty$
$L^2(p)$	A weighted square-integrable function space with the norm $\|\cdot\|_{L^2(p)} := \int_{R^n} (\cdot)^2 p\, dx$
$\delta_{a,b}$	The Kronecker notation, in which $\delta_{a,b} = 1$ if $a = b$, and $\delta_{a,b} = 0$, otherwise
$\int_0^T Y_s dX_s$	The Itô stochastic integration of Y_s with respect to the semi-martingale X_s
$\int_0^T Y_s \circ dX_s$	The Stratonovich stochastic integration of Y_s with respect to the semi-martingale X_s
Π	A partition $0 = t_0 < t_1 < \cdots t_n = T$ of the time interval $[0, T]$
$|\Pi|$	The maximal distance between the nodes of the partition Π
∇	The gradient operator
∇_x	The gradient operator with respect to x
$\nabla \cdot$	The divergence operator
Δ	The Laplacian operator
$\wedge$	The wedge product among differential forms
$Vol(U)$	The volume of a bounded subset U
$\mathbb{I}_A$	The indicator of the set A
$\mathscr{A}$	The second-order elliptic differential operator
E	Estimation algebra
$\Omega = (\omega_{ij})$	Wong's matrix
$[\cdot, \cdot]$	Lie bracket
$\mathfrak{g}$	Lie algebra
U_k	The linear space of differential operators with order no more than k, especially $U_0 := C^\infty(R^n)$
$(\cdot)\, mod\, U_l$	A member of the affine class of operators obtained by adding members of U_l to the argument
$Ad_A B$	The adjoint map between Lie algebra, i.e., $Ad_A B = [A, B]$
E_k	An Euler operator in $x_1, \cdots, x_k$ variables, i.e.,

$$E_k := \sum_{j=1}^{k} x_j \frac{\partial}{\partial x_j}$$

deg	The degree of a polynomial
$P_k(x_1, x_2, \cdots, x_j)$	The set of degree k polynomials in variable $x_1, \cdots, x_j$
$const$	Certain constant number
$X_{1:k}$	The discrete random processes $X_1, \cdots, X_k$
$p(X_k|Y_{1:k})$	The condition density function for X_k given $Y_1, \cdots, Y_k$
$\mu_{k|l}$	The mean of the condition density function $p(X_k|Y_{1:l})$
$P_{k|l}$	The covariance of the condition density function $p(X_k|Y_{1:l})$
I_n	The $n \times n$ identity matrix
$\delta(\cdot)$	The standard Dirac distribution

Part I
Preliminary Knowledge

Chapter 1
Probability Theory

In this chapter, we will introduce some basic concepts and conclusions about measure theoretical foundations of probability theory. The purpose of this chapter is to provide a background for readers who are not familiar with measure theory and probability theory before. We only introduce those topics to readers that are highly relative and necessary to subsequent chapters. Main results in this chapter come from the references [1–3]. This chapter is organized as follows. In Sect. 1.1, we define and introduce some basic concepts and properties about probability space. In Sect. 1.2, we introduce random variables defined on probability space. Then in Sect. 1.3, we give some numerical characteristics about random variables. In Sect. 1.4, we discuss conditional probability and expectation, which are important contents for subsequent filtering theory. Finally, in Sect. 1.5, we introduce two classical limit theorems.

1.1 Probability Space

A probability space is described by a triple $(\Omega, \mathcal{F}, P)$. Ω is a set, which represents possible "experimental outcomes" in trials. $\mathcal{F}$ is a set of "events," which will be defined explicitly later. $P : \mathcal{F} \to [0, 1]$ is a function that assigns probabilities to events. Next we shall introduce the important concept appeared in the probability triple.

Definition 1.1 (Algebra) $\mathcal{I}$, as a collection of subsets of Ω, is called an algebra if it satisfies:

(1) $\emptyset, \Omega \in \mathcal{F}$.
(2) If $A \in \mathcal{F}$, then the complement $A^c \in \mathcal{F}$.
(3) If $A, B \in \mathcal{F}$, then $A \cup B \in \mathcal{F}$.

© The Author(s), under exclusive license to Springer Nature Switzerland AG 2024
S. S.-T. Yau et al., *Principles of Nonlinear Filtering Theory*, Algorithms and Computation in Mathematics 33, https://doi.org/10.1007/978-3-031-77684-7_1

Definition 1.2 (σ-**algebra**) $\mathcal{F}$, as a collection of subsets of Ω, is called a σ-algebra if it satisfies:

1. $\emptyset, \Omega \in \mathcal{F}$.
2. If $A \in \mathcal{F}$, then the complement $A^c \in \mathcal{F}$.
3. If $A_1, A_2, \cdots \in \mathcal{F}$, then $\cup_{k=1}^{\infty} A_k \in \mathcal{F}$.

Example 1.1 Consider the experiment of rolling a die. The probability space can be defined as $\Omega = \{1, 2, 3, 4, 5, 6\}$. If we define $\mathcal{F}$ as the set of all subsets of Ω, then $\mathcal{F}$ is a σ-algebra. If we define $\mathcal{F} = \{\emptyset, \{1, 3, 5\}, \{2, 4, 6\}, \Omega\}$, then $\mathcal{F}$ is also a σ-algebra. However, if we take $\mathcal{F} = \{\emptyset, \{1, 3, 5\}, \{2, 4, 6\}, \{1, 2\}\}$, then $\mathcal{F}$ is not a σ-algebra.

Definition 1.3 Borel σ-algebra $\mathcal{B}$ is defined by the smallest σ-algebra containing all the open subsets of R^n.

If we do not consider P, $(\Omega, \mathcal{F})$ is called a measurable space, and we can put a measure on this space.

Definition 1.4 Let $\mathcal{F}$ be a σ-algebra of subsets of Ω. We call $P : \mathcal{F} \to [0, 1]$ a probability measure if P satisfies

(i) $P(\emptyset) = 0$, $P(\Omega) = 1$. For any set $A \in \mathcal{F}$, $P(A) \geq 0$.
(ii) If $A_1, A_2, \cdots$ are disjoint sets in $\mathcal{F}$, then

$$P\left(\bigcup_{k=1}^{\infty} A_k\right) = \sum_{k=1}^{\infty} P(A_k). \tag{1.1}$$

Next we give the exact definition of the probability space.

Definition 1.5 A triple $(\Omega, \mathcal{F}, P)$ is called a probability space provided Ω is any set, $\mathcal{F}$ is a σ-algebra of subsets of Ω, and P is a probability measure on $\mathcal{F}$.

Next we introduce some useful terminologies.

Definition 1.6

(i) A set $A \in \mathcal{F}$ is called an event; points $\omega \in \Omega$ are sample points.
(ii) $P(A)$ represents the probability of the event A.
(iii) An event is said to hold almost surely (abbreviated as "a.s.") if it happens with probability 1.

For a given physical experiment, at the beginning, probability space should be specified. And the probability space is not unique and not determined by the physical experiment itself. If probability space is not specified, sometimes, it will cause a paradox such as the well-known Bertrand's paradox [3]. Furthermore, we notice that $P(A) = 0$ does not imply that $A = \emptyset$. For instance, if we define Ω as the set of all real numbers in Example 1.1 and denote $P\{(a, b)\}$ as probability, that die lies in interval (a, b). Then, for example, we have $P\{(8, 9)\} = 0$. However, $(8, 9)$ itself is an open interval and is not an empty set. Similarly, $P(A) = 1$ does not represent

1.1 Probability Space

that $A = \Omega$. For example, we assume $A = (0, 9)$; then $P(A) = 1$. However, A is not a whole set Ω.

The next result gives some basic properties of the probability measure. In all cases, we assume that the sets we mention are in $\mathcal{F}$.

Theorem 1.1 *Let $(\Omega, \mathcal{F}, P)$ be a probability space, then*

(i) *monotonicity:* $A \subset B \Longrightarrow P(A) \leq P(B)$.
(ii) *subadditivity:* $P(\bigcup_{m=1}^{\infty} A_m) \leq \sum_{m=1}^{\infty} P(A_m)$.
(iii) *continuity from below.* $A_i \uparrow A$ *(i.e.,* $A_1 \subset A_2 \subset \cdots$ *and* $\bigcup_{i=1}^{\infty} A_i = A$*)* $\Longrightarrow$ $P(A_i) \uparrow P(A)$.
(iv) *continuity from above.* $A_i \downarrow A$ *(i.e.,* $A_1 \supset A_2 \supset \cdots$ *and* $\bigcap_{i=1}^{\infty} A_i = A$*)* $\Longrightarrow$ $P(A_i) \downarrow P(A)$.

Proof

(i) If $A \subset B$, then $B = A \cup (B \setminus A)$ and $A \cap (B \setminus A) = 0$. By definition, we obtain

$$P(B) = P(A) + P(B \setminus A) \geq P(A). \tag{1.2}$$

(ii) Let $D_n = \bigcup_{i=1}^{n} A_i$ and denote $F_n = D_n - D_{n-1}$ for all $n \geq 2$, then

$$\begin{aligned}
P\left(\bigcup_{i=1}^{n} A_i\right) &= P(D_n) \\
&= P(A_1) + P\left(\bigcup_{i=2}^{n} F_i\right) \\
&\leq P(A_1) + \sum_{i=2}^{n} P(A_i) \\
&= \sum_{i=1}^{n} P(A_i).
\end{aligned} \tag{1.3}$$

Subadditivity can be obtained if we let $n \to \infty$.

(iii) Let $A_0 = \emptyset$, $B_m = A_m \setminus A_{m-1}$ for $m \geq 1$. Since A_i satisfies that $A_1 \subset A_2 \subset \cdots$ and $\bigcup_{i=1}^{\infty} A_i = A$, then B_i's are disjoint and satisfy

$$\begin{aligned}
A &= \lim_{m \to \infty} \bigcup_{i=1}^{m} A_i = \lim_{m \to \infty} A_m \\
&= \lim_{m \to \infty} \bigcup_{i=1}^{m} B_i = \bigcup_{i=1}^{\infty} B_i.
\end{aligned} \tag{1.4}$$

Then we obtain

$$P(A) = P(\bigcup_{i=1}^{\infty} B_i)$$
$$= \sum_{i=1}^{\infty} P(B_i) \qquad (1.5)$$
$$= \lim_{m \to \infty} \sum_{i=1}^{m} P(B_i)$$
$$= \lim_{m \to \infty} P(A_m).$$

Notice that $A_m \subset A$; then $P(A_m) \leq P(A)$. Therefore, $P(A_m) \uparrow P(A)$.
(iv) Let $B_i = A_1 \setminus A_i$. Then $B_1 \subset B_2 \subset \cdots$ and

$$\bigcup_{i=1}^{m} B_i = \bigcup_{i=1}^{m} A_1 \setminus A_i = A_1 \bigcup \left(\bigcap_{i=1}^{m} A_i\right)^c = A_1 \setminus \bigcap_{i=1}^{m} A_i. \qquad (1.6)$$

Then $\cup_{i=1}^{\infty} B_i = A_1 \setminus A$. By using Theorem 1.1 (iii), we obtain $P(A_1 \setminus A) = \lim_{i \to \infty} P(B_i)$. Then $P(A_1) - P(A) = \lim_{i \to \infty}(P(A_1) - P(A_i))$. Therefore,

$$P(A) = \lim_{i \to \infty} P(A_i). \qquad (1.7)$$

Considering $P(A_i) \geq P(A)$, we have $P(A_i) \downarrow P(A)$.

□

1.2 Random Variables

1.2.1 Definition and Properties

In this section, we will introduce some basic concepts and results about random variables. In the previous section, we have defined probability space, which is a triple $(\Omega, \mathcal{F}, P)$. However, probability space itself is abstract and not "observable" for us. In order to observe "events" more clearly, we introduce a mapping X from sample points set Ω to R^n, i.e., $X : \Omega \to R^n$. With the help of mapping X, we can use subset of R^n to represent events in probability space.

Definition 1.7 Let $(\Omega, \mathcal{F}, P)$ be a probability space. A mapping

$$X : \Omega \to R^n \qquad (1.8)$$

1.2 Random Variables

is called an n-dimensional random variable if for each $B \in \mathcal{B}$, we have

$$X^{-1}(B) \in \mathcal{F}. \tag{1.9}$$

When we require to emphasize σ-algebra, equivalently, we say that X is $\mathcal{F}$-measurable.

In the following sections, we use capital letters to represent random variables. Lowercase letters denote sample values or realizations of the random variable. For example, X denotes a random variable, and $X(\omega) = x$ is realization of X. Traditionally, we denote $P(X^{-1}(B))$ as $P(X \in B)$, which represents the probability that X is in B. In many cases, the random variable is denoted by abbreviation "r.v."

Lemma 1.1 *Let $X : \Omega \to R^n$ be a random variable. Then*

$$\mathcal{F}(X) := \{X^{-1}(B) | B \in \mathcal{B}\} \tag{1.10}$$

is a σ-algebra, called the σ-algebra generated by X. This is the smallest sub-σ-algebra of $\mathcal{F}$ with respect to which X is measurable.

Proof First we notice for any $B, B_1, B_2, \cdots \in \mathcal{B}$,

$$\begin{aligned} X^{-1}\left(\bigcup_{i=1}^{\infty} B_i\right) &= \bigcup_{i=1}^{\infty} X^{-1}(B_i) \\ X^{-1}(B^c) &= (X^{-1}(B))^c. \end{aligned} \tag{1.11}$$

(i) Due to $X^{-1}(R^n) = \Omega \in \mathcal{F}(X)$, then $X^{-1}(\emptyset) = \emptyset \in \mathcal{F}(X)$.
(ii) If $B \in \mathcal{B}$, then $B^c \in \mathcal{B}$. Then

$$(X^{-1}(B))^c = X^{-1}(B^c) \in \mathcal{F}(X). \tag{1.12}$$

(iii) Due to $B_1, B_2, \cdots \in \mathcal{B}$, we have $\cup_{i=1}^{\infty} B_i \in \mathcal{B}$. Thus

$$\bigcup_{i=1}^{\infty} X^{-1}(B_i) = X^{-1}\left(\bigcup_{i=1}^{\infty} B_i\right) \in \mathcal{F}(X). \tag{1.13}$$

By using Definition 1.2, we deduce that $\mathcal{F}$ is a σ-algebra. Next we assume $\mathcal{G}$ is any sub-σ-algebra of $\mathcal{F}$ with respect to which X is measurable. Clearly, $\mathcal{G} \supset \mathcal{F}(X)$. Therefore, $\mathcal{F}(X)$ is smallest sub-σ-algebra of $\mathcal{F}$ with respect to which X is measurable.

□

Remark 1.1 In probability theory, σ-algebra $\mathcal{F}(X)$ essentially reflects "complete information" about random variable X.

Lemma 1.2 *Let X be a random variable and Y is a function of X, i.e., $Y = \Phi(X)$. If for any $B \in \mathcal{B}$, $\Phi^{-1}(B) \in \mathcal{B}$, then Y is $\mathcal{F}(X)$-measurable.*

Proof For any $B \in \mathcal{B}$, we calculate

$$Y^{-1}(B) = (\Phi \circ X)^{-1}(B) = X^{-1}(\Phi^{-1}(B)). \tag{1.14}$$

Due to $\Phi^{-1}(B) \in \mathcal{B}$, then $Y^{-1}(B) \in \mathcal{F}(X)$ for any $B \in \mathcal{B}$. Therefore, Y is $\mathcal{F}(X)$-measurable. □

Next we give some examples of random variables.

Example 1.2 Let $A \in \mathcal{F}$. We can define indicator function of A as below,

$$\chi_A(\omega) := \begin{cases} 1, & \text{if } \omega \in A, \\ 0, & \text{if } \omega \notin A, \end{cases} \tag{1.15}$$

which is a random variable.

Example 1.3 Let $A_1, A_2, \cdots, A_m \in \mathcal{F}$ and $A_i \cap A_j = \emptyset$ for any $1 \leq i, j \leq m$. Assume $a_1, a_2, \cdots, a_m$ are n real numbers. Then

$$X = \sum_{i=1}^{m} a_i \chi_{A_i}(\omega) \tag{1.16}$$

is a random variable and is called a simple function. A simple function is superposition of some indicator functions.

Let $(\Omega, \mathcal{F}, P)$ be a probability space and X is a random variable from Ω to R^n. Let $x = (x_1, x_2, \cdots, x_n) \in R^n$, $y = (y_1, y_2, \cdots, y_n) \in R^n$. We denote that $x \leq y$ means $x_i \leq y_i$, $1 \leq i \leq n$. Based on the definition of the random variable, we can define distribution function.

Definition 1.8 Distribution function of the random variable X is defined by

$$F_X(x) := P(X \leq x). \tag{1.17}$$

where $x \in R^n$.

In the following sections, if without special explanation, generally we consider the scalar random variable. Next we introduce some basic properties of distribution function.

Theorem 1.2 *Any distribution function F of scalar random variable has the following properties:*

(i) *F is monotone, non-decreasing.*
(ii) *$\lim_{x \to \infty} F(x) = 1$, $\lim_{x \to -\infty} F(x) = 0$*

1.2 Random Variables

(iii) F is right continuous, i.e., $\lim_{y \downarrow x} F(y) = F(x)$.
(iv) If $F(x-) = \lim_{y \uparrow x} F(y)$ then $F(x-) = P(X < x)$
(v) $P(X = x) = F(x) - F(x-)$

Proof

(i) Let $x_1 \leq x_2$. Then $\{X \leq x_1\} \subseteq \{X \leq x_2\}$. By applying Theorem 1.1 (i), we obtain $P(X \leq x_1) \leq P(X \leq x_2)$.

(ii) By using

$$\lim_{x \to \infty} P(X \leq x) = P(\Omega) = 1,$$
$$\lim_{x \to -\infty} P(X \leq x) = P(\emptyset) = 0, \tag{1.18}$$

we obtain (ii).

(iii) If $y \downarrow x$, then $\{X \leq y\} \downarrow \{X \leq x\}$. Then by Theorem 1.1 (iv), we obtain

$$\lim_{y \downarrow x} P(X \leq y) = P(X \leq x), \tag{1.19}$$

i.e.,

$$\lim_{y \downarrow x} F(y) = F(x). \tag{1.20}$$

(iv) If $y \uparrow x$, then $\{X \leq y\} \uparrow \{X < x\}$. Then

$$\lim_{y \uparrow x} P(X \leq y) = P(X < x), \tag{1.21}$$

i.e.,

$$P(X < x) = \lim_{y \uparrow x} F(y) = F(x-). \tag{1.22}$$

(v) We notice

$$P(X = x) = P(X \leq x) - P(X < x)$$
$$= F(x) - F(x-). \quad \text{(By definition and (iv))} \tag{1.23}$$

Then we obtain (v).

$\square$

Remark 1.2 (v) means that if $F_X(\cdot)$ is discontinuous at point x, then its jump is equal to probability $P(X = x)$.

Next we can do some basic calculations using Theorem 1.2.

$$\{x_1 \leq X \leq x_2\} = \{X = x_1\} \cup \{x_1 < X \leq x_2\}, \tag{1.24}$$

so that

$$P(x_1 \leq X \leq x_2) = F_X(x_1) - F_X(x_1-) + F_X(x_2) - F_X(x_1) \\ = F_X(x_2) - F_X(x_1-). \tag{1.25}$$

Similarly,

$$P(X < x_1) = F_X(x_1-), \\ P(X \geq x_1) = 1 - F_X(x_1-), \tag{1.26} \\ P(X > x_1) = 1 - F_X(x_1).$$

More similar results can be done by readers.

Definition 1.9 Suppose X is a scalar random variable and $F = F_X$ is its distribution function. If there exists a nonnegative, integrable function $p : R^n \to R$ such that

$$F_X(x) = \int_{-\infty}^{x} p_X(y) dy, \tag{1.27}$$

then p_X is called the density function of random variable X. Meanwhile, X is called a continuous random variable.

If the distribution function is absolutely continuous, the density function exists. In this case, the numbers of points at which $F_X(x)$ is not differentiable are at most countable. By Definition 1.9, if density function of random variable X exists, distribution function $F_X(x)$ is continuous at all x due to absolute continuity of integration. And

$$p_X(x) = \frac{d}{dx} F_X(x) \tag{1.28}$$

holds at all points at which the derivative exists.

Since $F_X(x)$ is continuous, by Theorem 1.2 (v), we deduce

$$P(X = x) = 0 \tag{1.29}$$

for any $x \in R$. Due to monotonicity of distribution function, by (1.28), we deduce

$$p_X(y) \geq 0. \tag{1.30}$$

In view of Theorem 1.2 (ii),

$$\int_{-\infty}^{\infty} p_X(y) dy = 1. \tag{1.31}$$

1.2 Random Variables

By using (1.25) and let $x_1 < x_2$,

$$P(x_1 \leq X \leq x_2) = F_X(x_2) - F_X(x_1)$$
$$= \int_{x_1}^{x_2} p_X(y) dy. \tag{1.32}$$

Above equation can be extended to vector case naturally,

$$P(X \in B) = \int_B p_X(y) dy \quad \text{for all } B \in \mathcal{B}, \tag{1.33}$$

where X is a n-dimensional random variable and $y = (y_1, \cdots, y_n)$. This formula is highly important. The left-hand side of equation is probability measure of event $X^{-1}(B)$ defined in probability space. The right-hand side of equation is integral defined on Euclidean space and can be calculated directly.

By Definition 1.9, the density function can determine distribution function, then density function can completely characterize properties of a continuous random variable. In the following sections, we only concern about continuous random variables. Therefore, we make assumption that the density function satisfies all smoothness properties that we require. All quantities that we are interested in can be calculated by density function. Equivalently, we shall specify a random variable with a density function.

Definition 1.10 The expectation of a continuous random variable X is defined by

$$EX := \int_{-\infty}^{\infty} x p_X(x) dx. \tag{1.34}$$

Example 1.4 Let $X : \Omega \to R^n$ be a random variable with distribution function F_X and density p_X. Suppose $f : R^n \to R$, and

$$Y = f(X) \tag{1.35}$$

is integrable. Then

$$EY = \int_{R^n} f(x) p(x) dx. \tag{1.36}$$

Definition 1.11

(i) The n-th moment of a random variable X is defined as

$$E(X^n) := \int x^n p_X(x) dx. \tag{1.37}$$

Especially, $E(X^2)$ is called mean square value of X.

(ii) The n-th central moment of a random variable X is defined as

$$E(X - E(X))^n := \int (x - EX)^n p_X(x) dx. \tag{1.38}$$

Especially, the second central moment is called variance of X:

$$var(X) := E(X - EX)^2 = E(X^2) - (EX)^2. \tag{1.39}$$

(iii) Standard deviation of X is defined as

$$\sigma(X) := \sqrt{var(X)}, \tag{1.40}$$

which measures dispersion between samples and its mean value.

Lemma 1.3 (Chebyshev's Inequality) *If X is a random variable and $1 \leq m < \infty$, then*

$$P(|X| \geq \lambda) \leq \frac{1}{\lambda^m} E(|X|^m) \quad \text{for all } \lambda > 0. \tag{1.41}$$

Proof We calculate

$$\begin{aligned} E(|X|^m) &= \int |x|^m p_X(x) dx \\ &\geq \int_{|x| \geq \lambda} |x|^m p_X(x) dx \\ &\geq \lambda^m \int_{|x| \geq \lambda} p_X(x) dx \\ &= \lambda^m P(|X| \geq \lambda). \end{aligned} \tag{1.42}$$

Then we obtain

$$P(|X| \geq \lambda) \leq \frac{1}{\lambda^m} E(|X|^m) \quad \text{for all } \lambda > 0. \tag{1.43}$$

$\square$

Lemma 1.4 *Let X be a random variable and c be a fixed constant. If*

$$E(X - c)^2 = 0, \tag{1.44}$$

then $X = c$, a.s..

1.2 Random Variables

Proof If $X = c, a.s.$ does not hold, we assume that $\exists\ \varepsilon > 0, P(|X - c| > \varepsilon) > 0$, i.e.,

$$\int_{(-\infty, c-\varepsilon) \cup (c+\varepsilon, \infty)} p_X(x) dx > 0. \tag{1.45}$$

Then

$$\begin{aligned} E(X - c)^2 &= \int_R (x - c)^2 p_X(x) dx \\ &\geq \int_{(-\infty, c-\varepsilon) \cup (c+\varepsilon, \infty)} (x - c)^2 p_X(x) dx \\ &> \varepsilon^2 \int_{(-\infty, c-\varepsilon) \cup (c+\varepsilon, \infty)} p_X(x) dx \\ &> 0. \end{aligned} \tag{1.46}$$

This is contradictory to $E(X - c)^2 = 0$. Therefore, $X = c, a.s.$ holds. □

Next we introduce some common probability distributions.

Example 1.5 Uniform distribution on $(0, 1)$. Density is $p(x) = 1$ for $x \in (0, 1)$ and 0 otherwise. Distribution function is shown below.

$$F(x) = \begin{cases} 0, & x < 0, \\ x, & 0 \leq x \leq 1, \\ 1, & x > 1. \end{cases} \tag{1.47}$$

Example 1.6 Exponential distribution with rate λ. Let density be $p(x) = \lambda e^{-\lambda x}$ for $x \geq 0$ and 0 otherwise. Distribution function is

$$F(X) = \begin{cases} 0, & x < 0, \\ 1 - e^{-x}, & x \geq 0. \end{cases} \tag{1.48}$$

Example 1.7 Gaussian distribution or, equivalently, normal distribution. A random variable is Gaussian if its density function is given by

$$p_X(x) = \frac{1}{\sqrt{2\pi}\sigma} \exp\left[-\frac{1}{2}\left(\frac{x - m}{\sigma}\right)^2\right], \tag{1.49}$$

where m and $\sigma > 0$ are mean value and standard deviation. m and σ completely characterize the properties of Gaussian distribution. If a random variable X is

Gaussian, we usually denote

$$X \sim N(m, \sigma^2). \tag{1.50}$$

Distribution function is given by

$$F_X(x) = \frac{1}{\sqrt{2\pi}\sigma} \int_{-\infty}^{x} \exp\left[-\frac{1}{2}\left(\frac{t-m}{\sigma}\right)^2\right] dt. \tag{1.51}$$

Example 1.8 Vector-valued Gaussian random variable. Let $X : \Omega \to R^n$ be a n-dimensional random variable. X is a vector-valued Gaussian random variable if its density function is given by

$$p_X(x) = \frac{1}{\sqrt{(2\pi)^n \det C}} \exp\left[-\frac{1}{2}(x-m)^T C^{-1}(x-m)\right], \tag{1.52}$$

where $m \in R^n$ and C is a symmetric positive definite matrix.

In probability theory, the random variable with finite second-order moment is a class of important random variables. We use $L^2(\Omega)$ to represent this class of random variables with finite second-order moment on probability space $\{\Omega, \mathcal{F}, P\}$. If $X \in L^2(\Omega)$, i.e.,

$$\int_\Omega X^2 dP < \infty, \tag{1.53}$$

we can introduce norm in this linear space,

$$\|X\| = \left(\int_\Omega X^2 dP\right)^{\frac{1}{2}}. \tag{1.54}$$

If we introduce inner product in $L^2(\Omega)$ as follows, for any $X, Y \in L^2(\Omega)$, we define

$$(X, Y) = \int_\Omega XY dP, \tag{1.55}$$

then $L^2(\Omega)$ will become a Hilbert space.

1.2.2 Jointly Distributed Random Variables

In this section, we consider n (continuous) scalar random variables $X_1, X_2, \cdots, X_n$.

1.2 Random Variables

Definition 1.12 Let $X_1, X_2, \cdots, X_n$ be n random variables. Jointly distribution function is defined as

$$F_{X_1,X_2,\cdots,X_n}(x_1, x_2, \cdots, x_n) := P(X_1 \leq x_1, X_2 \leq x_2, \cdots, X_n \leq x_n). \quad (1.56)$$

Definition 1.13 Suppose $X_1, X_2, \cdots, X_n$ are n random variables and $F_{X_1,X_2,\cdots,X_n}$ is its distribution function. If there exists a nonnegative, integrable function $p_{X_1,X_2,\cdots,X_n}: R^n \to R$ such that

$$F_{X_1,X_2,\cdots,X_n}(x_1, x_2, \cdots, x_n) = \int_{-\infty}^{x_1} \cdots \int_{-\infty}^{x_n} p_{X_1,X_2,\cdots,X_n}(\xi_1, \xi_2, \cdots, \xi_n) d\xi_1 \cdots d\xi_n. \quad (1.57)$$

then $p_{X_1,X_2,\cdots,X_n}$ is called the jointly density function of random variables $X_1, X_2, \cdots, X_n$.

Then we deduce

$$p_{X_1,X_2,\cdots,X_n}(x_1, \cdots, x_n) = \frac{\partial^n}{\partial x_1, \cdots, \partial x_n} F_{X_1,X_2,\cdots,X_n}(x_1, x_2, \cdots, x_n), \quad (1.58)$$

holds at all points at which the derivative exists. The probability properties of $X_1, X_2, \cdots, X_n$ can be completely characterized by their jointly distribution function or joint density function.

Sometimes, we only need to consider a part of random variables $X_1, X_2, \cdots, X_n$, i.e., $X_1, X_2, \cdots, X_m, m < n$. In order to describe the distribution of $X_1, X_2, \cdots, X_m$, we calculate marginal distribution function, i.e.,

$$F_{X_1,X_2,\cdots,X_m}(x_1, x_2, \cdots, x_m) = F_{X_1,X_2,\cdots,X_n}(x_1, x_2, \cdots, x_m, \infty, \cdots, \infty). \quad (1.59)$$

Corresponding marginal density function can be calculated by differentiating (1.59), then we have

$$p_{X_1,X_2,\cdots,X_m}(x_1, x_2, \cdots, x_m) = \int_{-\infty}^{\infty} p_{X_1,X_2,\cdots,X_n}(x_1, x_2, \cdots, x_n) dx_{m+1} \cdots dx_n. \quad (1.60)$$

Definition 1.14 For $1 \leq i \leq k$, the l-th moment of X_k is defined as

$$E(X_k^l) = \int \cdots \int x_k^l p_{X_1,X_2,\cdots,X_n}(x_1, x_2, \cdots, x_n) dx_1 \cdots dx_n. \quad (1.61)$$

where l is a positive integer.

Definition 1.15 For $1 \leq k, l \leq n$, the covariance of X_k, X_l is defined as

$$\begin{aligned} cov(X_k, X_l) &:= E(X_k - EX_k)(X_l - EX_l) \\ &= E(X_k X_l) - EX_k \cdot EX_l. \end{aligned} \quad (1.62)$$

Definition 1.16 For $1 \leq k, l \leq n$, the correlation coefficient of X_k, X_l is defined as

$$\rho(X_k, X_l) = \frac{cov(X_k, X_l)}{\sigma(X_k)\sigma(X_l)}, \quad (1.63)$$

where we assume that $\sigma(X_k), \sigma(X_l) < \infty$. If $EX_k^2, EX_l^2 < \infty$ and $cov(X_k, X_l) = 0$, X_k, X_l are called uncorrelated.

Finite estimation of moments is an important problem. Next we develop some inequalities as tools.

Lemma 1.5 *Let X and Y be two scalar random variables and a, b be two real numbers; then*

(i) (Linearity) $E(aX + bY) = aEX + bEY$.
(ii) (Schwarz's inequality)

$$E^2(XY) \leq E(X^2)E(Y^2). \quad (1.64)$$

(iii) (Triangle inequality)

$$E(|X + Y|) \leq E|X| + E|Y|. \quad (1.65)$$

(iv)

$$E(|X + Y|^2) \leq 2E|X|^2 + 2E|Y|^2. \quad (1.66)$$

Proof

(i)

$$\begin{aligned} E(aX + bY) &= \iint (ax + by) p_{XY}(x, y) dx dy \\ &= a \iint x p_{XY}(x, y) dx dy + b \iint y p_{XY}(x, y) dx dy \\ &= a \int x p_X(x) dx + b \int y p_Y(y) dy \\ &= aEX + bEY. \end{aligned} \quad (1.67)$$

1.2 Random Variables

(ii) We consider

$$E(a|X| - |Y|)^2 = a^2 E|X|^2 - 2a E(|XY|) + E|Y|^2 \geq 0. \tag{1.68}$$

This is a nonnegative quadratic function in terms of a. Discriminant is nonpositive, i.e.,

$$E^2(XY) - E(X^2)E(Y^2) \leq 0, \tag{1.69}$$

which leads to Schwarz's inequality.

(iii) It can be easily obtained by using triangle inequality of real numbers.

(iv)

$$E|X+Y|^2 = E|X|^2 + E|Y|^2 + 2E|XY|$$
$$\leq E|X|^2 + E|Y|^2 + 2\sqrt{E|X|^2 E|Y|^2} \tag{1.70}$$
$$\leq 2E|X|^2 + 2E|Y|^2.$$

□

Dealing with several scalar random variables is equivalent to dealing with a vector-valued random variable; if we arrange these scalar random variables $X_1, X_2, \cdots, X_n$ to a vector, then we have

$$X = \begin{bmatrix} X_1 \\ X_2 \\ \vdots \\ X_n \end{bmatrix}. \tag{1.71}$$

Definition 1.17 If X is a random vector, its expectation can be defined as

$$EX := \begin{bmatrix} EX_1 \\ EX_2 \\ \vdots \\ EX_n \end{bmatrix}, \tag{1.72}$$

which is usually called mean vector of X.

Definition 1.18 Let X be a random vector. Its covariance matrix, which is usually denoted by P_X, is defined as

$$P_X := [cov(X_i, X_j)]$$
$$= E(X - EX)(X - EX)^T$$
$$= \begin{bmatrix} var(X_1) & cov(X_1, X_2) & \cdots & cov(X_1, X_n) \\ cov(X_2, X_1) & var(X_2) & \cdots & cov(X_2, X_n) \\ \vdots & \vdots & \ddots & \vdots \\ cov(X_n, X_1) & cov(X_n, X_2) & \cdots & var(X_n) \end{bmatrix}. \quad (1.73)$$

By definition, covariance matrix is a symmetric, positive semidefinite matrix.

Let X, Y be two n-dimensional random variables. Y is a function of X. If the density function of X is known, we can derive density function of Y by the following theorem.

Theorem 1.3 *Let X, Y be two n-dimensional random variables with $Y = f(X)$. Suppose f^{-1} exists and both f and f^{-1} are continuously differentiable. Then*

$$p_Y(y) = p_X(f^{-1}(y)) \left\| \frac{\partial f^{-1}(y)}{\partial y} \right\|, \quad (1.74)$$

where $\left\| \frac{\partial f^{-1}(y)}{\partial y} \right\| > 0$ is the absolute value of the Jacobian determinant.

Proof For any set $S \in R^n$, we have

$$P(Y \in S) = P(f(X) \in S) = P(X \in f^{-1}(S)). \quad (1.75)$$

Next we calculate both sides in above equation by definition and obtain

$$P(Y \in S) = \int_S p_Y(y) dy, \quad (1.76)$$

and

$$P(X \in f^{-1}(S)) = \int_{f^{-1}(S)} p_X(x) dx$$
$$= \int_S p_X(f^{-1}(y)) \left\| \frac{\partial f^{-1}(y)}{\partial y} \right\| dy. \quad (1.77)$$

Then we have

$$\int_S \left[p_Y(y) - p_X(f^{-1}(y)) \left\| \frac{\partial f^{-1}(y)}{\partial y} \right\| \right] dy = 0 \quad (1.78)$$

holds for any set $S \in R^n$. Next we suppose (1.74) is not true. Without loss of generality, we assume there exists a point y_0 such that the integrand in (1.78) is positive. Due to continuity of integrand, there exists a small region in which integrand is always positive. Then (1.78) does not hold. Contradiction! □

1.3 Numerical Characteristics

1.3.1 Characteristic Function

In this section, we introduce a tool called characteristic function, which is helpful for us to calculate properties of random variables.

Definition 1.19 Let $X = [X_1, \cdots, X_n]^T$ be a R^n-valued random variable and $u = [u_1, \cdots, u_n]^T \in R^n$. Then characteristic function of X is defined by

$$\varphi_X(u) := E(e^{iu^T X}), \tag{1.79}$$

or equivalently,

$$\varphi_{X_1,\cdots,X_n}(u_1, \cdots, u_n) := E\left(\exp\left(i \sum_{j=1}^n u_j X_j\right)\right). \tag{1.80}$$

where i denotes imaginary unit.

Characteristic function always exists, since

$$|E(e^{iu^T X})| \leq E|e^{iu^T X}| = 1. \tag{1.81}$$

Following theorem gives some important properties of characteristic function.

Lemma 1.6

(i) If $X_1, \cdots, X_n$ are independent random variables, then for each $u \in R^n$,

$$\varphi_{X_1+\cdots+X_n}(u_1, \cdots, u_n) = \varphi_{X_1}(u_1) \cdots \varphi_{X_n}(u_n). \tag{1.82}$$

(ii) If X is a real-valued random variable and φ is differentiable, then

$$\varphi^{(k)}(0) = i^k E X^k, \quad k = 0, 1, \cdots. \tag{1.83}$$

(iii) If X and Y are two random variables and

$$\varphi_X(u) = \varphi_Y(u) \quad \text{for all } \lambda, \tag{1.84}$$

then

$$F_X(x) = F_Y(x) \quad \text{for all } x. \tag{1.85}$$

Assertion (iii) says that the characteristic function can completely determine the distribution of a random variable.

(iv) If density function exists, then Fourier transformation relation between density function and characteristic function is

$$\varphi_X(u) = \int_{-\infty}^{\infty} e^{iuX} p_X(x) dx, \tag{1.86}$$

and

$$p_X(x) = \frac{1}{2\pi} \int_{-\infty}^{\infty} e^{-iuX} \varphi_X(u) du. \tag{1.87}$$

Proof

(i) Due to independence of $X_1, \cdots, X_n$, we have

$$\begin{aligned}\varphi_{X_1+\cdots+X_n}(u_1,\cdots,u_n) &= E\left(\exp\left(i\sum_{j=1}^n u_j X_j\right)\right) \\ &= E\left(\prod_{j=1}^n e^{iu_j X_j}\right) \\ &= \prod_{j=1}^n E\left(e^{iu_j X_j}\right) \\ &= \prod_{j=1}^n \varphi_{X_j}(u_j).\end{aligned} \tag{1.88}$$

(ii) The derivative can be calculated by definition,

$$\frac{d^k}{du^k}\varphi_X(u) = i^k E(X^k e^{iuX}), k \geq 0. \tag{1.89}$$

Therefore

$$\varphi_X^{(k)}(0) = i^k E(X^k), k \geq 0. \tag{1.90}$$

(iii) See [1] for the proof of (iii).

1.3 Numerical Characteristics

(iv) By Definition 1.19, we deduce characteristic function is Fourier transform of density. Therefore, density function is the inverse Fourier transform of characteristic function.

□

1.3.2 Examples

Next we give some examples to show the properties of characteristic function.

Example 1.9

(i) Let X be a scalar Gaussian random variable with mean m and variance σ^2. Then its characteristic function is

$$\varphi_X(u) = \frac{1}{\sqrt{2\pi}\sigma} \int_{-\infty}^{\infty} \exp\left[iux - \frac{1}{2}\left(\frac{x-m}{\sigma}\right)^2\right] dx. \tag{1.91}$$

Let $y = \frac{x-m}{\sigma}$, then

$$\varphi_X(u) = \frac{1}{\sqrt{2\pi}} \exp(ium) \int_{-\infty}^{\infty} \exp\left(iu\sigma y - \frac{1}{2}y^2\right) dy$$

$$= \frac{1}{\sqrt{2\pi}} \exp\left(ium - \frac{1}{2}u^2\sigma^2\right) \int_{-\infty}^{\infty} \exp\left[-\frac{1}{2}(y - iu\sigma)^2\right] dy \tag{1.92}$$

$$= \exp\left(ium - \frac{1}{2}u^2\sigma^2\right).$$

(ii) Let X be a n-dimensional Gaussian random variable. Its characteristic function is

$$\varphi_X(u) = E(\exp(iu^T X)), \tag{1.93}$$

where $X^T = [X_1, \cdots, X_n]$ and $u^T = [u_1, \cdots, u_n]$. Similarly to the scalar case, characteristic function of X is

$$\varphi_X(u) = \exp\left(iu^T EX - \frac{1}{2}u^T P_X u\right), \tag{1.94}$$

where P_X is the covariance matrix of X.

1.4 Conditional Probability

1.4.1 Definition and Independence

Let $(\Omega, \mathcal{F}, P)$ be a probability space and $A, B \in \mathcal{F}$ be two events. We define $\tilde{\Omega} := B$, $\tilde{\mathcal{F}} = \{C \cap B | C \in \mathcal{F}\}$, $\tilde{P} = \frac{P}{P(B)}$, where $P(B) > 0$. Then $\{\tilde{\Omega}, \tilde{\mathcal{F}}, \tilde{P}\}$ consists a new probability space. The new probability measure $\tilde{P}$ is corresponding to conditional probability.

Definition 1.20 Conditional probability of event A given event B is defined by

$$P(A|B) := \tilde{P}(A \cap B) = \frac{P(A \cap B)}{P(B)} \quad \text{if } P(B) > 0. \tag{1.95}$$

Definition 1.21 Two events A and B are called independent if

$$P(A \cap B) = P(A)P(B), \tag{1.96}$$

or equivalently,

$$P(A|B) = P(A). \tag{1.97}$$

For more than two events, we have similar definition of independence.

Definition 1.22 Let $A_1, A_2, \cdots$ be events. These events are called mutually independent if for all choices $1 \leq k_1 < k_2 < \cdots < k_m$, we have

$$P(A_{k_1} \cap \cdots \cap A_{k_m}) = P(A_{k_1}) \cdots P(A_{k_m}). \tag{1.98}$$

Next we can extend this definition to σ-algebras:

Definition 1.23 Let $\mathcal{F}_i \in \mathcal{F}$ be σ-algebras, for $i = 1, \cdots$. $\{\mathcal{F}_i\}_{i=1}^{\infty}$ are called independent if for all choices of $1 \leq k_1 < k_2 < \cdots < k_m$ and of any events $A_{k_i} \in \mathcal{F}_{k_i}$, we have

$$P(A_{k_1} \cap \cdots \cap A_{k_m}) = P(A_{k_1}) \cdots P(A_{k_m}). \tag{1.99}$$

Furthermore, we define independence of two random variables.

Definition 1.24 Two continuous jointly distributed random variables X_1, X_2 are independent if

$$F_{X_1, X_2}(x_1, x_2) = F_{X_1}(x_1) F_{X_2}(x_2), \tag{1.100}$$

if density function exists, equivalently we have

$$p_{X_1, X_2}(x_1, x_2) = p_{X_1}(x_1) p_{X_2}(x_2). \tag{1.101}$$

1.4 Conditional Probability

Theorem 1.4 *Let X_1, X_2 are two independent, real-valued random variables, with*

$$E|X_1|, E|X_2| < \infty, \tag{1.102}$$

then $E|X_1 X_2| < \infty$ and

$$E(f(X_1)g(X_2)) = E(f(X_1))E(g(X_2)), \tag{1.103}$$

where f, g are fixed (non-random) functions.

Proof By definition of expectation, we calculated

$$\begin{aligned}
E(f(X_1)g(X_2)) &= \int\int f(x_1)g(x_2) p_{X_1 X_2}(x_1, x_2) dx_1 dx_2 \\
&= \int\int f(x_1)g(x_2) p_{X_1}(x_1) p_{X_2}(x_2) dx_1 dx_2 \\
&= \int f(x_1) p_{X_1}(x_1) dx_1 \cdot \int g(x_2) p_{X_2}(x_2) dx_2 \\
&= E(f(X_1))E(g(X_2)).
\end{aligned} \tag{1.104}$$

$\square$

Theorem 1.5 *Let X_1, X_2 be two independent, real-valued random variables, with*

$$var(X_1), var(X_2) < \infty, \tag{1.105}$$

then

$$var(X_1 + X_2) = var(X_1) + var(X_2). \tag{1.106}$$

Proof Let $m_1 = EX_1, m_2 = EX_2$. By definition of variance, we have

$$\begin{aligned}
var(X_1 + X_2) &= \int\int (X_1 + X_2 - m_1 - m_2)^2 p_{X_1 X_2}(x_1, x_2) dx_1 dx_2 \\
&= \int\int (X_1 - m_1)^2 p_{X_1 X_2}(x_1, x_2) dx_1 dx_2 \\
&\quad + \int\int (X_2 - m_2)^2 p_{X_1 X_2}(x_1, x_2) dx_1 dx_2 \\
&\quad + 2\int\int (X_1 - m_1)(X_2 - m_2) p_{X_1 X_2}(x_1, x_2) dx_1 dx_2 \\
&= var(X_1) + var(X_2) + 2E(X_1 - m_1)E(X_2 - m_2) \\
&= var(X_1) + var(X_2).
\end{aligned} \tag{1.107}$$

$\square$

Similarly, for more than two random variables, independence can be introduced.

Definition 1.25 Jointly distributed random variables $X_1, X_2, \cdots, X_n$ are mutually independent if

$$F_{X_1,\cdots,X_n}(x_1,\cdots,x_n) = F_{X_1}(x_1) \cdots F_{X_n}(x_n), \qquad (1.108)$$

or equivalently,

$$p_{X_1,\cdots,X_n}(x_1,\cdots,x_n) = p_{X_1}(x_1) \cdots p_{X_n}(x_n). \qquad (1.109)$$

Definition 1.26 Random variables $X_1, X_2, \cdots, X_n$ are jointly independent of random variables $Y_1, Y_2, \cdots, Y_n$ if

$$\begin{aligned} &F_{X_1,\cdots,X_n,Y_1,\cdots,Y_n}(x_1,\cdots,x_n,y_1,\cdots,y_n) \\ &= F_{X_1,\cdots,X_n}(x_1,\cdots,x_n) F_{Y_1,\cdots,Y_n}(y_1,\cdots,y_n), \end{aligned} \qquad (1.110)$$

or equivalently,

$$\begin{aligned} &p_{X_1,\cdots,X_n,Y_1,\cdots,Y_n}(x_1,\cdots,x_n,y_1,\cdots,y_n) \\ &= p_{X_1,\cdots,X_n}(x_1,\cdots,x_n) p_{Y_1,\cdots,Y_n}(y_1,\cdots,y_n). \end{aligned} \qquad (1.111)$$

Example 1.10 Let X, Y be independent, real-valued random variables, and if $X \sim N(m_1, \sigma_1^2)$, $Y \sim N(m_2, \sigma_2^2)$, then we calculate characteristic function of $X + Y$,

$$\begin{aligned} \varphi_{X+Y}(u) &= \varphi_X(u)\varphi_Y(u) \\ &= \exp\left(ium_1 - \frac{1}{2}u^2\sigma_1^2\right) \exp\left(ium_2 - \frac{1}{2}u^2\sigma_2^2\right) \\ &= \exp\left[iu(m_1 + m_2) - \frac{1}{2}u^2(\sigma_1^2 + \sigma_2^2)\right]. \end{aligned} \qquad (1.112)$$

Then we can deduce that $X + Y$ is a Gaussian random variable, with mean $m_1 + m_2$ and variance $\sigma_1^2 + \sigma_2^2$, i.e.,

$$X + Y \sim N(m_1 + m_2, \sigma_1^2 + \sigma_2^2). \qquad (1.113)$$

1.4.2 Conditional Probability Density

Let X, Y be two jointly distributed random variables. First, we formally derive the conditional probability density. Let events $A = \{X \leq x\}$ and $B = \{y \leq Y \leq y + \Delta y\}$. By Definition 1.20, we obtain

1.4 Conditional Probability

$$P(X \leq x | y \leq Y \leq y + \Delta y) = \frac{P(X \leq x, y \leq Y \leq y + \Delta y)}{P(y \leq Y \leq y + \Delta y)}. \quad (1.114)$$

By definition of distribution function, we obtain

$$F_{X, y \leq Y \leq y + \Delta y}(x | y \leq Y \leq y + \Delta y) = \frac{F_{X,Y}(x, y + \Delta y) - F_{X,Y}(x, y)}{F_Y(y + \Delta y) - F_Y(y)}$$

$$= \frac{[F_{X,Y}(x, y + \Delta y) - F_{X,Y}(x, y)]/\Delta y}{[F_Y(y + \Delta y) - F_Y(y)]/\Delta y}. \quad (1.115)$$

Formally let $\Delta y \to 0$, we deduce

$$F_{X,Y}(x|y) = \frac{\partial F_{X,Y}(x, y)}{\partial y} \Big/ \frac{\partial F_Y(y)}{\partial y}$$

$$= \frac{\int_{-\infty}^{x} p_{X,Y}(\xi, y) d\xi}{p_Y(y)}, \quad (1.116)$$

where

$$p_Y(y) = \int p_{X,Y}(x, y) dx \quad (1.117)$$

is the marginal density function. Then we differentiate (1.115) from both sides of (1.116) and obtain

$$p_{X|Y}(x|y) = \frac{p_{X,Y}(x, y)}{p_Y(y)}$$

$$= \frac{p_{X,Y}(x, y)}{\int p_{X,Y}(x, y) dx} \quad (1.118)$$

Therefore, we have

$$p_{X,Y}(x, y) = p_{X|Y}(x|y) p_Y(y). \quad (1.119)$$

Similarly, we exchange the role of X and Y and get

$$p_{X,Y}(x, y) = p_{Y|X}(y|x) p_X(x). \quad (1.120)$$

By combining these two equations, we obtain

$$p_{X|Y}(x|y) = \frac{p_{Y|X}(y|x) p_X(x)}{p_Y(y)}, \quad (1.121)$$

which is the well-known Bayes' formula.

Similarly, if X, Y are two n, m-dimensional random variables, respectively, then the conditional density (1.118) can be defined as

$$p_{X_1,\cdots,X_n|Y_1,\cdots,Y_m}(x_1,\cdots,x_n|y_1,\cdots,y_m)$$
$$= \frac{p_{X_1,\cdots,X_n,Y_1,\cdots,Y_m}(x_1,\cdots,x_n,y_1,\cdots,y_m)}{p_{Y_1,\cdots,Y_m}(y_1,\cdots,y_m)} \quad (1.122)$$
$$= \frac{p_{X_1,\cdots,X_n,Y_1,\cdots,Y_m}(x_1,\cdots,x_n,y_1,\cdots,y_m)}{\int\cdots\int p_{X_1,\cdots,X_n,Y_1,\cdots,Y_m}(\xi_1,\cdots,\xi_n,y_1,\cdots,y_m)d\xi_1\cdots d\xi_n}.$$

1.4.3 Conditional Expectation

In this subsection, we introduce the concept of conditional expectation, which is highly important to the later filtering theory. First we should give the definition.

Definition 1.27 Let X be a random variable and $B \in \mathcal{F}$ be an event. Conditional expectation of X given B is defined as

$$\begin{aligned} E(X|B) &= \int_B X d\tilde{P} \\ &= \frac{1}{P(B)} \int_B X dP. \end{aligned} \quad (1.123)$$

A more meaningful quantity is conditional expectation of random variable X given by random variable Y. If we give a realization $Y(\omega)$ for a sample point $\omega \in \Omega$, by Definition 1.27, $E(X|Y(\omega))$ will be a real number. Then we can extend this idea to the definition of $E(X|Y)$. If Y is a random variable, then $E(X|Y)$ should be a random variable that satisfies some properties. Actually, $E(X|Y)$ represents our best estimation for X by giving some information of Y. Next we give a reasonable definition of conditional expectation.

Definition 1.28 Let X, Y be random variables. Then $E(X|Y)$ is called conditional expectation of X given by Y, if:

(i) $E(X|Y)$ is any $\mathcal{F}(Y)$-measurable random variable.
(ii)

$$\int_A X dP = \int_A E(X|Y) dP \quad \text{for all } A \in \mathcal{F}(Y). \quad (1.124)$$

Example 1.11 Let $(\Omega, \mathcal{F}, P)$ be a probability space. Y is a simple random variable defined on this probability space, i.e.,

1.4 Conditional Probability

$$Y = \begin{cases} a_1 & \text{on } A_1, \\ a_2 & \text{on } A_2, \\ \cdots \\ a_m & \text{on } A_m, \end{cases} \quad (1.125)$$

where $A_1, A_2, \cdots, A_m$ are disjoint events in $\mathcal{F}$ and $a_1, a_2, \cdots, a_m$ are real numbers. For an any random variable X defined on Ω, our best estimation for X depends on realization of Y. Thus, we can define conditional expectation as

$$E(X|Y) := \begin{cases} \frac{1}{P(A_1)} \int_{A_1} X dP & \text{on } A_1, \\ \frac{1}{P(A_2)} \int_{A_2} X dP & \text{on } A_2, \\ \cdots \\ \frac{1}{P(A_m)} \int_{A_m} X dP & \text{on } A_m. \end{cases} \quad (1.126)$$

We can clearly find $E(X|Y)$ is a random variable satisfying Definition 1.28.

We notice $E(X|Y)$ essentially is defined by using σ-algebra generated by random variable Y. Furthermore, we can extend this definition to sub-σ-algebra of $\mathcal{F}$.

Definition 1.29 Let $(\Omega, \mathcal{F}, P)$ be a probability space and suppose $\mathcal{V}$ is a sub-σ-algebra, $\mathcal{V} \subset \mathcal{F}$. If $X : \Omega \to R^n$ is an integrable random variable, we define conditional expectation

$$E(X|\mathcal{V}) \quad (1.127)$$

to be any random variable on Ω satisfying:

(i) $E(X|\mathcal{V})$ is $\mathcal{V}$-measurable
(ii) $\int_A X dP = \int_A E(X|\mathcal{V}) dP$ for all $A \in \mathcal{V}$.

Remark 1.3 We can understand conditional expectation $E(X|\mathcal{V})$ as follows. Our goal is to find best estimation of X given $\mathcal{V}$. $\mathcal{V}$ can be regarded as observations. Then $E(X|\mathcal{V})$ should be constructed from the information of $\mathcal{V}$, which is exactly (i). Furthermore, X and our best estimation $E(X|\mathcal{V})$ should be consistent, at least on events in $\mathcal{V}$, which is (ii). Therefore, (i) and (ii) extract two basic properties of conditional expectation.

Conditional expectation has many nice properties, and some of them are listed in the following theorem.

Theorem 1.6 ([3])

(i) If X is $\mathcal{V}$-measurable, then $E(X|\mathcal{V}) = X$ a.s.
(ii) If a, b are constants, $E(aX + bY|\mathcal{V}) = aE(X|\mathcal{V}) + bE(Y|\mathcal{V})$ a.s.
(iii) If X is $\mathcal{V}$-measurable and XY is integrable, then $E(XY|\mathcal{V}) = XE(Y|\mathcal{V})$ a.s.
(iv) If X is independent of $\mathcal{V}$, then $E(X|\mathcal{V}) = E(X)$ a.s.

(v) If $\mathcal{W} \subset \mathcal{V}$, we have

$$E(X|\mathcal{W}) = E(E(X|\mathcal{V})|\mathcal{W}) = E(E(X|\mathcal{W})|\mathcal{V}) \text{ a.s.} \quad (1.128)$$

(vi) The inequality $X \le Y$ a.s. implies $E(X|\mathcal{V}) \le E(Y|\mathcal{V})$ a.s.

Proof

(i) It can be obtained by definition of conditional expectation.

(ii) For any $A \in \mathcal{V}$, we calculate

$$\int_A [aE(X|\mathcal{V}) + bE(Y|\mathcal{V})]dP = a\int_A E(X|\mathcal{V})dP + b\int_A E(Y|\mathcal{V})dP$$
$$= a\int_A XdP + b\int_A YdP$$
$$= \int_A (aX + bY)dP. \quad (1.129)$$

By definition, we have

$$\int_A E[aX + bY|\mathcal{V}]dP = \int_A (aX + bY)dP. \quad (1.130)$$

Hence by uniqueness, we obtain $E(aX+bY|\mathcal{V}) = aE(X|\mathcal{V})+bE(Y|\mathcal{V})$ a.s., which means condition expectation has linearity for first variable.

(iii) By uniqueness a.s. of $E(XY|\mathcal{V})$, we just need to prove, for any $A \in \mathcal{V}$,

$$\int_A XYdP = \int_A XE(Y|\mathcal{V})dP. \quad (1.131)$$

We first prove it by assuming X is a simple function, i.e.,

$$X = \sum_{i=1}^n b_i \chi_{B_i}, \quad (1.132)$$

where $B_i \in \mathcal{V}$. Then

$$\int_A XE(Y|\mathcal{V})dP = \sum_{i=1}^n b_i \int_A \chi_{B_i} E(Y|\mathcal{V})dP$$
$$= \sum_{i=1}^n b_i \int_{A \cap B_i} E(Y|\mathcal{V})dP$$
$$= \sum_{i=1}^n b_i \int_{A \cap B_i} YdP \quad (A \cap B_i \in \mathcal{V}) \quad (1.133)$$
$$= \int_A XYdP.$$

1.4 Conditional Probability

If X is a general measurable function in probability space $(\Omega, \mathcal{V}, P)$, we can approximate it by simple functions.

(iv) We just need to prove, for any $A \in \mathcal{V}$,

$$\int_A E(X)dP = \int_A XdP. \tag{1.134}$$

Next we calculate

$$\begin{aligned}\int_A XdP &= \int_\Omega \chi_A XdP \\ &= E(\chi_A X) \\ &= P(A)E(X) \\ &= \int_A E(X)dP.\end{aligned} \tag{1.135}$$

(v) We first try to prove $E(X|\mathcal{W}) = E(E(X|\mathcal{V})|\mathcal{W})$. The only thing is to prove, for any $A \in \mathcal{W}$,

$$\int_A E(E(X|\mathcal{V})|\mathcal{W})dP = \int_A XdP. \tag{1.136}$$

By definition, for any $A \in \mathcal{W}$, we have

$$\int_A E(E(X|\mathcal{V})|\mathcal{W})dP = \int_A E(X|\mathcal{V})dP. \tag{1.137}$$

Since $A \in \mathcal{W} \subset \mathcal{V}$, we have

$$\int_A E(X|\mathcal{V})dP = \int_A XdP. \tag{1.138}$$

Combining (1.137) and (1.138), we finish the proof. The proof of $E(X|\mathcal{W}) = E(E(X|\mathcal{W})|\mathcal{V})$ is similar.

(vi) Due to $X \leq Y$ a.s., for any $A \in \mathcal{V}$, we have

$$\begin{aligned}\int_A E(Y|\mathcal{V}) - E(X|\mathcal{V})dP &= \int_A E(Y-X|\mathcal{V})dP \\ &= \int_A (Y-X)dP \geq 0.\end{aligned} \tag{1.139}$$

Then we take $A = \{E(Y|\mathcal{V}) - E(X|\mathcal{V}) < 0\} \in \mathcal{V}$. By inequality (1.139), if $P(A) > 0$, then the left side of (1.139) will be negative contradiction! Therefore, $P(A) = 0$, which deduce $E(Y|\mathcal{V}) \geq E(X|\mathcal{V}), a.s.$

□

1.5 Limit Theorem

In this section, we consider a model of "repeated and independent experiments." We consider a sequence of random trials. In every trial, we will obtain a random "outcome." If we denote X_n as the n-th outcome, we will obtain a sequence of random variables, $X_1, X_2, \cdots, X_n, \cdots$.

Definition 1.30 A sequence $X_1, X_2, \cdots, X_n, \cdots$ of random variables are called independent and identically distributed if $X_1, X_2, \cdots, X_n, \cdots$ are mutually independent and

$$F_{X_1}(x) = F_{X_2}(x) = \cdots = F_{X_n}(x) = \cdots, \quad \text{for all } x. \tag{1.140}$$

1.5.1 Strong Law of Large Numbers

Following theorem shows that for almost all $\omega \in \Omega$, the average of $X_1, X_2, \cdots, X_n, \cdots$ will converge to real mean value.

Theorem 1.7 (Strong Law of Large Numbers) *Let $X_1, X_2, \cdots, X_n, \cdots$ be a sequence of independent, identically distributed, integrable real-valued random variables defined on the same probability space. Then*

$$P\left(\lim_{n \to \infty} \frac{X_1 + X_2 + \cdots + X_n}{n} = E(X_i)\right) = 1, \tag{1.141}$$

which means that $\frac{X_1 + X_2 + \cdots + X_n}{n} \to E(X_i)$ a.s. $n \to \infty$.

Proof By the Kolmogorov inequality, the following inequality holds for arbitrary $a > 0$,

$$P(\max_{1 \leq j \leq n} |\sum_{i=1}^{j} \frac{X_i - E(X_i)}{i}| \geq a) \leq \frac{1}{a^2} \sum_{i=1}^{n} \frac{DX_i}{i^2} \tag{1.142}$$

Then

$$P(\max_{j \geq 1} |\sum_{i=1}^{j} \frac{X_i - E(X_i)}{i}| \geq a) \leq \frac{1}{a^2} \sum_{i=1}^{\infty} \frac{DX_i}{i^2} \tag{1.143}$$

So we shall get

$$P(\sum_{i=1}^{\infty} \frac{X_i - E(X_i)}{i} < a) \geq P(\max_{j \geq 1} |\sum_{i=1}^{j} \frac{X_i - E(X_i)}{i}| < a) \geq 1 - \frac{1}{a^2} \sum_{i=1}^{\infty} \frac{DX_i}{i^2} \tag{1.144}$$

1.5 Limit Theorem

Due to $\sum_{i=1}^{\infty} \frac{DX_i}{i^2} < \infty$, let $a \to \infty$, which will lead to the following:

$$P(\sum_{i=1}^{\infty} \frac{X_i - E(X_i)}{i} < \infty) = 1 \qquad (1.145)$$

By the Kronecker lemma which states if $\sum_{k=1}^{\infty} \frac{c_k}{k} < \infty$, then $\frac{1}{n}\sum_{k=1}^{n} c_k \to 0$. Finally we shall get

$$\frac{1}{n}\sum_{i=1}^{n}(X_i - EX_i) \to 0, \quad a.s. \qquad (1.146)$$

which finishes the proof. □

1.5.2 Central Limit Theorem

In the previous subsection, we notice Strong Law of Large Numbers shows convergence of average value of the random sequence. In this subsection, we introduce Central Limit Theorem, which reflects the fluctuation of the random sequence $X_1, X_2, \cdots, X_n, \cdots$ from mean value.

Theorem 1.8 (Central Limit Theorem) *Let $X_1, X_2, \cdots, X_n$ be independent, identically distributed, real-valued random variables with*

$$E(X_i) = \mu, V(X_i) = \sigma^2 > 0. \qquad (1.147)$$

for $i = 1, \cdots, n$. Set

$$S_n := \sum_{k=1}^{n} X_i. \qquad (1.148)$$

Then for all $-\infty < a < b < \infty$

$$\lim_{n \to \infty} P\left(a \le \frac{S_n - n\mu}{\sqrt{n}\sigma} \le b\right) = \frac{1}{\sqrt{2\pi}} \int_{a}^{b} e^{-\frac{x^2}{2}} dx. \qquad (1.149)$$

Proof To begin with, we shall denote $Q_k = \frac{X_k - \mu}{\sigma\sqrt{n}}$, which has density function q_k. Let $T_n = \sum_{k=1}^{n} Q_k$ and its associated probability density f_n. The whole proof will contain three steps.

Step 1. It can be directly obtained that $f_n = q_1 * q_2 * \cdots * q_n$, where $*$ denotes the convolution. By the convolution formula, we shall get $f_n = \mathcal{F}^{-1} \Pi_{k=1}^{n} \mathcal{F}(q_k)$.

Step 2. The cumulative density function of Q_k is denoted as Q. By definition, we shall get

$$Q(q) = P(\frac{X_k - \mu}{\sigma\sqrt{n}} \leq q) = \int_{-\infty}^{\sigma\sqrt{n}q+\mu} f(t)dt = \int_{-\infty}^{q} \sigma\sqrt{n} f(\sigma\sqrt{n}t + \mu)dt \quad (1.150)$$

This leads to the probability density function $q(t) = \sigma\sqrt{n} f(\sigma\sqrt{n}t + \mu)$. Its Fourier transformation can be written down

$$\begin{aligned}\mathcal{F}(\omega) &= \int_{-\infty}^{\infty} q(t) e^{-j\omega t} dt \\ &= \int_{-\infty}^{\infty} f(x) e^{-j\omega \frac{x-\mu}{\sigma\sqrt{n}}} dx \\ &= 1 - \frac{\omega^2}{2n}\end{aligned} \quad (1.151)$$

Step 3. Finally, by convolution formula, we shall get

$$\begin{aligned}f(t) &= \lim_{n\to\infty} f_n(t) \\ &= \lim_{n\to\infty} \mathcal{F}^{-1}(\mathcal{F}^n(\omega)) \\ &= \frac{1}{2\pi} \int_{-\infty}^{\infty} e^{-\frac{\omega^2}{2}} e^{j\omega t} d\omega \\ &= \frac{1}{2\pi} e^{-\frac{t^2}{2}}\end{aligned} \quad (1.152)$$

That finishes the whole proof. □

Remark 1.4 The Central Limit Theorem is a significantly important theorem in the probability theory. Many practical models can be appropriated by a mathematical model "repeated and independent experiment." From the Central Limit Theorem, we can know that the average value of outcome of repeated trials will appropriately obey a Gaussian distribution. This provides useful guidance for physics and engineering technology.

1.6 Exercises

1. Let $\Omega = \{1, 2, 3, 4\}$ and denote $\mathcal{I} = \{\{1\}, \{3\}\}$. Write down the sigma algebra $\sigma(\mathcal{I})$ generated by $\mathcal{I}$.
2. Denote a sequence of subsets of Ω as $\mathcal{I}_1, \mathcal{I}_2, \cdots$, which satisfies $\mathcal{I}_n \subset \mathcal{I}_{n+1}$ for any n. If each of $\mathcal{I}_n$ is a σ-algebra, show that the infinite union $\cup_{n=1}^{\infty} \mathcal{I}_n$ may not be a σ-algebra.

3. Assume that $f(x) = a_0 + a_1 x + a_2 x^2$ is a quadratic polynomial in $\mathbb{R}$. Find the necessary and sufficient condition of coefficients $\{a_i\}$ so that the equation $E(f(\alpha X)) = \alpha^2 E(f(X))$ holds for arbitrary α and random variable X.
4. X and Y are general random variables. Prove inequality

$$|\text{Corr}(X, Y)| \leq 1. \tag{1.153}$$

5. Show that two discrete random variables with same mean and the same variance may not have the same distributions.
6. Consider two random variables X_1 and X_2 satisfy the following distributions:

$$P(X_1) = \begin{cases} \frac{1}{4}, & X_1 = -1 \\ \frac{1}{2}, & X_1 = 0 \\ \frac{1}{4}, & X_1 = 1 \end{cases} \tag{1.154}$$

and

$$P(X_2) = \begin{cases} \frac{1}{2}, & X_2 = 0 \\ \frac{1}{2}, & X_2 = 1 \end{cases} \tag{1.155}$$

and $P\{X_1 = 0, X_2 = 0\} = 0$. (1) Write down the joint distribution of X_1 and X_2. (2) Show whether X_1, X_2 are independent.
7. Let $\xi = (\xi_1, \cdots, \xi_n)^T$ satisfy n-dimensional Gaussian distribution $\mathcal{N}(\mu, \Sigma)$. C is arbitrary matrix with size $m \times n$. Prove that $\eta = C\xi$ satisfies m dimensional Gaussian distribution $\mathcal{N}(C\mu, C\Sigma C^T)$.
8. Prove that sufficient and necessary condition of Gaussian variables $\xi_1, \xi_2, \cdots, \xi_n$ being independent is that each two of them are irrelevant.
9. Characteristic function of n-dimensional Gaussian distribution is

$$f(t) = \exp\{i\mu^T t - \frac{1}{2} t^T \Sigma t\}. \tag{1.156}$$

10. Let ξ be a continuous random variable with density function $p(x)$. Assume $\eta = g(\xi)$ where g is strictly monotonically increasing and g^{-1} has continuous derivative. Prove that η has the following density function:

$$p(g^{-1}(y))|g^{-1}(y)'| \tag{1.157}$$

References

1. L. Breiman *Probability*. Massachusetts. 1968.
2. R. Durrett. *Probability: theory and examples*. Cambridge University Press, 2019.
3. L. C. Evans. *An introduction to stochastic differential equations*. American Mathematical Society, 2012.

Chapter 2
Stochastic Processes

This chapter provides a comprehensive overview of stochastic processes, an essential concept in probability theory and its applications. Beginning with an introduction to both discrete and continuous stochastic processes, we explore fundamental aspects of process similarity, filtrations, and the crucial concept of stopping times. The exposition then transitions to a detailed discussion on martingales, a class of stochastic processes with profound implications in financial mathematics, signal processing, and beyond. We delve into discrete martingales before extending our analysis to continuous martingales, where we introduce the core inequalities and the pivotal Doob-Meyer decomposition. The narrative further unfolds to encompass continuous local martingales, setting the stage for an in-depth examination of square-integrable martingales. Here, the focus sharpens on the quadratic variations of continuous martingales and local martingales, revealing the intricate behavior of stochastic processes over time. This chapter aims to equip the reader with a solid foundation in understanding and analyzing the dynamic nature of stochastic processes, underpinning much of modern probability theory and its diverse applications.

2.1 Introduction to Stochastic Processes and Filtrations

Throughout this chapter, we consider a probability space $(\Omega, \mathcal{F}, P)$. In this section, we shall provide several important results on the following topic:

- What is the discrete or continuous stochastic process?
- What is discrete or continuous martingale?
- What is square-integrable martingale?

The basic theorems used in filtering on stochastic process and martingale shall be introduced in detail in this section.

2.1.1 Discrete and Continuous Stochastic Processes

A stochastic process $X = (X_t), t \in T$ is a mathematical model for the occurrence, at each moment after the initial time, of a random phenomenon. The randomness is captured by the introduction of a measurable space $(\Omega, \mathcal{F})$, called the sample space, on which probability measures can be placed.

Definition 2.1 (Discrete and Continuous Stochastic Processes) The stochastic process is a collection of random variables $X = (X_t), t \in T$, where T is a set of real numbers, on $(\Omega, \mathcal{F})$. They take values in a second measurable space $(E, \mathcal{E})$, called the state space. Usually, we will consider that the state space $(E, \mathcal{E})$ will be the d-dimensional Euclidean space equipped with the σ-field of Borel sets. And the index $t \in T$ of the random variables X_t admits a convenient interpretation as times.

Especially, if the set T is a discrete set of R_+, the stochastic process is discrete; if the set T is an interval of R_+, the stochastic process is continuous.

Then for a fixed sample point $\omega \in \Omega$, the function $t \to X_t(\omega)$ is the sample path of the process X associated with ω. If the sample path is right continuous for almost $\omega \in \Omega$, then we call the process right continuous and left limits (RCLL) exists. So we have a new way to understand the stochastic process.

2.1.2 Sameness Between Two Processes

Similarly in probability theory, we need to define the sameness between two random processes. Let us consider two stochastic processes X_t and Y_t defined on the same probability space $(\Omega, \mathcal{F}, P, t \in T)$. When they are regarded as functions of t and ω, we say X_t and Y_t were the same if and only if $X_t(\omega) = Y_t(\omega)$ for all the $t \geq 0$ and all the $\omega \in \Omega$. However, in the presence of the probability measure P, we could weaken the requirement in at least three different ways to obtain three related concepts of sameness between the two processes. And we focus on the case $T = [0, \infty)$.

Definition 2.2 (Modification) Y_t is a modification of X_t if, for every $t \geq 0$, we have $P[X_t = Y_t] = 1$

Definition 2.3 (Same Finite-Dimensional Distributions) X_t and Y_t have the same finite-dimensional distributions if, for any integer $n \geq 1$, the real numbers $0 \leq t_1 < t_2, \cdots < t_n < \infty$, $A \in \mathcal{E}_1 \times \cdots \times \mathcal{E}_n$, where $\times$ denotes direct product and the $\mathcal{E}_i$ is the σ-algebra of X_{t_i} for $1 \leq i \leq n$, we have:

$$P[(X_{t_1}, \ldots, X_{t_n}) \in A] = P[(Y_{t_1}, \ldots, Y_{t_n}) \in A].$$

2.1 Introduction to Stochastic Processes and Filtrations

Definition 2.4 (Indistinguishable) X_t and Y_t are called indistinguishable if almost all their sample paths agree:

$$P[X_t = Y_t; 0 \leq t < \infty] = 1.$$

The third property is the strongest; it implies trivially the first one, which in turn yields the second. On the other hand, two processes can be modifications of one another and yet have completely different sample paths. Here is a standard example:

Example 2.1 Consider a positive random variable T with a continuous distribution, put $X_t = 0$, and let $Y_t = \begin{cases} 0; & t \neq T \\ 1; & t = T \end{cases}$. Y_t is a modification of X_t, since for every $t \geq 0$, we have $P[Y_t = X_t] = P[T \neq t] = 1$. But on the other hand, $P[Y_t(\omega) = X_t(\omega); 0 \leq t] = P[T \notin R_+] = 0$, the X_t is not indistinguishable with Y_t.

Proposition 2.1 *Let Y_t be a modification of X_t, and suppose that both processes have right-continuous sample paths. Then X_t and Y_t are indistinguishable.*

Proof Consider a dense countable subset $D = \{t_1, \ldots, t_n, \cdots\}$ of R_+. Because the sample path is right-continuous so for every $t \in [0, \infty)$, we have a sub-sequence $\{t_{i_m}\}_{m=1}^{\infty}$, which will have $\lim_{m \to \infty} t_{i_m} = t^+$. It is obvious that

$$\{\omega | X_t(\omega) = Y_t(\omega); t \in [0, \infty)\} \subset \{\omega | X_t(\omega) = Y_t(\omega); t \in D\}.$$

And from the above limit, we can have

$$\{\omega | X_t(\omega) = Y_t(\omega); t \in D\} \subset \{\omega | X_t(\omega) = Y_t(\omega); t \in [0, \infty)\}.$$

□

Remark 2.1 We will get the same result when the sample paths are left-continuous or continuous.

From Proposition 2.1, we can know that the modification is equivalent to indistinguishable under the regularity on the sample path (here we need the sample path to be right/left continuous).

2.1.3 Filtrations and Measurement of Processes

As a function of two variables based on measurable space, we can naturally define measurable random processes.

Definition 2.5 (Measurable) A stochastic process $X = (X_t)$, $t \geq 0$ with values in a measurable space $(E, \mathcal{E})$ is said to be measurable if the mapping

$$(\omega, t) \to X_t(\omega)$$

defined on $\Omega \times R_+$ equipped with the product σ-field $\mathcal{F} \otimes \mathcal{B}(R_+)$ is measurable. (We recall that $\mathcal{B}(R_+)$ stands for the Borel σ-field of R_+.)

As the index t represents the time. The random process at a fixed time must have its past history, so the X_t should be measurable in any σ-field $\mathcal{E}_s$ associated with the time $0 \leq s \leq t$. In this way, it is natural to define the filtration of the random process.

Definition 2.6 (Filtration) A filtration on $(\Omega, \mathcal{F}, P)$ is a collection $(\mathcal{F}_t)_{0 \leq t \leq \infty}$ indexed by $[0, \infty]$ of sub σ-field of $\mathcal{F}$, such that $\mathcal{F}_s \subset \mathcal{F}_t$ for every $s \leq t \leq \infty$.

We have thus, for every $0 \leq s < t$,

$$\mathcal{F}_0 \subset \mathcal{F}_s \subset \mathcal{F}_t \subset \mathcal{F}_\infty \subset \mathcal{F}$$

We also say that $(\Omega, \mathcal{F}, (\mathcal{F}_t), \mathbf{P})$ is a filtered probability space.

Let $(\mathcal{F}_t)_{0 \leq t \leq \infty}$ be a filtration on $(\Omega, \mathcal{F}, \mathbf{P})$. We set, for every $t \geq 0$,

$$\mathcal{F}_{t^+} = \bigcap_{s > t} \mathcal{F}_s.$$

and $\mathcal{F}_{\infty^+} = \mathcal{F}_\infty$. Note that $\mathcal{F}_t \subset \mathcal{F}_{t^+}$ for every $t \in [0, \infty]$. The collection $(\mathcal{F}_{t^+})_{0 \leq t \leq \infty}$ is also a filtration. We say that the filtration $(\mathcal{F}_t)$ is right-continuous if

$$\mathcal{F}_{t^+} = \mathcal{F}_t, \qquad t \geq 0.$$

By construction, the filtration $(\mathcal{F}_{t^+})_{0 \leq t \leq \infty}$ is right-continuous.

Let $(\mathcal{F}_t)_{0 \leq t \leq \infty}$ be a filtration and let $\mathcal{N}$ be the class of all $(\mathcal{F}_\infty, P)$-negligible sets (i.e., $A \in \mathcal{N}$ if there exists an $A' \in \mathcal{F}_\infty$ such that $A \subset A'$ and $P(A') = 0$). The filtration is said to be complete if $\mathcal{N} \subset \mathcal{F}_0$.

If $(\mathcal{F}_t)_{0 \leq t \leq \infty}$ is not complete, it can be completed by setting $\mathcal{F}'_t = \mathcal{F}_t \vee \sigma(\mathcal{N})$, for every $t \in [0, \infty)$, using the notation $\mathcal{F}_t \vee \sigma(\mathcal{N})$ for the smallest σ-field that contains bath $\mathcal{F}_t$ and $\sigma(\mathcal{N})$. We will often apply this completion procedure to the canonical filtration of a random process X_t and call the resulting filtration the completed canonical filtration of X.

If we have an increasing flow of information, naturally we can require that the random variable at each moment be measurable in the corresponding σ-field, which is adaption.

2.1 Introduction to Stochastic Processes and Filtrations

Definition 2.7 (Adaption) A random process $(X_t)_{t \geq 0}$ with values in a measurable space $(E, \mathcal{E})$ is called adapted to filtration $(\mathcal{F}_t)_{0 \leq t \leq \infty}$ if, for every $t \geq 0$, X_t is $\mathcal{F}_t$-measurable.

More efficiently, we can define stronger measurability for our processes.

Definition 2.8 (Progressively Measurable) This process is said to be progressively measurable if, for every $t \geq 0$, the mapping

$$(\omega, s) \to X_s(\omega)$$

defined on $\Omega \times [0, t]$ is measurable for the σ-field $\mathcal{F}_t \otimes \mathcal{B}([0, t])$.

Note that a progressively measurable process is both adapted and measurable. So naturally, if the process is measurable and adapted, how do we make sure the process is progressively measurable?

Proposition 2.2 *Let $(X_t)_{t \geq 0}$ be a random process with values in a metric space $(E, \mathcal{E})$(equipped with its Borel σ-field). Suppose that X_t is adapted and that the sample paths of X_t are right-continuous (i.e., for every $\omega \in \Omega, t \to X_t(\omega)$ is right-continuous). Then X_t is progressively measurable. The same conclusion holds if one replaces right-continuous with left-continuous.*

Proof We treat the case of right continuity. With $t > 0, n \geq 1, k = 0, 1, \ldots, 2^n - 1$, and $0 \leq s \leq t$, we define:

$$X_s^{(n)}(\omega) = X_{\frac{(k+1)t}{2^n}}(\omega) \quad for \quad \frac{kt}{2^n} < s \leq \frac{(k+1)t}{2^n}$$

as well as $X_0^{(n)}(\omega) = X_0(\omega)$. The so-constructed map $(s, \omega) \to X_s^{(n)}(\omega)$ from $[0, t] \times \Omega$ into R^d is demonstrably $\mathcal{B}([0, t]) \otimes \mathcal{F}_t$-measurable. Besides, by right-continuity, we have: $\lim_{n \to \infty} X_s^{(n)}(\omega) = X_s(\omega)$, $(s, \omega) \in [0, t] \times \Omega$. Therefore, the map $(s, \omega) \to X_s(\omega)$ is also $\mathcal{B}([0, t]) \otimes \mathcal{F}_t$-measurable. □

Without using the regularity of the sample paths, we have an important result

Proposition 2.3 *Let $(X_t)_{t \geq 0}$ be a random process with values in a metric space $(E, \mathcal{E})$(equipped with its Borel σ-field). Then, X_t is progressively measurable.*

Proof To show that $(X_t)_{t \geq 0}$ is progressively measurable, we need to demonstrate that for each $t \geq 0$, the mapping $X : [0, t] \times \Omega \to E$, defined by $(s, \omega) \mapsto X_s(\omega)$, is $\mathcal{B}([0, t]) \otimes \mathcal{F}_t$-measurable, where $\mathcal{B}([0, t])$ is the Borel σ-field on $[0, t]$ and $\mathcal{F}_t$ is the filtration to which X_t is adapted.

Since X_t is a random process with values in $(E, \mathcal{E})$, by definition, for each fixed t, X_t is $\mathcal{F}_t$-measurable. This means that for any $B \in \mathcal{E}$, the set $\{\omega \in \Omega : X_t(\omega) \in B\}$ is in $\mathcal{F}_t$.

To extend this to the progressive measurability over $[0, t] \times \Omega$, we consider the product σ-field $\mathcal{B}([0, t]) \otimes \mathcal{F}_t$. For each $s \in [0, t]$, since X_s is $\mathcal{F}_s$-measurable and

$\mathcal{F}_s \subseteq \mathcal{F}_t$ for $s \leq t$ due to the filtration being increasing, X_s is also $\mathcal{F}_t$-measurable. Therefore, the pre-image of a Borel set $B \in \mathcal{E}$ under X_s is an element of $\mathcal{F}_t$, making the mapping $X : [0, t] \times \Omega \to E$ measurable with respect to $\mathcal{B}([0, t]) \otimes \mathcal{F}_t$.

Thus, $(X_t)_{t \geq 0}$ is progressively measurable, completing the proof. □

2.1.4 Stopping Time and Associated σ-Fields

Definition 2.9 (Stopping Time and Optional Time) A random variable $T : \Omega \to [0, \infty]$ is a stopping time of the filtration $(\mathcal{F}_t)$ if $\{T \leq t\} \in \mathcal{F}_t$, for every $t \geq 0$. The σ-field of the past before T is then defined by

$$\mathcal{F}_T = \{A \in \mathcal{F}_\infty : t \geq 0, A \cap \{T \leq t\} \in \mathcal{F}_t\}.$$

A random variable $T : \Omega \to [0, \infty]$ is an optional time of the filtration $(\mathcal{F}_t)$ if $\{T < t\} \in \mathcal{F}_t$, for every $t \geq 0$. The σ-field of the past before T is then defined by

$$\mathcal{F}_T = \{A \in \mathcal{F}_\infty : t \geq 0, A \cap \{T < t\} \in \mathcal{F}_t\}.$$

In what follows, "stopping time" will mean the stopping time of the filtration $(\mathcal{F}_t)$ otherwise specified. If T is a stopping time, we also have $\{T < t\} \in \mathcal{F}_t$ for every $t > 0$.

Proposition 2.4 *Every random time equal to a non-negative constant is a stopping time. Every stopping time is optional, and the two concepts coincide if the filtration is right-continuous.*

Proof To prove this proposition, we need to establish two main points:

1. Every constant time is a stopping time: Let τ be a random time equal to a non-negative constant c. By definition, a stopping time with respect to a filtration $\{\mathcal{F}_t\}_{t \geq 0}$ is a random time τ such that for every $t \geq 0$, the event $\{\tau \leq t\}$ is in $\mathcal{F}_t$. Since τ is a constant, $\{\tau \leq t\}$ is trivially in $\mathcal{F}_t$ for all $t \geq 0$ (it is either the entire sample space or the empty set). Therefore, τ is a stopping time.

2. Every stopping time is optional; the concepts coincide if the filtration is right-continuous: A stopping time τ is said to be optional if the stochastic process $\mathbf{1}_{\{\tau \leq t\}}$ is adapted to the filtration $\{\mathcal{F}_t\}_{t \geq 0}$. Since $\{\tau \leq t\} \in \mathcal{F}_t$ for a stopping time τ, it follows that $\mathbf{1}_{\{\tau \leq t\}}$ is $\mathcal{F}_t$-measurable, making τ optional.

If the filtration $\{\mathcal{F}_t\}_{t \geq 0}$ is right-continuous, i.e., $\mathcal{F}_t = \mathcal{F}_{t+}$ for all t, then the concepts of stopping times and optional times coincide. This is because the right-continuity of the filtration ensures that the measurability conditions for stopping times and optional times are equivalent.

Thus, we have shown that every random time equal to a non-negative constant is a stopping time and that every stopping time is optional. Moreover, these two concepts coincide if the filtration is right-continuous. □

2.1 Introduction to Stochastic Processes and Filtrations

Let us establish some simple properties of stopping times.

Lemma 2.1 *If T and S are stopping times, then so are TS, $T \vee S$, and $T + S$.*

Proof The first two assertions are trivial by using the definition. For the third one, start with the decomposition, valid for $t > 0$:

$$\{T + S > t\}$$
$$= \{T = 0, S > t\} \cup \{0 < T < t, T + S > t\} \cup \{T > t, S = 0\} \cup \{T \geq t, S > 0\}$$

The $\{T = 0, S > t\}$, $\{T > t, S = 0\}$ and $\{T \geq t, S > 0\}$ events in this decomposition are in $\mathcal{F}_t$, either trivially. As for the second event, we rewrite it as:

$$\bigcup_{r \in Q^+, 0 < r < t} \{t > T > r, S > t - r\}$$

where Q^+ is the set of rational numbers in $[0, \infty)$. Membership in $\mathcal{F}_t$ is now obvious. □

Lemma 2.2 *Let $\{T_n\}_{n=1}^{\infty}$ be a sequence of optional times; then so are*

$$\sup_{n \geq 1} T_n, \quad \inf_{n \geq 1} T_n, \quad \overline{\lim_{n \to \infty}} T_n, \quad \underline{\lim_{n \to \infty}} T_n$$

Furthermore, if the T_ns are stopping times, then so is $\sup_{n \geq 1} T_n$.

Proof From the identities

$$\left\{\sup_{n \geq 1} T_n \leq t\right\} = \bigcap_{n=1}^{\infty} \{T_n \leq t\} \text{ and } \left\{\inf_{n \geq 1} T_n < t\right\} = \bigcup_{n=1}^{\infty} \{T_n < t\},$$

we can obtain the desired results. □

Then we consider the associated σ-fields.

Proposition 2.5 *Consider a complete filtration $(\mathcal{F}_t, t \in [0, \infty])$, and we define $\mathcal{G}_t$, which satisfies $\mathcal{G}_t = \mathcal{F}_{t^+}$ for $t \in [0, \infty]$.*
Let T be a stopping time of the filtration $(\mathcal{G}_t)$; then

$$\mathcal{F}_T = \{A \in \mathcal{F}_\infty : t \geq 0, A \cap \{T \leq t\} \in \mathcal{F}_t\}$$

is a σ-field, and

$$\mathcal{G}_T = \{A \in \mathcal{F}_\infty | t > 0, A \cap \{T < t \in \mathcal{F}_t\}\}.$$

Here, we can have $\mathcal{F}_{T^+} := \mathcal{G}_T$.

Proof To prove that $\mathcal{F}_T$ is a σ-field, we need to show that it is non-empty, closed under complementation, and closed under countable unions.

- Non-emptiness is guaranteed since $\emptyset \in \mathcal{F}_T$ as $\emptyset \cap \{T \leq t\} = \emptyset$ belongs to $\mathcal{F}_t$ for all t.
- Closure under complementation: If $A \in \mathcal{F}_T$, then $A^c \in \mathcal{F}_\infty$, and for all $t \geq 0$, $(A^c \cap \{T \leq t\}) = (A \cap \{T \leq t\})^c \in \mathcal{F}_t$ since $\mathcal{F}_t$ is a σ-field.
- Closure under countable unions: If $A_i \in \mathcal{F}_T$ for all $i \in \mathbb{N}$, then $\bigcup_i A_i \in \mathcal{F}_\infty$, and for all $t \geq 0$, $\left(\bigcup_i A_i\right) \cap \{T \leq t\} = \bigcup_i (A_i \cap \{T \leq t\}) \in \mathcal{F}_t$.

Thus, $\mathcal{F}_T$ satisfies all the conditions of a σ-field.

Now, considering $\mathcal{G}_T$, by definition, for any $A \in \mathcal{G}_T$, and for all $t > 0$, we have $A \cap \{T < t\} \in \mathcal{F}_{t^+}$. Since $\mathcal{F}_{t^+} = \mathcal{G}_t$ by the filtration's definition, and T is a stopping time with respect to $\mathcal{G}_t$, it follows that $\mathcal{G}_T$ is equivalent to $\mathcal{F}_{T^+}$.

Therefore, $\mathcal{F}_{T^+} = \mathcal{G}_T$ as required. □

In the following, some properties of stopping times and of associated σ-fields are given, which are necessary for the rest of the chapter.

Proposition 2.6 (Properties of Stopping Times and of the Associated σ-fields [2])

(a) For every stopping time T, we have $\mathcal{F}_T \subset \mathcal{F}_{T^+}$. If the filtration $(\mathcal{F}_t)$ is right-continuous, we have $\mathcal{F}_T = \mathcal{F}_{T^+}$.
(b) If $T = t$ is a constant stopping time, then $\mathcal{F}_T = \mathcal{F}_t$, and $\mathcal{F}_{T^+} = \mathcal{F}_{t^+}$.
(c) Let T be a stopping time. Then T is $\mathcal{F}_T$-measurable.
(d) Let T be a stopping time and $A \in \mathcal{F}_\infty$. Set:

$$T^A(\omega) = \begin{cases} T(\omega) & if\ \omega \in A. \\ +\infty & if\ \omega \notin A \end{cases}$$

Then $A \in \mathcal{F}_T$ if and only if T^A is a stopping time.
(e) Let S, T be two stopping times such that $S \leq T$. Then $\mathcal{F}_S \subset \mathcal{F}_T$, and $\mathcal{F}_{S^+} \subset \mathcal{F}_{T^+}$.
(f) If (S_n) is a monotone decreasing sequence of stopping times, then $S = \lim_{n\to\infty} S_n$ is a stopping time of the filtration $(\mathcal{F}_{t^+})$ (the definition of $\mathcal{F}_{t^+}$ is given in Proposition 2.5).

$$\mathcal{F}_{S^+} = \bigcap_n \mathcal{F}_{S_n^+}$$

(g) If (S_n) is a monotone decreasing sequence of stopping times, which is also stationary (in the sense that, for every ω, there exists an integer $N(\omega)$ such that $S_n(\omega) = S(\omega)$ for every $n \geq N(\omega)$), then $S = \lim_{n\to\infty} S_n$ is also a stopping time.

$$\mathcal{F}_S = \bigcap_{1 \leq n} \mathcal{F}_{S_n}$$

2.1 Introduction to Stochastic Processes and Filtrations

(h) Let T be a stopping time. A function $\omega \to Y(\omega)$ defined on the set $\{T < \infty\}$ and taking values in the measurable set $(E, \mathcal{E})$ is $\mathcal{F}_T$-measurable if and only if, for every $t \geq 0$, the restriction of Y to the set is $\mathcal{F}_T$-measurable.

Proof

(a), (b), and (c) are almost immediately from our definitions. Let us prove the other statements.

(d) For every $t \geq 0$,

$$\{T^A \leq t\} = A \cap \{T \leq t\}$$

and the result follows from the definition of $\mathcal{F}_T$.

(e) It is enough to prove that $\mathcal{F}_S \subset \mathcal{F}_T$. If $A \in \mathcal{F}_S$, we have

$$A \cap \{T \leq t\} = (A \cap \{S \leq t\}) \cap \{T \leq t\} \in \mathcal{F}_t,$$

hence $A \in \mathcal{F}_T$.

(f) We have for every $t \geq 0$,

$$\{S \leq t\} = \bigcap_{1 \leq n} \{S_n \leq t\} \in \mathcal{F}_t.$$

(g) Similarly

$$\{S < t\} = \bigcup_{1 \leq n} \{S_n < t\} \in \mathcal{F}_t.$$

Then by (e), we have $\mathcal{F}_{S^+} \subset \mathcal{F}_{S_n^+}$ for every n. And conversely, if $A \in \bigcap_n \mathcal{F}_{S_n^+}$,

$$A \cap \{S < t\} = \bigcup_{1 \leq n}(A \cap \{S_n < t\}) \in \mathcal{F}_t,$$

hence $A \in \mathcal{F}_{S^+}$.

(h) First assume that, for every $t \geq 0$, the restriction of Y to $\{T \leq t\}$ is $\mathcal{F}_t$-measurable subset A of E,

$$\{Y \in A\} \cap \{T \leq t\} \in \mathcal{F}_t$$

Letting $t \to \infty$, we first obtain that $\{Y \in A\} \in \mathcal{F}_\infty$, and then we deduce from the previous display that $\{Y \in A\} \in \mathcal{F}_T$ and thus $\{Y \in A\} \cap \{T \leq t\} \in \mathcal{F}_t$, giving the desired result.

□

2.2 Discrete Martingales

We use the notation $\mathbb{N} = \{0, 1, 2, 3, \ldots\}$. Let us start by recalling the basic definitions. We consider a probability space $(\Omega, \mathcal{F}, P)$, and we fix a discrete filtration, that is, an increasing sequence $(\mathcal{G}_n)_{n \in \mathbb{N}}$ of sub-σ-fields of $\mathcal{F}$. We also let

$$\mathcal{G}_\infty = \bigcup_{n=0}^{\infty} \mathcal{G}_n$$

Definition 2.10 A sequence $(Y_n)_{n \in \mathbb{N}}$ of integrable random variables, such that Y_n is $\mathcal{G}_n$-measurable for every $n \in \mathbb{N}$, is called:

- a martingale if, whenever $0 \leq m < n$, $E[Y_n | \mathcal{G}_m] = Y_m$;
- a supermartingale if, whenever $0 \leq m < n$, $E[Y_n | \mathcal{G}_m] \leq Y_m$;
- a submartingale if, whenever $0 \leq m < n$, $E[Y_n | \mathcal{G}_m] \geq Y_m$.

All these notions depend on the choice of the filtration $\mathcal{G}_n$, which is fixed in what follows.

Theorem 2.1 (Maximal Inequality) *If $(Y_n)_{n \in \mathbb{N}}$ is a submartingale, then, for every $\lambda > 0$ and every $k \in \mathbb{N}$,*

$$\lambda P\left(\sup_{n \leq k} |Y_n| > \lambda\right) \leq E[|Y_k|]$$

Proof Define the stopping time $\tau = \min\{n \leq k : |Y_n| > \lambda\}$, with the convention that $\min \emptyset = \infty$. Consider the stopped process $Y_n^\tau = Y_{\min(n, \tau)}$. By the optional stopping theorem, since (Y_n) is a supermartingale and τ is a bounded stopping time, we have $E[Y_\tau] \leq E[Y_0]$.

For any $n \leq k$, if $\tau \leq k$, then $|Y_\tau| > \lambda$. Thus,

$$\lambda P\left(\sup_{n \leq k} |Y_n| > \lambda\right) = \lambda P(\tau \leq k) \leq E[|Y_\tau|].$$

Since $|Y_\tau| \leq |Y_k|$ for $\tau \leq k$, and $E[|Y_\tau|] \leq E[|Y_k|]$ for $\tau > k$ (as (Y_n) is a supermartingale), we conclude

$$E[|Y_\tau|] \leq E[|Y_k|].$$

Combining the inequalities, we get

$$\lambda P\left(\sup_{n \leq k} |Y_n| > \lambda\right) \leq E[|Y_k|].$$

This completes the proof of the maximal inequality for supermartingales. □

2.2 Discrete Martingales

Theorem 2.2 (Doob's Inequality in L^p) *If $(Y_n)_{n \in \mathbb{N}}$ is a martingale, and every $k \in \mathbb{N}$, $p > 1$, we have*

$$E\left[\sup_{n \leq k} |Y_n|^p\right] \leq \left(\frac{p}{p-1}\right)^p E[|Y_k|^p]$$

Proof To prove this inequality, we use the Doob's maximal inequality for martingales and the concept of conjugate exponents. Let q be such that $\frac{1}{p} + \frac{1}{q} = 1$. By integration by part and for any random variable X, $r > 1$, we can have

$$E[|X|^r] = \int_0^\infty r x^{r-1} P(|X| \geq x) dx.$$

Now, consider the supermartingale $Y_k^* := \sup_{n \leq k} |Y_n|$, and we choose $r = p$, which yields

$$E[|Y_k^*|^p] = p \int_0^\infty \lambda^{p-1} p\left(|Y_k^*| \geq \lambda\right) d\lambda \leq p \int_0^\infty \lambda^{p-1} \cdot \frac{1}{\lambda} \left(\int_{\{|Y_k^*| \geq \lambda\}} |Y_k| d\lambda\right), \tag{2.1}$$

where the last inequality holds that by Doob's maximal inequality.

Now, we consider to estimate the $p \int_0^\infty \lambda^{p-1} \cdot \frac{1}{\lambda} \left(\int_{\{|Y_k^*| \geq \lambda\}} |Y_k^*| d\lambda\right)$. Using the switch two integration, we have

$$p \int_0^\infty \lambda^{p-1} \cdot \frac{1}{\lambda} \left(\int_{\{|Y_k| \geq \lambda\}} |Y_k| d\lambda\right) = p \int_\Omega Y_k \int_0^\infty \lambda^{p-2} 1_{\{|Y_k^*| \geq \lambda\}} d\lambda dP$$

$$= p \int_\Omega \left(|Y_k| \int_0^{n+2} \lambda^{p-2} d\lambda\right) dP = p \cdot E\left[Y_k \cdot \int_0^{|Y_k|} \lambda^{p-2} d\lambda\right] \tag{2.2}$$

$$= \frac{p}{p-1} E[Y_k |Y_k|^{(p-1)}] \leq \frac{p}{p-1} E[|Y_k|^p]^{\frac{1}{p}} \cdot E[|Y_k^*|^{(p-1)q}]^{\frac{1}{q}}$$

where the inequality uses the Holder inequality and $\frac{1}{p} + \frac{1}{q} = 1$. Here, we divide the $E[|Y_k^*|^{(p-1)q}]^{\frac{1}{q}} = E[|Y_k^*|^p]^{\frac{1}{q}}$, and combining (2.1) and (2.2), we can finish the proof. □

If $y = (y_n)_{n \in \mathbb{N}}$ is a sequence of real numbers, and a< b, the crossing number of this sequence along $[a, b]$ before time n, denoted by $U_F(a, b, n)$, is the largest integer k such that there exists a strictly increasing finite sequence

$$m_1 < n_1 < m_2 < n_2 < \cdots < m_k < n_k$$

of nonnegative integers smaller than or equal to n with the properties $y_{m_i} \leq a$ and $y_{n_i} \geq b$, for every $i \in \{1, 2, \ldots k\}$. In what follows, we consider a sequence $Y = (Y_n)_{n \in \mathbb{N}}$ of real random variables, and the associated upcrossing number $U_F(a, b, n)$ is then an integer-valued random variable.

Theorem 2.3 (Doob's Upcrossing Inequality for Continuous-Time Submartingales) *If $(X_t)_{t \in \mathbb{R}}$ is a submartingale with right-continuous sample paths, then for every $a < b$ and letting $[\sigma, \tau]$ be a subinterval of $[0, \infty)$,*

$$E[D_{[\sigma,\tau]}(a,b,n)] \leq \left(\frac{1}{b-a}\right) E[(X_\tau - a)^+],$$

$$E[U_{[\sigma,\tau]}(a,b,n)] \leq \left(\frac{E[X_\tau^+] + \|a\|}{b-a}\right).$$

Proof To extend the discrete version of Doob's upcrossing inequality to the continuous setting, we first define the continuous-time analogues of upcrossings and downcrossings within the interval $[\sigma, \tau]$.

Step 1: Approximation by discretization. Consider a sequence of partitions of the interval $[\sigma, \tau]$ with mesh going to zero. For each partition, define a discrete-time process by sampling X_t at the partition points. Each sampled process is a submartingale by the submartingale property of X_t.

Step 2: Application of discrete upcrossing inequality. For each discrete approximation, apply the discrete version of Doob's upcrossing inequality to obtain:

$$E[D_F(a,b,n)] \leq \left(\frac{1}{b-a}\right) E[(X_\tau - a)^+] \text{ for downcrossings,}$$

$$E[U_F(a,b,n)] \leq \left(\frac{E[X_\tau^+] + \|a\|}{b-a}\right) \text{ for upcrossings.}$$

Step 3: Passing to the limit. As the mesh of the partition goes to zero, the expected number of upcrossings and downcrossings in the discrete approximations converge to the expected number of upcrossings and downcrossings in the continuous-time process. The right-continuous sample paths of X_t ensure that the limits of the expectations on the right-hand side converge to the corresponding expectations for the continuous-time process.

Thus, we conclude that:

$$E[D_{[\sigma,\tau]}(a,b,n)] \leq \left(\frac{1}{b-a}\right) E[(X_\tau - a)^+],$$

$$E[U_{[\sigma,\tau]}(a,b,n)] \leq \left(\frac{E[X_\tau^+] + \|a\|}{b-a}\right).$$

2.2 Discrete Martingales

This completes the proof by showing that the continuous-time upcrossing and downcrossing expectations are bounded as stated. □

This inequality is a crucial tool for proving the convergence theorems for discrete-time martingales and supermartingales. Let us recall two important instances of these theorems.

Theorem 2.4 (Convergence Theorem for Discrete-Time Submartingales [1]) *If $(Y_n)_{n \in \mathbb{N}}$ is a submartingale, and if the sequence $(Y_n)_{n \in \mathbb{N}}$ is bounded in L^1, then then there exists a random variable $Y_\infty \in L^1$ such that*

$$\lim_{n \to \infty} Y_n = Y_\infty \quad a.s.$$

Proof Given that $(Y_n)_{n \in \mathbb{N}}$ is bounded in L^1, it follows that $\sup_n E[|Y_n|] < \infty$. This implies, by the Markov inequality, that for any $\epsilon > 0$, $P(|Y_n| > K) \leq \frac{E[|Y_n|]}{K}$ for any $K > 0$, which can be made arbitrarily small by choosing K large enough. Hence, $(Y_n)_{n \in \mathbb{N}}$ is uniformly integrable.

By Doob's upcrossing inequality, for any $a < b$ and $n \in \mathbb{N}$,

$$E[U_F(a, b, n)] \leq \left(\frac{1}{b-a}\right) E[(Y_n - a)^+].$$

As $(Y_n)_{n \in \mathbb{N}}$ is uniformly integrable and bounded in L^1, the expected number of upcrossings is finite. This implies that the number of upcrossings of any interval $[a, b]$ by the sequence (Y_n) is almost surely finite. Therefore, (Y_n) must converge almost surely to a limit Y_∞.

To show $Y_\infty \in L^1$, we use the fact that $(Y_n)_{n \in \mathbb{N}}$ is uniformly integrable. This implies that the limit Y_∞ is also integrable, and hence $Y_\infty \in L^1$. This completes the proof of the convergence theorem for discrete-time submartingales. □

Theorem 2.5 (Convergence Theorem for Uniformly Integrable Discrete-Time Martingales [1])

If $(Y_n)_{n \in \mathbb{N}}$ is a martingale, then the following assertions are equivalent:

(i) *The martingale $(Y_n, \mathcal{G}_n, P)_{n \in \mathbb{N}}$ is closed, in the sense that there exists a random variable $Z \in L^1(\Omega, \mathcal{F}, P)$ such that $Y_n = E[Z|\mathcal{G}_n]$ for every $n \in \mathcal{N}$.*
(ii) *The sequence $(Y_n)_{n \in \mathbb{N}}$ converges a.s. and in L^1.*
(iii) *The sequence $(Y_n)_{n \in \mathbb{N}}$ is uniformly integrable.*

Proof (i) ⇒ (ii): If there exists a random variable $Z \in L^1(\Omega, \mathcal{F}, P)$ such that $Y_n = E[Z|\mathcal{G}_n]$ for every $n \in \mathbb{N}$, then by the Martingale Convergence Theorem, $(Y_n)_{n \in \mathbb{N}}$ converges almost surely and in L^1 to $E[Z|\mathcal{G}_\infty]$, where $\mathcal{G}_\infty$ is the sigma-algebra generated by $\bigcup_{n=1}^\infty \mathcal{G}_n$.

(ii) ⇒ (iii): If $(Y_n)_{n \in \mathbb{N}}$ converges almost surely and in L^1, then by Vitali's convergence theorem, the sequence is uniformly integrable.

(iii) ⇒ (i): If $(Y_n)_{n\in\mathbb{N}}$ is uniformly integrable, then again by the Martingale Convergence Theorem, there exists a random variable Y_∞ such that $(Y_n)_{n\in\mathbb{N}}$ converges to Y_∞ almost surely and in L^1. Define $Z = Y_\infty$, which belongs to $L^1(\Omega, \mathcal{F}, P)$; hence, for each n, $Y_n = E[Z|\mathcal{G}_n]$. This completes the proof of the equivalence of the three conditions. □

We now recall two versions of the optional stopping theorem in discrete time. A (discrete) stopping time is a random variable T with values in $\mathcal{N} \cap \{\infty\}$, such that $\{T = n\} \in \mathcal{G}_n$ for every $n \in \mathbb{N}$. The σ-field of the past before T is then $\mathcal{G}_T = \{A \in \mathcal{G}_\infty | A \cap \{T \leq n\} \in \mathcal{G}_n$ for every $n \in \mathbb{N}\}$.

Theorem 2.6 (Optional Stopping Theorem for Uniformly Integrable Discrete-Time Martingales [1]) *Let $(Y_n)_{n\in\mathbb{N}}$ be a uniformly integrable martingale, and let Y_∞ be the a.s. limit of Y_n when $n \to \infty$. Then, for every choice of the stopping times S and T such that $S \leq T$, we have $Y_T \in L^1$ and*

$$Y_S = E[Y_T|\mathcal{G}_S]$$

with the convention that $Y_T = Y_\infty$ on the event $\{T = \infty\}$, and similarly for Y_S.

Proof Since $(Y_n)_{n\in\mathbb{N}}$ is uniformly integrable, it converges both almost surely and in L^1 to Y_∞. This uniform integrability also ensures that for any stopping time T, the stopped process Y_T is integrable, i.e., $Y_T \in L^1$.

For stopping times S and T with $S \leq T$, the martingale property and uniform integrability imply that for any $n \geq S$,

$$E[Y_{n\wedge T}|\mathcal{G}_S] = Y_S.$$

Taking the limit as $n \to \infty$ and using the Dominated Convergence Theorem, we obtain

$$E[Y_T|\mathcal{G}_S] = Y_S,$$

where we have used the fact that $Y_{n\wedge T}$ converges to Y_T as $n \to \infty$, and the conditional expectation is well-defined due to the uniform integrability of the martingale sequence.

This establishes that $Y_S = E[Y_T|\mathcal{G}_S]$, completing the proof. The convention that $Y_T = Y_\infty$ on the event $\{T = \infty\}$ ensures that the theorem also applies in cases where the stopping time T may be infinite. □

Theorem 2.7 (Optional Stopping Theorem for Discrete-Time Supermartingales (Bounded Case))

Let $(Y_n)_{n\in\mathbb{N}}$ be a supermartingale, and for every choice of the bounded stopping times S and T such that $S \leq T$, we have

$$Y_S \geq E[Y_T|\mathcal{G}_S].$$

2.3 Continuous Martingales

Proof Given that $(Y_n)_{n \in \mathbb{N}}$ is a supermartingale, it follows that for all $n \geq m$,

$$E[Y_n | \mathcal{G}_m] \leq Y_m.$$

Now, consider two bounded stopping times S and T with $S \leq T$. We aim to show that

$$Y_S \geq E[Y_T | \mathcal{G}_S].$$

Since S and T are bounded, there exists a maximum time N such that $S, T \leq N$ for all $\omega \in \Omega$. We proceed by conditioning on the filtration $\mathcal{G}_S$ and utilizing the supermartingale property as follows:

$$E[Y_T | \mathcal{G}_S] = E[E[Y_T | \mathcal{G}_N] | \mathcal{G}_S] \leq E[Y_S | \mathcal{G}_S] = Y_S,$$

where the inequality $E[Y_T | \mathcal{G}_N] \leq Y_S$ for $S \leq N$ uses the supermartingale property and the fact that $S \leq T \leq N$. This inequality holds because conditioning on a later time $(\mathcal{G}_N)$ and then taking the conditional expectation given an earlier time $(\mathcal{G}_S)$ preserve the supermartingale inequality due to the tower property of conditional expectation.

Thus, we have shown that for any two bounded stopping times S and T with $S \leq T$, the value of the supermartingale at time S is greater than or equal to the expected value of the supermartingale at time T, conditioned on the information available at time S. This completes the proof of the optional stopping theorem for discrete-time supermartingales in the bounded case. □

2.3 Continuous Martingales

2.3.1 Introduction of Continuous Martingales

Recall that we have fixed a filtered probability space $(\Omega, \mathcal{F}, (\mathcal{F}_t), \mathbf{P})$. In the remaining part of this chapter, all processes take values in R. The following is an obvious analog of the corresponding definition in discrete time.

Definition 2.11 A sequence $(X_t)_{t \in R}$ of integrable random variables, such that X_t is $\mathcal{F}_n$-measurable for every $n \in \mathcal{N}$, is called:

(1) a martingale, if, whenever $0 \leq s < t$, $E[X_t | \mathcal{F}_s] = Y_s$;
(2) a supermartingale, if, whenever $0 \leq s < t$, $E[Y_t | \mathcal{F}_s] < Y_s$;
(3) a submartingale, if, whenever $0 \leq s < t$, $E[Y_t | \mathcal{F}_s] \geq Y_s$;

Example 2.2 We say that a process $(Z_t)_{t \geq 0}$ with values in R or in R^d has independent increments with respect to the filtration $(\mathcal{F}_t)$ if Z is adapted and if, for every $0 \leq s < t$, $Z_t - Z_s$ is independent of $(\mathcal{F}_s)$ (for instance, a Brownian motion has independent increments with respect to its canonical filtration). If Z is a real-valued process having independent increments with respect to $(\mathcal{F}_t)$, then

(i) if $Z_t \in L^1$ for every $t \geq 0$, then $\tilde{Z}_t = Z_t - E[Z_t]$ is a martingale.
(ii) if $Z_t \in L^2$ for every $t \geq 0$, then $Y_t = \tilde{Z}_t^2 - E[\tilde{Z}_t^2]$ is a martingale.
(iii) if, for some $\theta \in R$, we have $E[e^{\theta Z_t}] < \infty$ for every $t \geq 0$, then

$$X_t = \frac{e^{\theta Z_t}}{E[e^{\theta Z_t}]}$$

is a martingale.

Our next goal is to study the regularity properties of sample paths of martingales and supermartingales. We first establish continuous time analogs of classical inequalities in the discrete-time setting.

Theorem 2.8 (Maximal Inequality [1]) *If $(X_t)_{t \in R_+}$ is a submartingale with right-continuous sample paths, then, for every $\lambda > 0$ and every $t > 0$,*

$$\lambda P\left(\sup_{0 \leq s \leq t} |X_s| > \lambda\right) \leq E[X_t^+]$$

Proof Let the finite set F consist of $0, t$, and a finite subset of $[0, t] \cap \mathbb{Q}$. We obtain from [1]: $\mu P[\max_{t \in F} X_t > \mu] \leq E(X_t^+)$ By considering an increasing sequence $\{F_n\}_{n=1}^{\infty}$ of finite sets whose union is the whole of $([0, t] \cap \mathbb{Q}) \cup \{0, t\}$ we may replace F by this union in the preceding inequalities. The right-continuity of sample paths implies that $\mu P[\sup_{0 \leq s \leq t} X_s > \mu] \leq E(X_t^+)$. □

Theorem 2.9 (Doob's Inequality in L^p) *If $(X_t)_{t \in R_+}$ is a martingale with right-continuous sample paths, then, for every $t > 0$, $p > 1$,*

$$E\left[\sup_{0 \leq s \leq t} |X_s|^p\right] \leq \left(\frac{p}{p-1}\right)^p E[|X_t|^p]$$

Proof From the discrete martingales result, we can have, for every $m \geq 1$,

$$E\left[\sup_{s \in D_m} |X_s|^p\right] \leq \left(\frac{p}{p-1}\right)^p E[|X_t|^p].$$

where the $D_m = \{t_0, t_1, \ldots, t_m\}$. It is an increasing union of a sequence $(D_m)_{m \geq 1} \subset D$, in which D is a countable dense subset of R_+.

Then, we just have to let m tend to infinity, using the monotone convergence theorem and then the identity

$$\sup_{s \in D \cap [0,t]} |X_s| = \sup_{s \in [0,t]} |X_s|.$$

□

2.3.2 The Important Inequalities for Martingales

Let $X = \{X_t; 0 \leq t < \infty\}$ be a real-valued stochastic process. As $a < b$, the up-crossing number of this sequence along $[a, b]$ with a finite subset F of $[0, \infty]$, which is denoted by $U_F(a, b, X(\omega))$.

$$\tau_1(\omega) = \min\{t \in F; X_t(\omega) \leq a\}$$

and define for $j = 1, 2, \ldots$.

$$\sigma_j = \min\{t \in F; X_t(\omega) > b\}.$$

$$\tau_{j+1} = \min\{t \in F; X_t(\omega) \leq a\}.$$

The convention here is that the minimum of the empty set is $+\infty$, and we denote by $U_F(a, b, X(\omega))$ the largest integer j for which $\sigma_j(\omega) < \infty$. If $I \subset [0, \infty)$ is not necessarily finite, we define

$$U_I(a, b; X(\omega)) = \sup\{U_F(a, b; X(\omega; F \subset I, F\ is\ finite))\}.$$

Theorem 2.10 *Let $(X_t)_{t \geq 0}$ be a supermartingale, and let D be a countable dense subset of R_+.*

(i) *For almost every $\omega \in \Omega$, the restriction of the function $s \to X_s(\omega)$ to the set D has a right-limit*

$$X_{t^+}(\omega) := \lim_{s \in D \to t^+} X_s(\omega)$$

at every $t \in [0, \infty)$, and a left-limit

$$X_{t^-}(\omega) := \lim_{s \in D \to t^-} X_s(\omega).$$

(ii) *For every $t \in R_+$, $X_{t^+} \in L^1$ and*

$$X_t \geq E[X_{t^+}|\mathcal{F}_t]$$

with equality, if the function $t \to E[X_t]$ is right-continuous (in particular if X is a martingale). The process X_t is a supermartingale with respect to the filtration. It is a martingale if X is a martingale.

Proof

(i) Fix D. We have:

$$\sup_{s \in D \cap [0,t]} |X_s| < \infty. \quad a.s.$$

We can choose a sequence D_m, $m \geq 1$ of finite subsets of D that increase to $D \cap [0, T]$ and are such that $0, T \in D_m$. Doob's upcrossing inequality for discrete supermartingales gives, for every $a < b$ and every $m \geq 1$,

$$E[U_{D_m}(a,b)] \leq \frac{1}{b-a} E[(X_T - a)^-].$$

We let $m \to \infty$ and get by monotone convergence

$$E[U_{\{[0,T] \cap D\}}(a,b)] \leq \frac{1}{b-a} E[(X_T - a)^-] < \infty.$$

We thus have

$$U_{[0,T] \cap D}(a,b) \leq \infty.$$

Set

$$N = \bigcup_{T \in D} \left\{ \sup_{t \in D \cap [0,T]} |X_t| = \infty \right\} \cup \left(\bigcup_{a < b \in Q} \{U_{D \cap [0,T]} = \infty\} \right). \quad (2.3)$$

Then $P(N) = 0$ by the preceding considerations. On the other hand, if $\omega \notin N$, the function $t \to X_t(\omega)$ helps us easily get the conclusion.

(ii) To define $X_{t+}(\omega)$ for every $\omega \in \Omega$ and not only on Ω/N, we set

$$X_{t+}(\omega) = \begin{cases} \lim_{s \to t, s \in D} & \text{if the limit exits} \\ 0 & \text{otherwise} \end{cases}$$

By definition, X_{t+} is $\mathcal{F}_{t+}$-measurable. Fix $t \geq 0$ and choose a sequence $(t_n)_{n \geq 0}$ in D such that t_n decreases strictly to t as $n \to \infty$. Then, by construction, we have a.s.

$$X_{t+} = \lim_{n \to \infty} X_{t_n}.$$

Set $Y_k = X_{t-k}$ for every integer $k \leq 0$; then Y is a backward supermartingale concerning the (backward) discrete filtration, and we have $\sup_{k \leq 0} E[|Y_k|] < \infty$.

2.3 Continuous Martingales

Then the convergence theorem for backward supermartingales implies that the sequence X_{t_n} converges to X_{t^+} in L^1. In particular, $X_{t^+} \in L^1$. Thanks to the L^1-convergence, we can pass to the limit $n \to \infty$ in the inequality $X_t \geq E[X_{t_n}|\mathcal{F}_t]$ and get

$$X_t \geq E[X_{t^+}|\mathcal{F}_t]$$

□

The number of downcrossings $D_I(a, b; X(\omega))$ is defined similarly.

Theorem 2.11 (Doob's Upcrossing Inequality [3]) *If $(X_t)_{t \in R}$ is a submartingale with right-continuous sample paths, then for every $a < b$ and letting $[\sigma, \tau]$ be a subinterval of $[0, \infty)$,*

$$E[D_{[\sigma,\tau]}(a,b,n)] \leq \left(\frac{1}{b-a}\right) E[(X_\tau - a)^+],$$

$$E[U_{[\sigma,\tau]}(a,b,n)] \leq \left(\frac{E[X_\tau^+] + \|a\|}{b-a}\right).$$

Proof The proof of Doob's upcrossing inequality in a continuous setting can be conceptually tied back to its discrete counterpart. However, given the statement involves a continuous-time process, we will sketch an approach that leverages a discretization argument, which indirectly invokes the essence of the discrete upcrossing inequality.

1. **Discretization:** Approximate the continuous-time submartingale $(X_t)_{t \in [\sigma, \tau]}$ by a discrete-time submartingale $(X_{t_n})_{n=0}^N$, where $\sigma = t_0 < t_1 < \cdots < t_N = \tau$ and each t_n is a rational number in $[\sigma, \tau]$. The right-continuous sample paths ensure that we can make this approximation arbitrarily close by refining the partition.
2. **Applying discrete upcrossing inequality:** For each discrete approximation, apply the known discrete version of Doob's upcrossing inequality to obtain

$$E[D_{[\sigma,\tau]}(a,b,n)] \leq \left(\frac{1}{b-a}\right) E[(X_{t_N} - a)^+],$$

$$E[U_{[\sigma,\tau]}(a,b,n)] \leq \left(\frac{E[X_{t_N}^+] + \|a\|}{b-a}\right),$$

where $D_{[\sigma,\tau]}(a, b, n)$ and $U_{[\sigma,\tau]}(a, b, n)$ are the downcrossing and upcrossing numbers for the discretized process.

3. **Taking limits:** As the partition gets finer, $t_N \to \tau$, and by the right-continuity of the sample paths and the Dominated Convergence Theorem, we can pass the limits inside the expectations

$$E[D_{[\sigma,\tau]}(a,b,n)] \leq \left(\frac{1}{b-a}\right) E[(X_\tau - a)^+],$$

$$E[U_{[\sigma,\tau]}(a,b,n)] \leq \left(\frac{E[X_\tau^+] + \|a\|}{b-a}\right).$$

This approach effectively bridges the discrete and continuous realms by leveraging the right-continuous nature of the sample paths to apply discrete martingale results and then transitioning back to the continuous setting. The key step is the discretization and the careful limit process, which ensures the inequalities hold in the continuous setting as well. □

We shall find that RCLL is very important. Next we will give a valid criterion for RCLL.

Theorem 2.12 *Let $X = \{X_t; \mathcal{F}_t, 0 \leq t < \infty\}$ be a submartingale, and assume the filtration $(\mathcal{F}_t)$ satisfies the usual conditions. Then the process X has a right-continuous modification if and only if the function $t \to EX_t$ from $[0,\infty)$ to R is right-continuous. If this right-continuous modification exists, it can be chosen to be RCLL and adapted to $(\mathcal{F}_t)$, hence a submartingale concerning $(\mathcal{F}_t)$.*

Proof Please refer to Theorem 3.18 in [3].
Necessity: Assume X has a right-continuous modification Y. Since Y is right-continuous and adapted to $(\mathcal{F}_t)$, for every t, Y_t is $\mathcal{F}_t$-measurable and $E|Y_t| < \infty$. As X_t and Y_t are modifications of each other, $E[X_t] = E[Y_t]$ for all t. The right-continuity of $t \to EY_t$ follows from the right-continuity of Y and Dominated Convergence Theorem, showing that $t \to EX_t$ is right-continuous.
Sufficiency: Suppose the function $t \to EX_t$ is right-continuous. We construct a right-continuous modification of X by considering the regular conditional probability distribution given $\mathcal{F}_t$. Due to the submartingale property and the right-continuity of EX_t, we can apply a version of the Doob-Dynkin lemma to obtain a sequence of simple processes that converge uniformly to a limit process Y, which is right-continuous.

For the RCLL property, we utilize the fact that every right-continuous function can be approximated by functions that are right-continuous with left limits. Specifically, we can define

$$Y_t = \lim_{s \downarrow t} \sup_{u \in \mathbb{Q}, u > s} X_u,$$

where the limit is taken over the rationals. This definition ensures that Y_t is right-continuous with left limits. The adaptiveness of Y to $(\mathcal{F}_t)$ follows from

2.3 Continuous Martingales

the measurability of the supremum of a countable set of $\mathcal{F}_t$-measurable random variables.

Finally, to show that Y is a submartingale with respect to $(\mathcal{F}_t)$, we need to verify that $E[Y_t|\mathcal{F}_s] \geq Y_s$ for $s \leq t$. This follows from the construction of Y as a limit of $\mathcal{F}_t$-measurable random variables and the submartingale property of X.

Thus, we have shown that if $t \to EX_t$ is right-continuous, then X has an RCLL modification that is a submartingale with respect to $(\mathcal{F}_t)$. □

We start with a convergence theorem for martingales.

Theorem 2.13 (Submartingale Convergence [2]) *Let $X = \{X_t; \mathcal{F}_t, 0 \leq t < \infty\}$ be a right-continuous submartingale, and assume $C := \sup_{t \geq 0} E(X_t^+) < \infty$. Then $X_\infty(\omega) := \lim_{t \to \infty} X_t(\omega)$ exists, and $E|X_\infty| < \infty$.*

Proof The proof involves several steps and utilizes key properties of submartingales and the assumption on the uniform integrability criterion.

Step 1: Uniform integrability. The condition $C := \sup_{t \geq 0} E(X_t^+) < \infty$ implies that the positive part of the submartingale is uniformly integrable. This is because, for any $\epsilon > 0$, there exists a constant K such that for all t, $E[X_t^+ 1_{\{X_t^+ > K\}}] < \epsilon$, due to the uniform integrability criterion.

Step 2: Doob's martingale convergence theorem. By Doob's martingale convergence theorem, since (X_t) is a right-continuous submartingale and is uniformly integrable, it converges almost surely and in L^1 to a limit X_∞ as $t \to \infty$.

Step 3: Existence of X_∞. The almost-sure convergence guarantees that $X_\infty(\omega) = \lim_{t \to \infty} X_t(\omega)$ exists for almost every $\omega \in \Omega$.

Step 4: Finiteness of $E|X_\infty|$. The convergence in L^1 implies that $E|X_\infty| = \lim_{t \to \infty} E|X_t| \leq \limsup_{t \to \infty} E[X_t^+] \leq C < \infty$, ensuring that the expected value of the limit is finite.

Thus, under the condition that $\sup_{t \geq 0} E(X_t^+) < \infty$, a right-continuous submartingale (X_t) converges almost surely and in L^1 to a limit X_∞, for which $E|X_\infty| < \infty$. □

Theorem 2.14 ([2]) *If $(X_t)_{t \in R_+}$ is a martingale with right-continuous sample paths, then the following assertions are equivalent:*

(i) *The martingale $(X_t)_{t \in \mathbb{R}_+}$ is closed.*
(ii) *The sequence $(X_t)_{t \in R_+}$ converges a.s. and in L^1.*
(iii) *The sequence $(X_t)_{t \in R_+}$ is uniformly integrable.*

Proof We prove the equivalence by showing (i) $\Rightarrow$ (ii), (ii) $\Rightarrow$ (iii), and (iii) $\Rightarrow$ (i).

(i) $\Rightarrow$ (ii): If the martingale $(X_t)_{t \in \mathbb{R}_+}$ is closed, it means there exists an integrable random variable X_∞ such that for every t, $X_t = E[X_\infty|\mathcal{F}_t]$. By the Martingale Convergence Theorem, (X_t) converges almost surely and in L^1 to X_∞, proving (ii).

(ii) $\Rightarrow$ (iii): Assuming (ii) that (X_t) converges almost surely and in L^1 to some limit X_∞, we can argue that (X_t) is uniformly integrable. Convergence in L^1 implies $\sup_t E[|X_t|] < \infty$, and almost-sure convergence along with Vitali's convergence theorem ensures that the family (X_t) is uniformly integrable.

(iii) ⇒ (i): Assuming (iii) that (X_t) is uniformly integrable, we apply the Uniform Integrability Convergence Theorem. This theorem states that if a family of integrable random variables is uniformly integrable and converges almost surely, then it also converges in L^1. By the Doob's Martingale Convergence Theorem, (X_t) converges almost surely to some limit X_∞. The uniform integrability of (X_t), combined with its almost-sure convergence, implies that X_∞ is integrable and $E[X_\infty|\mathcal{F}_t] = X_t$ for all t. Thus, (X_t) is closed, proving (i).

Therefore, we have shown that (i), (ii), and (iii) are equivalent for a martingale $(X_t)_{t\in\mathbb{R}_+}$ with right-continuous sample paths. □

We will now use the optional stopping theorems for discrete martingales and supermartingales in order to establish similar results in the continuous-time setting. Let $(X_t)_{t\in R}$ be a martingale or a supermartingale with right-continuous sample paths, such that X_t converges a.s. as $t \to \infty$ to a random variable denoted by X_∞. Then, for every stopping time T, we write X_T for the random variable

$$X_T(\omega) = 1_{\{T(\omega)<\infty\}} X_{T(\omega)}(\omega) + 1_{\{T(\omega)=\infty\}} X_\infty(\omega).$$

The random variable X_T was only defined on the subset $\{T(\omega) < \infty\} X_{T(\omega)}$ of ω. With this definition, the random variable X_T is still $\mathcal{F}_T$-measurable: Use Theorem 3.7 and we can easily verify the fact that $1_{T(\omega)=\infty} X_\infty(\omega)$ is $\mathcal{F}_T$-measurable.

Theorem 2.15 (Optional Sampling) *Let $X = \{X_t; \mathcal{F}_t, 0 \le t < \infty\}$ be a right-continuous submartingale, and let the last time element X_∞ and $S \le T$ be two optional times of the filtration $(\mathcal{F}_t)$. We have*

$$E(X_T|\mathcal{F}_{S^+}) \ge X_S \quad a.s.P.$$

If S is a stopping time, then $\mathcal{F}_S$ can replace $\mathcal{F}_{S^+}$ above. In particular, $EX_T \ge EX_0$.

Proof We divide the proof into two parts: the case where S and T are bounded stopping times and the general case.

Part 1: Bounded stopping times. Assume S and T are bounded by some constant $N < \infty$. By the submartingale property, for any $t \ge 0$, $E(X_{t+\epsilon}|\mathcal{F}_t) \ge X_t$ a.s. for any $\epsilon > 0$. Letting $\epsilon \to 0$, and using the right-continuity of X, we have $E(X_{t^+}|\mathcal{F}_t) \ge X_t$ a.s., where $X_{t^+} = \lim_{\epsilon\to 0^+} X_{t+\epsilon}$. Since S and T are optional times, they are $\mathcal{F}_t$-measurable, and we can apply the tower property of conditional expectation and the submartingale property to get $E(X_T|\mathcal{F}_{S^+}) \ge E(X_S|\mathcal{F}_{S^+}) = X_S$ a.s., where the equality follows from the fact that X_S is $\mathcal{F}_{S^+}$-measurable.

Part 2: General optional times. For the general case, approximate S and T by bounded stopping times $S_n = \min\{S, n\}$ and $T_n = \min\{T, n\}$, which converge to S and T, respectively, as $n \to \infty$. Apply the result from Part 1 to S_n and T_n, yielding $E(X_{T_n}|\mathcal{F}_{S_n^+}) \ge X_{S_n}$ a.s. By the right-continuity of X and the dominated convergence theorem, we can take limits as $n \to \infty$ to conclude $E(X_T|\mathcal{F}_{S^+}) \ge X_S$ a.s.

2.3 Continuous Martingales 57

If S is a stopping time, $\mathcal{F}_{S^+} = \mathcal{F}_S$ because $\mathcal{F}_S$ contains all null sets of $\mathcal{F}$, and any event in $\mathcal{F}_{S^+}$ depends on events strictly before S and thus is measurable with respect to $\mathcal{F}_S$.

Finally, by taking $S = 0$ and noting that X_0 is $\mathcal{F}_0$-measurable, we obtain $E(X_T | \mathcal{F}_0) \geq X_0$ a.s., which implies $EX_T \geq EX_0$ since X_0 is constant. □

2.3.3 The Doob-Meyer Decomposition

Definition 2.12 Consider a probability space $(\Omega, \mathcal{F}, P)$ and a random sequence $(A_n)_{n=0}^{\infty}$ adapted to the discrete filtration $(\mathcal{F}_n)_{n=0}^{\infty}$. The sequence is called increasing, if for P a.e. $\omega \in \Omega$ we have $0 = A_0(\omega) \leq A_1(\omega) \leq \ldots$, and $E[A_n] < \infty$ holds for every $n \geq 1$.

Definition 2.13 An increasing sequence $\{A_n, \mathcal{F}_n; n = 0, 1, \ldots\}$ is called natural if for every bounded martingale $\{M_n, \mathcal{F}_n; n = 0, 1, \ldots\}$ we have

$$E(M_n A_n) = E \sum_{k=1}^{n} M_{k-1}(A_k - A_{k-1}), \quad n \geq 1$$

Proposition 2.7 *An increasing sequence $\{A_n, \mathcal{F}_n; n = 0, 1, \ldots\}$ is predictable if and only if it is natural.*

Proof We prove this proposition in two parts: showing that every predictable sequence is natural and then showing that every natural sequence is predictable.

Predictable ⇒ Natural: Assume $\{A_n, \mathcal{F}_n; n = 0, 1, \ldots\}$ is predictable. This means that for all n, A_n is $\mathcal{F}_{n-1}$-measurable. Let $X = \{X_n, n = 0, 1, \ldots\}$ be a bounded martingale. We need to show that the sequence $\{A_n X_n\}$ is a martingale, which would imply that $\{A_n\}$ is natural.

Since A_n is $\mathcal{F}_{n-1}$-measurable and X_n is a martingale, we have

$$E[A_n X_n | \mathcal{F}_{n-1}] = A_n E[X_n | \mathcal{F}_{n-1}] = A_n X_{n-1}.$$

This shows that $E[A_n X_n] = E[A_{n-1} X_{n-1}]$, proving that $\{A_n X_n\}$ is a martingale, and thus $\{A_n\}$ is natural.

Natural ⇒ Predictable: Assume $\{A_n, \mathcal{F}_n; n = 0, 1, \ldots\}$ is natural. This implies that for any bounded martingale $\{X_n\}$, the product $\{A_n X_n\}$ is a martingale. To show predictability, we must demonstrate that A_n is $\mathcal{F}_{n-1}$-measurable for all n.

Consider a martingale $\{X_n\}$ where $X_n = \mathbf{1}_{\{A_n > a\}}$ for some $a \in \mathbb{R}$. Since $\{A_n\}$ is natural, $\{A_n X_n\}$ is a martingale. Observe that

$$E[A_n X_{n+1} | \mathcal{F}_n] = A_n E[X_{n+1} | \mathcal{F}_n] = A_n X_n,$$

which implies $A_n = E[A_n|\mathcal{F}_{n-1}]$ for all n. Hence, A_n is $\mathcal{F}_{n-1}$-measurable, proving that $\{A_n\}$ is predictable.

Therefore, we conclude that an increasing sequence $\{A_n, \mathcal{F}_n; n = 0, 1, \ldots\}$ is predictable if and only if it is natural. □

Definition 2.14 An adapted process A is called increasing if for $\omega \in \Omega$ we have:

1. $A_0(\omega) = 0$
2. $t \to A_t(\omega)$ is a nondecreasing, right-continuous function, and $E(A_t) < \infty$ holds for every $t \in [0, \infty)$. An increasing process is called integrable if $E(A_\infty) < \infty$, where $A_\infty = \lim_{t \to \infty} A_t$.

Definition 2.15 An increasing process A is called natural if for every bounded right-continuous martingale $\{M_t, \mathcal{F}_t; 0 \le t < \infty\}$ we have

$$E \int_{(0,t]} M_s dA_s = E \int_{(0,t]} M_{s^-} dA_s, \; for \; every \; 0 < t < \infty. \tag{2.4}$$

Remark 2.2

(i) if A is an increasing measurable process and X is a measurable process, then with $\omega \in \Omega$ fixed, the sample path $\{X_t(\omega); 0 \le t < \infty\}$ is a measurable function from $[0, \infty)$ into R. It follows that the Lebesgue-Stieltjes integrals

$$I_t^\pm(\omega) := \int_{(0,.t]} X_s^\pm(\omega) dA_s(\omega)$$

are well-defined. If X is progressively measurable, and if $I_t = I_t^+ - I_t^-$ is well defined and finite for all $t \ge 0$, then I is right-continuous and progressively measurable.

(ii) Every continuous, increasing process is natural. Indeed then, for P-a.e $\omega \in \Omega$, we have

$$\int_{(0,t]} (M_s(\omega) - M_{s^-}(\omega)) dA_s(\omega) = 0 \; for \; every \; 0 < t < \infty,$$

because every path $\{M_s(\omega); 0 \le s < \infty\}$ has only countably many discontinuities points due to that M_s is RCLL.

(iii) It can be shown that every natural increasing process is adapted to the filtration $\{\mathcal{F}_{t^-}\}$.

Lemma 2.3 *In Definition 2.15, Eq. (2.4) is equivalent to*

$$E(M_t A_t) = E \int_{(0,t]} M_{s^-} dA_s. \tag{2.5}$$

2.3 Continuous Martingales

Proof To show the equivalence, we start with the left-hand side of (2.5) and apply the integration by parts formula for semimartingales, which gives

$$M_t A_t = \int_{(0,t]} M_{s-} dA_s + \int_{(0,t]} A_{s-} dM_s + [M, A]_t, \tag{2.6}$$

where $[M, A]_t$ denotes the quadratic covariation of M and A.

Since A is an increasing process, it is of finite variation, and thus the quadratic covariation $[M, A]_t$ is zero. Moreover, because A is increasing, $A_{s-} = A_s$ for all s, making the second integral on the right-hand side of (2.6) equal to zero when M is a martingale (due to the martingale property $E(dM_s|\mathcal{F}_{s-}) = 0$).

Therefore, Eq. (2.6) simplifies to

$$M_t A_t = \int_{(0,t]} M_{s-} dA_s. \tag{2.7}$$

Taking expectations on both sides, and noting that M_t is bounded and A_t is of finite variation, we get

$$E(M_t A_t) = E\left(\int_{(0,t]} M_{s-} dA_s\right), \tag{2.8}$$

which shows that Eq. (2.4) is equivalent to Eq. (2.5). $\square$

Definition 2.16 Let us consider the class $\mathcal{P}_a$ of stopping times T of the filtration $\{\mathcal{F}_t\}$ which satisfy $P(T < \infty) = 1$. The right-continuous process $\{X_t, \mathcal{F}_t; 0 \leq t < \infty\}$ is said to be of class D, if the family $\{X_T\}_{T \in \mathcal{P}}$ is uniformly integrable; of class DL, if the family $\{X_T\}_{T \in \mathcal{P}_a}$ is uniformly integrable, for every $0 < a < \infty$.

Theorem 2.16 (Doob-Meyer Decomposition) *Let $\{\mathcal{F}_t\}$ satisfy the usual condition. If the right-continuous submartingale $X = \{X_t; \mathcal{F}_t, 0 \leq t < \infty\}$ is of class DL, then it admits the decomposition $X_t = M_t + A_t$ (M_t is a right-continuous martingale and A_t is an increasing process). The latter can be taken to be natural; under this additional condition, the decomposition is unique. Further, if X is of class D, then M is a uniformly integrable martingale and A is integrable.*

Proof For the uniqueness, let us assume that X admits both decompositions $X_t = M'_t + A'_t = M_t + A_t$, then $\left\{B_t = A'_t - A_t = M_t - M'_t; 0 \leq t < \infty\right\}$ is martingale, and for all bounded and right-continuous martingale $(\delta)_t$, we have

$$E[\delta_t(A'_t - A_t)] = E\int_{(0,t]} \delta_{s-} dB_s = \lim_{n \to \infty} \sum_{j=1}^{m_n} \delta_{t_{j-1}^{(n)}} [B_{t_j^{(n)}} - B_{t_{j-1}^{(n)}}],$$

where $\Pi_n = \left\{t_0^{(n)}, \ldots, t_{m_n}^{(n)}\right\}$, $n \geq 1$ is a sequence of partitions of $[0, t]$ with $\|\Pi_n\| = \max_{1 \leq j \leq m_n}(t_j^{(n)} - t_{j-1}^{(n)})$ converging to zero as $n \to \infty$, and as such, each Π_{n+1} is a refinement of Π_n. But now

$$E\left[\delta_{t_{j-1}^{(n)}}[B_{t_j^{(n)}} - B_{t_{j-1}^{(n)}}]\right] = 0, \quad \text{and thus} \quad E[\delta_t(A_t' - A_t)] = 0,$$

For an arbitrary bounded random variable δ_t, we can select $\{\delta, \mathcal{F}_t\}$ to be a right-continuous modification of $(E[\delta|\mathcal{F}_t], \mathcal{F}_t)$, and we obtain that $E[\delta(A_t' - A_t)] = 0$ because of the right-continuity now gives us their indistinguishability.

For the existence of the decomposition, please find a detailed proof in book [3]. □

Definition 2.17 A submartingale $\{X_t(\omega); 0 \leq t < \infty\}$ is called regular if for every $a > 0$ and every nondecreasing sequence of stopping times $(T_n)_{n=1}^{\infty} \subset \mathcal{P}_a$ with $T = \lim_{n \to \infty} T_n$, we have $\lim_{n \to \infty} E(X_{T_n}) = E(X_T)$.

Theorem 2.17 *Suppose that $\{X_t(\omega); 0 \leq t < \infty\}$ is a right-continuous submartingale of class DL with respect to the $(\mathcal{F}_t)$, which satisfies the usual conditions, and let $\{A_t, 0 \leq t < \infty\}$ be the natural increasing process in the Doob-Meyer decomposition. The process A is continuous if and only if X is regular.*

Proof Given the Doob-Meyer decomposition $X_t = M_t + A_t$, where X is a right-continuous submartingale of class DL, M_t is a right-continuous martingale, and A_t is a natural increasing process.

To prove that A is continuous if and only if X is regular, we proceed as follows:

($\Rightarrow$) Assume A is continuous. By the definition of a regular submartingale, for every predictable stopping time τ, it holds that $E[X_\tau|\mathcal{F}_{\tau-}] \geq X_{\tau-}$. Since A is continuous, $A_\tau = A_{\tau-}$ for every predictable stopping time τ, implying that the jumps of X at predictable stopping times come solely from M. However, M being a martingale implies $E[M_\tau|\mathcal{F}_{\tau-}] = M_{\tau-}$. Thus, the continuity of A ensures that the jumps of X are controlled, making X regular.

($\Leftarrow$) Conversely, assume X is regular. This implies that for any predictable stopping time τ, the potential discontinuities of X are controlled, which in turn restricts the behavior of A, given that M is a martingale and thus has predictable jumps that are compensated. The regularity of X ensures that its discontinuities are such that they do not contribute to an increase in A in a way that would result in A being discontinuous. Therefore, the only way to maintain the regularity of X through all predictable stopping times is if A itself is continuous, as any discontinuity in A would contradict the submartingale property of X in the context of its predictability and the martingale property of M.

Hence, we conclude that A is continuous if and only if X is regular, completing the proof. □

2.4 Continuous Local Martingale

We consider again a filtered probability space $(\Omega, \mathcal{F}, (\mathcal{F}_t), P)$. If T is a stopping time, and if $X = (X_t)_{t \geq 0}$ is an adapted process with continuous sample paths, we will write X_T for process X stopped at T, defined by $X_t^T = X_{t \wedge T}$ for every $t \geq 0$. It is useful to observe that if S is another stopping time, then

$$(X^T)^S = (X^S)^T = X^{S \wedge T}.$$

Definition 2.18 An adapted process $M = (M_t)_{t \geq 0}$ with continuous sample paths and such that $M_0 = 0$ a.s. is called a continuous local martingale if there exists a non-decreasing sequence $(T_n)_{n \geq 0}$ of stopping times such that $T_n \to \infty (i.e., _n \to \infty)$ for every ω) and, for every n, the stopped process M_{T_n} is a uniformly integrable martingale.

Remark 2.3

(i) We do not require M_t is in L^1 in the definition of a continuous local martingale. (compare with the definition of martingales). In particular, the variable M_0 may be any $\mathcal{F}_0$-measurable random variable.
(ii) Any martingale with continuous sample paths is a continuous local martingale (see property (a) below), but the converse is false, and for this reason, we will sometimes speak of "true martingales" to emphasize the difference with local martingales. Let us give a few examples of continuous local martingales that are not (true) martingales. If B is an $\mathcal{F}_t$-Brownian motion started from 0, and Z is an $\mathcal{F}_0$-measurable random variable, the process $M_t = Z + B_t$ is always a continuous local martingale, but is not a martingale if $E[|Z|] = 1$. If we require the property $M_0 = 0$, we can also consider $M_t = ZB_t$, which is always a continuous local martingale.
(iii) One can define a notion of local martingale with RCLL sample paths. In this course, however, we consider only continuous local martingales.

Proposition 2.8 ([4]) *Let M be a martingale. Then the following properties hold:*

(a) A martingale with continuous sample paths is a continuous local martingale.
(b) In the definition of a continuous local martingale starting from 0, one can replace "uniformly integrable martingale" with "martingale" (indeed, one can then observe that $M^{T_n \wedge n}$ is uniformly integrable, and we still have $T_n \wedge n \to \infty$).
(c) If M is a continuous local martingale, then, for every stopping time T, M_T is a continuous local martingale.
(d) If (T_n) reduces M and if (S_n) is a sequence of stopping times such that $S_n \to \infty$, then the sequence $T_n \wedge S_n$ also reduces M.
(e) The space of all continuous local martingales is a vector space.

Proof

(a) By definition, a martingale M with continuous sample paths satisfies for all $t \geq 0$, $E[M_t|\mathcal{F}_s] = M_s$ for $s \leq t$, and the continuity of paths implies it is also a continuous local martingale since it can be approximated locally by bounded martingales due to its continuous paths.

(b) Starting from 0, a continuous process M is a local martingale if there exists a sequence of stopping times (T_n) increasing to infinity such that M^{T_n} is a martingale for each n. If M is a martingale, then $M^{T_n \wedge n}$ is also a martingale, and since $T_n \wedge n \to \infty$, it implies that M can be considered a continuous local martingale. The uniform integrability follows from the fact that $M^{T_n \wedge n}$ is bounded for each n by the martingale property and Doob's martingale convergence theorem.

(c) Given a continuous local martingale M and a stopping time T, the stopped process M_T is defined by $M_T(t) = M(t \wedge T)$. Since the stopping at T preserves the martingale properties and continuity, M_T remains a continuous local martingale.

(d) If (T_n) reduces M and (S_n) is such that $S_n \to \infty$, then for each n, $M^{T_n \wedge S_n}$ is a martingale because both T_n and S_n are stopping times, and their minimum also constitutes a stopping time. The fact that $T_n \wedge S_n$ increases to infinity ensures that M is locally a martingale with respect to this new sequence, thus reducing M.

(e) To prove that continuous local martingales form a vector space, we need to show that the sum of two continuous local martingales is again a continuous local martingale and that scalar multiples of continuous local martingales are also continuous local martingales. This follows from the linearity of expectation and the preservation of the martingale property under linear operations. □

Proposition 2.9

(i) A nonnegative continuous local martingale M such that $M_0 \in L^1$ is a supermartingale.

(ii) A continuous local martingale M such that there exists a random variable $Z \in L^1$ with $|M_t| \leq Z$ for every $t \geq 0$ (in particular a bounded continuous local martingale) is a uniformly integrable martingale.

(iii) If M is a continuous local martingale and $M_0 = 0$ (or more generally $M_0 \in L^1$, the sequence of stopping times

$$T_n = \inf\{t \geq 0;\ |M_t| \geq n\}$$

reduces M.

2.4 Continuous Local Martingale

Proof

(i) Write $M_t = M_0 + N_t$. By definition, there exists a sequence (T_n) of stopping times that reduce N. Then, if $s \leq t$, we have for every n

$$N_{s \wedge T_n} = E[N_{t \wedge T_n} | \mathcal{F}_s].$$

We can add on both sides the random variable M_0 (which is $\mathcal{F}_0$-measurable and in L^1 by assumption), and we get

$$M_{s \wedge T_n} = E[M_{t \wedge T_n} | \mathcal{F}_s],$$

since M takes nonnegative values, we can now let n tend to ∞ and apply the version of Fatou's lemma for conditional expectations, which gives

$$M_s \geq E[M_t | \mathcal{F}_s].$$

Taking $s = 0$, we get $E[M_t] \leq E[M_0] < \infty$; hence, $M_t \in L^1$ for every $t \geq 0$. The previous inequality now shows that M is a supermartingale.

(ii) By the same argument as in (i), we get for $0 \leq s \leq t$,

$$M_{s \wedge T_n} = E[M_{t \wedge T_n} | \mathcal{F}_s], \qquad (2.9)$$

Since $|M_{t \wedge T_n}| \leq Z$, we can use dominated convergence to obtain that the sequence $M_{t \wedge T_n}$ converges to M_t in L^1. We can thus pass to the limit $n \to \infty$ in Theorem 2.9 and get that $M_s = E[M_t | \mathcal{F}_s]$.

(iii) Suppose that $M_0 = 0$. The random times T_n are stopping times. The desired result is an immediate consequence of (ii) since M^{T_n} is a continuous local martingale and $|M^{T_n}| \leq n$. If we only assume that $M_0 \in L^1$, we observe that M_{T_n} is dominated by $n + |M_0|$. □

Theorem 2.18 *Let M be a continuous local martingale. Assume that M is also a finite variation process (in particular $M_0 = 0$). Then $M_t = 0$ for every $t \geq 0$, a.s.*

Proof Set

$$\tau_n = \inf\left\{ t \geq 0 : \int_0^t |dM_s| \geq n \right\}.$$

For every integer $n \geq 0$, τ_n is a stopping time (recall that $\int_0^t |dM_s|$ is an increasing process if M is a finite variation process). Fix $n \geq 0$ and set $N = M^{\tau_n}$. Note that, for every $t \geq 0$,

$$|N_t| = |M_{t \wedge \tau_n}| \leq \int_0^{t \wedge \tau_n} |dM_s| \leq n.$$

By Proposition 2.8, N is a martingale. Let $t \geq 0$ and $0 = t_0 < t_1 < \cdots < t_p = t$ be any subdivision of $[0, t]$. Then, we have

$$E[N_t^2] = \sum_{i=1}^{p} E[(N_{t_i} - N_{t_{i-1}})^2] \tag{2.10}$$

$$\leq E[(\sup_{1 \leq i \leq p} |N_{t_i} - N_{t_{i-1}}|) \sum_{i=1}^{p} |N_{t_i} - N_{t_{i-1}}|] \tag{2.11}$$

$$\leq n E[\sup_{1 \leq i \leq p} |N_{t_i} - N_{t_{i-1}}|], \tag{2.12}$$

noting that $\int_0^t |dN_s| \leq n$ by the definition of τ.

We now apply the preceding bound to a sequence $0 = t_0^k < t_1^k < \cdots < t_{p_k}^k = t$ of subdivisions of $[0, t]$ whose mesh tends to 0. Using the continuity of sample paths, and with the fact that N is bounded (to justify dominated convergence), we get

$$\lim_{k \to \infty} E[\sup_{1 \leq i \leq p_k} |N_{t_i^k} - N_{t_{i-1}^k}|] = 0.$$

We then conclude that $E[N_t^2] = 0$; hence, $M_{t \wedge n} = 0$. Letting n tend to ∞, we get that $M_t = 0$. □

2.5 Square-Integrable Martingale

2.5.1 The Quadratic Variation of a Continuous Martingale

Definition 2.19 Let $X = \{X_t; \mathcal{F}_t, 0 \leq t < \infty\}$ be a right-continuous martingale. We say that X is square-integrable if $EX_t^2 < \infty$ for every $t \geq 0$. If, in addition, $X_0 = 0$, we write $X \in \mathcal{M}_2$ (or $X \in \mathcal{M}_2^c$, if X is also continuous).

Definition 2.20 For $X \in \mathcal{M}_2$, we define the quadratic variation of X to be the process $\langle X \rangle_t := A_t$, where A is the natural increasing process in the Doob-Meyer decomposition of X^2.

Definition 2.21 For two martingales X, Y of $\mathcal{M}_2$, we define the cross variation of X to be the process $\langle X, Y \rangle_t$ by

$$\langle X, Y \rangle_t = \frac{1}{4}(\langle X + Y \rangle_t - \langle X - Y \rangle_t)$$

and observe that $X_t Y_t - \langle X, Y \rangle_t$ is a martingale. Two elements X, Y of $\mathcal{M}_2$ are called orthogonal if $\langle X, Y \rangle_t = 0$ for every $0 \leq t < \infty$.

Easily we can find the important properties for $\langle \cdot, \cdot \rangle$

2.5 Square-Integrable Martingale

Proposition 2.10 ([2]) $\langle \cdot, \cdot \rangle$ *is a bilinear form on* $\mathcal{M}_2$, *i.e., for any members* X, Y, Z *of* $\mathcal{M}_2$ *and real numbers* a, b, *we have:*

(i) $\langle aX + bY, Z \rangle = a \langle X, Z \rangle + b \langle Y, Z \rangle$,
(ii) $\langle X, Y \rangle = \langle Y, X \rangle$,
(iii) $|\langle X, Y \rangle|^2 \leq \langle X, X \rangle \langle Y, Y \rangle$.

Proof They follow from $(aX + bY)Z = aXZ + bYZ$, $XY = YX$, $X^2 \geq 0$ and Cauchy-Schwarz inequality. □

The use of the term quadratic variation may appear to be unfounded. Indeed, a more conventional use of this term is the following: Let $X = \{X_t; \mathcal{F}_t, 0 \leq t < \infty\}$ be a process, and it is a partition of $[0, t]$. Define the p-th variation of X of over the partition Π to be

$$V_t^{(p)}(\Pi) = \sum_{k=1}^{m} |X_{t_k} - X_{t_{k-1}}|^p.$$

Now define the mesh of the partition Π as $\|\Pi\| = \max_{1 \leq k \leq m} |t_k - t_{k-1}|$. If $V_t^{(2)}(\Pi)$ converges in some sense as $\|\Pi\| \to 0$, the limit is called the quadratic variation of X on $[0, t]$.

Theorem 2.19 ([3]) *Let X be in $\mathcal{F}_2$, and it is continuous. For Π of $[0, t]$, we have* $\lim_{\|\Pi\| \to 0} V_t^{(2)}(\Pi) = \langle X \rangle$, *(in probability); for every $\epsilon > 0, \eta > 0$, there exists $\delta > 0$ such that $\|\Pi\| < \delta$ implies*

$$P[|V_t^{(2)}(\Pi) - \langle X, X \rangle_t| > \epsilon] < \eta.$$

Proof Let $X = \{X_t; \mathcal{F}_t, 0 \leq t < \infty\}$ be a continuous process in $\mathcal{F}_2$. Consider a partition $\Pi = \{0 = t_0 < t_1 < \ldots < t_m = t\}$ of the interval $[0, t]$.

The p-th variation of X over the partition Π is defined as

$$V_t^{(p)}(\Pi) = \sum_{k=1}^{m} |X_{t_k} - X_{t_{k-1}}|^p.$$

We focus on $p = 2$, which gives us the quadratic variation.

Assume that $\|\Pi\| = \max_{1 \leq k \leq m} |t_k - t_{k-1}| \to 0$. We want to show that $V_t^{(2)}(\Pi)$ converges to $\langle X \rangle_t$ in probability.

Given any $\epsilon > 0$ and $\eta > 0$, we will demonstrate that there exists $\delta > 0$ such that if $\|\Pi\| < \delta$, then $P[|V_t^{(2)}(\Pi) - \langle X, X \rangle_t| > \epsilon] < \eta$.

By definition of quadratic variation for continuous semimartingales, we know that as $\|\Pi\| \to 0$, the sums $\sum_{k=1}^{m} (X_{t_k} - X_{t_{k-1}})^2$ approximate the increasing process $\langle X \rangle_t$, which is the limit of these sums as the mesh of the partition goes to zero, reflecting the accumulated "energy" of the process X over $[0, t]$.

The convergence in probability follows from the fact that, for continuous processes, the variance of these sums (which measure the mean quadratic fluctuation of the process around its path) tends to zero as the mesh of the partition goes to zero. This is due to the continuity of the sample paths of X and the properties of the predictable quadratic variation process $\langle X \rangle_t$ associated with X.

Therefore, by taking δ sufficiently small, we ensure that the probability that the quadratic variation $V_t^{(2)}(\Pi)$ deviates from $\langle X, X \rangle_t$ by more than ϵ is less than η, which proves the theorem. $\square$

Definition 2.22 For any $X \in \mathcal{M}_2$, and $0 \leq t < \infty$, we define

$$\|X\|_t := \sqrt{E[X_t^2]}.$$

We also set

$$\|X\| := \sum_{n=1}^{\infty} \frac{\|X\|_n \wedge 1}{2^n}.$$

Proposition 2.11 *Under the preceding metric, $\mathcal{M}_2$ is a complete metric space, and $\mathcal{M}_2^c$ a closed subspace of $\mathcal{M}_2$.*

Proof Let $\mathcal{M}_2$ denote the space of square-integrable martingales and $\mathcal{M}_2^c$ the subspace consisting of continuous square-integrable martingales. We define the metric on $\mathcal{M}_2$ by $d(X, Y) = (\mathbb{E}[\langle X - Y, X - Y \rangle_\infty])^{1/2}$, where $\langle \cdot, \cdot \rangle$ denotes the quadratic variation.

Completeness of $\mathcal{M}_2$: To show that $\mathcal{M}_2$ is complete, consider a Cauchy sequence $(X^n)_{n \in \mathbb{N}}$ in $\mathcal{M}_2$. For every $\epsilon > 0$, there exists $N \in \mathbb{N}$ such that for all $m, n \geq N$, $d(X^n, X^m) < \epsilon$. This implies that $(X^n)_{n \in \mathbb{N}}$ converges in L^2 to some limit process X. Since the space of square-integrable martingales is closed under L^2 convergence, $X \in \mathcal{M}_2$, proving that $\mathcal{M}_2$ is complete.

Closedness of $\mathcal{M}_2^c$: To prove that $\mathcal{M}_2^c$ is a closed subspace, let $(X^n)_{n \in \mathbb{N}}$ be a sequence in $\mathcal{M}_2^c$ converging to some limit X in the metric of $\mathcal{M}_2$. The convergence in L^2 implies that $X^n \to X$ in probability, and since each X^n is continuous, it follows that X is also continuous almost surely. Therefore, $X \in \mathcal{M}_2^c$, proving that $\mathcal{M}_2^c$ is a closed subspace of $\mathcal{M}_2$. $\square$

2.5.2 The Quadratic Variation of a Continuous Local Martingale

From now on until the end of this chapter, we assume that the filtration $\mathcal{F}_t$ is complete. The next theorem will play a very important role in forthcoming developments.

2.5 Square-Integrable Martingale

Let $M = (M_t)_{t \geq 0}$ be a continuous local martingale. There exists an increasing process denoted by $(\langle M, M \rangle_t)_{t \geq 0}$, which is unique up to indistinguishability, such that $M_t^2 - \langle M, M \rangle_t$ is a continuous local martingale. Furthermore, for every fixed $t > 0$, if $0 = t_0^n < t_1^n < \ldots t_{p_n}^n = t$ is an increasing sequence of subdivisions of $[0, t]$ with mesh tending to 0, we have

Proposition 2.12 *For a martingale M, the quadratic variation of M is defined as*

$$\langle M, M \rangle_t = \lim_{n \to \infty} \sum_{i=1}^{p_n} (M_{t_i^n} - M_{t_{i-1}^n})^2 \tag{2.13}$$

in probability, where $(\langle M, M \rangle_t)_{t \geq 0}$ is called the quadratic variation of M.

Proof The definition of the quadratic variation $\langle M, M \rangle_t$ directly leads to its representation as the limit of the sum of squared increments of the martingale M as the partition gets finer, in the sense of probability.

Given a sequence of partitions of the interval $[0, t]$ with the n-th partition given by $0 = t_0^n < t_1^n < \ldots < t_{p_n}^n = t$ and the mesh of the partition $\|\Pi_n\| = \max_{1 \leq i \leq p_n} |t_i^n - t_{i-1}^n| \to 0$ as $n \to \infty$, we consider the sums $S_n = \sum_{i=1}^{p_n} (M_{t_i^n} - M_{t_{i-1}^n})^2$.

By Theorem 2.19, for every martingale M that is continuous and belongs to $\mathcal{F}_2$, it has been established that

$$\lim_{\|\Pi_n\| \to 0} S_n = \langle M, M \rangle_t$$

in probability. This convergence implies that, for any $\epsilon > 0$ and $\eta > 0$, there exists an N such that for all $n \geq N$,

$$P\left(\left|S_n - \langle M, M \rangle_t\right| > \epsilon\right) < \eta,$$

demonstrating that the sum of the squared increments of M over increasingly finer partitions converges in probability to the quadratic variation $\langle M, M \rangle_t$. This convergence in probability is what defines the quadratic variation of the martingale M over the interval $[0, t]$.

Thus, the proposition is proved using the definition and Theorem 2.19 on the convergence of the quadratic variation. □

Proposition 2.13 *Let M be a continuous local martingale and let T be a stopping time. Then we have a.s. for every $t \geq 0$,*

$$\left\langle M^T, M^T \right\rangle_t = \langle M, M \rangle_{t \wedge T}.$$

Proof This follows from the fact that $M_{t \wedge T}^2 - (\langle M, M \rangle_{t \wedge T}$ is a continuous local martingale (cf. property (c) of continuous local martingales). □

Proposition 2.14 *Let M be a continuous local martingale and $M_0 = 0$. Then we have $\langle M, M \rangle = 0$ if and only if $M = 0$.*

Proof Suppose that $\langle M, M \rangle = 0$. Then M_t^2 is a nonnegative continuous local martingale, M_t^2 is a supermartingale, and hence $E[M_t^2] \leq E[M_0^2] = 0$, so that $M_t = 0$ for every t. The converse is obvious. □

Theorem 2.20 *Let M be a continuous local martingale with $M_0 \in L^2$.*

(i) *The following assertions are equivalent:*

  (a) *M is a martingale bounded in L^2.*
  (b) *$E[\langle M, M \rangle_\infty] < \infty$. Furthermore, if these properties hold, the process $M_t^2 - \langle M, M \rangle_t$ is a uniformly integrable martingale, and in particular $E[M_\infty^2] = E[M_0^2] + E[\langle M, M \rangle_\infty]$.*

(ii) *The following assertions are equivalent:*

  (a) *M is a martingale and $M_t \in L^2$ for every $t \geq 0$.*
  (b) *$E[\langle M, M \rangle_t] < \infty$ for every $t \geq 0$.*

Furthermore, if one of (i) and one of (ii) are held, the process $M_t^2 - \langle M, M \rangle_t$ is a martingale.

Proof Replacing M by $M - M_0$, we may assume that $M_0 = 0$ in the proof. Let us first assume that M is a martingale bounded in L^2. Doob's inequality in L^2. Proposition 2.9(ii) shows that, for every $T > 0$,

$$E\left[\sup_{0 \leq t \leq T} M_t^2\right] \leq 4 E[M_T^2].$$

By letting T go to ∞, we have

$$E\left[\sup_{t \geq 0} M_t^2\right] \leq 4 \sup_{t \geq 0} E[M_t^2] := C < \infty.$$

Set $S_n = \inf\{t \geq 0 : \langle M, M \rangle_t \geq n\}$. Then the continuous local martingale $M_{t \wedge S_n}^2 - \langle M, M \rangle_{t \wedge S_n}$ is dominated by the variable

$$\sup_{s \geq 0} M_s^2 + n$$

which is integrable. From Proposition 2.9, we get that this continuous local martingale is a uniformly integrable martingale; hence,

$$E[\langle M, M \rangle_{t \wedge S_n}] = E[M_{t \wedge S_n}^2] \leq E\left[\sup_{s \geq 0} M_s^2\right] \leq C.$$

By letting n and t tend to infinity, and using monotone convergence, we get $E[\langle M, M \rangle_\infty] \leq C < \infty$.

2.6 Exercises

Conversely, assume that $E\left[\langle M, M\rangle_\infty\right] < \infty$. Set $T_n = \inf\{t \geq 0 : |M_t| \geq n\}$. Then the continuous local martingale $M^2_{t\wedge T_n} - \langle M, M\rangle_{t\wedge T_n}$ is dominated by the variable

$$n^2 + \langle M, M\rangle_\infty,$$

which is integrable. From Proposition 2.9(ii) again, this continuous local martingale is a uniformly integrable martingale; hence, for every $t \geq 0$,

$$E[M^2_{t\wedge T_n}] = E[\langle M, M\rangle_{t\wedge T_n}] < 1$$

By letting $n \to \infty$ and using Fatou's lemma, we get $E[M_t^2] \leq C'$, so that the collection $(M_t)_{t\geq 0}$ is bounded in L^2. We have not yet verified that $(M_t)_{t\geq 0}$ is a martingale. However, the previous bound on $E[M^2_{t\wedge T_n}]$ shows that the sequence $(M_{t\wedge T_n})$ is uniformly integrable and therefore converges both a.s. and in L^1 to M_t, for every t_0. Recalling that M_{T_n} is a martingale (Proposition 2.9(iii)), the L^1-convergence allows us to pass to the limit $n \to \infty$ in the martingale property $E[M_{t\wedge T_n}|\mathcal{F}_s] = M_{s\wedge T_n}$, for $0 \leq s < t$, and to get that M is a martingale.

Finally, if properties (a) and (b) hold, the continuous local martingale $M^2 - \langle M, M\rangle$, M_i is dominated by the integrable variable

$$\sup_{t\geq 0} M_t^2 + \langle M, M\rangle_\infty,$$

and is therefore (by Proposition 2.9(ii)) a uniformly integrable martingale.

(ii) It suffices to apply (i) to $(M_{t\wedge a})_{t\geq 0}$ for every choice of $a \geq 0$. □

2.6 Exercises

1. Show that a sum of martingales is a martingale
2. (1) Is any Markovian process a martingale? If yes, prove it. Otherwise, construct a counterexample.
   (2) Is any martingale Markovian? If yes, prove it. Otherwise, construct a counterexample.
3. Given that $\{X_n, n \geq 1\}$ are independent random variables with $E[X_i] = m_i$, $\mathrm{Var}(X_i) = \sigma_i^2$, and $i \geq 1$, let $S_n = \sum_{i=1}^n X_i$ and $\mathcal{F}_n = \sigma(X_1, \ldots, X_n)$.

   (1) Find sequences (b_n), (c_n) of real numbers such that $S_n^2 + b_n S_n + c_n$ is an $\mathcal{F}_n$-martingale.
   (2) Assume moreover that $\lambda \in \mathbb{R}$ such that $\exp(\lambda X_i) \in L^1$ for any $i \geq 1$, and set $G_i(\lambda) = E[\exp(\lambda X_i)]$, $i \geq 1$. Find a sequence $(a_{\lambda_n})_{n\geq 0}$ such that $\left\{\exp(\lambda S_n - a_{\lambda_n})\right\}_{n\geq 0}$ is an $\mathcal{F}_n$-martingale.

4. Let $M = \{M_t : t \in \{0, 1, 2, \ldots\}\}$ be a square-integrable martingale existing on a filtered probability space $(\Omega, \mathcal{F}, \mathcal{F}_t, P)$, where $\mathcal{F} = \{\mathcal{F}_t : t \in \{0, 1, 2, \ldots\}\}$ denotes the filtration.

The predictable process $\Theta = \{\Theta_t : t \in \{0, 1, 2, \ldots\}\}$ is constructed on the same space. For all $t \in \{1, 2, \ldots\}$, Θ_t is $\mathcal{F}_{t-1}$-measurable, and Θ_0 is $\mathcal{F}_0$-measurable. We also assume that for all $t \in \{0, 1, 2, \ldots\}$, the random variable Θ_t is square-integrable, meaning $E[|\Theta_t|^2] < \infty$.

We aim to show that the process $N = \{N_t : t \in \{0, 1, 2, \ldots\}\}$ defined as

$$N_t = N_0 + \sum_{k=1}^{t} \Theta_k (M_k - M_{k-1}),$$

is a martingale, provided that N_0 is $\mathcal{F}_0$-measurable.

5. Let X be a square-integrable random variable constructed on a filtered probability space $(\Omega, \mathcal{F}, \{\mathcal{F}_t\}, P)$ with $\mathbb{E}[|X|] < \infty$. We aim to prove that the stochastic process $\{M_t : t \in \{0, 1, 2, \ldots\}\}$ defined as

$$M_t = \mathbb{E}_P[X \mid \mathcal{F}_t], \quad t \geq 0,$$

is a martingale.

6. Consider a probability space $(\Omega, \mathcal{F}, P)$ on which two filtrations are constructed, $\{\mathcal{F}_t : t \geq 0\}$ and $\{\mathcal{G}_t : t \geq 0\}$, satisfying $\mathcal{F}_t \subseteq \mathcal{G}_t$ for all t.

   (1) Let $M = \{M_t : t \geq 0\}$ be a $\{\mathcal{F}_t\}$-martingale, and let $N = \{N_t : t \geq 0\}$ be a $\{\mathcal{G}_t\}$-martingale. We are to determine if M is a $\{\mathcal{G}_t\}$-martingale and if N is a $\{\mathcal{F}_t\}$-martingale.
   (2) Let τ be a $\{\mathcal{F}_t\}$-stopping time and σ be a $\{\mathcal{G}_t\}$-stopping time. We need to check if σ is a $\{\mathcal{F}_t\}$-stopping time and if τ is a $\{\mathcal{G}_t\}$-stopping time.

7. Let $\{\mathcal{F}_n\}$ be a filtration and $\{M_n\}$ a uniformly integrable (UI) $\{\mathcal{F}_n\}$-martingale. Show that $\{M_n, n \geq 0\}$ converges almost surely (a.s.) and in L^1 toward a limiting M_∞. Also, show that for any $n \in \mathbb{N}$, $M_n = E[M_\infty | \mathcal{F}_n]$.

8. Set $X_0 = 0$, and for $k \geq 0$, define the transition probabilities:

$$P(X_{k+1} = 1 \mid X_k = 0) = P(X_{k+1} = -1 \mid X_k = 0) = \frac{1}{2^k},$$

$$P(X_{k+1} = 0 \mid X_k = 0) = 1 - \frac{1}{2^k},$$

$$P(X_{k+1} = kX_k \mid X_k \neq 0) = \frac{1}{k},$$

$$P(X_{k+1} = 0 \mid X_k \neq 0) = 1 - \frac{1}{k}.$$

Show that $\{X_n, n \geq 0\}$ is a martingale. Does it converge almost surely? In probability? In L^1?

9. Let $\{V_i, i \geq 1\}$ be nonnegative i.i.d. random variables, such that $E[V_i] = 1$ and $P(V_i = 1) < 1$. We define $X_0 = 1$, $X_n = \prod_{i=1}^n V_i$, and $F_n = \sigma(V_i, i \leq n)$.

   (1) Show that $\{X_n\}$ is a $\{F_n\}$-martingale.
   (2) Does $\{X_n\}$ converge? In what sense?

References

1. K. L. Chung. *A course in probability theory*. Academic Press, 2001.
2. L. Gall and J. Francois. *Brownian motion, martingales, and stochastic calculus*. Springer, 2016.
3. I. Karatzas and S. E. Shreve. *Brownian motion and stochastic calculus*. Springer, New York, 1991.
4. B. Oksendal. *Stochastic differential equations: an introduction with applications*. Springer Science & Business Media, 2013.

Chapter 3
Stochastic Differential Equations

In this chapter, we will give a very general review to the main results in the subject of stochastic differential equation (SDE), which are important in stochastic filtering theory. We will start from the theory of stochastic integral, which serves to be a foundation in the development of SDEs. The famous Itô's formula and Girsanov's change-of-measure method, as well as the Burkholder-Davis-Gundy inequality, are presented here and will be applied in the derivation of filtering equations in Chap. 5. Next, different kinds of formulations to the SDEs will be summarized, and the relationship between SDEs and partial differential equations (PDEs) will also be studied. As a special reminder, for most theorems mentioned in this chapter, we will only give a brief sketch of the proofs or simply omit the proofs. We highly recommend interested readers to refer to monographs such as [5] and [6] for detailed proofs. Also, advanced readers may skip this chapter or regard it as a quick reference.

3.1 Stochastic Integral

The concept of stochastic integral is quite common in natural science, engineering, and finance, when it comes to representing cumulative effects of random phenomena in the real world. Historically, the rigorous mathematical formulation of stochastic integral was proposed by K. Itô in 1942, and fruitful theoretical and practical results based on this formulation have been achieved during the last decades.

In comparison with traditional Riemann-Stieltjes integral and Lebesgue integral, a major difference for stochastic integral is that the paths of most important stochastic processes that we are interested in, such as Brownian motion, are almost surely nowhere differentiable. If we make an attempt to construct stochastic integrals based on traditional deterministic ones, we will soon find out that the Riemann-Stieltjes sums do not converge and the paths of the process cannot generate a measure as functions with bounded variation do in the measure theory. Therefore, K. Itô introduced a novel definition of stochastic integral quite different from traditional, deterministic ones.

In this section, we will first summarize the construction procedure of stochastic integral based on the work of Karatzas and Shreve [6] and then introduce the famous Itô's change-of-variable formula. After that, we will focus on some basic properties and results of stochastic integral that are useful and important in SDE and stochastic filtering theory. Finally, we will give a brief introduction of another definition of stochastic integral, Stratonovich's integral, and its relationship with that of Itô sense.

3.1.1 Construction of Stochastic Integral

In the construction of Lebesgue integral in measure theory, we first define simple functions from characteristic functions, $1_B(x)$, $B \in \mathcal{B}(R^d)$, and define the integral of simple functions. Next, with the approximation of Lebesgue measurable functions by simple functions, we generate the definition of integration to a broader case.

Similarly, when constructing the stochastic integrals, we also first introduce the concept of simple processes defined below, which play the role of simple functions in the construction of Lebesgue integral in measure theory.

Definition 3.1 A process X is called simple if there exists a strictly increasing sequence of real numbers $\{t_n\}_{n=0}^{\infty}$ with $t_0 = 0$ and $\lim_{n \to \infty} t_n = \infty$, as well as a sequence of random variables $\{\xi_n\}_{n=0}^{\infty}$ and a nonrandom constant $C < \infty$ with $\sup_{n \geq 0} |\xi_n(\omega)| \leq C$, for every $\omega \in \Omega$, such that ξ_n is $\mathcal{F}_{t_n}$-measurable for every $n \geq 0$ and

$$X_t = \xi_0 1_{\{0\}}(t) + \sum_{i=0}^{\infty} \xi_i 1_{(t_i, t_{i+1}]}(t), \quad 0 \leq t < \infty. \tag{3.1}$$

The next Lemma shows that a large number of processes can be approximated by a series of simple processes.

Lemma 3.1 *Let X be a bounded, measurable, $\{\mathcal{F}_t\}$-adapted process. Then there exists a series of simple processes $\{X^{(m)}\}_{m=1}^{\infty}$, such that*

$$\sup_{T>0} \lim_{m \to \infty} E \int_0^T \left| X_t^{(m)} - X_t \right|^2 dt = 0. \tag{3.2}$$

Proof We will briefly present the procedure of constructing the series of simple processes here, and for detailed proof, interested readers can refer to Karatzas and Shreve [6].

3.1 Stochastic Integral

For each $T > 0$, we can first define a sequence of simple processes $\{X^{(n,T)}\}_{n=1}^{\infty}$ by

$$X_t^{(n,T)} = X_0 1_{\{0\}}(t) + \sum_{k=0}^{2^n-1} X_{kT/2^n} 1_{(kT/2^n, (k+1)T/2^n]}(t); \quad n \geq 1.$$

Then for each bounded, adapted, and continuous process X, we have from the bounded convergence theorem that

$$\lim_{n \to \infty} E \int_0^T \left| X_t^{(n,T)} - X_t \right|^2 dt = 0$$

Thus, for each $m \in \mathbb{N}$, there exists $n_m \in \mathbb{N}$, such that

$$E \int_0^m \left| X_t^{(n_m, m)} - X_t \right|^2 dt < \frac{1}{m}$$

and the series $\{X^{(n_m, m)}\}$ can be used to approximate the continuous process X in the sense of Eq. (3.2).

For a general bounded measurable and adapted process X, we can first approximate it with a continuous process and then approximate the continuous process with simple ones. Therefore, each bounded measurable and adapted process can be approximated by a series of simple processes in the sense of Eq. (3.2). □

For simple process X defined in (3.1), we can define its stochastic integral with respect to a continuous, square-integrable martingale as follows.

Definition 3.2 Let $M = \{M_t, \mathcal{F}_t; 0 \leq t < \infty\} \in \mathcal{M}_2^c$ be a continuous, square-integrable martingale, and X is a simple process defined by (3.1). Then the stochastic integral of X with respect to martingale M is defined by the following martingale transform $I^M(X) = \{I_t^M(X); 0 \leq t < \infty\}$:

$$I_t^M(X) = \int_0^t X_s dM_s \triangleq \sum_{i=0}^{\infty} \xi_i (M_{t \wedge t_{i+1}} - M_{t \wedge t_i}), \quad 0 \leq t < \infty. \tag{3.3}$$

Definition 3.2 shows that the stochastic integral of simple processes is defined by a form of Riemann-Stieltjes sum. However, instead of choosing the value of integrand arbitrarily in each small interval $[t_i, t_{i+1}]$, as it does in standard Riemann-Stieltjes sum, the value of integrand in the sum in the stochastic integral is fixed to be the value at the left point of each interval.

Just as we did in the Lebesgue integration theory, stochastic integrals of a larger group of processes are defined to be the limits of integrals of simple processes.

Definition 3.3 For a bounded, measurable, $\{\mathcal{F}_t\}$-adapted process X and a continuous square-integrable martingale $M = \{M_t, \mathcal{F}_t; 0 \leq t < \infty\} \in \mathcal{M}_2^c$, the stochastic integral $I^M(X)$ of X with respect to M is defined to be the unique process that satisfies $\lim_{n\to\infty}|I(X^{(n)}) - I(X)| = 0$, for every series of simple processes $\{X^{(n)}\}_{n=1}^{\infty}$ satisfying (3.2). We denote

$$I_t(X) = \int_0^t X_s dM_s; \quad 0 \leq t < \infty.$$

With the definitions above, the stochastic integral we introduce in this section enjoys many good properties, which are summarized in the following theorem.

Theorem 3.1 *Let* $M = \{M_t, \mathcal{F}_t; 0 \leq t < \infty\}$, $N = \{N_t, \mathcal{F}_t; 0 \leq t < \infty\}$ *be two continuous, square-integrable martingales, and* $X = \{X_t; 0 \leq t < \infty\}$ *and* $Y = \{Y_t; 0 \leq t < \infty\}$ *are two bounded measurable,* $\{\mathcal{F}_t\}$-*adapted process. Then, the following properties hold for the stochastic integral process defined above:*

(1) $I_0^M(X) = 0$, *a.s.* P;
(2) $I^M(X) = \{I_t^M(X), \mathcal{F}_t; 0 \leq t < \infty\}$ *is a continuous square-integrable martingale, i.e.*

$$E\left[I_t^M(X)\Big|\mathcal{F}_s\right] = I_s^M(X), \quad a.s.\ P,\ 0 \leq s \leq t < \infty; \quad (3.4)$$

(3) $\langle I^M(X), N\rangle_t = \int_0^t X_s d\langle M, N\rangle_s$, *a.s.* P, $0 \leq t < \infty$;
(4) $\langle I^M(X), I^N(Y)\rangle_t = \int_0^t X_s Y_s d\langle M, N\rangle_s$, *a.s.* P, $0 \leq t < \infty$;
(5) $I^M(\alpha X + \beta Y) = \alpha I^M(X) + \beta I^M(Y)$.

Proof Here, we only need to prove the above properties for any simple process X defined by (3.1), and for the general case, we can get these properties through approximation.

(1) is obvious, and (5) can be obtained directly from the definition of stochastic integral.

For properties (2) to (4), we first notice that the stochastic integral process $\{I_t^M(X); 0 \leq t < \infty\}$ is an adapted, continuous process. In order to prove the martingale property, we can assume that s and t are two positive numbers with $0 \leq s < t < \infty$ and $t_m \leq s < t_{m+1}$ and $t_n \leq t < t_{n+1}$, where $m, n \in \mathbb{N}$, with $m \leq n$; then, because ξ_i is $\mathcal{F}_{t_i}$- measurable and M is a martingale, we have

$$E\left[(I_t^M(X) - I_s^M(X))\Big|\mathcal{F}_s\right] = \sum_{i=0}^{\infty} \xi_i E\left[M_{t\wedge t_{i+1}} + M_{s\wedge t_{i+1}} - M_{t\wedge t_i} - M_{s\wedge t_i}\Big|\mathcal{F}_s\right].$$

The right-hand side of the above equation is equal to zero because of the martingale property of M. Thus, the process $I^M(X)$ is a continuous martingale.

3.1 Stochastic Integral

For property (3), let's first calculate the following conditional expectation:

$$E\left[(I_t^M(X) - I_s^M(X))(N_t - N_s)\Big|\mathcal{F}_s\right]. \tag{3.5}$$

The two terms in the brackets can be expanded as follows:

$$I_t^M(X) - I_s^M(X) = \xi_{m-1}(M_{t_m} - M_s) + \sum_{i=m}^{n-1}\xi_i(M_{t_{i+1}} - M_{t_i}) + \xi_n(M_t - M_{t_n}),$$

$$N_t - N_s = (N_{t_m} - N_s) + \sum_{i=m}^{n-1}(N_{t_{i+1}} - N_{t_i}) + (N_t - N_{t_n}).$$

Next, notice that when we substitute the expansion form back to Eq. (3.5), because of the martingale property of M and N, the following terms

$$E\left[\xi_i(M_{t_{i+1}} - M_{t_i})(N_{t_{j+1}} - N_{t_j})\Big|\mathcal{F}_s\right]$$

with $i \neq j$ all become zero. Therefore, we have

$$E\left[(I_t^M(X) - I_s^M(X))(N_t - N_s)\Big|\mathcal{F}_s\right]$$

$$= E\left[\sum_{i=m-1}^{n}\xi_i(M_{t_{i+1}} - M_{t_i})(N_{t_{i+1}} - N_{t_i})\Big|\mathcal{F}_s\right]$$

with the convenient notation $t_{n+1} = t$ and $t_{m-1} = s$. Because of the definition of cross variation, we have

$$E\left[(I_t^M(X) - I_s^M(X))(N_t - N_s)\Big|\mathcal{F}_s\right]$$

$$= E\left[\sum_{i=m-1}^{n}\xi_i(\langle M, N\rangle_{t_{i+1}} - \langle M, N\rangle_{t_i})\Big|\mathcal{F}_s\right]$$

$$= E\left[\int_s^t X_u d\langle M, N\rangle_u\Big|\mathcal{F}_s\right].$$

Therefore,

$$\langle I^M(X), N\rangle_t = \int_0^t X_s d\langle M, N\rangle_s.$$

With similar calculations, we can prove property (4). Take $N = M$ and $X = Y$, and the square-integrability of $I^M(X)$ is obtained. □

For the detailed proof of Theorem 3.1 and further properties of stochastic integrals, one can refer to monographs about stochastic calculus theory, and for us, the above discussion on the construction of stochastic integrals is enough for the development of stochastic differential equation and stochastic filter theory. We would like to finish the construction of stochastic integrals with the following two remarks.

Remark 3.1 In fact, Theorem 3.1 (3) can also be regarded as a characterization of the stochastic integral, that is, the stochastic integral $I^M(X)$ is the unique continuous, square-integrable martingale $\Phi \in \mathcal{M}_2^c$, such that

$$\langle \Phi, N \rangle_t = \int_0^t X_s d\langle M, N \rangle_s; \quad 0 \le t < \infty, \quad \text{a.s. P},$$

for every $N \in \mathcal{M}_2^c$

Remark 3.2 The construction of stochastic integral can be generated to a broader class of processes. We denote by $\mathcal{L}(M)$ the set of all progressively measurable processes satisfying

$$E \int_0^T X_s^2 d\langle M \rangle_s < \infty, \quad \forall T > 0.$$

It can be proved that each process in class $\mathcal{L}(M)$ can be approximated by a series of simple processes in the sense of Eq. (3.2). Therefore, the stochastic integral of $X \in \mathcal{L}(M)$ with respect to M can be defined similarly, and the stochastic integral also satisfies the properties summarized in Theorem 3.1.

Moreover, according to the standard localization procedure, the definition of stochastic integral can be further broadened for continuous local martingale $M \in \mathcal{M}^{c,loc}$ and processes $X \in \mathcal{L}(M)$, satisfying

$$P\left[\int_0^T X_s^2 d\langle M \rangle_s < \infty\right] = 1, \quad \forall T \ge 0.$$

However, in this case, the stochastic integral process $I^M(X)$ is a local martingale, rather than a martingale. Nevertheless, properties (1), (3), (4), and (5) also hold with $\langle \cdot, \cdot \rangle$ denoting the cross-variation of local martingales.

3.1.2 Itô's Formula

One of the most important theorem in calculus is the famous Newton-Leibnitz formula, which serves to be a connection between differentiation and integration and makes a great number of integrals computable.

Itô's formula, which will be introduced next, enjoys the same status in stochastic calculus as Newton-Leibnitz formula in calculus. It is a stochastic version of change-

3.1 Stochastic Integral

of-variable formula and introduces the "chain rule" to a large group of processes, called semimartingale. In order to introduce the Itô's formula, we first give the definition of semimartingales.

Definition 3.4 A continuous semimartingale $X = \{X_t, \mathcal{F}_t; 0 \leq t < \infty\}$ is an adapted process that has the decomposition, P a.s.,

$$X_t = X_0 + M_t + B_t; \quad 0 \leq t < \infty, \tag{3.6}$$

where $M = \{M_t, \mathcal{F}_t; 0 \leq t < \infty\} \in \mathcal{M}^{c,loc}$ and $B = \{M_t, \mathcal{F}_t; 0 \leq t < \infty\}$ is the difference of continuous, nondecreasing, adapted processes $\{A_t^\pm, \mathcal{F}_t; 0 \leq t < \infty\}$, with $A_0^\pm = 0$, P a.s., i.e.

$$B_t = A_t^+ - A_t^-, \quad 0 \leq t < \infty$$

With the concept of semimartingales, we can introduce the famous Itô's formula as follows.

Theorem 3.2 Let $f : R \to R$ be a C^2 function and let $X = \{X_t, \mathcal{F}_t; 0 \leq t < \infty\}$ be a continuous semimartingale with decomposition (3.6). Then, P a.s.,

$$\begin{aligned} f(X_t) = f(X_0) &+ \int_0^t f'(X_s) dM_s + \int_0^t f'(X_s) dB_s \\ &+ \frac{1}{2} \int_0^t f''(X_s) d\langle M \rangle_s, \quad 0 \leq t < \infty, \end{aligned} \tag{3.7}$$

where $\langle \cdot \rangle$ means the square variation of martingales.

Proof We will summarize the main idea of the above theorem, and for the detailed proof, readers can also refer to monographs in stochastic calculus.

The main idea of the proof is to expand the difference $f(X_t) - f(X_0)$ as a Riemann sum and approximate the sum by Taylor's expansion.

Firstly, let $n \in \mathbb{N}$ and $0 = t_0 \leq t_1 \leq \cdots \leq t_n = t$ be a partition of $[0, t]$; then, we can write

$$f(X_t) - f(X_0) = \sum_{i=0}^{n-1} f(X_{t_{i+1}}) - f(X_{t_i}) \tag{3.8}$$

For each term in the right-hand summation, because f is a C^2 function, we can do the Taylor's expansion as follows:

$$\begin{aligned} f(X_{t_{i+1}}) - f(X_{t_i}) = f'(X_{t_i})(X_{t_{i+1}} - X_{t_i}) &+ \frac{1}{2} f''(X_{t_i})(X_{t_{i+1}} - X_{t_i})^2 \\ &+ o((X_{t_{i+1}} - X_{t_i})^2). \end{aligned} \tag{3.9}$$

Let's denote $\lambda = \max_{1 \leq i \leq n}(t_i - t_{i-1})$; then as $\lambda \to 0$, from the definition of stochastic integral, the summation of the first two terms will converge to the right hand of Eq. (3.5),

$$\sum_{i=0}^{n-1} f'(X_{t_i})(X_{t_{i+1}} - X_{t_i}) \to \int_0^t f'(X_s) dM_s + \int_0^t f'(X_s) dB_s,$$

$$\sum_{i=0}^{n-1} \frac{1}{2} f''(X_{t_i})(X_{t_{i+1}} - X_{t_i})^2 \to \frac{1}{2} \int_0^t f''(X_s) d\langle M \rangle_s.$$
(3.10)

The convergence above is at least in the sense of convergence in probability. Besides, the summation of the third term in the right hand of Eq. (3.9) will converge to zero. Therefore, the right-hand side of Eq. (3.8) will converge to the right-hand side of Eq. (3.5) as $\lambda \to 0$, and this proves Itô's formula. □

Exercise 3.1 Provide a mathematically rigorous proof for the two formulae in (3.10) in the sense of convergence in probability, i.e., for an arbitrary $\epsilon > 0$, we have

$$\lim_{\lambda \to 0} P\left[\left|\sum_{i=0}^{n-1} f'(X_{t_i})(X_{t_{i+1}} - X_{t_i}) - \int_0^t f'(X_s) dM_s - \int_0^t f'(X_s) dB_s\right| > \epsilon\right] = 0,$$

$$\lim_{\lambda \to 0} P\left[\left|\sum_{i=0}^{n-1} \frac{1}{2} f''(X_{t_i})(X_{t_{i+1}} - X_{t_i})^2 - \frac{1}{2} \int_0^t f''(X_s) d\langle M \rangle_s\right| > \epsilon\right] = 0.$$
(3.11)

with $\lambda = \max_{1 \leq i \leq n}(t_i - t_{i-1})$ defined in the proof of Theorem 3.2

Remark 3.3 Formally, we can denote formula (3.7) in differential form as

$$df(X_t) = f'(X_t) dX_t + \frac{1}{2} f''(X_t) d\langle X \rangle_t.$$
(3.12)

Itô's formula can also be promoted to multi-dimensional vectors of semimartingales, where we only need to change the derivatives into partial derivatives and square variations into cross variations.

Theorem 3.3 Let $\{M_t \triangleq (M_t^{(1)}, M_t^{(2)}, \cdots, M_t^{(d)}), \mathcal{F}_t; 0 \leq t < \infty\}$ be a vector of continuous local martingales and $\{B_t \triangleq (B_t^{(1)}, B_t^{(2)}, \cdots, B_t^{(d)})\}$ a vector of adapted process of bounded variation with $B_0 = 0$, and set $X_t = X_0 + M_t + B_t$, $0 \leq t < \infty$, where X_0 is an $\mathcal{F}_0$-measurable random vector in R^d. Let $f(t, x) : [0, \infty) \times R^d \to R$ be a $C^{1,2}$ function. Then, P a.s.,

$$f(t, X_t) = f(0, X_0) + \int_0^t \frac{\partial}{\partial t} f(s, X_s) ds + \sum_{i=1}^d \int_0^t \frac{\partial}{\partial x_i} f(s, X_s) dX_s^i \quad (3.13)$$

3.1 Stochastic Integral

$$+ \frac{1}{2} \sum_{i=1}^{d} \sum_{j=1}^{d} \int_0^t \frac{\partial^2}{\partial x_i \partial x_j} f(s, X_s) d\langle X^i, X^j \rangle_s, \quad 0 \le t < \infty. \quad (3.14)$$

where $\langle \cdot, \cdot \rangle$ means the cross variation of the two processes.

Proof The idea to prove this theorem is quite similar to the one-dimensional case. We can get formula (3.13) from the same partition and Taylor expansion procedure as in the proof of Theorem 3.2. Therefore, we skip the proof of this theorem. □

Exercise 3.2 Imitate the proof in Theorem 3.2, and give a sketch of the proof of Itô's formula for multi-variate case (Theorem 3.3).

Theorems 3.2 and 3.3 also show that smooth functions of semimartingales are also semimartingales. Therefore, a large number of processes are contained in the group of semimartingales, and from now on, we will focus our attention to this kind of processes.

3.1.3 Girsanov's Theorem and Novikov Condition

With the definition of stochastic integral and Itô's formula, we can develop fundamental additional feature about semimartingales that are useful in solving stochastic differential equations and filtering theory.

Here, we will introduce Girsanov's change-of-measure method. The main purpose of this method is to construct a new probability measure, under which a semimartingale or a solution to a stochastic equation can be converted to be a Brownian motion.

Let $W = \{W_t = (W_t^1, W_t^2, \cdots, W_t^d), \mathcal{F}_t; 0 \le t < \infty\}$ be a d-dimensional standard Brownian motion defined on probabilistic space $\{\Omega, \mathcal{F}, P\}$ and $X = \{X_t = (X_t^1, X_t^2, \cdots, X_t^d), \mathcal{F}_t; 0 \le t < \infty\}$ a vector of measurable, adapted processes satisfying

$$P\left[\int_0^T (X_s^i)^2 ds < \infty\right] = 1, \quad \forall T \ge 0. \quad (3.15)$$

Therefore, according to Remark 3.2, for each i, the stochastic integral $I^{W^{(i)}}(X^{(i)})$ is well defined and is a continuous local martingale.

We define the process $Z = \{Z_t; 0 \le t < \infty\}$ with

$$Z_t = \exp\left[\sum_{i=1}^{d} \int_0^t X_s^i dW_s^i - \frac{1}{2} \int_0^t |X_s|^2 ds\right]. \quad (3.16)$$

According to Itô's formula, we have

$$Z_t = 1 + \sum_{i=1}^{d} \int_0^t Z_s X_s^i dW_s^i. \tag{3.17}$$

Therefore, $Z = \{Z_t, \mathcal{F}_t; 0 \le t < \infty\}$ is a continuous local martingale with $Z_0 = 1$.

Let us first assume that Z is in fact a martingale and the required conditions will be discussed later. Since $EZ_t = EZ_0 = 1$, $\forall t \ge 0$, we can define, for each $T > 0$, a probability measure $\widetilde{P}_T$ on $\mathcal{F}_T$ as follows:

$$\widetilde{P}_T(A) = E[1_A Z_T]; \quad A \in \mathcal{F}_T. \tag{3.18}$$

Because of the martingale property, for $0 \le t \le T$, we have

$$\widetilde{P}_T(A) = E[1_A Z_T] = E[E[1_A Z_T | \mathcal{F}_t]] = E[1_A Z_t] = \widetilde{P}_t(A)$$

The following theorem shows that under the probability measure $\widetilde{P}_T$, we can construct a new Brownian motion from the process W and X.

Theorem 3.4 (Girsanov [4]) *Assume that Z defined by (3.16) is a martingale, then the process $\widetilde{W} = \{\widetilde{W}_t = (\widetilde{W}_t^{(1)}, \widetilde{W}_t^{(2)}, \cdots, \widetilde{W}_t^{(d)}), \mathcal{F}_t; 0 \le t \le T\}$ defined by*

$$\widetilde{W}_t^{(i)} = W_t^{(i)} - \int_0^t X_s^{(i)} ds; \quad 1 \le i \le d, \quad 0 \le t \le T$$

is a standard d-dimensional Brownian motion on probabilistic space $(\Omega, \mathcal{F}_T, \widetilde{P}_T)$, $\forall T > 0$.

In order to prove the above theorem, we need to first introduce two lemmas. The first one is about a Bayesian-like formula, which presents relationships between conditional expectations under probability measures P and $\widetilde{P}_T$.

Lemma 3.2 *For a fixed $0 \le T < \infty$, if Z defined by (3.16) is a martingale and, thus, $\widetilde{P}_T$ defined in (3.18) is a probability measure, then, for any $0 \le s \le t \le T$, and $\mathcal{F}_t$-measurable random variable Y, satisfying $\widetilde{E}_T |Y| < \infty$, we have the following equation:*

$$\widetilde{E}_T[Y|\mathcal{F}_s] = \frac{1}{Z_s} E[Y Z_s | \mathcal{F}_s], \quad a.s.\ P \text{ and } \widetilde{P}_T. \tag{3.19}$$

Here, $\widetilde{E}_T$ denotes the expectation operator with respect to $\widetilde{P}_T$.

Proof The result in Lemma 3.2 can be derived directly from the definition of conditional expectations.

3.1 Stochastic Integral

Notice that the right-hand side in Eq. (3.19) is $\mathcal{F}_s$-measurable. Besides, for any $A \in \mathcal{F}_s$,

$$\tilde{E}_T\left[1_A \frac{1}{Z_s} E[YZ_t|\mathcal{F}_s]\right] = E\left[1_A \frac{Z_T}{Z_s} E[YZ_s|\mathcal{F}_s]\right]$$

$$= E\left[\left[1_A \frac{Z_T}{Z_s} E[YZ_t|\mathcal{F}_s]\right] | \mathcal{F}_s\right]$$

$$= E\left[1_A \frac{1}{Z_s} E[YZ_t|\mathcal{F}_s] E[Z_T|\mathcal{F}_s]\right]$$

$$= E[1_A Y Z_t] = \tilde{E}_T[1_A Y].$$

According to the definition of conditional expectations, Eq. (3.19) holds a.s. P and $\tilde{P}_T$. □

The next lemma gives another characterization of Brownian motion using the concept of cross variations.

Lemma 3.3 *Let* $X = \{X_t = (X_t^{(1)}, X_t^{(2)}, \cdots, X_t^{(d)}), \mathcal{F}_t, 0 \leq t < \infty\}$ *be a vector of continuous local martingales, and cross variations are given by*

$$\langle X^{(k)}, X^{(j)} \rangle_t = \delta_{kj} t; \; 1 \leq k, j \leq d$$

where δ_{kj} *is the Kronecker-delta notation. Then,* $\{X_t, \mathcal{F}_t; 0 \leq t < \infty\}$ *is a d-dimensional Brownian motion.*

Proof According to the definition of Brownian motion, we only need to show that increments $X_t - X_s$ and $(0 \leq s < t < \infty)$ are independent of $\mathcal{F}_s$ and have the d-dimensional normal distribution $\mathcal{N}(0, (t-s)I)$.

Besides, since the distribution can be totally determined by its characteristic function, we only need to prove the following equation for every $u \in R^d$,

$$E\left[e^{i\langle u, X_t - X_s\rangle} \Big| \mathcal{F}_s\right] = e^{\frac{1}{2}|u|^2(t-s)}. \tag{3.20}$$

To this end, we can apply Itô's formula in Theorem 3.3 to function $f(x) = e^{i\langle u, x \rangle}$.

$$e^{i\langle u, X_t\rangle} = e^{i\langle u, X_s\rangle} + i \sum_{j=1}^d u_j \int_s^t e^{i\langle u, X_v\rangle} dM_v^{(j)} - \frac{1}{2} \sum_{j=1}^d u_j^2 \int_s^t e^{i\langle u, X_v\rangle} dv. \tag{3.21}$$

Here we use the fact that $\langle X^{(k)}, X^{(j)} \rangle_t = \delta_{kj} t; \; 1 \leq k, j \leq d$.

Because the stochastic integral in Eq. (3.21) is a martingale, the expectation of that term is zero. If we multiply (3.21) by $e^{-i(u,X_s)}1_A$ for every $A \in \mathcal{F}_s$, we have

$$E\left[e^{i(u,X_t-X_s)}1_A\right] = P(A) - \frac{1}{2}|u|^2 \int_s^t E\left[e^{i(u,X_v-X_s)}1_A\right]dv, \qquad (3.22)$$

which is an integral equation for $t \to E[e^{i(u,X_t-X_s)}1_A]$. Therefore, we have

$$E\left[e^{i(u,X_t-X_s)}1_A\right] = P(A)e^{\frac{1}{2}|u|^2(t-s)}. \qquad (3.23)$$

Because $A \in \mathcal{F}_s$ is arbitrarily chosen, we have proved Eq. (3.20), and thus X is a d-dimensional Brownian motion. □

With the above two lemmas, we can start to prove Girsanov's theorem.

Proof We will use Lemma 3.3 to prove that the new process $\widetilde{W}$ is a Brownian motion under probability measure $\widetilde{P}_t$.

Firstly, each $\widetilde{W}^{(i)}$, $1 \leq i \leq d$ is a continuous local martingale under $\widetilde{P}_t$. This is because from Itô's formula, the product $Z_t \widetilde{W}_t^{(i)}$ is a local martingale under P:

$$d(Z_t \widetilde{W}_t^{(i)}) = Z_t d\widetilde{W}_t^{(i)} + \widetilde{W}_t^{(i)} dZ_t + d\langle Z, \widetilde{W}^{(i)}\rangle_t$$

$$= Z_t dW_t^{(i)} - Z_t X_t^{(i)} dt + \widetilde{W}_t^{(i)} \sum_{j=1}^d Z_t X_t^{(j)} dW_t^{(j)} + Z_t X_t^{(i)} dt$$

$$= Z_t dW_t^{(i)} + \widetilde{W}_t^{(i)} \sum_{j=1}^d Z_t X_t^{(j)} dW_t^{(j)},$$

where we use the fact that $dZ_t = \sum_{j=1}^d Z_t X_t^{(j)} dW_t^{(j)}$.

With the Bayesian rule we derived in Lemma 3.2, if $\widetilde{W}^{(i)}$ is bounded, we have

$$\widetilde{E}_T\left[\widetilde{W}_t^{(i)}\Big|\mathcal{F}_s\right] = \frac{1}{Z_s}E\left[Z_t \widetilde{W}_t^{(i)}\Big|\mathcal{F}_s\right] = \frac{1}{Z_s}Z_s \widetilde{W}_s^{(i)} = \widetilde{W}_s^{(i)}. \qquad (3.24)$$

This shows that the new process $\widetilde{W}$ is a martingale. Through a standard localization procedure, we can show that $\widetilde{W}$ is a local martingale under $\widetilde{P}_T$.

Next, we need to compute cross variations between $\widetilde{W}^{(k)}$ and $\widetilde{W}^{(j)}$. Since W is a Brownian motion under P, we only need to show that

$$\langle \widetilde{W}^{(k)}, \widetilde{W}^{(j)}\rangle_t = \langle W^{(k)}, W^{(j)}\rangle_t. \qquad (3.25)$$

3.1 Stochastic Integral

Here, we use Itô's formula again,

$$\widetilde{W}_t^{(k)} \widetilde{W}_t^{(j)} - \langle W^{(k)}, W^{(j)} \rangle_t = \int_0^t W_s^{(k)} dW_s^{(j)} + \int_0^t W_s^{(j)} dW_s^{(k)}$$
$$- \int_0^t \left(W_s^{(k)} X_s^{(j)} + W_s^{(j)} X_s^{(k)} \right) ds.$$

Then

$$Z_t [\widetilde{W}_t^{(k)} \widetilde{W}_t^{(j)} - \langle W_t^{(k)}, W_t^{(j)} \rangle_t] = \int_0^t Z_s W_s^{(k)} dW_s^{(j)} + \int_0^t Z_s W_s^{(j)} dW_s^{(k)}$$
$$+ \int_0^t [\widetilde{W}_s^{(k)} \widetilde{W}_s^{(j)} - \langle W^{(k)}, W^{(j)} \rangle_s] \sum_{i=1}^d Z_s X_s^{(i)} dW_s^{(i)},$$

which is also a martingale under P. Therefore, the Bayesian rule in Lemma 3.3 also implies that

$$\widetilde{E}_T \left[\widetilde{W}_t^{(k)} \widetilde{W}_t^{(j)} - \langle W_t^{(k)}, W_t^{(j)} \rangle_t \Big| \mathcal{F}_s \right] = [\widetilde{W}_s^{(k)} \widetilde{W}_s^{(j)} - \langle W_s^{(k)}, W_s^{(j)} \rangle_t], \quad (3.26)$$

which means that $[\widetilde{W}_t^{(k)} \widetilde{W}_t^{(j)} - \langle W_t^{(k)}, W_t^{(j)} \rangle_t]$ is a $\widetilde{P}_T$-martingale. According to the definition of cross variation, we have

$$\langle \widetilde{W}^{(k)}, \widetilde{W}^{(j)} \rangle_t = \langle W^{(k)}, W^{(j)} \rangle_t, \ a.s. \ P \text{ and } \widetilde{P}_T \quad (3.27)$$

We have now proved that $\widetilde{W}$ is a d-dimensional standard Brownian motion under probability measure $\widetilde{P}_T$. □

In comparison with an arbitrary semimartingales, Brownian motion is simpler and easier to analyze. The above theorem is useful in solving stochastic differential equations and has an important application in the development of Duncan-Mortensen-Zakai equation (DMZ equation, for short), which will be introduced in the next part.

However, in order to use Theorem 3.4, we must check that the local martingale Z is indeed a martingale. Novikov's condition, which is summarized in the next theorem, is a useful approach to prove the martingale property of process Z.

Theorem 3.5 (Novikov [7]) $W = \{W_t = (W_t^1, W_t^2, \cdots, W_t^d), \mathcal{F}_t; 0 \leq t < \infty\}$ is a d-dimensional standard Brownian motion and $X = \{X_t = (X_t^1, X_t^2, \cdots, X_t^d), \mathcal{F}_t; 0 \leq t < \infty\}$ a vector of measurable, adapted processes satisfying (3.15). If

$$E \left[\exp \left(\frac{1}{2} \int_0^T |X_s|^2 ds \right) \right] < \infty; \quad 0 \leq T < \infty, \quad (3.28)$$

then Z defined by (3.16) is a martingale. Condition (3.28) is called Novikov's condition.

For the proof of this theorem, we need to use tools of stopping time as well as the famous martingale representation theorems for Brownian motion. We would like to concisely summarize the proof of this theorem and leave some technical procedures as exercises. In the meantime, interested readers can find the proof of the theorem in the original papers as well as many monographs in stochastic calculus.

Proof We first show that the process $\{Z_t : t \geq 0\}$ is a supermartingale. Consider the stopping times

$$T_n = \inf\left\{t \geq 0 : \max_{1 \leq i \leq d} \int_0^t (Z_s X_s^{(i)})^2 ds = n\right\}. \quad (3.29)$$

Let $Z^{(n)} = \{Z_t^{(n)} = Z_{t \wedge T_n} : t \geq 0\}$; then $Z^{(n)}$ are martingales, and we have

$$E[Z_{t \wedge T_n} | \mathcal{F}_s] = Z_{s \wedge T_n}, \ 0 \leq s \leq t, \ n \geq 1. \quad (3.30)$$

According to Fatou's lemma, as $n \to \infty$, we have

$$E[Z_t | \mathcal{F}_s] \leq Z_s, \ 0 \leq s \leq t, \quad (3.31)$$

which shows that $\{Z_t : t \geq 0\}$ is a supermartingale.

In order to prove that $\{Z_t : t \geq 0\}$ is indeed a martingale, we only need to show that $EZ_t = 1$, for all $0 \leq t < \infty$.

The procedure of showing $EZ_t = 1$ for all $0 \leq t < \infty$ is technical. The idea is to construct a new Brownian motion according to the martingale representation theorem.

Let us denote by

$$M_t := \sum_{i=1}^d \int_0^t X_s^i dW_s^i \quad (3.32)$$

the local martingale in the exponential of Z_t, and let $T(s) = \inf\{t \geq 0 : \langle M \rangle_t > s\}$ be a set of stopping times. Then the time-changed process $(B_s, \mathcal{G}_s, s \geq 0)$ given by

$$B_s = M_{T(s)}, \ \mathcal{G}_s = \mathcal{F}_{T(s)}, \ s \geq 0, \quad (3.33)$$

is a Brownian motion.

For $b < 0$, consider the stopping time for $\{\mathcal{G}_s\}$ given by

$$S_b = \inf\{s \geq 0 : B_s - s = b\}. \quad (3.34)$$

3.1 Stochastic Integral

Then,

$$EZ_t = E\left[\exp\left(M_t - \frac{1}{2}\langle M\rangle_t\right)\right]$$

$$= E\left[1_{\{S_b \leq \langle M\rangle_t\}} \exp\left(M_t - \frac{1}{2}\langle M\rangle_t\right)\right] + E\left[1_{\{S_b > \langle M\rangle_t\}} \exp\left(M_t - \frac{1}{2}\langle M\rangle_t\right)\right]$$

$$= E\left[1_{\{S_b \leq \langle M\rangle_t\}} \exp\left(b + \frac{1}{2}S_b\right)\right] + E\left[1_{\{S_b > \langle M\rangle_t\}} \exp\left(M_t - \frac{1}{2}\langle M\rangle_t\right)\right], \tag{3.35}$$

where the last equality holds because $\{\exp(B_s - \frac{1}{2}s) : s \geq 0\}$ is a martingale and $M_t = B_{\langle M\rangle_t}$. We can also check that $\langle M\rangle_t$ is a stopping time of $\{\mathcal{G}_s\}$, and therefore, according to the optimal sampling theorem,

$$E\left[1_{\{S_b \leq \langle M\rangle_t\}} \exp\left(M_t - \frac{1}{2}\langle M\rangle_t\right)\right]$$

$$= E\left[1_{\{S_b \leq \langle M\rangle_t\}} \exp\left(B_{\langle M\rangle_t} - \frac{1}{2}\langle M\rangle_t\right)\right]$$

$$= E\left[1_{\{S_b \leq \langle M\rangle_t\}} \exp\left(B_{S_b} - \frac{1}{2}S_b\right)\right] \tag{3.36}$$

$$= E\left[1_{\{S_b \leq \langle M\rangle_t\}} \exp\left(b + \frac{1}{2}S_b\right)\right]$$

The first term in (3.35) is bounded above by

$$e^b E[\exp(\frac{1}{2}\langle M\rangle_t)] = e^b E\left[\exp\left(\frac{1}{2}\int_0^t |X_s|^2 ds\right)\right].$$

The expectation is finite for all $t \geq 0$ according to the Novikov condition (3.28), and thus the first term tends to zero as $b \to -\infty$.

For the second term, as $b \to -\infty$, it tends to EZ_t according to the monotone convergence theorem, and therefore, we have proved that $EZ_t = 1$, for all $t \geq 0$. □

Exercise 3.3 Check that $B_{\langle M\rangle_t} = M_t$, for all $t \geq 0$.

Exercise 3.4 Prove that $\langle M\rangle_t$ is a stopping time of $\{\mathcal{G}_s\}$. *(Hint: check that the event $\{\langle M\rangle_t > s\}$ is $\mathcal{G}_s$-measurable for all $s \geq 0$.)*

Exercise 3.5 Prove that the time-changed process $(B_s, \mathcal{G}_s, s \geq 0)$ defined by (3.33) is a Brownian motion. *(Hint: check the conditions in Lemma 3.3 hold for $(B_s, \mathcal{G}_s, s \geq 0)$)*

Sometimes, the above Novikov's condition is hard to verify. To this end, we will use another useful condition when using Girsanov's theorem.

Theorem 3.6 (Bain and Crisan [1]) *Let $X = \{X_t, \mathcal{F}_t; \leq t < \infty\}$ be a continuous, d-dimensional, adapted process such that*

$$E\left[\int_0^T |X_s|^2 ds\right] < \infty \tag{3.37}$$

and Z_t is defined by

$$Z_t = \exp\left[\sum_{i=1}^d \int_0^t X_s^i dW_s^i - \frac{1}{2}\int_0^t |X_s|^2 ds\right] \tag{3.38}$$

If for all $T \geq 0$,

$$E\left[\sum_{i=1}^d \int_0^T Z_s(X_s^{(i)})^2 ds\right] < \infty, \tag{3.39}$$

then $Z = \{Z_t, \mathcal{F}_t; 0 \leq t < \infty\}$ is a martingale.

Proof Similar to the proof of Theorem 3.5, it suffices to show that $EZ_t = 1$ for all $t \geq 0$.

Consider the auxiliary process $\frac{Z_t}{1+\epsilon Z_t}$ for a given $\epsilon > 0$, according to Itô's formula,

$$\begin{aligned}\frac{Z_t}{1+\epsilon Z_t} &= \frac{1}{1+\epsilon} + \sum_{i=1}^d \int_0^t \frac{Z_s}{(1+\epsilon Z_s)^2} X_s^i dW_s^i \\ &\quad - \sum_{i=1}^d \int_0^t \frac{\epsilon Z_s^2}{(1+\epsilon Z_s)^3}(X_s^i)^2 ds.\end{aligned} \tag{3.40}$$

According to condition (3.39), the second term on the right-hand side of (3.40) is a martingale, and thus we have

$$E\left[\frac{Z_t}{1+\epsilon Z_t}\right] = \frac{1}{1+\epsilon} - E\sum_{i=1}^d \int_0^t \frac{\epsilon Z_s^2}{(1+\epsilon Z_s)^3}(X_s^i)^2 ds \tag{3.41}$$

The desired result $EZ_t = 1$ follows from the dominated convergence theorem with $\epsilon \to 0$. □

3.1 Stochastic Integral

Exercise 3.6 Prove that the second term on the right-hand side of (3.40) is a martingale by showing that the quadratic variation term is finite, i.e.,

$$E\left[\sum_{i=1}^{d}\int_{0}^{t}\left(\frac{Z_s}{(1+\epsilon Z_s)^2}\right)^2 (X_s^i)^2 ds\right] < \infty, \tag{3.42}$$

for all $\epsilon > 0$ and $t \geq 0$.

3.1.4 Burkholder-Davis-Gundy Inequality

In this subsection, we provide a useful inequality called Burkholder-Davis-Gundy inequality (BDG inequality, for short) [2, 3]. The inequality is useful in moment estimation of local martingales and the development of existence and uniqueness theory of stochastic (partial) differential equations.

Let $M \in \mathcal{M}^{c,loc}$ be a continuous local martingale. We can define a nondecreasing process

$$M_t^* = \max_{0 \leq s \leq t} |M_s|; \quad 0 \leq t < \infty. \tag{3.43}$$

The following theorem shows that the moments of M_t^* can be bounded by the moments of the square variation $\langle M \rangle_t$.

Theorem 3.7 *Let* $M \in \mathcal{M}^{c,loc}$, *and* M_t^* *is defined as in (3.43). For every* $m > 0$, *there exist universal positive constants* k_m *and* K_m *(depending only on m), such that*

$$k_m E[\langle M \rangle_t^m] \leq E[(M_t^*)^{2m}] \leq K_m E[\langle M \rangle_t^m] \tag{3.44}$$

holds for every $t \geq 0$.

Proof In order to prove Theorem 3.7, we first consider the following non-negative process:

$$Y_t = \delta + \epsilon \langle M \rangle_t + M_t^2, \tag{3.45}$$

where $\delta > 0$ and $\epsilon > 0$ are non-negative constants to be determined later.

Applying Itô's formula to $f(x) = x^m$ and process $\{Y_t; 0 \leq t < \infty\}$, we have

$$Y_t^m = \delta^m + m(1+\epsilon)\int_0^t Y_s^{m-1} d\langle M \rangle_s + 2m(m-1)\int_0^t Y_s^{m-2} M_s^2 d\langle M \rangle_s$$
$$+ 2m\int_0^t Y_s^{m-1} M_s dM_s, \quad 0 \leq t < \infty. \tag{3.46}$$

According to a standard localization procedure, we can, without loss of generality, only consider the case where M and $\langle M \rangle$ are bounded processes, and therefore, M and stochastic integrals with respect to M are all martingales. In this case, according to Doob's maximal inequality, we only need to prove

$$k_m E[\langle M \rangle_t^m] \leq E[(M_t)^{2m}] \leq K_m E[\langle M \rangle_t^m]. \tag{3.47}$$

Taking expectations to both sides of Eq. (3.46), we have

$$EY_T^m = \delta^m + m(1+\epsilon)E\int_0^T Y_s^{m-1} d\langle M \rangle_s \\ + 2m(m-1)E\int_0^T Y_s^{m-2} M_s^2 d\langle M \rangle_s. \tag{3.48}$$

Take $\delta \downarrow 0$ in Eq. (3.48), and we have

$$E\left[\epsilon\langle M \rangle_T + M_T^2\right]^m = m(1+\epsilon)E\int_0^T \left[\epsilon\langle M \rangle_s + M_s^2\right]^{m-1} d\langle M \rangle_s \\ + 2m(m-1)E\int_0^T \left[\epsilon\langle M \rangle_s + M_s^2\right]^{m-2} M_s^2 d\langle M \rangle_s. \tag{3.49}$$

Next, we need to divide the proof into five cases according to the value of m.

Case 1: $0 < m \leq 1$, the upper bound of $E(M_T^{2m})$.

In this case, the second term on the right-hand side of (3.49) is non-positive. Therefore,

$$E\left[\epsilon\langle M \rangle_T + M_T^2\right]^m \leq m(1+\epsilon)E\int_0^T \left[\epsilon\langle M \rangle_s + M_s^2\right]^{m-1} d\langle M \rangle_s \\ \leq m(1+\epsilon)\epsilon^{m-1} E\int_0^T \langle M \rangle_s^{m-1} d\langle M \rangle_s \\ = (1+\epsilon)\epsilon^{m-1} E(\langle M \rangle_T^m).$$

Also, when $0 < m \leq 1$, for every $x, y \geq 0$,

$$(x+y)^m \geq 2^{m-1}(x^m + y^m)$$

Then,

$$\epsilon^m E(\langle M \rangle_T^m + E(M_T^{2m})) \leq E\left[\epsilon\langle M \rangle_T + M_T^2\right]^m \leq (1+\epsilon)\left(\frac{\epsilon}{2}\right)^{m-1} E(\langle M \rangle_T^m),$$

$$E(M_T^{2m}) \leq \left((1+\epsilon)\left(\frac{\epsilon}{2}\right)^{m-1} - \epsilon^m\right) E(\langle M \rangle_T^m). \tag{3.50}$$

3.1 Stochastic Integral

Case 2: $m > 1$, the lower bound of $E(M_T^{2m})$.

This case is quite similar to *Case 1*. Here, we can repeat the estimation in *Case 1* and only change some symbols to fit the condition $m > 1$. Then we have the lower bound

$$E(M_T^{2m}) \geq \left((1+\epsilon) \left(\frac{\epsilon}{2}\right)^{1-m} - \epsilon^m \right) E(\langle M \rangle_T^m). \tag{3.51}$$

Case 3: $\frac{1}{2} < m \leq 1$, the lower bound of $E(M_T^{2m})$.

If we take $\epsilon = 0$ and let $\delta \downarrow 0$ in Eq. (3.45), we have from Eq. (3.46) that

$$E(M_T^{2m}) = 2m \left(m - \frac{1}{2} \right) E \int_0^T M_s^{2(m-1)} d\langle M \rangle_s. \tag{3.52}$$

Meanwhile, the discussion in *Case 1* shows that

$$2^{m-1} \left[\epsilon^m E(\langle M \rangle_T^m) + E(M_T^{2m}) \right] \leq m(1+\epsilon) E \int_0^T M_s^{2(m-1)} d\langle M \rangle_s. \tag{3.53}$$

Combining Eqs. (3.52) and (3.53), we have

$$E(M_T^{2m}) \geq \epsilon^m \left(\frac{(1+\epsilon)2^{1-m}}{2m-1} - 1 \right)^{-1} E(\langle M \rangle_T^m). \tag{3.54}$$

Case 4: $m > 1$, the upper bound of $E(M_T^{2m})$.

This case is also quite similar to *Case 3*, except that the inequality (3.53) is reversed; thus, we have

$$E(M_T^{2m}) \leq \epsilon^m \left(\frac{(1+\epsilon)2^{1-m}}{2m-1} - 1 \right)^{-1} E(\langle M \rangle_T^m). \tag{3.55}$$

for all $\epsilon > 0$ such that the right-hand side of (3.55) is positive.

Case 5: $0 < m \leq \frac{1}{2}$, the lower bound of $E(M_T^{2m})$.

For the proof of this case, we need to first introduce the following lemma.

Lemma 3.4 *Let $X = \{X_t, \mathcal{F}_t; 0 \leq t < \infty\}$ be a continuous, non-negative process with $X_0 = 0$ a.s. and $A = \{A_t, \mathcal{F}_t; 0 \leq t < \infty\}$ a continuous increasing process for which*

$$E(X_T) \leq E(A_T)$$

for every bounded stopping time T. With the same notation in (3.43), for every continuous increasing function $F : [0, \infty) \to [0, \infty)$ with $F(0) = 0$, and $G(x) =

$2F(x) + x \int_x^\infty \frac{1}{u} dF_u$, we have

$$EF(X_T^*) \leq EG(A_T) \qquad (3.56)$$

for any stopping time T.

Proof of the Lemma For every $\epsilon > 0$ and $\delta > 0$, we first define two stopping times as follows:

$$H_\epsilon = \inf\{t \geq 0; X_t \geq \epsilon\}; \quad S_\delta = \inf\{t \geq 0; A_t \geq \delta\}.$$

Define $T_n := T \wedge n \wedge H_\epsilon = \min\{T, n, H_\epsilon\}$ to be a bounded stopping time; then

$$\epsilon P[X_{T_n}^* \geq \epsilon] \leq E[X_{T_n} 1_{\{X_{T_n}^* \geq \epsilon\}}] \leq E[X_{T_n}] \leq E[A_{T_n}] \leq E[A_T]$$

as $n \to \infty$, we have

$$P[X_T^* \geq \epsilon] \leq \frac{1}{\epsilon} E(A_T).$$

Therefore,

$$P[X_T^* \geq \epsilon, A_T < \delta] \leq P[X_{T \wedge S_\delta}^* \geq \epsilon] \leq \frac{1}{\epsilon} E(A_{T \wedge S_\delta}) = \frac{1}{\epsilon} E(\delta \wedge A_T).$$

Using the fact that

$$F(x) = \int_0^\infty 1_{x \geq u} dF_u,$$

we have

$$E[F(X_T^*)] = \int_0^\infty P[X_T^* \geq u] dF_u \leq \int_0^\infty (P[X_T^* \geq u, A_T < u] + P[A_T \geq u]) dF_u$$

$$\leq \int_0^\infty \left\{\frac{E(u \wedge A_T)}{u} + P(A_T \geq u)\right\} dF_u$$

$$= \int_0^\infty \left[2P(A_T \geq u) + \frac{1}{u} E[A_T 1_{\{A_T < u\}}]\right] dF_u$$

$$= E\left[2F(A_T) + A_T \int_0^\infty \frac{1}{u} dF_u\right] = E[G(A_T)].$$

Back to the Proof of Case 5: Since we have proved the theorem for Case 1 and Case 3, we can use the fact that

$$k_1 E\langle M \rangle_t \leq E[(M_t^*)^2]. \qquad (3.57)$$

3.1 Stochastic Integral

Taking $X = k_1 \langle M \rangle$, $A = (M^*)^2$, and $F(x) = x^m$, with $0 < m \leq \frac{1}{2}$ in Lemma 3.4, we have

$$G(x) = \frac{1-m}{2-m} x^m$$

and

$$\frac{1-m}{2-m} k_1^m E \langle M \rangle_t^m \leq E[(M_t^*)^{2m}]. \tag{3.58}$$

Combining (3.50), (3.51), (3.54), (3.55), and (3.58), we can get the desired result (3.47) for every $m > 0$ and thus conclude the proof. □

3.1.5 Stratonovich's Integral

In the construction of Itô's stochastic integral, we mentioned that Itô's integral roots from the Riemann-Stieltjes sum where the left point values of the integrand are involved in each small interval.

Although Itô's integral satisfies many notable features as we introduced in Theorem 3.1 and is widely used in many areas, it also suffers from a drawback that Itô's formula, i.e., the change-of-variable formula, is different from the traditional chain rules in calculus. In this subsection, we will introduce another construction of stochastic integral, called Stratonovich's integral, which satisfies the chain rule and is, to some extent, more convenient in modeling.

The definition of Stratonovich's integral is as follows, where we have to restrict the integrands to semimartingales.

Definition 3.5 Let X and Y be two semimartingales with the following decompositions:

$$X_t = X_0 + M_t + B_t, \quad Y_t = Y_0 + N_t + C_t; \quad 0 \leq t < \infty.$$

Stratonovich's integral of Y with respect to X is defined by

$$\int_0^t Y_s \circ dX_s \triangleq \int_0^t Y_s dX_s + \frac{1}{2} \langle X, Y \rangle_s, \quad 0 \leq t < \infty, \tag{3.59}$$

where the first term of the right-hand side means the stochastic integral in Itô's sense.

Just as in the above definition, Stratonovich's integral is defined based on Itô's integral. However, the next theorem shows that the change-of-variable formula of Stratonovich's version is formally the same as the chain rules in calculus.

Theorem 3.8 *Let* $X = (X^{(1)}, X^{(2)}, \cdots, X^{(d)})$ *be a vector of continuous semi-martingales with decompositions*

$$X_t^{(i)} = X_0^{(i)} + M_t^{(i)} + B_t^{(i)}; \quad 1 \leq i \leq d. \tag{3.60}$$

If $f : R^d \to R$ *is a* C^3 *function, then*

$$f(X_t) = f(X_0) + \sum_{i=1}^{d} \int_0^t \frac{\partial}{\partial x_i} f(X_s) \circ dX_s^{(i)}. \tag{3.61}$$

Proof According to the multi-dimensional Itô's formula in Theorem 3.3, we have

$$f(X_t) = f(X_0) + \sum_{i=1}^{d} \int_0^t \frac{\partial}{\partial x_i} f(X_s) dX_s^i$$

$$+ \frac{1}{2} \sum_{i=1}^{d} \sum_{j=1}^{d} \int_0^t \frac{\partial^2}{\partial x_i \partial x_j} f(X_s) d\langle X^i, X^j \rangle_s.$$

With the relationship between Stratonovich's integral and Itô's integral, each terms in the summation of Eq. (3.61) can be expanded as:

$$\int_0^t \frac{\partial}{\partial x_i} f(X_s) \circ dX_s^{(i)} = \int_0^t \frac{\partial}{\partial x_i} f(X_s) dX_s^{(i)} + \frac{1}{2} \langle \frac{\partial}{\partial x_i} f(X), X \rangle_t.$$

Since $f \in C^3$, we can apply Itô's formula to $\frac{\partial}{\partial x_i}(X_t)$. Combining this with the properties of stochastic integral in Itô's sense we derive before, we can obtain

$$\left\langle \frac{\partial}{\partial x_i} f(X), X \right\rangle_t = \sum_{j=1}^{d} \int_0^t \frac{\partial^2}{\partial x_i \partial x_j} f(X_s) d\langle X^i, X^j \rangle_s.$$

Thus, we have finished the proof of the above theorem. □

At the beginning of this subsection, we mentioned that Itô's integral roots from the Riemann-Stieltjes sum with the left point values of integrands at each small interval. In fact, the Stratonovich's integral defined here can also by regarded as a limit of Riemann-Stieltjes sum. The only difference is that we use the middle-point values of integrands at each interval.

Proposition 3.1 *Let X and Y be two continuous semimartingales and $\Pi = \{t_0, t_1$ and $\cdots, t_n\}$ a partition of $[0, t]$. Denote $|\Pi| = \max_{1 \leq i \leq n}(t_i - t_{i-1})$. Then,*

$$\lim_{|\Pi| \to 0} \sum_{i=0}^{n-1} \left(\frac{1}{2} Y_{t_{i+1}} + \frac{1}{2} Y_{t_i} \right) (X_{t_{i+1}} - X_{t_i}) = \int_0^t Y_s \circ dX_s,$$

where the limit is taken in the sense of convergence in probability.

3.2 Formulations of Stochastic Differential Equations

Proof From the construction of Itô's integral, we can see that Itô's integral can be regarded as the limitation of a kind of Riemann-Stieltjes sum, that is,

$$\lim_{|\Pi| \to 0} \sum_{i=0}^{n-1} Y_{t_i} (X_{t_{i+1}} - X_{t_i}) = \int_0^t Y_s dX_s.$$

According to the relationship between Itô's integral and Stratonovich's one, we only need to show that

$$\lim_{|\Pi| \to 0} \sum_{i=0}^{n-1} \left(Y_{t_{i+1}} - Y_{t_i}\right)(X_{t_{i+1}} - X_{t_i}) = \langle X, Y \rangle_t, \quad (3.62)$$

and Eq. (3.62) can be derived directly from the properties of cross variation. □

At the end of this subsection, we have to remark that although Stratonovich's integral seems to be closer to the traditional deterministic calculus, the more popular stochastic integral is the one by Itô's. Two of the reasons are that Itô's integral can be defined for a broader class of stochastic processes, while Stratonovich's integral can only be defined for semimartingales; besides, Stratonovich's integral does not provide any new insights in mathematics. The definition of Stratonovich's integral is based on Itô's integral, and the relationship between two integrals can be summarized by a close formula (3.59).

Up to now, we have reviewed the basic results in stochastic integral theory, and after all this preparation, we can now give the formulations of stochastic differential equations.

3.2 Formulations of Stochastic Differential Equations

Brownian motion is one of the most important processes studied in stochastic calculus fields, and as we mentioned in the previous section, semimartingales can be converted into a standard Brownian motion by Girsanov's change-of-measure method. Therefore, stochastic integrals with respect to Brownian motion are mainly concerned in the theory of stochastic differential equations. In this section, we will formulate the stochastic differential equation with respect to Brownian motion as follows:

$$\begin{cases} dX_t = b(t, X_t)dt + \sigma(t, X_t)dW_t, & t > 0, \\ X_0 = \xi, \end{cases} \quad (3.63)$$

where $W = \{W_t; 0 \leq t < \infty\}$ is an r-dimensional standard Brownian motion and $b_i(t, x)$, $\sigma_{ij}(t, x)$, $1 \leq i \leq d$, and $1 \leq j \leq r$ are Borel-measurable functions from $[0, \infty) \times R^d$ to R.

The vector $\{b_i(t, x); 1 \leq i \leq d\}$ is called the *drift vector*; the matrix function $\{\sigma_{ij}(t, x); 1 \leq i \leq d, 1 \leq j \leq r\}$ is called the *dispersion matrix*; and the matrix $a(t, x) = \sigma(t, x)\sigma^\top(t, x)$, with elements

$$a_{ij}(t, x) = \sum_{k=1}^{r} \sigma_{ik}(t, x)\sigma_{jk}(t, x), \quad 1 \leq i, j \leq d$$

is called the *diffusion matrix*.

In comparison to ordinary differential equations in deterministic sense, the solution of stochastic differential equations can have different meanings because of the introduction of probability space and lead to different formulations. In this section, we will introduce two kinds of formulations, the strong solution, where the probability space and the Brownian motions are fixed ahead of time, and the weak solution, where the probability space and even the Brownian motions are a part of the solution. In the formulation of weak solutions, apart from solving the SDE directly, we can also apply stochastic integral theory and transform the SDE into a martingale problem, which is more convenient for analyzing the properties of the solution.

3.2.1 Strong Solutions

In this subsection, we will study the stochastic differential equation (3.63) in its strong form. After giving the definition of strong solution, we will consider the existence and uniqueness of solutions under the strong formulation.

Definition 3.6 For a given probability space $(\Omega, \mathcal{F}, P)$, a fixed Brownian motion W, and initial condition ξ, a strong solution of the stochastic differential equation (3.63) is a process $X = \{X_t; 0 \leq t < \infty\}$ with continuous sample paths and with the following properties:

(1) X is adapted to the filtration $\{\mathcal{F}_t\}$;
(2) $P[X_0 = \xi] = 1$;
(3) $P\left[\int_0^t |b_i(s, X_s)| + \sigma_{ij}^2(s, X_s)ds < \infty\right] = 1$, for every $0 \leq t < \infty, 1 \leq i \leq d$, and $1 \leq j \leq r$,
(4) $X_t = X_0 + \int_0^t b(s, X_s)ds + \int_0^t \sigma(s, X_s)dW_s, 0 \leq t < \infty$, a.s. P.

In the formulation of strong sense, the probability space and the Brownian motion are both fixed ahead of time. What we need to do is just find a stochastic process that makes the stochastic integral in the equations sensible and satisfies the corresponding stochastic differential equations. The next theorem shows that with relatively strong restrictions on the parameters (global Lipschitz continuous), stochastic differential equation (3.63) has a strong solution.

3.2 Formulations of Stochastic Differential Equations

Theorem 3.9 *Suppose that coefficients $b(t, x)$ and $\sigma(t, x)$ satisfy the global Lipschitz and linear growth conditions, that is, there exists a positive constant K, such that for every $0 \le t < \infty$, $x, y \in R^d$,*

$$|b(t, x) - b(t, y)| + |\sigma(t, x) - \sigma(t, y)| \le K|x - y|, \quad (3.64)$$

$$|b(t, x)|^2 + |\sigma(t, x)|^2 \le K^2(1 + |x|^2). \quad (3.65)$$

For a fixed probability space $(\Omega, \mathcal{F}, P)$, let ξ be an R^d-valued random vector, independent of the r-dimensional Brownian motion $W = \{W_t; 0 \le t < \infty\}$, and with finite second moment

$$E[|\xi|^2] < \infty.$$

Then, there exists a continuous, adapted process $X = \{X_t, 0 \le t < \infty\}$, which is a strong solution of Eq. (3.63).

Proof The main idea to prove the existence is similar to the deterministic case, where we use the method of Peano iteration.

We can first define iteratively a sequence of processes by $X_t^{(0)} \equiv \xi$ and

$$X_t^{(k+1)} = \xi + \int_0^t b(s, X_s^{(k)}) ds + \int_0^t \sigma(s, X_s^{(k)}) dW_s; \ 0 \le t < \infty, \ k \ge 0. \quad (3.66)$$

Next, we only need to show that this sequence $\{X^{(k)}\}_{k=0}^{\infty}$ will converge to the solution of (3.63).

To this end, we first notice that $X_t^{(k+1)} - X_t^{(k)}$ is also a semimartingale and have the decomposition:

$$X_t^{(k+1)} - X_t^{(k)} = B_t + M_t,$$

where

$$B_t = \int_0^t (b(s, X_s^{(k+1)}) - b(s, X_s^{(k)})) ds, \ M_t = \int_0^t (\sigma(s, X_s^{(k+1)}) - \sigma(s, X_s^{(k)})) dW_s.$$

The Burkholder-Davis-Gundy inequality in Theorem 3.7 guarantees the validation of the following moment estimation:

$$E\left[\max_{0 \le s \le t} |M_s|^2\right] \le \Lambda_1 E \int_0^t |\sigma(s, X_s^{(k+1)}) - \sigma(s, X_s^{(k)})|^2 ds$$

$$\le \Lambda_1 K^2 E \int_0^t |X_s^{(k+1)} - X_s^{(k)}|^2 ds$$

for some $\Lambda_1 \ge 0$.

Besides, according to (3.64), we can also get a moment estimation of B_t:

$$E|B_t|^2 \leq K^2 E \int_0^t |X_s^{(k+1)} - X_s^{(k)}|^2 ds.$$

Thus, there exists an $L \geq 0$, such that

$$E\left[\max_{0 \leq s \leq t} |X_s^{(k+1)} - X_s^{(k)}|^2\right] \leq L \int_0^t E[|X_s^{(k+1)} - X_s^{(k)}|^2] ds.$$

Iteratively, we have

$$E\left[\max_{0 \leq s \leq t} |X_s^{(k+1)} - X_s^{(k)}|^2\right] \leq C \times \frac{(Lt)^k}{k!}; \quad 0 \leq t \leq T \quad (3.67)$$

for every $T > 0$. By Chebyshev's inequality, we have

$$P\left[\max_{0 \leq s \leq t} |X_s^{(k+1)} - X_s^{(k)}|^2 \geq \frac{1}{2^{k+1}}\right] \leq 4C \times \frac{(4LT)^k}{k!} \quad (3.68)$$

and the right-hand side of (3.68) is the general term of a convergence series. By Borel-Cantelli lemma, we can conclude that $t \to X_t^{(k)}$, $t \in [0, T]$ is a uniform Cauchy sequence a.s. and, thus, will converge to a process with continuous path, which is a strong solution to the stochastic differential equation (3.63). □

In this strong formulation of SDE, we need to find a solution in a fixed probability space. This means that strong restrictions are needed to guarantee the existence of a strong solution. Meanwhile, we do not need so strong restrictions on parameters for uniqueness. In this formulation, the uniqueness of solution means that two solutions are indistinguishable with respect to the same initial condition.

Definition 3.7 Consider the stochastic differential equation (3.63). If X and $\widetilde{X}$ are two strong solutions with respect to Brownian motion W and initial condition ξ and $P[X_t = \widetilde{X}_t, 0 \leq t < \infty] = 1$, then we say that strong uniqueness holds for the given stochastic differential equation.

The next theorem shows that under weaker conditions, we can guarantee the uniqueness of strong solution to a stochastic differential equation.

Theorem 3.10 *Suppose that $b(t, x)$ and $\sigma(t, x)$ are both locally Lipschitz-continuous in the space variable, that is, for every $n \geq 1$, there exists a positive number K_n, such that*

$$|b(t, x) - b(t, y)| + |\sigma(t, x) - \sigma(t, y)| \leq K_n |x - y|$$

for all $t \geq 0$ and $x, y \in R^d$, with $|x| \leq n$ and $|y| \leq n$. Then strong uniqueness holds for this stochastic differential equation.

3.2 Formulations of Stochastic Differential Equations

Proof The key point in the proof of uniqueness is also similar to that in the deterministic case. We will also do moment estimation and apply Gronwall's inequality to get the uniqueness result.

Suppose that X and $\widetilde{X}$ are two strong solutions to the same probability space, initial value ξ, and Brownian motion W. According to a standard localization procedure, we can assume without loss of generality that X_t is bounded. Besides, as in the deterministic case, we only need to show the uniqueness for every bounded interval $[0, T]$, $T \geq 0$.

With the properties of stochastic integral and cross variation, we can obtain the following moment estimation:

$$\begin{aligned}
E|X_t - \widetilde{X}_t|^2 &\leq C \left\{ E \int_0^t |\sigma(s, X_s) - \sigma(s, \widetilde{X}_s)|^2 ds \right. \\
&\quad \left. + E \left[\int_0^t |b(s, X_s) - b(s, \widetilde{X}_s)| ds \right]^2 \right\} \\
&\leq C(T+1) E \int_0^t \left(|b(s, X_s) - b(s, \widetilde{X}_s)|^2 \right. \\
&\quad \left. + |\sigma(s, X_s) - \sigma(s, \widetilde{X}_s)|^2 \right) ds \\
&\leq C(T+1) K_n^2 \int_0^t E|X_s - \widetilde{X}_s|^2 ds.
\end{aligned}$$

Now, we can apply the Gronwall's inequality to $t \to E|X_t - \widetilde{X}_t|^2$ and we can get the desired result. $\square$

3.2.2 Weak Solutions

In the formulation of a strong solution, we need a quite strict condition on the parameters to guarantee the existence of a solution. However, for practical use, many stochastic equations do not satisfy those conditions. Therefore, we need another formulation of solution so that it is easier to find a solution. In this subsection, we will introduce the definition of weak solution, in which probability space and the Brownian motion are also a part of the solution.

Definition 3.8 A weak solution of Eq. (3.63) is $(\Omega, \mathcal{F}, \{\mathcal{F}_t\} P, X, W)$, where

(1) $(\Omega, \mathcal{F}, P)$ is a probability space and $\{\mathcal{F}_t\}$ is a filtration of sub-σ-fields of $\mathcal{F}$ satisfying the usual conditions;
(2) $X = \{X_t, \mathcal{F}_t; 0 \leq t < \infty\}$ is a continuous, adapted R^d-valued process and $W = \{W_t, \mathcal{F}_t; 0 \leq t < \infty\}$ is a r-dimensional standard Brownian motion;

(3) $P\left[\int_0^t |b_i(s, X_s)| + \sigma_{ij}^2(s, X_s)ds < \infty\right] = 1$, for every $0 \leq t < \infty$, $1 \leq i \leq d$, and $1 \leq j \leq r$,

(4) $X_t = X_0 + \int_0^t b(s, X_s)ds + \int_0^t \sigma(s, X_s)dW_s$, $0 \leq t < \infty$, a.s. P.

In this weak formulation, because the requirement of a solution is weaken, the uniqueness of a solution is relatively more important compared with the existence. We would like to leave the discussion of the existence in the next subsection and focus on the uniqueness of solution here.

In the strong formulation, the concept of uniqueness is quite clear, which means that the path of two solutions with respect to the same initial value must be indistinguishable. In the weak formulation, we can give the concept of uniqueness two different meanings.

Definition 3.9 Suppose that $(\Omega, \mathcal{F}, \{\mathcal{F}_t\}P, X, W)$ and $(\widetilde{\Omega}, \widetilde{\mathcal{F}}, \{\widetilde{\mathcal{F}}_t\}\widetilde{P}, \widetilde{X}, \widetilde{W})$ are two arbitrarily chosen, weak solutions of (3.63), with the same initial distribution, i.e.,

$$P[X_0 \in U] = \widetilde{P}[\widetilde{X}_0 \in U], \quad \forall U \in \mathcal{B}(R^d).$$

If the process X and $\widetilde{X}$ have the same law, we say that the uniqueness in the sense of probability law holds for the stochastic differential equation.

Definition 3.10 Suppose that $(\Omega, \mathcal{F}, \{\mathcal{F}_t\}P, X, W)$ and $(\Omega, \mathcal{F}, \{\mathcal{F}_t\}P, \widetilde{X}, W)$ are two arbitrarily chosen, weak solutions of (3.63), with common Brownian motion W on a common probabilistic space $(\Omega, \mathcal{F}, P)$, and common initial value. If two processes X and $\widetilde{X}$ are indistinguishable, i.e., $P[X_t = \widetilde{X}_t, 0 \leq t < \infty] = 1$, then we say that pathwise uniqueness holds for the stochastic equation.

Although we have given two meanings to the uniqueness of weak solutions, these two meanings are not entirely irrelevant. In fact, the next proposition shows that pathwise uniqueness is a stronger property for a stochastic differential equation than the uniqueness in the sense of probability law.

Proposition 3.2 *Pathwise uniqueness implies uniqueness in the sense of probability law.*

Proof The proof of this theorem requires some basic knowledge about measure theory in infinite dimensional space. Therefore, we just summarize the key points in the proof here, and for a detailed proof, one can refer to the original paper by Yamada and Watanabe [9].

Suppose that $(\Omega^j, \mathcal{F}^j, \{\mathcal{F}_t^j\}P^j, X^j, W^j)$ and $j = 1, 2$ are two weak solutions to (3.63). Then, we can construct another two weak solutions with the same law of the above two processes on the canonical probability space for Brownian motion, that is,

$$(\Omega, \mathcal{F}, \{\mathcal{F}_t\}_{t \geq 0}) = (C[0, \infty)^d, \mathcal{B}(C[0, \infty)^d)).$$

3.2 Formulations of Stochastic Differential Equations

On this probability space, under the Wiener measure P, the coordinate process

$$W_t(\omega) = \omega(t), \ 0 \leq t < \infty, \ \omega \in \Omega,$$

is a standard Brownian motion.

We denote by x^1 and x^2 two new constructed weak solutions on $(\Omega, \mathcal{F}, P)$. Then, according to the assumption that the pathwise uniqueness holds for Eq. (3.63), we have $P[x_t^1 = x_t^2, 0 \leq t < \infty] = 1$ and, thus, have the same law. Therefore, the two processes X^1 and X^2, though defined on a different probability space, also have the same probability law. We then proved the uniqueness in the sense of probability law for Eq. (3.63). □

Meanwhile, although we have different sets for strong and weak solutions, the existence of strong and weak solutions are also relevant by the following theorem, which shows that a weak solution with the property of pathwise uniqueness implies a strong solution with respect to a certain Brownian motion in a certain probability space.

Theorem 3.11 *Suppose that Eq. (3.63) has a weak solution $(\Omega, \mathcal{F}, \{\mathcal{F}_t\}, P, X, W)$ with initial distribution μ, and suppose that pathwise uniqueness holds for Eq. (3.63). Then, given any probability space $(\widetilde{\Omega}, \widetilde{\mathcal{F}}, \widetilde{P})$ rich enough to support an R^d-valued random variable ξ with distribution μ and an independent r-dimensional Brownian motion $\widetilde{W} = \{\widetilde{W}_t, \mathcal{F}_t^{\widetilde{W}}; 0 \leq t < \infty\}$, ($\mathcal{F}_t^{\widetilde{W}}$ denotes the σ-field generated by Brownian motion $\widetilde{W}$), a strong solution of Eq. 3.63 exists with initial condition ξ.*

Remark 3.4 The proof of Theorem 3.11 also involves the application of canonical probability space and other probability tools such as regular conditional probabilities. Therefore, we would like to skip the proof here and refer interested readers to monographs with respect to stochastic differential equations [5].

For filtering equations we will focus on later, it is hard for the parameters to meet the requirement for the existence of strong solutions. Therefore, from now on, we will pay more attention on the weak formulation of stochastic differential equations. Without special instructions, the solution of a stochastic differential equation means a weak solution.

3.2.3 The Martingale Problem of Stroock and Varadhan

In this subsection, we will view the weak formulation of a stochastic differential equation from a different aspect. The main result in this subsection is that in some sense, a weak solution to an SDE is equivalent to a probability measure that guarantees a group of semimartingales the martingale property.

Suppose that $(\Omega, \mathcal{F}, \{\mathcal{F}_t\}, P, X, W)$ is a weak solution of Eq. (3.63). For every $t \geq 0$, we introduce the second-order differential operator

$$\mathscr{A}_t f \triangleq \frac{1}{2} \sum_{i,j=1}^{d} a_{ij}(t,x) \frac{\partial^2}{\partial x_i \partial x_j} f + \sum_{i=1}^{d} b(t,x) \frac{\partial}{\partial x_i} f; \quad f \in C^2(R^d). \quad (3.69)$$

The motivation of the introduction of the martingale problem is that, according to Itô's formula, we can construct a group of local martingales from the solution to a stochastic differential equation.

Theorem 3.12 *For every $f \in C\big([0,\infty) \times R^d\big)$, the process $M^f = \{M_t^f, \mathcal{F}_t; 0 \leq t < \infty\}$ given by*

$$M_t^f = f(t, X_t) - f(0, X_0) - \int_0^t \left(\frac{\partial f}{\partial s} + \mathscr{A}_s f \right)(s, X_s) ds \quad (3.70)$$

is a continuous local martingale.

Proof This result is a direct corollary of Itô's formula. According to Itô's formula, we have

$$f(t, X_t) = f(0, X_0) + \int_0^t \left(\frac{\partial f}{\partial s} + \mathscr{A}_s f \right)(s, X_s) ds$$

$$+ \sum_{i=1}^{d} \sum_{j=1}^{d} \int_0^t \frac{\partial f}{\partial x_i}(s, X_s) \sigma_{ij}(s, X_s) dW_s^{(j)}.$$

Therefore,

$$M_t^f = \sum_{i=1}^{d} \sum_{j=1}^{d} \int_0^t \frac{\partial f}{\partial x_i}(s, X_s) \sigma_{ij}(s, X_s) dW_s^{(j)}.$$

is a continuous, local martingale. □

Later on, we will see that heuristically, a process that guarantees the martingale property of M_t^f is just the solution of the corresponding stochastic differential equation. Before that, we will first give the definition of a martingale problem.

Definition 3.11 (Stroock and Varadhan [8]) A probability measure P on $\big(C[0,\infty)^d, \mathcal{B}(C[0,\infty)^d)\big)$ is called a solution to local martingale problem associated with $\{\mathscr{A}_t\}$, if for every $f \in C\big([0,\infty) \times R^d\big)$, under P, the process

$$M_t^f(\omega) = f(t, \omega(t)) - f(0, \omega(0)) - \int_0^t \left(\frac{\partial f}{\partial s} + \mathscr{A}_s f \right)(s, \omega(s)) ds \quad (3.71)$$

is a continuous local martingale.

3.2 Formulations of Stochastic Differential Equations

The next theorem illustrates the equivalence of a solution to the martingale problem and a weak solution to the corresponding stochastic differential equation.

Theorem 3.13 *The existence of a solution P to the local martingale problem associated with $\{\mathscr{A}_t\}$ is equivalent to the existence of a weak solution, $(\widetilde{\Omega}, \widetilde{\mathcal{F}}, \{\widetilde{\mathcal{F}}_t\}, \widetilde{P}, X, W)$, to Eq. (3.63). The two solutions are related by $P = \widetilde{P} X^{-1}$.*

The uniqueness of the solution P of the local martingale problem with initial distribution μ:

$$P[\omega \in C[0,\infty)^d; \omega(0) \in U] = \mu(U), \quad U \in \mathcal{B}(R^d)$$

is equivalent to the uniqueness in the sense of probability law for Eq. (3.63).

Remark 3.5 The proof of Theorem 3.13 also involves a lengthy but fundamental discussion on measure theory in infinite dimensional spaces. Again, we would like to omit all these discussions that are not used very often in the development of filter theory. We only need to remember that the solution of the martingale problem is relevant to a weak solution of stochastic differential equation by $P = \widetilde{P} X^{-1}$.

With the equivalence of the martingale problem and the weak solution, we can obtain a weaker condition for the existence of weak solutions from the existence of a solution to the martingale problem. Here, we consider the time-homogeneous version of the stochastic differential equations.

Theorem 3.14 *Consider the time-homogeneous stochastic differential equation*

$$dX_t = b(X_t)dt + \sigma(X_t)dW_t, \tag{3.72}$$

where $b_i, \sigma_{ij} : R^d \to R$ are bounded and continuous functions. Then, there exists a weak solution to (3.72) with respect to every initial distribution μ on $\mathcal{B}(R^d)$ with

$$\int_{R^d} |x|^{2m} \mu(dx) < \infty, \quad \text{for some } m > 1.$$

Proof The main idea for the proof of this theorem is quite similar to the case where we proved the existence of a strong solution. However, instead of the construction of a Peano-like iteration procedure, we will make an approximation to the solution by a sequence of processes that are a solution to a simpler stochastic differential equation.

For each $n \in \mathbb{N}$, we define the process $\{X_t^{(n)}\}$ as follows:

$$X_0^{(n)} = \xi; \quad X_t^{(n)} = X_{t_j^{(n)}} + b(X_{t_j^{(n)}})(t - t_j^{(n)}) + \sigma(X_{t_j^{(n)}})(W_t - W_{t_j^{(n)}}) \tag{3.73}$$

for each $t_j^{(n)} < t \le t_{j+1}^{(n)}$, where $t_j^{(n)} = j/2^n$ are the dyadic rationals.

If we define the ladder functions $\psi_n(t) = t_j^{(n)}$; $t_j^{(n)} \le t < t_{j+1}^{(n)}$ and the new coefficients

$$b^{(n)}(t, y(t)) = b(y(\psi_n(t))), \quad \sigma^{(n)}(t, y(t)) = \sigma(y(\psi_n(t))), \tag{3.74}$$

then $X^{(n)}$ solves the stochastic integral equation

$$X_t^{(n)} = \xi + \int_0^t b^{(n)}(s, X_s^{(n)}) ds + \int_0^t \sigma^{(n)}(s, X_s^{(n)}) dW_s. \tag{3.75}$$

According to Theorem 3.13, the existence of a solution to (3.75) implies a solution to the corresponding local martingale problem $P^{(n)} = P(X^{(n)})^{-1}$.

Since coefficients of each (3.75) are bounded, the series of probability measure $\{P^{(n)}\}_{n=1}^\infty$ is tight and thus converges weakly to a probability measure P^* on the canonical probability space.

The limitation theory of integrals leads to the result that P^* is a solution to the local martingale problem corresponding to stochastic differential equation (3.72), which then implies the existence of a solution to that equation. □

Starting from the martingale problem, we can also give another condition that guarantees the uniqueness of a weak solution. The next theorem shows that the uniqueness of a weak solution can be obtained from the existence of a solution to the Cauchy problem of a parabolic partial differential equation.

Theorem 3.15 *Suppose that with coefficients $b(x)$ and $\sigma(x)$ in (3.72), the Cauchy problem*

$$\begin{cases} \dfrac{\partial u}{\partial t} - \mathscr{A} u = 0; & \text{in } [0, \infty) \times R^d \\ u(0, \cdot) = f; & \text{in } R^d \end{cases} \tag{3.76}$$

has a solution $u_f \in C\left([0, \infty) \times R^d\right) \cap C^{1,2}\left((0, \infty) \times R^d\right)$, which is bounded on each $[0, T] \times R^d$, for every $f \in C_0^\infty(R^d)$. Then for every $x \in R^d$, there exists at most one solution to the time-homogeneous martingale problem with initial distribution

$$P[\omega \in C[0, \infty)^d; \omega(0) = 0] = 1$$

Proof For every fixed $T > 0$, we define $g(t, x) = u_f(T-t, x)$. Then, g is bounded and satisfies

$$\frac{\partial g}{\partial t} + \mathscr{A} g = 0, \quad g(T, \cdot) = f(\cdot).$$

Assume that X and $\tilde{X}$ are two weak solutions to the stochastic differential equation (3.72) and P and $\tilde{P}$ are corresponding solutions to the local martingale problem; then, under P and $\widetilde{P}$, $Z_t(\omega) = \omega(t)$, $\omega \in C[0, \infty)^d$ is a solution to the same stochastic differential equation.

Thus, according to Itô's formula, $\{g(t, Z_t), 0 \leq t < \infty\}$ is a local martingale under P and $\widetilde{P}$. However, since g is a bounded function, the local martingale is actually a martingale, and thus, we have

$$E[f(Z_T)] = E[g(T, Z_T)] = E[g(0, Z_0)]$$
$$= \widetilde{E}[g(0, Z_0)] = \widetilde{E}[g(T, Z_T)] = \widetilde{E}[f(Z_T)]. \tag{3.77}$$

Since Eq. (3.77) holds for every $f \in C_0^\infty$, we have $P = \widetilde{P}$, and there exists at most one solution to Eq. (3.72). □

3.3 Connections Between Stochastic Differential Equations and Partial Differential Equations

In the previous section, Theorem 3.15 reveals one of the connections between stochastic differential equations and partial differential equations. The uniqueness of solution to an SDE is related to the existence of solution to a corresponding PDE, in the context of the martingale problem.

In this section, we will review two other links between SDEs and PDEs. The first one is given by the Feynman-Kac formula, where the solution of a PDE can be represented by the expectation of the solution to an SDE. The solution of an SDE can also be represented by its probability law, and its probability density function satisfies the Kolmogorov's equation.

3.3.1 Feynman-Kac Representation

In order to illustrate the Feynman-Kac representation formula, we first consider the following stochastic differential equation:

$$\begin{cases} dX_s = b(s, X_s)ds + \sigma(s, X_s)dW_s, \ t \leq s < \infty, \\ X_t = x, \end{cases} \tag{3.78}$$

where coefficients $b_i(t, x)$ and $\sigma_{ij}(t, x)$ are continuous and satisfy the linear growth condition (3.65). Meanwhile, Eq. (3.78) has a weak solution, and the weak solution is unique in the sense of probability law.

We denote the solution to the above differential equation as $X^{(t,x)} = \{X_s^{(t,x)}; 0 \leq s < \infty\}$. The equation $X^{(t,x)}$ can also be expressed in the following integration form:

$$X_s^{(t,x)} = x + \int_t^s b(\theta, X_\theta^{(t,x)})d\theta + \int_t^s \sigma(\theta, X_\theta^{(t,x)})dW_\theta; \quad t \leq s < \infty. \quad (3.79)$$

Besides, we will consider the Cauchy problem of the following parabolic differential equation:

$$\begin{cases} -\dfrac{\partial v}{\partial t} + kv = \mathcal{A}_t v + g; \text{ in } [0, T) \times R^d, \\ v(T, x) = f(x); \text{ in } x \in R^d, \end{cases} \quad (3.80)$$

where $T > 0$ is arbitrary but fixed; functions $f(x) : R^d \to R$, $g(t, x) : [0, T] \times R^d \to R$, and $k(t, x) : [0, T] \times R^d \to [0, \infty)$ are continuous and satisfy

(1) $|f(x)| \leq L(1 + |x|^{2\lambda})$ or $f(x) \geq 0$; $x \in R^d$,
(2) $|g(t, x)| \leq L(1 + |x|^{2\lambda})$ or $g(t, x) \geq 0$; $0 \leq t \leq T$, $x \in R^d$,

with appropriate constants $L > 0$ and $\lambda \geq 1$.

The following theorem shows that the solution to the Cauchy problem with polynomial growth condition can be characterized by the process $X^{(t,x)}$.

Theorem 3.16 *Suppose that $v(t, x) \in C\left([0, T] \times R^d\right) \cap C^{1,2}\left([0, T) \times R^d\right)$ is a solution to the Cauchy problem, which satisfies the polynomial growth condition*

$$\max_{0 \leq t \leq T} |v(t, x)| \leq M(1 + |x|^{2\mu}); \quad x \in R^d \quad (3.81)$$

for some $M > 0$ and $\mu \leq 1$. Then $v(t, x)$ admits the stochastic representation

$$v(t, x) = E^{t,x}\left[f(X_T)exp\left\{-\int_t^T k(\theta, X_\theta)d\theta\right\} \right. \\ \left. + \int_t^T g(s, X_s)exp\left\{-\int_t^s k(\theta, X_\theta)d\theta\right\}ds\right] \quad (3.82)$$

on $[0, T] \times R^d$; in particular, such a solution is unique.

Proof The key point of the proof is to apply Itô's formula to the process $v(s, X_s)exp\{-\int_t^s k(\theta, X_\theta)d\theta\}$, and Eq. (3.82) can then be derived from the martingale property.

In the process, conditions on coefficients guarantee those local martingales we get from Itô's formula are actually martingales and the validation of applying localization procedure. For simplicity, we skip the proof of this part and admit that those local martingales mentioned here are martingales, so that the value of expectations won't change through time.

3.3 Connections Between Stochastic Differential Equations and Partial...

According to Itô's formula, the process

$$M_s = v(s, X_s) exp\{-\int_t^s k(\theta, X_\theta)d\theta\} + \int_t^s g(u, X_u) exp\{-\int_t^u k(\theta, X_\theta)d\theta\}du$$

is a martingale. Therefore, $EM_T = EM_t$, and the desired result is directly from the fact that

$$v(t, X_t) = v(t, x), \quad v(T, X_T) = f(X_T).$$

□

3.3.2 Kolmogorov Equation

Apart from Feynman-Kac formula presented in the previous subsection, the solution to an SDE and the solution to a PDE can also be connected by the probability density function. In fact, the probability density function, as a function of time variable t and space variables x, satisfies a parabolic partial differential equation.

To give the above discussion a rigorous mathematical context, we first define the concept of "well-pose" for a time-homogeneous stochastic differential equation.

Definition 3.12 The time-homogeneous stochastic differential equation in integral form

$$X_t = x + \int_0^t b(X_s)ds + \int_0^t \sigma(X_s)dW_s \tag{3.83}$$

is called well-posed, if for every initial condition $x \in R^d$, it admits a weak solution that is unique in the sense of probability law.

The following theorem shows that the solution to a well-posed time-homogeneous equation is a strong Markov process and has a strong relationship to an elliptic differential operator.

Theorem 3.17 *Suppose that coefficients b and σ are bounded on compact subsets of R^d and that the time-homogeneous stochastic integral equation (3.83) is well-posed. Then, strong Markov property holds for process X.*

Besides, if we further assume that b and σ are bounded and continuous, then the relation

$$\lim_{t \downarrow 0} \frac{1}{t} \left[E^x f(X_t) - f(x) \right] = (\mathscr{A} f)(x); \quad \forall x \in R^d \tag{3.84}$$

holds for every $f \in C^2(\mathbb{R}^d)$, where $\mathscr{A}$ is the time-homogeneous version of operator (3.69)

$$(\mathscr{A}f)(x) \triangleq \frac{1}{2}\sum_{i,j=1}^{d} a_{ij}(x)\frac{\partial^2 f}{\partial x_i \partial x_j}(x) + \sum_{i=1}^{d} b_i(x)\frac{\partial f}{\partial x_i}(x) \qquad (3.85)$$

and $a(x) = \sigma(x)\sigma^\top(x)$. Moreover, the process X is a diffusion process.

Proof The proof of strong Markov property requires a long journey in the theory of Markov process, and interested readers may refer to monographs for help. For us, we would like to admit the strong Markov property and give a proof of Eq. (3.84).

In fact, under the conditions in Theorem 3.17, it is a straight corollary of Itô's formula. Since X is a solution to the stochastic integral equation, thus

$$f(X_t) = f(x) + \sum_{i=1}^{d}\int_0^t \frac{\partial f}{\partial x_i}(X_s)b^i(X_s)ds + \sum_{i,j=1}^{d}\int_0^t \frac{\partial f}{\partial x_i}(X_s)\sigma_{ij}(X_s)dW_s^{(j)}$$

$$+ \frac{1}{2}\sum_{i,j=1}^{d}\int_0^t \frac{\partial^2 f}{\partial x_i \partial x_j}(X_s)\left(\sum_{k=1}^{d}\sigma_{ik}(X_s)\sigma_{jk}(X_s)\right)ds.$$

(3.86)

Taking expectations to both sides of Eq. (3.86) and noticing that the stochastic integral is a martingale, we have

$$E^x f(X_t) = f(x) + \sum_{i=1}^{d} E\int_0^t \frac{\partial f}{\partial x_i}(X_s)b^i(X_s)ds$$

$$+ \frac{1}{2}\sum_{i,j=1}^{d} E\int_0^t \frac{\partial^2 f}{\partial x_i \partial x_j}(X_s)\left(\sum_{k=1}^{d}\sigma_{ik}(X_s)\sigma_{jk}(X_s)\right)ds.$$

(3.87)

Calculate $\lim_{t\downarrow 0}\frac{1}{t}[E^x f(X_t) - f(x)]$ based on (3.87), and note that $X_0 \equiv x$. We have derived Eq. (3.84). □

Since X is a diffusion process under the conditions in Theorem 3.17, the probability law of X is totally determined by initial values and the **transition probability density function**, $\Gamma(t; x, y)$, which is defined as follows:

$$P^x[X_t \in dy] = \Gamma(t; x, y)dy; \quad \forall x \in \mathbb{R}^d, \ t > 0. \qquad (3.88)$$

The next theorem shows that the transition density function $\Gamma(t; x, y)$ satisfies the following parabolic partial differential equations.

3.3 Connections Between Stochastic Differential Equations and Partial...

Theorem 3.18 *The transition density $\Gamma(t; x, y)$ satisfies the forward Kolmogorov equation, for every fixed $x \in R^d$:*

$$\frac{\partial}{\partial t}\Gamma(t; x, y) = \mathscr{A}^*\Gamma(t; x, y); \quad (t, x) \in (0, \infty) \times R^d, \tag{3.89}$$

and the backward Kolmogorov equation, for every fixed $y \in R^d$:

$$\frac{\partial}{\partial t}\Gamma(t; x, y) = \mathscr{A}\Gamma(t; x, y); \quad (t, y) \in (0, \infty) \times R^d, \tag{3.90}$$

where the operator $\mathscr{A}^*$ is given by

$$(\mathscr{A}^* f)(x) \triangleq \frac{1}{2}\sum_{i,j=1}^{d} \frac{\partial^2}{\partial x_i \partial x_j}[a_{ij}(x)f(x)] - \sum_{i=1}^{d}\frac{\partial}{\partial x_i}[b_i(x)f(x)]. \tag{3.91}$$

Proof Here, we only give a formal proof of the above theorem. A rigid proof can be obtained by applying the approximation of smooth functions.

For each $(t, y) \in (0, \infty) \times R^d$, we first calculate

$$\frac{\partial \Gamma}{\partial t}(t, x, y) = \lim_{\Delta t \to 0} \frac{1}{\Delta t}(\Gamma(t + \Delta t, x, y) - \Gamma(t, x, y)). \tag{3.92}$$

Meanwhile, according to the strong Markov property of X,

$$\Gamma(t + \Delta t, x, y) = \int_{R^d} \Gamma(\Delta t, x, z)\Gamma(t, z, y)dz = E^x[\Gamma(t, X_{\Delta t}, y)]. \tag{3.93}$$

Therefore,

$$\lim_{\Delta t \to 0} \frac{1}{\Delta t}(\Gamma(t + \Delta t, x, y) - \Gamma(t, x, y))$$
$$= \lim_{\Delta t \to 0} \frac{1}{\Delta t}(E^x[\Gamma(t, X_{\Delta t}, y)] - \Gamma(t, x, y)). \tag{3.94}$$

According to Theorem 3.17, the right-hand side in (3.94) is equal to $\mathscr{A}\Gamma(t, x, y)$, and thus, we have proved Kolmogorov's backward equation (3.90).

In order to prove Kolmogorov's forward equation, we only need to show that

$$\int_{[0,\infty)\times R^d} \frac{\partial}{\partial t}\Gamma(t; x, y)\varphi(t, y)dtdy = \int_{[0,\infty)\times R^d} \mathscr{A}^*\Gamma(t; x, y)\varphi(t, y)dtdy \tag{3.95}$$

holds for every $\varphi \in C_0^{1,2}([0, \infty) \times R^d)$.

With the method of integration by part, it is equivalent to

$$\int_{[0,\infty)\times R^d} \left(\frac{\partial \varphi}{\partial t} + \mathscr{A}\varphi\right)(t,y)\Gamma(t,x,y)dtdy = 0 \qquad (3.96)$$

holds for every $\varphi \in C_0^{1,2}([0,\infty) \times R^d)$.

Notice that the left-hand side in (3.96) is equal to

$$\int_0^\infty E^x\left[\frac{\partial \varphi}{\partial t}(t,X_t) + \mathscr{A}\varphi(t,X_t)\right]dt \qquad (3.97)$$

and Itô's formula shows that the expectation is equal to 0, for every $t > 0$.

Thus, Eq. (3.96) holds, and we have proved Kolmogorov's forward equation (3.89). □

3.4 Exercises

In this chapter, some in-line exercises are provided during the main text, which serves as complements of proofs of theorems and lemmas. Besides, some extra exercises are also listed below.

1. Let $\{W_t : t \geq 0\}$ be a standard one-dimensional Brownian motion. Use Itô's formula to prove that

$$\int_0^t W_s dW_s = \frac{1}{2}\left(W_t^2 - t\right). \qquad (3.98)$$

2. Let $\{Z_t : t \geq 0\}$ be defined in Sect. 3.1.3

$$Z_t = \exp\left[\sum_{i=1}^d \int_0^t X_s^i dW_s^i - \frac{1}{2}\int_0^t |X_s|^2 ds\right]. \qquad (3.99)$$

Use Itô's formula to prove that

$$Z_t = 1 + \sum_{i=1}^d \int_0^t Z_s X_s^i dW_s^i. \qquad (3.100)$$

3. Let $\{X_t : t \geq 0\}$ and $\{Y_t : t \geq 0\}$ be two semimartingales. Prove the integration by parts formula for stochastic integral in Itô's sense.

$$\int_0^t X_s dY_s = X_t Y_t - X_0 Y_0 - \int_0^t Y_s dX_s - \langle X, Y\rangle_t. \qquad (3.101)$$

4. Let $\{X_t : t \geq 0\}$ and $\{Y_t : t \geq 0\}$ be two semimartingales. Prove the integration by parts formula for stochastic integral in Stratonovich sense.

$$\int_0^t X_s \circ dY_s = X_t Y_t - X_0 Y_0 - \int_0^t Y_s \circ dX_s. \tag{3.102}$$

Notice that the integration by parts formula for stochastic integral in Stratonovich sense is identical with the normal integrals in calculus.

5. Let $\{W_t : t \geq 0\}$ be a standard one-dimensional Brownian motion. Show that $X_t = e^{W_t}$ is the solution of the stochastic differential equation

$$dX_t = \frac{1}{2} X_t dt + X_t dW_t. \tag{3.103}$$

6. Let $A, B \in R^{d \times d}$ be $d \times d$ matrices and $\{W_t : t \geq 0\}$ be a standard d-dimensional Brownian motion. Solve the linear stochastic differential equation

$$dX_t = AX_t dt + B dW_t \tag{3.104}$$

References

1. A. Bain and D. Crisan. *Fundamentals of stochastic filtering*. Springer-Verlag, New York, 2009.
2. D. L. Burkholder. Distribution function inequalities for martingales. *The Annals of Probability*, 1:19–42, 1973.
3. B. Davis. Brownian motion and analytic functions. *The Annals of Probability*, 7:913–932, 1979.
4. I. V. Girsanov. On transforming a certain class of stochastic processes by absolutely continuous substitution of measures. *Theory of Probability and Its Applications*, 5:285–301, 1960.
5. N. Ikeda and S. Watanabe. *Stochastic differential equations and diffusion processes*. North-Holland, Amsterdam, 1981.
6. I. Karatzas and S. E. Shreve. *Brownian motion and stochastic calculus*. Springer, New York, 1991.
7. A. A. Novikov. On an identity for stochastic integrals. *Theory of Probability and Its Applications*, 17:717–720, 1972.
8. D. W. Stroock and S. R. S. Varadhan. *Multidimensional diffusion processes*. Springer, New York, 1979.
9. T. Yamada and S. Watanabe. On the uniqueness of solutions of stochastic differential equations. *Journal of Mathematics of Kyoto University*, 11:155–167, 1971.

Chapter 4
Optimization

In this chapter, we will start from basic concepts and examples, which contain some well-known optimization problems used in a wide fields. Then, for general optimization, we introduce optimal condition satisfied by local or global minimum. Well-known Karush-Kuhn-Tucker condition is introduced in detail. In the following, dual optimization is discussed, which includes weak and strong forms. Condition satisfied by strong duality is formulated through the framework of constraint qualification. Finally, we stress on the convex optimization and introduce the corresponding version of optimal condition. In terms of numerical algorithm of convex optimization, we mention active set method to solve quadratic convex problem. For nonconvex optimization, we include sequential quadratic programming for quadratic optimization system.

4.1 Background

At the beginning of this chapter, we will briefly describe the reasons for the popularity of optimization and its history and then give some examples of optimization problems. Optimization is denoted to solve the minimum of a function on a given set:

$$\begin{aligned} \min \quad & f(x) \\ \text{s.t.} \quad & x \in \mathcal{F}, \end{aligned} \tag{4.1}$$

where $\mathcal{F} \subset R^n$ is the feasible region, $x \in \mathcal{F}$ is a feasible solution, and f is a real function called objective function.

Optimization has been developing quickly as a field and a hot topic. An important reason is due to requirement of machine learning. A machine learning problem can finally be transformed into an optimization problem.

The development of optimization can be traced back to seventeenth century. At the beginning, Newton and Rapson transformed the problem of solving nonlinear equations to a minimization problem as follows:

$$f(x) = 0 \Rightarrow \min f^2(x). \tag{4.2}$$

Then we try to find x through optimization methods. Similar methods are also used by Gauss-Seidel and Jacobi; they are used to solve such a problem as follows:

$$\begin{cases} f_1(x) &= 0 \\ \ldots & \\ f_n(x) &= 0 \end{cases} \Rightarrow \min \sum_{i=1}^n f_i^2(x).$$

Therefore, an important idea is the root of the solution equation and the minimum value of the optimization function are often equivalent.

Over time, before and after World War II, there were actually many interesting developments. For example, in 1940, Bellman developed a dynamic programming algorithm [2] (Dynamic Programming, DP). The basic idea is to decompose the problem to be solved into several sub-problems, first solve the sub-problems, and then obtain the solution of the original problem from the solutions of these sub-problems. A classic example of dynamic programming is Dijkstra algorithm of shortest path [5]. Speaking of convex optimization, we generally mention the simplex method [7]. When it comes to numerical optimization, the interior point method [6] will be mentioned. The interior point method was proposed by Karmarkar in 1984. Afterward, many optimization developments focused on many details of the interior point method.

4.1.1 Basic Concepts

Definition 4.1 (Convex Set) For the set S and any two points x, y in the set, if there is $\theta x + (1-\theta) y \in S$ for any $\theta \in [0, 1]$, then the set is said to be a convex set.

Definition 4.2 (Convex Hull) The smallest convex set containing set S is defined as convex hull of S, and we denote $conv(S)$.

Next we introduce some properties about convex set, which can be easily obtained by properties of set calculation.

Proposition 4.1 *If S_i is a convex set, then $\bigcap_{i=1}^{\infty} S_i$ is also a convex set.*

Remark 4.1 The union of convex sets in general will not be convex.

Definition 4.3 (Cone) Set $\mathcal{K} \subset R^n$ is a cone if $\lambda x \in \mathcal{K}$ for any $x \in \mathcal{K}$ and $\lambda \geq 0$.

Definition 4.4 For a set $S \subset R^n$, dual set is defined as

4.1 Background

$$S^* = \{y \in R^n | y^T x \geq 0, \forall x \in S\}. \quad (4.3)$$

If $S = S^*$, then it is self-dual.

Definition 4.5 (Convex Function) If the domain of the function $f : R^n \to R$ is a convex set and f satisfies $f(\theta x + (1-\theta)y) \leq \theta f(x) + (1-\theta) f(y)$ for any $\theta \in [0, 1]$, then f is said to be a convex function.

Remark 4.2 We say a function is strictly convex if Definition 4.5 holds with strict inequality for $x \neq y$ and $0 < \theta < 1$. We say that f is concave if $-f$ is convex, and f is strictly concave if $-f$ is strictly convex.

Convex functions give rise to a particularly important type of convex set called an α-sublevel set, which is defined as follows.

Definition 4.6 (α-Sublevel Set) Given a convex function $f : R^n \to R$ and a real number $\alpha \in R$, the α-sublevel set is defined as

$$\{x \in \mathcal{D}(f) : f(x) \leq \alpha\},$$

where $\mathcal{D}(f)$ is the domain of definition.

Remark 4.3 The α-sublevel set is the set of all points x such that $f(x) \leq \alpha$. To see it is a convex set, we consider any $x, y \in \mathcal{D}(f)$ such that $f(x) \leq \alpha$ and $f(y) \leq \alpha$. Then

$$f(\theta x + (1-\theta)y) \leq \theta f(x) + (1-\theta) f(y) \leq \theta \alpha + (1-\theta)\alpha = \alpha.$$

Definition 4.7 (Subgradient) For function $f(x)$, subgradient at point $\bar{x}$ is defined as vector $d \in R^n$ satisfying the following condition:

$$f(x) \geq f(\bar{x}) + d^T(x - \bar{x}), \quad \forall x \in D(f) \quad (4.4)$$

and the set of all subgradients at $\bar{x}$ is denoted as $\partial f(\bar{x})$.

Definition 4.8 (Conjugate Function) Conjugate function of a function $f : D(f) \to R$ is defined as $h : \mathcal{Y} \to R$, where

$$h(y) = \sup_{x \in D(f)} \{y^T x - f(x)\}, \quad (4.5)$$

and

$$\mathcal{Y} = \{y \in R^n | h(y) < \infty\}. \quad (4.6)$$

In the following section, conjugate function will be used to establish conjugate dual model.

4.1.2 Examples

Now we will introduce some optimization problems from different fields.

Example 4.1 (Shortest-Path Problem [4]) The shortest path problem is the problem of finding a path between two vertices (or nodes) in a graph such that the sum of the weights of its constituent edges is minimized, which is an important application in graph theory.

Consider the shortest path problem. Given a directed graph (V, A) with source node s, target node t, and cost w_{ij} for each edge (i, j) in A, consider the program with variables x_{ij}

$$\min \sum_{i,j \in A} w_{ij} x_{ij}$$
$$\text{s.t.} \quad x_{ij} \in \{0, 1\}, \sum_j x_{ij} - \sum_j x_{ji} = \begin{cases} 1 & i = s, \\ -1 & i = t, \\ 0 & \text{else}, \end{cases}$$

where the x_{ij} here means whether to choose the path from i to j. So for a path, if it is the starting point, its out-degree is 1 larger than the in-degree; if it is the end point, its in-degree must be 1 greater than the out-degree (i.e., the out-degree must be greater than the in-degree-1). The out-degree and in-degree of the passing point in the middle are the same.

Remark 4.4 This optimization problem is not a very easy problem. Because its constraints are discrete points, this constraint condition is not convex. Therefore, the common technique of convex relaxation will be considered, i.e.,

$$x_{ij} = 0, 1 \Rightarrow x_{ij} \geq 0.$$

It is needed to note that the new problem may not be equivalent to original problem after convex relaxation.

Example 4.2 (LASSO [8]) LASSO is a regression analysis method that performs both variable selection and regularization in order to enhance the prediction accuracy and interpretability of the resulting statistical model, which is an important method in statistics and machine learning.

Consider a m-dimensional Fused LASSO problem, which can be formulated as the following optimization problem:

$$\min_{\theta} \frac{1}{2} \sum_{i=1}^{n} \|y_i - x_i \theta^\top\|_2^2 + \lambda \|\theta\|_1,$$

where $\{(x_i, y_i)\}_{i=1}^{n}$ is the observation set with $x_i \in R^m$, $y_i \in R^m$, and $\theta \in R^{m \times m}$ is parameter matrix. If the penalty term $\lambda \|\theta\|$ is removed, then it is a classic least squares regression problem.

4.1 Background

Remark 4.5 In view of the penalty term, it is to punish the gap between adjacent points. So, the large λ will make the fitting result tend to be segmented straight.

Example 4.3 (Principal Components Analysis (PCA) [9]) Principal component analysis (PCA) is a technique for reducing the dimensionality of large datasets, increasing interpretability but at the same time minimizing information loss. It is a type of projection method. Notice the variance of data reflects the contained information. The aim of PCA is to find a projective subspace that can maximize variance of projected data. In the following, we can provide the formulation of PCA.

We assume that we have dataset $\{x^{(1)}, x^{(2)}, \ldots, x^{(m)}\}$ that has been centralized, i.e., $\sum_{i=1}^{m} x^{(i)} = 0$. After projection transform, we get new coordinate system $\{w_1, w_2, \ldots, w_n\}$ with $\|w_i\|_2 = 1$, $w_i^T w_j = 0$ for $i \neq j$.

If we want to reduce dimension of dataset from n to n', that is, discard some of the coordinates in the new coordinate system, the new coordinate system is $\{w_1, w_2, \ldots, w_{n'}\}$, and the sample point $x^{(i)}$ in the projection in the dimensional coordinate system is $z^{(i)} = (z_1^{(i)}, z_2^{(i)}, \ldots, z_{n'}^{(i)})^T$. Among them, $z_j^{(i)} = w_j^T x^{(i)}$ is the coordinate of the j-th dimension of $x^{(i)}$ in the low-dimensional coordinate system.

If we use $z^{(i)}$ to restore the original data $x^{(i)}$, the obtained restored data $\bar{x}^{(i)} = \sum_{j=1}^{n'} z_j^{(i)} w_j = W z^{(i)}$, where W is a matrix composed of standard orthogonal bases.

Now we consider the entire dataset. We hope that the distance between all data and this hyperplane is close enough, that is, minimize the following formula:

$$\sum_{i=1}^{m} \|\bar{x}^{(i)} - x^{(i)}\|_2^2$$

Expanding the norm in the previous summation, we can get:

$$\begin{aligned}
\sum_{i=1}^{m} \left\|\bar{x}^{(i)} - x^{(i)}\right\|_2^2 &= \sum_{i=1}^{m} \left\|W z^{(i)} - x^{(i)}\right\|_2^2 \\
&= \sum_{i=1}^{m} \left(W z^{(i)}\right)^T \left(W z^{(i)}\right) - 2 \sum_{i=1}^{m} \left(W z^{(i)}\right)^T x^{(i)} \\
&\quad + \sum_{i=1}^{m} x^{(i)T} x^{(i)} \\
&= -\operatorname{tr}\left(W^T X X^T W\right) + \sum_{i=1}^{m} x^{(i)T} x^{(i)}.
\end{aligned} \quad (4.7)$$

Notice that $\sum_{i=1}^{m} x^{(i)} x^{(i)T}$ is the covariance matrix of the dataset, and each vector w_j of W is an orthonormal basis. And $\sum_{i=1}^{m} x^{(i)T} x^{(i)}$ is a constant. Minimizing the above equation is equivalent to:

$$\min_W -\operatorname{tr}\left(W^T X X^T W\right) \text{ s.t. } W^T W = I.$$

This minimization is not difficult; direct observation can also find that W corresponding to the minimum value is composed of eigenvectors corresponding to the largest n' eigenvalues of the covariance matrix XX^T. Of course, mathematical derivation is also very easy. Using the Lagrangian function, we can get

$$J(W) = -\mathrm{tr}\left(W^T X X^T W\right) + \lambda(W^T W - I).$$

The derivative of W has $-XX^T W + \lambda W = 0$, which means that $XX^T W = \lambda W$. In this way, it can be seen more clearly that W is a matrix composed of n' eigenvectors of XX^T, and λ is a matrix composed of several eigenvalues of XX^T; the eigenvalues are on the main diagonal, and the remaining positions are 0. When we reduce the dataset from n-dimensional to n'-dimensional, we need to find eigenvectors corresponding to the largest n' eigenvalues. The matrix W composed of n' eigenvectors is the matrix we need. For the original dataset, we only need to use $z^{(i)} = W^T x^{(i)}$ to reduce the dimensionality of the original dataset to the n'-dimensional dataset with the minimum projection distance. In the general form, we can conclude PCA as the following optimization problem:

$$\min_{R} \quad \|X - R\|^2 \qquad (4.8)$$
$$\mathrm{s.t.} \quad r(R) = k$$

where $X \in R^{n \times p}$.

4.2 Optimal Condition and Duality Theory

In this section, we will introduce some sufficient and necessary conditions for a feasible solution to be local or global minimum. We will mainly focus on the first-order optimal condition and simply mention the second-order optimal condition as a supplement. In the following, we will introduce typical constraint qualification (CQ) and the relations between each other. Finally, duality theory will be discussed in detail. Main reference comes from [3].

4.2.1 Optimal Condition

Fundamental optimization problem can be expressed as below.

$$\min \quad f(x) \qquad (4.9)$$
$$\mathrm{s.t.} \quad x \in \mathcal{F} = \mathcal{C} \cap \mathcal{D},$$

4.2 Optimal Condition and Duality Theory

where $\mathcal{F} \subset R^n$ is called a feasible region, C and $\mathcal{D}$ are constraint set and domain of definition, $x \in \mathcal{F}$ is a feasible solution, and f is a real function and is called objective function. Without special notations, f is assumed to be smooth.

In typical research studies, the nonlinear programming problem is commonly formulated as follows:

$$\begin{aligned} \min \quad & f(x) \\ \text{s.t.} \quad & g(x) \le 0, \\ & x \in R^n, \end{aligned} \quad (4.10)$$

where $g(x) = (g_1(x), g_2(x), \cdots, g_m(x))^T \in R^m$ is a real value vector function. $g_i(x) \le 0$ is called i-th constraint condition. Without special notations, $g(x)$ is assumed to be smooth.

For optimization problem (4.9), when $\mathcal{F} = R^n$, we call this optimization an unconstrained problem. When $\mathcal{F} \ne \emptyset$, we call this optimization is a feasible problem. When there exists a feasible solution $x^* \in \mathcal{F}$ such that $f(x^*)$ attains its minimal value, this problem is called attainable. When an optimization problem is both feasible and attainable, it is called a solvable problem.

Definition 4.9 (Local Minimizer) A point x is a local minimizer if it is feasible (i.e., it satisfies the constraints of the optimization problem) and if there exists some $R > 0$ such that all feasible points z with $\|x - z\|_2 \le R$, satisfy $f(x) \le f(z)$.

Definition 4.10 (Global Minimizer) A point x is a global minimizer if it is feasible and for all feasible points $z \in \mathcal{F}$ such that $f(x) \le f(z)$ holds.

In previous two definitions, if corresponding inequality conditions strictly hold for any feasible solution $x \ne x^*$, we call x^* a strictly local minimizer/global minimizer.

Next we introduce a first-order necessary optimal condition.

Theorem 4.1 *If $\bar{x} \in \mathcal{F}$ is a local minimizer of optimization problem (4.9), then*

$$\nabla f(\bar{x})^T d \ge 0, \quad \forall d \in \mathcal{D}(\bar{x}), \quad (4.11)$$

where $\mathcal{D}(x)$ is a set of feasible direction at point $x \in \mathcal{F}$.

Proof By Taylor's expansion, we get

$$f(x) = f(\bar{x}) + \nabla f(\bar{x})^T (x - \bar{x}) + o(\|x - \bar{x}\|). \quad (4.12)$$

If there exists feasible direction $d \in \mathcal{D}(\bar{x})$ such that $\nabla f(\bar{x})^T d < 0$, then by the previous Taylor's expansion, we get that there exists $x = \bar{x} + \delta d \in \mathcal{F}$ satisfying $f(x) < f(\bar{x})$. This is a contradiction! □

Remark 4.6 Condition (4.11) is a necessary condition but not a sufficient condition. A simple counter-example is $f(x) = x^3$ at $\bar{x} = 0$.

Definition 4.11 (Active Constraint Set) An active constraint set at $x \in \mathcal{F}$ is as follows:
$$\mathcal{I} := \{i | g_i(x) = 0\}. \tag{4.13}$$

Definition 4.12 (Set of Locally Constrained Directions) A set of locally constrained directions at $x \in \mathcal{F}$ is as follows:
$$\mathcal{L}(x) := \{d \in R^n | \nabla g_i(x)^T d \leq 0, \forall i \in \mathcal{I}(x)\}. \tag{4.14}$$

Remark 4.7 Directions in $\mathcal{L}(x)$ are not all feasible directions for constraint conditions. A simple counter-example is $g(x) = x^2 \leq 0$ at point $x = 0$.

Set $\mathcal{L}(x)$ has been proven to satisfy some important conditions and play a significant role.

Theorem 4.2 *If problem (4.10) is feasible, then for any $x \in \mathcal{F}$, $\mathcal{L}(x)$ is a non-empty, closed, convex cone and*
$$\mathcal{D}(x) \subset \mathcal{L}(x). \tag{4.15}$$

Proof It is direct to verify $0 \in \mathcal{L}(x)$ and $\mathcal{L}(x)$ is a cone. For a sequence $d^k \to d^*$ satisfying $\nabla g_i(x)^T d^k \leq 0, i \in \mathcal{I}(x)$, by the property of limit, we get $\nabla g_i(x)^T d^* \leq 0, i \in \mathcal{I}(x)$. So $\mathcal{L}(x)$ is closed. Convex property is easy to verify by definition.

For any $d \in \mathcal{D}(x)$, there exists $\delta_0 > 0$ such that $g_i(\hat{x}) \leq 0, \hat{x} = x + \delta d$ for any $0 < \delta < \delta_0$ holds. Thus, we get
$$\begin{aligned} g_i(\hat{x}) &= g_i(x) + \nabla g_i(x)^T d \cdot \delta + o(\delta) \\ &= \nabla g_i(x)^T d \cdot \delta + o(\delta) \\ &\leq 0, \quad \forall i \in \mathcal{I}(x). \end{aligned} \tag{4.16}$$

It derives that $\nabla g_i(x)^T d \leq 0, \forall i \in \mathcal{I}(x)$ holds, i.e., $d \in \mathcal{L}(x)$. It follows $\mathcal{D}(x) \subset \mathcal{L}(x)$. □

In the following, we introduce a well-known sufficient condition for local minimizer.

Theorem 4.3 (Karush-Kuhn-Tucker Condition) *Let $\bar{x} \in \mathcal{F}$ be a local minimizer of problem (4.10). If $\mathcal{L}(\bar{x}) \subset cl(conv(\mathcal{D}(\bar{x})))$ holds, then there exists $\bar{\lambda} \in R_+^m$ such that*
$$\nabla f(\bar{x}) + \sum_{i=1}^m \bar{\lambda}_i \nabla g_i(\bar{x}) = 0, \tag{4.17}$$
$$\bar{\lambda}_i g_i(\bar{x}) = 0, \quad i = 1, 2, \cdots, m.$$

4.2 Optimal Condition and Duality Theory

The detailed proof of KKT theorem can be found in [3]. In Theorem 4.3, pair $(\bar{x}, \bar{\lambda})$ is called KKT pair, and $\bar{x}$ is called KKT point. $\bar{\lambda}_i g_i(\bar{x}) = 0$ is called complementary slackness condition, which shows that there exists one zero at least between $\bar{\lambda}_i$ and $g_i(\bar{x})$. $\bar{\lambda}_i$ is called Lagrange multiplier.

KKT condition is a first-order optimal condition that provides searching scope for local optimal solution. In actual examples, the first problem is to verify whether the preliminary condition $\mathcal{L}(\bar{x}) \subset cl(conv(\mathcal{D}(\bar{x})))$ holds. This is not a straightforward verification. In order to simplify the verification of such preliminary condition, some important sufficient conditions are proposed called constraint qualifications, which will be introduced in the next subsection.

Remark 4.8 In the following, we will introduce some geometrical explanation for KKT condition. We notice $\nabla g_i(x)$ is a normal vector at x of hyperplane $g_i(x) = c$ in R^n. Along direction $\nabla g_i(x)$, the value of $\nabla g_i(x)$ will increase, which will lead to those constraints not satisfied. We can call direction $\nabla g_i(x)$ "Tending unfeasible direction." From KKT condition, we find

$$-\nabla f(\bar{x}) = \sum_{i=1}^{m} \bar{\lambda}_i \nabla g_i(\bar{x}). \tag{4.18}$$

It shows that possible descent direction $-\nabla f(\bar{x})$ is composed by some "Tending unfeasible direction" $\nabla g_i(x)$.

If we consider information of twice derivatives furthermore, second-order optimal condition can be proposed.

Theorem 4.4 Let $(\bar{x}, \bar{\lambda})$ be a KKT pair, and denote $L(x, \lambda) = f(x) + \sum_{i=1}^{m} \lambda_i g_i(x)$. If Hessian matrix of $L(x, \lambda)$ satisfies

$$d^T \nabla_x^2 L(\bar{x}, \bar{\lambda}) d > 0, \quad \forall d \in \bar{\mathcal{L}}(\bar{x}), d \neq 0, \tag{4.19}$$

where

$$\bar{\mathcal{L}}(\bar{x}) := \{d \in R^n | \nabla g_i(\bar{x})^T d = 0, i \in \bar{I}(\bar{x}); \nabla g_i(\bar{x})^T d \leq 0, i \in I(\bar{x}) \setminus \bar{I}(\bar{x})\},$$

and

$$\bar{I}(\bar{x}) := \{i | i \in I(\bar{x}), \bar{\lambda}_i > 0\}.$$

Then $\bar{x}$ is a strictly local minimizer of (4.10).

Proof If $\bar{x}$ is not a strictly local minimizer. Then there exists a sequence $\{x^k\} \subset \mathcal{F}$ such that $x^k \to \bar{x}$ and $f(x^k) \leq f(\bar{x})$. We denote $d^k = \frac{x^k - \bar{x}}{\|x^k - \bar{x}\|}, \delta_k = \|x^k - \bar{x}\|$. By Bolzano theorem, there exists a convergent subsequence in a bounded sequence. Without loss of generality, we assume $d_k \to d$. Next by Taylor expansion of $g_i(x^k)$ at point $\bar{x}$, we can get $\nabla g_i(\bar{x})^T d \leq 0$. Similarly, we get $\nabla f(\bar{x})^T d \leq 0$.

In the following, we claim $\nabla g_i(\bar{x})^T d = 0, i \in \bar{I}(\bar{x})$. Otherwise, there exists $i \in \bar{I}(\bar{x})$ such that $\nabla g_i(\bar{x})^T d < 0$. Then KKT condition follows

$$\nabla f(\bar{x})^T d = -\sum_{i=1}^{m} \bar{\lambda}_i \nabla g_i(\bar{x})^T d > 0. \tag{4.20}$$

This is contradictory to $\nabla f(\bar{x})^T d \leq 0$. Then $\nabla g_i(\bar{x})^T d = 0, i \in \bar{I}(\bar{x}) \Longrightarrow d \in \bar{\mathcal{L}}(\bar{x})$.

Next

$$\begin{aligned} L(\bar{x}, \bar{\lambda}) &= f(\bar{x}) \\ &\geq f(x^k) \\ &\geq L(x^k, \bar{\lambda}) \\ &= L(\bar{x}, \bar{\lambda}) + \frac{1}{2}\delta_k^2 (d^k)^T \nabla_x^2 L(\bar{x}, \bar{\lambda}) d^k + o(\delta_k^2) \end{aligned} \tag{4.21}$$

Letting $d^k \to 0$, it follows $d^T \nabla_x^2 L(\bar{x}, \bar{\lambda}) d \leq 0$—a contradiction! □

4.2.2 Constraint Qualification

KKT condition restricts the scope of searching local minimizer and provides a possible method. In Theorem 4.3, we require to verify the condition

$$\mathcal{L}(\bar{x}) \subset cl(conv(\mathcal{D}(\bar{x}))) \tag{4.22}$$

So far, there exist many different types of constraint qualifications, which makes it difficult for researchers from different fields learn to learn the background and relations between them. Some direct sufficient conditions are proposed to guarantee condition (4.22), which is called constraint qualification.

In this subsection, we will summarize some constraint qualifications in known literatures.

- Linearly independent constraint qualification, LICQ: $\{\nabla g_i(x), i \in I(x)\}$ are linearly independent.
- Slater's constraint qualification: $\{g_i(x), i \in I(x)\}$ are convex functions, and x^0 is strictly interior point, i.e., $g_i(x^0) < 0, i = 1, 2, \cdots, m$.
- Cottle's constraint qualification: There exists one direction such that $\nabla g_i(x)^T d < 0, \forall i \in I(x)$.
- Zangwill's constraint qualification: $\mathcal{L}(x) \subset cl(\mathcal{D}(x))$.

4.2.3 Duality Theory

Duality method is an effective method to study optimization problem. For a constrained optimization problem (4.10), Lagrange function is defined as

$$L(x, \lambda) = f(x) + \sum_{i=1}^{m} \lambda_i g_i(x), \tag{4.23}$$

where $\lambda_i \geq 0$ is a Lagrange multiplier.

Denote feasible set $\mathcal{F} = \{x \in R^n | g_i(x) \leq 0, i = 1, 2, \cdots, m\}$. It is obvious that

$$\max_{\lambda \in R_+^m} L(x, \lambda) = \begin{cases} f(x), & x \in \mathcal{F} \\ +\infty, & x \notin \mathcal{F} \end{cases} \tag{4.24}$$

Then we get

$$v_p = \min_{x \in \mathcal{F}} f(x) = \min_{x \in R^n} \max_{\lambda \in R_+^m} L(x, \lambda) \tag{4.25}$$

A basic idea of establishing Lagrange duality problem is to try to get help from duality problem when original problem is difficult.

Next we consider to solve the following problem:

$$v_d = \max_{\lambda \in R_+^m} \min_{x \in R^n} L(x, \lambda) = \max_{\lambda \in R_+^m} v(\lambda) \tag{4.26}$$

where

$$v(\lambda) = \min_{x \in R^n} L(x, \lambda). \tag{4.27}$$

is an unconstrained optimization and is easy to solve. We denote (4.26) as a Lagrange dual problem of prime problem (4.10).

Theorem 4.5 (Weak Duality) *For optimization problem (4.10), $v_p \geq v_d$.*

Proof By definition of Lagrange function, we have

$$L(x, \lambda) = f(x) + \sum_{i=1}^{m} \lambda_i g_i(x) \leq f(x), \forall x \in \mathcal{F}, \lambda \geq 0 \tag{4.28}$$

Then $v(\lambda) \leq f(x), \forall x \in \mathcal{F}, \lambda \geq 0$ holds. Therefore, $v_d \leq v_p$. □

Then we can find that Lagrange dual problem always provides a lower bound for prime problem. This conclusion is called weak duality principle. If $v_p = v_d$, strong duality holds.

In solving (4.26), we first require to solve subproblem (4.27). If subproblem satisfies some specific condition, we can directly obtain the optimal solution of prime problem.

Theorem 4.6 (Strong Duality) *For a given $\bar{\lambda} \geq 0$, let $\bar{x}$ denote the optimal solution of subproblem (4.27), and $(\bar{x}, \bar{\lambda})$ satisfies complementary condition $\bar{\lambda}_i g_i(\bar{x}) = 0$. When $\bar{x} \in \mathcal{F}$, $\bar{x}$ is the optimal solution of (4.26).*

Proof If $\bar{x}$ is the optimal solution of subproblem, then

$$v_d \geq v(\bar{\lambda}) = L(\bar{x}, \bar{\lambda}) = f(\bar{x}) + \sum_{i=1}^{m} \bar{\lambda}_i g_i(\bar{x}) = f(\bar{x}) \geq v_p. \tag{4.29}$$

However, we notice weak duality $v_d \leq v_p$ holds naturally. Then above inequalities become equalities. Then $\bar{x}$ is the optimal solution of (4.26). □

4.3 Convex Optimization

In the following, we classify optimization problem to convex problem and non-convex problem based on convexity of objective function and feasible region.

In this section, we will focus on the convex optimization. The research on it is very complete mathematically. What follows is largely based on [1], which is the well-known textbook for convex optimization.

4.3.1 Basic Properties

We introduce the first-order and second-order condition for convexity.

Proposition 4.2 *Suppose a function $f : R^n \to R$ is differentiable (i.e., the gradient $\nabla_x f(x)$ exists at all points x in the domain of f). Then f is convex if and only if $\mathcal{D}(f)$ is a convex set, and for all $x, y \in \mathcal{D}(f)$,*

$$f(y) \geq f(x) + \nabla_x f(x)^T (y - x). \tag{4.30}$$

Proof We only consider the case $n = 1$. Assume that f is convex and $x, y \in \mathcal{D}(f)$. Since $\mathcal{D}(f)$ is convex, we conclude that for all $0 < t \leq 1$, $x + t(y - x) \in \mathcal{D}(f)$, and by convexity of f,

$$f(x + t(y - x)) \leq (1 - t)f(x) + tf(y).$$

If we divide both sides by t, we obtain

4.3 Convex Optimization

$$f(y) \geq f(x) + \frac{f(x + t(y - x)) - f(x)}{t},$$

Taking the limit as $t \to 0$, we obtain (4.30).

To show sufficiency, assume the function satisfies (4.30) for all x and y in $\mathcal{D}(f)$. Choose any $x \neq y$, and $0 \leq \theta \leq 1$, and let $z = \theta x + (1 - \theta)y$. Applying (4.30) twice yields

$$f(x) \geq f(z) + f'(z)(x - z), \quad f(y) \geq f(z) + f'(z)(y - z).$$

Multiplying the first inequality by θ and the second by $1 - \theta$ and adding them yields

$$\theta f(x) + (1 - \theta) f(y) \geq f(z),$$

which proves that f is convex. $\square$

The function $f(x) + \nabla_x f(x)^T (y - x)$ is called the **first-order approximation** to the function f at the point x. Intuitively, this can be thought of as approximating f with its tangent line at the point x. The first-order condition for convexity says that f is convex if and only if the tangent line is a global underestimator of the function f. In other words, if we take a function f and draw a tangent line at any point, then every point on this line will lie below the corresponding point on f.

Remark 4.9 Similarly to the definition of convexity, f will be strictly convex if this holds with strict inequality, concave if the inequality is reversed, and strictly concave if the reverse inequality is strict.

Proposition 4.3 *Suppose a function $f : R^n \to R$ is twice differentiable (i.e., the Hessian $\nabla_x^2 f(x)$ is defined for all points x in the domain of f). Then f is convex if and only if $\mathcal{D}(f)$ is a convex set and its Hessian is positive semidefinite: i.e., for any $x \in \mathcal{D}(f)$*

$$\nabla_x^2 f(x) \geq 0. \tag{4.31}$$

Proof We still consider the case $n = 1$. Suppose $f : R \to R$ is convex. Let $x, y \in \mathcal{D}(f)$ with $y > x$. By the first-order condition,

$$f'(x)(y - x) \leq f(y) - f(x) \leq f'(y)(y - x).$$

Subtracting the right-hand side from the left-hand side and dividing by $(y - x)^2$ gives

$$\frac{f'(y) - f'(x)}{y - x} \geq 0.$$

Taking the limit for $y \to x$ yields $f''(x) \geq 0$. Conversely, suppose $f''(z) \geq 0$ for all $z \in \mathcal{D}(f)$. Consider two arbitrary points $x, y \in \mathcal{D}(f)$ with $x < y$. We have

$$\begin{aligned} 0 &\leq \int_x^y f''(z)(y-z)dz \\ &= \left(f'(z)(y-z)\right)\big|_{z=x}^{z=y} + \int_x^y f'(z)dz \\ &= -f'(x)(y-x) + f(y) - f(x) \end{aligned}$$

i.e., $f(y) \geq f(x) + f'(x)(y-x)$. This shows that f is convex. □

Remark 4.10 Again analogous to both the definition and the first-order conditions for convexity, f is strictly convex if its Hessian is positive definite, concave if the Hessian is negative semidefinite, and strictly concave if the Hessian is negative definite. However, we should notice that a property of strictly convex cannot deduce the Hessian matrix if f is positive definite. For example, $f(x) = |x|^3$.

At last, we will give an important property for convex function. Suppose we start with the inequality in the basic definition of a convex function:

$$f(\theta x + (1-\theta)y) \leq \theta f(x) + (1-\theta)f(y) \text{ for } 0 \leq \theta \leq 1.$$

Using induction, this can be fairly easily extended to convex combinations of more than single variable,

$$f\left(\sum_{i=1}^k \theta_i x_i\right) \leq \sum_{i=1}^k \theta_i f(x_i) \text{ for } \sum_{i=1}^k \theta_i = 1, \theta_i \geq 0 \quad \forall i.$$

In fact, this can also be extended to infinite sums or integrals. In the latter case, the inequality can be written as

$$f\left(\int p(x)x dx\right) \leq \int p(x)f(x)dx \text{ for } \int p(x)dx = 1, p(x) \geq 0 \quad \forall x.$$

Since the integration of $p(x)$ is 1, it is common to consider it as a probability density, in which case, the previous equation can be written in terms of expectations,

$$f(E[x]) \leq E[f(x)], \tag{4.32}$$

where $E[\cdot]$ refers the expectation with respect to $p(x)$. (4.32) is well-known as Jensen's inequality in probability theory.

4.3.2 Optimal Condition

In this subsection, we first give the formal form of convex optimization problem:

$$\begin{aligned} \min \ & f(x) \\ \text{s.t.} \ & x \in C, \end{aligned} \quad (4.33)$$

where f is a convex function, C is a convex set, and x is the optimization variable. More specifically, we can write it as follows:

$$\begin{aligned} \min \ & f(x) \\ \text{s.t.} \ & h_i(x) \leq 0, \quad i = 1, \ldots, m \\ & \ell_j(x) = 0, \quad j = 1, \ldots, r, \end{aligned} \quad (4.34)$$

where f is a convex function, h_i are convex functions, ℓ_j are affine functions, and x is the optimization variable.

The optimal value of an optimization problem is denoted as p^* and is equal to the minimum possible value of the objective function in the feasible region

$$p^* = \min \left\{ f(x) : h_i(x) \leq 0, i = 1, \ldots, m, \ell_j(x) = 0, j = 1, \ldots, r \right\}.$$

We allow p^* to take on the values $+\infty$ and $-\infty$ when the problem is either infeasible (the feasible region is empty) or unbounded below (there exists feasible points such that $f(x) \to -\infty$), respectively. We say that x^* is an optimal point if $f(x^*) = p^*$.

One of the most important reasons for considering convex optimization problem is the following proposition.

Proposition 4.4 *The local optimal solution of the convex optimization problem is the global optimal solution.*

Proof By contradiction, we assume there is a local minimum at the point x and there is a point $z \in D$ such that $f(z) < f(x)$.

Now we assume $y = tx + (1-t)z$; then, at this time, because $h_i(x) \leq 0$ still holds (this is because of the function convexity), $l_j(y) = 0$ still holds (this is because of linearity), so y is still a feasible solution. But at this time, notice that the objective function is also convex, so

$$f(y) \leq tf(x) + (1-t)f(z) < f(x).$$

The key point here is to explain that if the local minimum is not the global minimum, then we can modify the t so that the y is close to the x, which will become a competitive "local minimum" with x. Due to

$$\|y - x\|_2 = (1-t)\|z - x\|_2.$$

this shows that we can choose the appropriate t so that $\|y - x\|_2$ can be arbitrarily small. In other words, even if y is in the neighborhood of x, there is $f(y) < f(x)$, which violates the definition of local minimum. □

4.3.3 Examples

- **Linear Programming**. We say that a convex optimization problem is a linear program (LP) if both the objective function f and inequality constraints g_i are affine functions. In other words, these problems have the form:

$$\min c^T x + d$$
$$\text{s.t. } Gx \leq h$$
$$Ax = b,$$

  where $x \in R^n$ is the optimization variable, $c \in R^n, d \in R, G \in R^{m \times n}, h \in R^m$ $A \in R^{p \times n}, b \in R^p$.

- **Quadratic Programming**. We say that a convex optimization problem is a quadratic program (QP) if the inequality constraints g_i are still all affine, but if the objective function f is a convex quadratic function. In other words, these problems have the following form:

$$\min \tfrac{1}{2} x^T P x + c^T x + d$$
$$\text{s.t. } Gx \leq h$$
$$Ax = b,$$

  where again $x \in R^n$ is the optimization variable and $c \in R^n, d \in R, G \in R^{m \times n}, h \in R^m$ $A \in R^{p \times n}, b \in R^p$ are defined by the problem, and we also have $P \in \mathbb{S}_+^n$ as a symmetric positive semidefinite matrix.

- **Quadratically Constrained Quadratic Programming(QCQP)**. We say that a convex optimization problem is a quadratically constrained quadratic program if both the objective f and the inequality constraints g_i are convex quadratic functions:

$$\min \tfrac{1}{2} x^T P x + c^T x + d$$
$$\text{s.t. } \tfrac{1}{2} x^T Q_i x + r_i^T x + s_i \leq 0, \quad i = 1, \ldots, m$$
$$Ax = b,$$

  where, as before, $x \in R^n$ is the optimization variable, $c \in R^n, d \in R, A \in R^{p \times n}, b \in R^p$ $P \in \mathbb{S}_+^n$, and we also have $Q_i \in \mathbb{S}_+^n, r_i \in R^n, s_i \in R$, for $i = 1, \ldots, m$.

4.3 Convex Optimization

- **Semidefinite Programming**. This last example is more complex than the previous ones, so don't worry if it doesn't make much sense at first. However, semidefinite programming is becoming more prevalent in many areas of machine learning research, so you might encounter these at some point, and it is good to have an idea of what they are. We say that a convex optimization problem is a semidefinite program (SDP) if it is of the form

$$\begin{aligned} \min \ & \operatorname{tr}(CX) \\ \text{s.t.} \ & \operatorname{tr}(A_i X) = b_i, \quad i = 1, \ldots, p \\ & X \succeq 0, \end{aligned}$$

where the symmetric matrix $X \in S^n$ is the optimization variable, symmetric matrices $C, A_1, \ldots, A_p \in S^n$ are defined by the problem, and the constraint $X \succeq 0$ means that we are constraining X to be positive semidefinite. This looks a bit different than the problems we have seen previously, since the optimization variable is now a matrix instead of a vector.

Remark 4.11 It is needed to note that quadratic programs are more general than linear programs since a linear program is just a special case of a quadratic program and likewise that quadratically constrained quadratic programs are more general than quadratic programs. However, what is not obvious is that semidefinite programs are in fact more general than all the previous types, that is, any quadratically constrained quadratic program (and hence any quadratic program or linear program) can be expressed as a semidefinite program.

4.3.4 Quadratic Convex Optimization

Quadratic optimization (QP) plays an important role in many fields such as mathematics, economy, biology, etc. Generally, we can express the quadratic optimization problem in the following form:

Prime problem :
$$\begin{aligned} \min \ & F(x) = \frac{1}{2} x^T G x + c^T x \\ \text{s.t.} \ & a_i^T x - b_i = 0, i \in E = \{1, \ldots, l\} \\ & a_i^T x - b_i \geq 0, i \in I = \{l+1, \ldots, q\}, \end{aligned} \quad (4.35)$$

where $G \in S^n$ and $c \in R^n$. Feasible region is $\Omega = \{x \in R^n | a_i^T x - b_i = 0, i \in E = \{1, \ldots, l\}, a_i^T x - b_i \geq 0, i \in I = \{l+1, \ldots, q\}$. If $G \in S_+^n$, it shows that problem (4.35) is a convex optimization. For a convex optimization problem, because local minimum of convex optimization is global minimum, KKT condition becomes a necessary and sufficient condition for global minimum.

An effective method is active-set method for convex quadratic problem. First we give definition of active set.

Definition 4.13 (Active Set) For a feasible point $x \in \Omega$, active set of x is

$$I(x) = \{i \mid a_i^T x - b_i = 0, i \in I\}. \tag{4.36}$$

By using active set, we can transform prime problem to problem with only equality constraints.

Theorem 4.7 *Assume x^* is global minimizer of problem (4.35) and active set of x^* is denoted by I^*; then x^* is global minimizer of the following QP constrained by equalities.*

$$\text{(II)}: \quad \begin{aligned} \min \quad & \frac{1}{2} x^T G x + c^T x \\ s.t. \quad & a_i^T x - b_i = 0, i \in I^* \cup E. \end{aligned} \tag{4.37}$$

Proof If x^* is global minimizer of (min QP), then x^* satisfies KKT equations of (min QP).

$$\begin{cases} G x^* + c - \sum_{i \in E} \lambda_i a_i - \sum_{i \in I} \lambda_i a_i = 0 \\ a_i^T x - b_i = 0, i \in E \\ \lambda_i \geq 0, i \in I^*; \lambda_i = 0, i \in I \setminus I^*, \end{cases} \tag{4.38}$$

which is equivalent to equation

$$\begin{cases} G x^* + c - \sum_{i \in E \cup I^*} \lambda_i a_i = 0 \\ a_i^T x - b_i = 0, i \in E \\ \lambda_i \geq 0, i \in I^*, \end{cases} \tag{4.39}$$

First two equations show x^* is a global minimizer of (II). □

Theorem 2.1 shows that if we know an active set of x^*, we can directly solve (II) and obtain the global minimizer of prime problem. Next we design an active set method to approach an accurate active set of x^*. Our consideration is that given initial point x_0, we calculate active set I_0, and solve the corresponding subproblem. By iteration, we change I_k and hope $I_k \to I^*$.

Algorithm:
Step [1]. Construct subproblem and obtain searching direction.
Assume x_k is a feasible point and $I_k = I(x_k) \cup E$. Subproblem is

4.3 Convex Optimization

$$\begin{cases} \min & \frac{1}{2}x^T G x + c^T x \\ s.t. & a_i^T x = b_i, i \in I_k. \end{cases} \quad (4.40)$$

Assume $x = x_k + d$ and above suboptimization can be transformed in terms of d:

$$\begin{cases} \min & \frac{1}{2}d^T G d + g_k^T d \\ s.t. & a_i^T d = 0, i \in I_k, \end{cases} \quad (4.41)$$

where $g_k = \nabla f(x_k) = Gx_k + c$. Assume global optimal point is d_k. λ_k is a corresponding Lagrangian multiplier.

Step [2]. If $d_k \neq 0$, we make a line search.

Step [2.1]. If $x_k + d_k$ is feasible point of (min QP), then we assume $x_{k+1} = x_k + d_k$. Maximum step of line search is denoted by α_k. In Step [2.1], we assume $\alpha_k = 1$.

Step [2.2]. If $x_k + d_k$ is not a feasible point of prime problem.

Next we fix d_k and assume $d = \alpha d_k$, $\alpha \in (0, 1)$. We want to regard α as an optimization variable. By substituting $d = \alpha d_k$ into (4.41), We obtain the following problem in terms of α.

$$\begin{cases} \min & F(\alpha) := \left(\frac{1}{2}d_k^T G d_k\right)\alpha^2 + (g_k^T d_k)\alpha \\ s.t. & 0 < \alpha < 1. \end{cases} \quad (4.42)$$

It can be verified that F is a decreasing function.

In order to get a feasible point x, we require to solve the following problem:

$$\begin{aligned} \bar{\alpha}_k := \mathrm{argmax} \quad & \alpha \\ s.t. \quad & x_k + \alpha d_k \in \Omega. \end{aligned} \quad (4.43)$$

By detailed calculation, we get

$$\bar{\alpha}_k = \min_{i \notin I_k} \left\{ \frac{b_i - a_i^T x_k}{a_i^T d_k} \middle| a_i^T d_k < 0 \right\}. \quad (4.44)$$

Combining **Step [2.1]** and **Step [2.2]**, we get

$$\begin{aligned} \alpha_k &= \min\{1, \bar{\alpha}_k\} \\ x_{k+1} &= x_k + \alpha_k d_k. \end{aligned} \quad (4.45)$$

Step [3]. Update active set I_k.

For $\alpha_k < 1$, $\alpha_k = \bar{\alpha}_k = \frac{b_t - a_t^T x_k}{a_t^T d_k}$ ($t \in (I \cup E) \setminus I_k$). Then $a_t^T(x_k + \alpha_k d_k) = b_t$. This is equivalent to increase in active constraint in x_{k+1}, $I_{k+1} = I_k \cup t$. When $\alpha_k = 1$, active set is unchanged, $I_{k+1} = I_k$.

Step [4]. Convergence test.

If $d_k = 0$, x_k is global minimizer of subproblem.

Furthermore, we focus on Lagrange multipliers of the subproblem. If Lagrange multipliers corresponding to inequality are non-negative, x_k satisfies KKT equations of the prime problem. It derives that x_k is the global optimal solution, and iteration is stopped.

If there are negative Lagrange multipliers corresponding to inequality, we need to choose a new feasible decreasing direction for the prime problem. Assume $\lambda_s^{(k)} < 0$, $s \in I_k \setminus E$. Next we solve the following optimization problem:

$$\min \quad \frac{1}{2} d^T G d + g_k^T d$$
$$\text{s.t.} \quad a_j^T d = 0, j \in I_k' = I_k \setminus \{s\} \quad (4.46)$$

It can be verified that the optimal minimizer d^* of (4.46) is not equal to 0, and we continue to proceed **Step [2]**.

End of algorithm.

4.4 Non-convex Optimization

In this section, we consider non-convex optimization problem. We introduce a type of effective method for a general optimization problem: sequential quadratic programming (SQP). The general optimization problem is described below.

$$\min \quad f(x)$$
$$\text{s.t.} \quad h_i(x) = 0, i \in E = \{1, 2, \cdots, l\}, \quad (4.47)$$
$$g_i(x) \geq 0, i \in I = \{1, 2, \cdots, m\},$$

where f, h_i, g_j are assumed to be differentiable functions. The basic thought of SQP is that in each iteration step, we will solve a QP subproblem to determine a decreasing direction. Followed by reducing cost function, we will determine step size. Finally, by repeating this process, we will obtain an approximate local minimizer.

4.4 Non-convex Optimization

4.4.1 Newton-Lagrange Method

First we consider a problem constrained only by equalities.

$$\begin{cases} \min \quad f(x) \\ s.t. \quad h_i(x) = 0, i \in E = \{1, 2, \cdots, l\}. \end{cases} \quad (4.48)$$

We denote $h(x) = (h_1(x), h_2(x), \cdots, h_l(x))^T$. Then the Lagrangian function is $L(x, \mu) = f(x) - \mu^T h(x)$, where $\mu = (\mu_1, \mu_2, \cdots, \mu_l)^T$ are Lagrangian multipliers. KKT conditions (Lagrangian multiplier method) is

$$\nabla L(x, \mu) = \begin{pmatrix} \nabla_x L(x, \mu) \\ \nabla_\mu L(x, \mu) \end{pmatrix} = \begin{pmatrix} \nabla f(x) - A(x)^T \mu \\ -h(x) \end{pmatrix} = 0, \quad (4.49)$$

where A is a Jacobian matrix of $h(x)$, i.e., $A = (\nabla h_1, \cdots, \nabla h_n)^T :\triangleq \nabla h(x)^T$.

Next we use the Newton iterative method to solve nonlinear equation (4.49). First we calculate Jacobian matrix of $\nabla L(x, \mu)$.

$$N(x, \mu) = \begin{pmatrix} W(x, \mu) & -A(x)^T \\ -A(x) & 0, \end{pmatrix} \quad (4.50)$$

where $W(x, \mu) = \nabla_{xx} L(x, \mu) = \nabla^2 f(x) - \sum_{i=1}^{l} \mu_i \nabla^2 h_i(x)$.

Then Newton iterative format is given below. We denote $z_k = (x_k, \mu_k)^T$, $z_{k+1} = z_k + p_k$, where $p_k = (d_k.v_k)^T$ satisfies $N(x_k, \mu_k) p_k = -\nabla L(x_k, \mu_k)$, i.e.,

$$\begin{pmatrix} W(x, \mu) & -A(x)^T \\ -A(x) & 0 \end{pmatrix} \begin{pmatrix} d_k \\ v_k \end{pmatrix} = \begin{pmatrix} -\nabla f(x_k) + A(x_k)^T \mu_k \\ h(x_k) \end{pmatrix}. \quad (4.51)$$

Essentially, in the above method, we use Newton iterative method to solve KKT conditions. Thus, we call it Newton-Lagrange method. The disadvantage of this method is numerical instability when solving nonlinear equations.

4.4.2 SQP Constrained by Equalities

In this subsection, we still consider optimization (4.48). In order to improve numerical stability, we introduce sequential quadratic programming. Basic thought is to transform (4.49) to a convex quadratic optimization problem.

We assume local minimizer x^* in (4.48) satisfying second-order optimality sufficient condition (OSOSC), i.e., $\forall 0 \neq d \in R^n$ satisfying $A(x^*)d = 0$, $d^T W(x^*, \mu^*)d > 0$ holds. OSOSC can guarantee x^* is strictly local minimizer. Next we introduce a lemma.

Lemma 4.1 *Let $U \in R^{n \times n}$, $S \in R^{n \times m}$. Then $\forall x \neq 0$ s.t. $S^T x = 0$, $x^T U x > 0 \iff$ there exists a positive constant $\sigma^* > 0$, $\forall \sigma \geq \sigma^*$, $U + \sigma S S^T > 0$ holds on.*

Proof Sufficiency: For $S^T x = 0$, we have

$$x^T U x = x^T (U + \sigma^* S S^T) x > 0. \tag{4.52}$$

Necessaty: If we can find a number $\sigma^* > 0$ such that $x^T(U + \sigma^* S S^T)x > 0$, then for any $\sigma \geq \sigma^*$, it is easy to get $x^T(U + \sigma S S^T)x > 0$. Next, we will prove there exists σ^* satisfying the above condition. Otherwise, there exists a sequence $\{x_k\}$ and $\|x_k\| = 1$ such that $x_k^T(U + k S S^T)x_k \leq 0$. By Bolzano theorem, $\{x_k\}$ has a convergent subsequence $\{x_{k_i}\} \to \bar{x}$ such that

$$x_{k_i}^T(U + k_i S S^T)x_{k_i} \leq 0. \tag{4.53}$$

By putting limit $k_i \to \infty$, we get

$$\bar{x}^T U \bar{x} + \lim_{k_i \to \infty} k_i x_{k_i}^T S S^T x_{k_i} \leq 0 \implies S^T x = 0 \tag{4.54}$$

and $\bar{x}^T U \bar{x} \leq 0$—a contradiction! □

Then by Lemma 4.1, for sufficiently small $r > 0$, we have

$$W(x^*, \mu^*) + \frac{1}{2r} A(x^*)^T A(x^*) > 0. \tag{4.55}$$

Then we can rewrite Eq. (4.51).

$$[W(x^k, \mu_k) + \frac{1}{2r} A^T(x_k) A(x_k)] d_k - A(x_k)^T [\mu_k + v_k + \frac{1}{2r} A(x_k) d_k] = -\nabla f(x_k)$$

$$A(x_k) d_k = -h(x_k). \tag{4.56}$$

If we denote $B(x_k, \mu_k) :\triangleq W(x_k, \mu_k) + \frac{1}{2r} A^T(x_k) A(x_k)$ and $\bar{\mu}_k :\triangleq \mu_k + v_k + \frac{1}{2r} A(x_k) d_k$, Eq. (4.51) will be equivalent to the below equation:

$$\begin{pmatrix} B(x_k, \mu_k) & -A(x_k)^T \\ A(X_k) & 0 \end{pmatrix} \begin{pmatrix} d_k \\ \bar{\mu}_k \end{pmatrix} = - \begin{pmatrix} \nabla f(x_k) \\ h(x_k) \end{pmatrix}. \tag{4.57}$$

Since $B(x^*, \mu^*) > 0$, when $(x_k, \mu_k) \to (x^*, \mu^*)$, $B(x_k, \mu_k) > 0$ holds on. We assume $A(x_k)$ is a row nonsingular matrix so that the above equation is solvable. Furthermore, we can transform Eq. (4.57) to a convex quadratic programming by the following lemma.

4.4 Non-convex Optimization

Lemma 4.2 *Let $B(x_k, \mu_k) > 0$ be positive definite. Thus d_k satisfies Eq. (4.57) if and only if d_k is a global minimum point of the following strictly convex quadratic programming:*

$$\begin{cases} \min_d & q_k(d) = \frac{1}{2} d^T B(x_k, \mu_k) d + \nabla f(x_k)^T d, \\ s.t. & h(x_k) + A(x_k) d = 0. \end{cases} \quad (4.58)$$

Proof We use equivalence form to prove this lemma. Thus,

d_k is global minimum

$\iff (d_k, \lambda_k)$ satisfies KKT condition, where λ_k is KKT multiplier.

(Since (4.58) is a convex quadratic optimization)

$$\iff \nabla L(d_k, \lambda_k) = \begin{pmatrix} \nabla_d L(d_k, \lambda_k) \\ \nabla_\lambda L(d_k, \lambda_k) \end{pmatrix} = 0, \quad (4.59)$$

where $L(d, \lambda) = q_k(d) - \lambda^T (h(x_k) + A(x_k) d)$.

$$\iff \begin{cases} B(x_k, \lambda_k) d_k + \nabla f(x_k) - A(x_k)^T \lambda_k = 0 \\ h(x_k) + A(x_k) d_k = 0 \end{cases}$$

$\iff d_k$ is solution of Eq. (4.57).

$\square$

4.4.2.1 SQP in General Case

In this section, we consider to extend thought to solve the general optimization problem (4.47). Given point (x_k, μ_k, λ_k), we make linearization for constrained functions, calculate Hessian matrix of Lagrangian function, and obtain the following QP suboptimization problem by analogy of (4.58):

$$\begin{aligned} \min_d \quad & \frac{1}{2} d^T B_k d + \nabla f(x_k)^T d \\ s.t. \quad & h_i(x_k) + \nabla h_i(x_k)^T d = 0, i \in E = \{1, 2, \cdots, l\} \\ & g_i(x_k) + \nabla g_i(x_k)^T d \geq 0, i \in I = \{1, 2, \cdots, m\}, \end{aligned} \quad (4.60)$$

where B_k is a positive definite approximation of $\nabla_{xx} L(x_k, \mu_k, \lambda_k)$ and $L(x_k, \mu_k, \lambda_k) = f(x) - \sum_{i \in E} \mu_i h_i(x) - \sum_{i \in I} \lambda_i g_i(x)$ is Lagrangian function. Thus, we can define optimal solution d^* of (4.60) as update direction d_k. For convenience of discussion, we assume that in suboptimization problem (4.60), there exists a feasible point.

Table 4.1 General SQP algorithm

Step 0	Give initial point $(x_0, \mu_0, \lambda_0) \in R^n \times R^l \times R^m$, symmetric positive definite matrix $B_0 \in R^{n \times n}$ Calculate $A_0^E = \nabla h(x_0)^T$, $A_0^I = \nabla g(x_0)^T$, $A_0 = \begin{pmatrix} A_0^E \\ A_0^I \end{pmatrix}$. Select parameter $\eta \in (0, 0.5)$, $\rho \in (0, 1)$ Tolerance error $0 \leq \varepsilon_1, \varepsilon_2 \ll 1$, assume $k := 0$
Step 1	Solve suboptimization problem $$\begin{cases} \min_d & \frac{1}{2} d^T B_k d + \nabla f(x_k)^T d \\ s.t. & h(x_k) + A_k^E d = 0 \\ & g(x_k) + A_k^I d \geq 0 \end{cases}$$ and obtain optimal solution d_k
Step 2	If $\|d_k\|_1 \ll \varepsilon_1$, $\|h(x_k)\|_1 + \|g(x_k)_-\| \ll \varepsilon_2$, we stop algorithm and obtain an approximate KKT point (x_k, μ_k, λ_k)
Step 3	For some cost function $\phi(x, \sigma)$, we select penalty parameter σ_k so that d_k is a decreasing direction at x_k
Step 4	Armijo searching. Assume m_k is minimal nonnegative integer m satisfying below inequality $\phi(x_k + \rho^m d_k, \sigma_k) \leq \phi(x_k, \sigma_k) + \eta \rho^m \phi'(x_k, \sigma_k; d_k)$ Assume step size $\alpha_k := \rho^{m_k}$ and update $x_{k+1} := x_k + \alpha_k d_k$
Step 5	Calculate $A_{k+1}^E = \nabla h(x_{k+1})^T$, $A_{k+1}^I = \nabla g(x_{k+1})^T$ $A_{k+1} = \begin{pmatrix} A_{k+1}^E \\ A_{k+1}^I \end{pmatrix}$. And KKT multiplier $\begin{pmatrix} \mu_{k+1} \\ \lambda_{k+1} \end{pmatrix} = [A_{k+1}, A_{k+1}^T]^{-1} A_{k+1} \nabla f(x_{k+1})$
Step 6	Update matrix B_k to B_{k+1}. Assume $s_k = x_{k+1} - x_k$, $y_k = \nabla_x L(x_{k+1}, \mu_{k+1}, \lambda_{k+1}) - \nabla_x L(x_k, \mu_{k+1}, \lambda_{k+1})$ $B_{k+1} = B_k - \frac{B_k s_k s_k^T B_k}{s_k^T B_k s_k} + \frac{z_k z_k^T}{s_k^T z_k}$, where $z_k = \theta_k y_k + (1 - \theta_k) B_k s_k$ and $\theta_k = \begin{cases} 1, & \text{if } s_k^T y_k \geq 0.2 s_k^T B_k s_k \\ \frac{0.8 s_k^T B_k s_k}{s_k^T B_k s_k - s_k^T y_k}, & \text{if } s_k^T y_k < 0.2 s_k^T B_k s_k \end{cases}$
Step 7	Assume $k := k + 1$ and return to Step 1

In summary, we conclude with the SQP algorithm of a general constrained optimization in Table 4.1.

4.5 Exercises

1. Let $\mathcal{X} \subset \mathbb{R}^n$ be a non-empty, closed, and convex set and $z \in \mathbb{R}^n$. Prove there exists a unique point $\bar{x} \in \mathcal{X}$ such that

$$\text{dist}(z, \mathcal{X}) := \min\{\|z - x\| \,|\, x \in \mathcal{X}\} \\ = \|z - \bar{x}\| \tag{4.61}$$

   and $(z - \bar{x})^\top (x - \bar{x}) \leq 0, \forall x \in \mathcal{X}$.

2. Let $\mathcal{X} \subset \mathbb{R}^n$ be a non-empty and convex set and $z \notin cl(\mathcal{X})$, where $cl(\mathcal{X})$ denotes the closure of $\mathcal{X}$. Prove that there exists a hyperplane $\mathcal{H} := \{y \in \mathbb{R}^n \,|\, a^\top y = b\}$ determined by a, b such that following statement holds:

$$a^\top x \geq b > a^\top z \tag{4.62}$$

   for arbitrary $x \in \mathcal{X}$.

3. Dual set of $\mathcal{X} \subset \mathbb{R}^n$ is defined as $\mathcal{X}^* = \{y \,|\, y^\top x \geq 0, \forall x \in \mathcal{X}\}$. Prove that (1) $(\mathbb{R}_+^n)^* = \mathbb{R}_+^n$ and (2) $(\mathcal{S}_+^n)^* = \mathcal{S}_+^n$.

4. Given a linear programming optimization as follows:

$$\begin{aligned} \min \quad & c^T x \\ \text{s.t.} \quad & Ax \geq b, \\ & x \in R_+^n. \end{aligned} \tag{4.63}$$

   Prove that its corresponding Lagrange duality problem is also the following linear programming:

$$\begin{aligned} \max \quad & b^T \lambda \\ \text{s.t.} \quad & A^T \lambda \leq c, \\ & \lambda \in R_+^m. \end{aligned} \tag{4.64}$$

5. Given a quadratic programming optimization as follows:

$$\begin{aligned} \min \quad & \frac{1}{2} x^T A x \\ \text{s.t.} \quad & \frac{1}{2} x^T B x \leq 1, \\ & x \in R_+^n. \end{aligned} \tag{4.65}$$

where $A \in \mathcal{S}^n$ is symmetric and $B \in \mathcal{S}^n_{++}$ is positive definite. Prove that its Lagrange duality problem is

$$\max \quad -\sigma$$
$$\text{s.t.} \quad A + \sigma B \in \mathcal{S}^n_+, \tag{4.66}$$
$$\sigma \geq 0.$$

6. Given function $f(x) = -2\sqrt{-x}$, $\mathcal{X} = \{x \in \mathbb{R}^n | x \leq 0\}$, verify its corresponding conjugate function is $h(y) = \frac{1}{y}$, $y > 0$.
7. Assume function f defined on $\mathcal{X}$ with its conjugate h defined on $\mathcal{Y}$. Prove that

$$x^\top y \leq f(x) + h(y), \forall x \in \mathcal{X} \text{ and } y \in \mathcal{Y} \tag{4.67}$$

and $x^\top y = f(x) + h(y)$ is equivalent to $y \in \partial f(x)$, where $\partial f(x)$ denotes the subgradient.
8. Show that for optimization problem, its optimal value might not be attainable.
9. Let $A \in \mathcal{S}^n$ with its associated eigenvalues $\lambda_1, \lambda_2, \cdots, \lambda_n$. Prove that if $x \neq 0$, then

$$\min_{1 \leq i \leq n} \{\lambda_i\} \leq \frac{x^\top A x}{x^\top x} \leq \max_{1 \leq i \leq n} \{\lambda_i\} \tag{4.68}$$

References

1. S. Boyd, S. P. Boyd, and L. Vandenberghe. *Convex optimization*. Cambridge University Press, 2004.
2. R. Bellman. Dynamic programming. *Science*, 153(3731):34–37, 1966.
3. S. Fang and W. Xing. *Linear cone optimization* North Carolina State University, 2013.
4. G. Gallo and S. Pallottino. Shortest path algorithms. *Annals of Operations Research*, 13(1):1–79, 1988.
5. D. B. Johnson. (1973). A note on Dijkstra's shortest path algorithm. *Journal of the ACM*, 20(3):385–388, 1973.
6. N. Karmarkar. A new polynomial-time algorithm for linear programming. *Proceedings of the 16th Annual ACM Symposium on Theory of Computing*, 302–311, 1984.
7. J. A. Nelder and R. Mead. A simplex method for function minimization. *The Computer Journal*, 7(4):308–313, 1965.
8. R. Tibshirani. Regression shrinkage and selection via the Lasso. *Journal of the Royal Statistical Society: Series B (Methodological)*, 58(1):267–288, 1996.
9. S. Wold, K. Esbensen and P. Geladi. Principal component analysis. *Chemometrics and Intelligent Laboratory Systems*, 2(1–3):37–52, 1987.

Part II
Filtering Theory

Chapter 5
Filtering Equations

In this chapter, we shall introduce the most important results for continuous filtering problem. we will introduce stochastic partial differential equations satisfied by the filtered posterior distribution, which is well-known as the Kushner-Stratonovich equation. We will use two different methods to derive this equation, namely, the change-measure method and innovation process method. In the change-measure method, we will introduce the conditional density formula for the continuous type, and we will derive an equation satisfied by a unnormalized conditional density function, which is well-known as Duncan-Mortensen-Zakai equation. In application, observations are not a sequence of probability distributions but a specific sampled path. At the end of this chapter, we present the robust DMZ equation designed for path-observation, which is used as a special change-measure method.

5.1 Introduction to Filtering Equations

A filtering problem is about to compute the conditional distribution of a state process X_t, which is somehow not easy to observe directly, given an observation process Y_t, which provides the information of the state process X_t, and can be observed. In practice, the state process X_t and the observation process Y_t are often modeled by two diffusion processes on a given probability space.

Let $(\Omega, \mathcal{F}, P)$ be a probability space with a filtration $(\mathcal{F}_t)$ satisfying the usual conditions. The state process $X = (X^i)_{i=1}^d$ is given by the solution of a d-dimensional stochastic differential equation driven by a p-dimensional Brownian motion $V = (V^j)_{j=1}^p$:

$$X_t^i = X_0^i + \int_0^t f^i(X_s)ds + \sum_{j=1}^p \int_0^t \sigma^{ij}(X_s)dV_s^j, \tag{5.1}$$

where in order to guarantee the existence of a solution to the stochastic differential equation (5.1), we assume that $f = (f^i)_{i=1}^d : R^d \to R^d$ and $\sigma = (\sigma^{ij})_{i,j=1}^d : R^d \to R^{d \times d}$ are Lipschitz continuous, that is, there exists a positive constant $K \in R_+$, such that for all $x, y \in R^d$, we have

$$\|f(x) - f(y)\| \leq K\|x - y\|; \quad \|\sigma(x) - \sigma(y)\| \leq K\|x - y\|. \tag{5.2}$$

Define a matrix-valued function $a = (a^{ij})_{i,j=1}^d$ as follows:

$$a^{ij} = \frac{1}{2}\sum_{k=1}^p \sigma^{ik}\sigma^{jk},$$

i.e., $a = \frac{1}{2}\sigma\sigma^\top$.

Consider the second-order differential operator

$$\mathscr{A} = \sum_{i=1}^d f^i \frac{\partial}{\partial x_i} + \sum_{i,j=1}^d a^{ij} \frac{\partial^2}{\partial x_i \partial x_j}.$$

The theory in Sect. 3.2 shows that for any twice differentiable, bounded, continuous function $\varphi \in C_b^2(R^d)$, the process $M^\varphi = \{M_t^\varphi : t \geq 0\}$ defined by

$$M_t^\varphi = \varphi(X_t) - \varphi(X_0) - \int_0^t \mathscr{A}\varphi(X_s)ds \tag{5.3}$$

is a martingale.

The observation process is modeled by the following equation:

$$Y_t = Y_0 + \int_0^t h(X_s)ds + W_t, \tag{5.4}$$

where $h = (h_i)_{i=1}^m : R^d \to R^m$ is a measurable function such that

$$P\left(\int_0^t \|h(X_s)\|ds < \infty\right) = 1.$$

in order to guarantee that the integration in (5.4) makes sense. W_t is another $\mathcal{F}_t$ adapted Brownian motion on $(\Omega, \mathcal{F}, P)$ independent of X. Denote the filtration generated by the observation process Y_t by

5.2 The Change of Probability Measure Method

$$\mathcal{Y}_t = \sigma(Y_s, 0 \leq s \leq t),$$

which represents the information we can obtain through the observations until time t.

The filtering problem is about to give an estimation to the state process at each time t based on the observations up to time t. Mathematically, at each time t, we would like to compute the conditional expectation of X_t given the information $\mathcal{Y}_t$ in order to get a minimal mean square error estimate of X_t.

In this chapter, we will derive the evolution equations satisfied by the conditional distribution π_t defined by

$$\pi_t(\varphi) = E[\varphi(X_t)|\mathcal{Y}_t], \quad \forall \varphi \in C_b(R^d). \tag{5.5}$$

In the next two sections, we will deduce the evolution equations through two different approaches, i.e., the change of probability measure approach and the innovation process approach. After that, we will consider the existence and evolution equations satisfied by the density function of the conditional distribution. At last, we will consider a robust form of the evolution equation of density function, which is widely used in numerical computations.

5.2 The Change of Probability Measure Method

As the beginning, we shall start with an easy example. If the observation equation is $h(x) = 0$, the $Y_t = W_t$ holds, and the Y_t is independent with X_t, which makes the $E[X_t|\mathcal{Y}_t] = E[X_t]$.

So for any $0 \leq t \leq T$, how to transform the general Y_t to be independent with X_t becomes an important starting point to derive the equation satisfied by the posterior density function. Girsanov's theorem provides a wonderful solution for our start point. Basically, we can define a new measure $\tilde{P}$ equivalent to the original measure P. And the Y_t is a standard Brownian motion under $\tilde{P}$, and it is independent with X_t.

Next, a powerful tool called Girsanov's theorem is introduced as follows.

Theorem 5.1 *Let M_t be a continuous martingale on a Probability space $(\Omega, \mathcal{F}_t, P)$ for $t \geq 0$, and let T_t satisfy the following SDE:*

$$dT_t = T_t dM_t \text{ with } T_0 = 1, \text{ for } t \in [0, \infty). \tag{5.6}$$

Or T_t can be the associated exponential martingale,

$$T_t = \exp(M_t - \frac{1}{2}\langle M \rangle_t). \tag{5.7}$$

If T_t is an uniformly integrable martingale, there is new measure Q equivalent to P, which is defined by

$$\frac{dQ}{dP} := T_\infty. \tag{5.8}$$

Furthermore, if X is a continuous P local martingale, then $X_t - \langle X, M \rangle_t$ is a Q local martingale.

Proof Because the T_t is a a uniformly integrable martingale, then there is a random variable T_∞ and $T_t = E[T_\infty | \mathcal{F}_t]$. Furthermore, the Q is a probability measure that is equivalent to P.

Now consider X_t, which is a P local martingale. Consider the process Y_t defined via

$$Y_t^n := X_{t \wedge s_n} - \langle X, M \rangle_{t \wedge s_n}, \tag{5.9}$$

where the $s_n := \inf\{t \geq 0 : |X_t| > n \text{ or } |\langle X, M \rangle| \geq n\}$.

By applying Itô's formula to $T_t Y_t^n$,

$$\begin{aligned} d(T_t Y_t^n) &= T_t dY_t^n + Y_t^n dT_t + d\langle T, Y^n \rangle_t \\ &= T_t (dX_t - d\langle X, M \rangle_t) + Y_t^n T_t dM_t + d\langle T, Y^n \rangle_t \\ &= (X_t - \langle X, M \rangle_t) T_t dM_t + T_t dX_t, \end{aligned} \tag{5.10}$$

where the result $\langle T, Y^n \rangle_t = T_t \langle X, M \rangle_t$ holds due to the Kunita-Watanabe identity and $t \leq s_n$. So, $T_t Y_t^n$ is a P local martingale. Furthermore, T_t is uniformly integrable, and Y_t^n is bounded, which makes the $T_t Y_t^n$ P martingale.

Next we verify that Y_t^n is Q martingale, for any $s < t$ and $A \in \mathcal{F}_s$, there is

$$E_Q[(Y_t^n - Y_s^n) 1_A] = E_P[T_\infty (Y_t^n - Y_s^n) 1_A] = E[(T_t Y_t^n - T_s Y_s^n) 1_A] = 0. \tag{5.11}$$

Hence, Y_t^n is a Q-martingale. Take $n \to \infty$; the $X - \langle X, M \rangle_t$ is a Q local martingale. □

Now we can get back to our filtering problem. It is clear that the observation process can be decomposed as follows:

$$Y_t = W_t + \int_0^t h(X_s) ds \text{ for } t \geq 0, \tag{5.12}$$

where W_t is a P−Brownian motion. Define $\tilde{P} = Q$ as in the Theorem 5.1, and $\tilde{W}_t = W_t - \langle W, M \rangle_t$. Then, since $\langle W \rangle_t = t I_m$ for all $t \geq 0$ and $\langle W \rangle_t = \langle \tilde{W} \rangle_t$ where I_m is m dimensional identity matrix, it follows from Levy's characterization of Brownian motion that $\tilde{W}_t$ is a Q−Brownian motion.

5.2 The Change of Probability Measure Method

If there is a M_t process that makes $\tilde{W}_t = W_t + \int_0^t h(X_s)ds$ hold, then we can reduce the general filtering problem to the trivial cases we give at the beginning of this section, which is that Y_t is independent with X_t. It is easy to find that

$$dM_t = -\sum_{i=1}^{m} h(X_t)dW_t^i$$

satisfies the conditions we need. Now, there are only two assumptions that are needed in constructing the new measure $\tilde{P}$ by using Theorem 5.1:

- The T_t associated with $M_t := -\sum_{i=1}^{m} \int_0^t h^i(X_s)dW_s^i ds$ in Theorem 5.1 need to be a uniformly integrable martingale.
- The T_t need to be an $\mathcal{F}_t$-adapted martingale.

Firstly, Let $T = \{T_t, t > 0\}$ be the process defined by

$$T_t = \exp(-\sum_{i=1}^{m} \int_0^t h^i(X_s)dW_s^i - \frac{1}{2}\sum_{i=1}^{m} \int_0^t h^i(X_s)^2 ds), \quad t \geq 0. \tag{5.13}$$

There is the condition to guarantee that T_t is a martingale, which is well-known as Novikov's condition as follows:

$$E\left[\exp(\frac{1}{2}\sum_{i=1}^{m} \int_0^t h^i(X_s)^2 ds)\right] < \infty \text{ for all } t \geq 0. \tag{5.14}$$

By using Lemma 3.9 in [1], the condition can be changed into the following:

$$E\left[\sum_{i=1}^{m} \int_0^t h^i(X_s)^2 ds\right] < \infty \text{ for all } t \geq 0. \tag{5.15}$$

Similarly, expectations of $\int_0^t h^i(X_s)^2 ds$ should also be finite under the new measure, which is

$$E\left[\sum_{i=1}^{m} \int_0^t T_s h^i(X_s)^2 ds\right] < \infty \text{ for all } t \geq 0. \tag{5.16}$$

A complete proof of the required properties in 5.2 obtained by condition (5.15), (5.16) is given by the following Lemma.

Lemma 5.1 *Let $T = \{T_t, t > 0\}$ defined in (5.13) be an RCLL m-dimensional process such that condtions (5.15), (5.16) hold and then T_t is an integrable martingale and is an $\mathcal{F}_t$-adapted martingale.*

Proof From (5.15), we see that the process

$$t \to -\sum_{i=1}^{m} \int_0^t h^i(X_s)dW_s^i, \qquad (5.17)$$

is a continuous (square-integrable) martingale with quadratic variation process

$$\sum_{i=1}^{m} \int_0^t (h(X_t)^i)^2 ds.$$

By Itô's formula, the process T_t satisfies the equation

$$T_t = 1 - \sum_{i=1}^{m} \int_0^t T_s h^i(X_s)dW_s^i. \qquad (5.18)$$

So, T_t is a non-negative, continuous, local martingale. It is a continuous supermartingale by using Fatou's lemma. In order to prove that T_t is a martingale, it is enough to show that it has a constant expectation.

Firstly, the following equation is held by the supermartingale property:

$$E[T_t] \leq E[T_0] = 1.$$

And we only need to prove that for any $t \geq 0$

$$E\left[\int_0^t \sum_{i=1}^m T_s^2 (h^i(X_s))^2\right] < \infty.$$

If T_t is bounded, the

$$E\left[\int_0^t \sum_{i=1}^m T_s^2 (h^i(X_s))^2\right] < \|T_t\|_\infty E\left[\int_0^t \sum_{i=1}^m T_s (h^i(X_s))^2\right] < \infty,$$

by the condition (5.15). So we need to deal with the cases T_t is unbounded. We can construct the $\frac{T_t}{1+\epsilon T_t}$, for some $\epsilon > 0$. By Itô's formula,

$$\begin{aligned}\frac{T_t}{1+\epsilon T_t} &= \frac{1}{1+\epsilon} + \sum_{i=1}^m \int_0^t \frac{T_t}{(1+\epsilon T_t)^2} h^i(X_s)dW_s^i \\ &\quad - \sum_{i=1}^m \int_0^t \frac{\epsilon T_t^2}{(1+\epsilon T_t)^3} h^i(X_t)^2 ds.\end{aligned} \qquad (5.19)$$

5.2 The Change of Probability Measure Method

The second part of (5.19), which is

$$E\left[\sum_{i=1}^{m}\int_0^t \frac{T_t^2}{(1+\epsilon T_t)^4}(h^i(X_s))^2 ds\right] \leq \frac{1}{\epsilon^2}E\left[\sum_{i=1}^{m}\int_0^t h^i(X_s)^2 ds\right] < \infty,$$

where the last inequality holds by (5.15).

From the third part in (5.19), it follows that

$$E\left[\sum_{i=1}^{m}\int_0^t \frac{T_s^2}{(1+\epsilon T_s)^3}(h^i(X_s))^2 ds\right]$$

$$\leq E\left[\sum_{i=1}^{m}\int_0^t (\frac{\epsilon T_s}{(1+\epsilon T_s)})T_s(h^i(X_s))^2 ds\right]$$

$$< \infty. \tag{5.20}$$

By taking the limit which is that ϵ tends to 0. And the proof is completed. □

The above change of probability measure method (CPMM) can be summarized as follows:

$$\text{Measure } P \text{ The Brownian motion } W_t \xrightarrow{\text{CPMM}} \text{Measure } \tilde{P}_t \text{ The Brownian motion } Y_t \tag{5.21}$$

where $\frac{d\tilde{P}_t}{dP} = T_t$ holds for any $t \geq 0$, and it is well defined since Radon-Nikodym derivative with respect to P to be given by T_t.

There is a clear clue in the following, which is to represent the filtering equations into the new measure $\tilde{P}_t$, which required the reverse Radon-Nikodym derivative $\frac{dP}{d\tilde{P}_t} = T_t^{-1}$.

For Itô's formula,

$$d(T_t^{-1}) = -(T_t^{-1})^2 dT_t = T_t^{-1}\left(\sum_{i=1}^{m} h^i(X_t)dW_t^i\right) \text{ with } T_0^{-1} = 1, \tag{5.22}$$

where the measure is P.

The measure P for W_t is equal to the measure $\tilde{P}_t$ for Y_t by (5.21). Furthermore, we donate the $T_t^{-1} = \tilde{T}_t$, and (5.22) will become

$$d\tilde{T}_t = \sum_{i=1}^{m}\tilde{T}_t h^i(X_t)dY_t^i \text{ with } \tilde{T}_0 = 1. \tag{5.23}$$

The well-defines for $\tilde{T}_t$ is similar with T_t.

5.3 The DMZ Equation

In this section, we will introduce the Kallianpur-Striebel formula, which can be considered as the continuous version of Baye's formula.

In the following, we denote that

$$E[\xi] = \int_\Omega \xi dP, \quad \tilde{E}[\xi] = \int_\Omega \xi d\tilde{P}_t, \qquad (5.24)$$

where ξ is a random variable and P, $\tilde{P}$ are defined in (5.21).

Proposition 5.1 (Kallianpur and Striebel [5]) *Let us assume that conditions (5.15) and (5.16) hold. For every $\varphi \in L^\infty(R^d, R)$, which is a bounded and measurable function, for any fixed $t \in [0, \infty)$,*

$$\pi_t(\varphi) = \frac{\tilde{E}[\tilde{T}_t \varphi(X_t)|\mathcal{Y}]}{\tilde{E}[\tilde{T}_t|\mathcal{Y}]} \quad \tilde{P}-a.s. \qquad (5.25)$$

where the $\tilde{T}_t$ is defined in (5.23).

Proof It is clear from the (5.23) that $\tilde{T}_t \geq 0$, and it is observed that

$$0 = \tilde{E}[1_{\{\tilde{T}_t=0\}}\tilde{T}_t] = E[1_{\{\tilde{T}_t=0\}}] = P(\tilde{T}_t = 0) = 0. \qquad (5.26)$$

So, it follows that $\tilde{T}_t > 0$, $P-a.s.$ as a consequence of which $\tilde{E}[\tilde{T}_t|\mathcal{Y}] > 0$, $P-a.s.$.

So we only need to prove the following statements:

$$\pi_t(\varphi)\tilde{E}[\tilde{T}_t|\mathcal{Y}] = \tilde{E}[\tilde{T}_t\varphi(X_t)|\mathcal{Y}] \quad \tilde{P}-a.s. \qquad (5.27)$$

As both left- and right-hand sides of this equation are $\mathcal{Y}_t$-measurable, we can prove this equation with a standard functional point of view, which is for any test bounded and $\mathcal{Y}_t$-measurable random variable α,

$$\tilde{E}[\pi_t(\varphi)\tilde{E}[\tilde{T}_t|\mathcal{Y}]\alpha] = \tilde{E}[\tilde{E}[\tilde{T}_t\varphi(X_t)|\mathcal{Y}]\alpha] \quad \tilde{P}-a.s. \qquad (5.28)$$

A consequence of the definition of the process π_t is that $\pi_t(\varphi) = E[\varphi(X_t)|\mathcal{Y}_t]$ $\tilde{P}-a.s.$, so from the definition of Kolmogorov conditional expectation

$$E[\pi_t(\varphi)\alpha] = E[\varphi(X_t)\alpha]. \qquad (5.29)$$

Writing this under the measure $\tilde{P}_t$,

$$\tilde{E}[\pi_t(\varphi)\alpha\tilde{T}_t] = \tilde{E}[\varphi(X_t)\alpha\tilde{T}_t]. \qquad (5.30)$$

5.3 The DMZ Equation

By the tower property of the conditional expectation, since by assumption the function α is $\mathcal{Y}_t$-measurable. □

Definition 5.1 Define the un-normalized conditional distribution of X to be the measure-valued process $\rho = \{\rho_t, t \geq 0\}$, which is determined by the values of $\rho_t(\varphi)$ for any φ is bound and measure, which are given for $t \geq 0$ by

$$\rho_t(\varphi) := \pi_t(\varphi)\zeta_t. \tag{5.31}$$

Lemma 5.2 The process $\{\rho_t, t \geq 0\}$ is a RCLL and $\mathcal{Y}_t$-adapted. Furthermore, for any $t \geq 0$,

$$\rho_t(\varphi) = \tilde{E}[\tilde{T}_t \varphi(X)_t | \mathcal{Y}_t] \quad \tilde{P} - a.s.. \tag{5.32}$$

Proof Both $\pi_t(\varphi)$ and ζ_t are $\mathcal{Y}_t$-adapted. By construction, $\{\zeta, t \geq 0\}$ is also RCLL. By Sect. 3.2, $\{\pi_t, t \geq 0\}$ is RCLL and $\mathcal{Y}_t$-adapted; therefore, the process $\{\rho_t, t \geq 0\}$ is also RCLL and $\mathcal{Y}_t$-adapted. For the second part, it follows that

$$\pi_t(\varphi)\tilde{E}[\tilde{T}_t | \mathcal{Y}_t] = \tilde{E}[\tilde{T}_t \varphi(X_t) | \mathcal{Y}_t] \quad \tilde{P} - a.s. \tag{5.33}$$

□

Corollary 5.1 Assume that conditions (5.15) and (5.16) hold. For every $\varphi \in L^\infty(R^d, R)$, which is a bounded and measurable function,

$$\pi_t(\varphi) = \frac{\rho_t(\varphi)}{\rho(1)} \quad \forall t \in [0, \infty). \tag{5.34}$$

Proof The result is a direct consequence of Definition 5.1. □

In the following, we further assume that for all $t \geq 0$,

$$\tilde{P}\left[\int_0^t \rho_s(\|h\|)^2 < \infty\right] = 1. \tag{5.35}$$

Theorem 5.2 If conditions (5.15) and (5.16) hold and (5.35) are satisfied, then the process ρ_t satisfies the following evolution equation, called the DMZ equation [4, 8, 10]:

$$\rho_t(\varphi) = \pi_0(\varphi) + \int_0^t \rho_s(\mathscr{A}\varphi)ds + \int_0^t \rho_s(\varphi h^T)dY_s, \quad \tilde{P} - a.s. \ \forall t \geq 0, \tag{5.36}$$

for all $\varphi \in C^2(R^d, R)$.

Proof The following equation holds by Itô's formula:

$$d(\tilde{T}_t \varphi(X_t)) = \tilde{T}_t(\mathscr{A}\varphi(X_t))dt + \tilde{T}_t dM_t^\varphi + \tilde{T}_t h^T(X_t)\varphi(X_t)dY_t. \tag{5.37}$$

If the $\tilde{T}_t$ is bounded, then the first term in the right-hand side can take the expectation,

$$\tilde{E}\left[\tilde{T}_t(\mathscr{A}\varphi(X_t))|\mathcal{Y}_t\right] = \rho(\mathscr{A}\varphi). \tag{5.38}$$

Furthermore, the second term is a well-defined martingale by the boundness of $\tilde{T}_t$:

$$\tilde{E}\left[\int_0^t \tilde{T}_t dM_s^\varphi |\mathcal{Y}_t\right] = 0. \tag{5.39}$$

The third term is well-defined since the quadratic variation of the process is

$$\tilde{E}\left[\int_0^t \tilde{T}_s^2 \|h(X_s)\|^2 ds\right] \le \|T_s\|_\infty \tilde{E}\left[\int_0^t \tilde{T}_s \|h(X_s)\|^2 ds\right]$$
$$= \|T_s\|_\infty E\left[\int_0^t \|h(X_s)\|^2 ds\right] < \infty, \tag{5.40}$$

where the last inequality holds by the condition (5.15).

For general cases, we can use a bounded trick, which is to approximate $\tilde{T}_t$ with $\tilde{T}_t^\epsilon$ given by

$$\tilde{T}_t^\epsilon = \frac{\tilde{T}_t}{1 + \epsilon \tilde{T}_t}.$$

Using Itô's rule and integration by parts, we find

$$d\left(\tilde{T}_t^\epsilon \varphi(X_t)\right) = \tilde{T}_t^\epsilon \mathscr{A}\varphi(X_t) dt + \tilde{T}_t^\epsilon dM_t^\varphi$$
$$- \epsilon \varphi(X_t)(1 + \epsilon \tilde{T}_t)^{-3} \tilde{T}_t^2 \|h(X_t)\|^2 dt$$
$$+ \varphi(X_t)(1 + \epsilon \tilde{T}_t)^{-2} \tilde{T}_t h^T(X_t) dY_t.$$

Since $\tilde{T}_t^\epsilon$ is bounded, the following equation holds:

$$\tilde{E}\left[\int_0^t \tilde{T}_t^\epsilon dM_s^\varphi |\mathcal{Y}\right] = 0.$$

And similarly with the above methods, by taking conditional expectation, the following equation holds:

$$\tilde{E}[\tilde{T}_t^\epsilon \varphi(X_t)|\mathcal{Y}] = \frac{\pi_0(\varphi)}{1+\epsilon} + \int_0^t \tilde{E}[\tilde{T}_t^\epsilon \mathscr{A}\varphi(X_s)|\mathcal{Y}] ds$$

5.3 The DMZ Equation

$$-\int_0^t \tilde{E}\left[\epsilon\varphi(X_s)(\tilde{T}_t^\epsilon)^2 \frac{1}{(1+\epsilon\tilde{T}_t)} \|h(X_s)\|^2 | \mathcal{Y}\right] ds$$

$$+\int_0^t \tilde{E}\left[\tilde{T}_t^\epsilon \frac{1}{1+\epsilon\tilde{T}_t} \varphi(X_s) h^T(X_s) | \mathcal{Y}\right] dY_s. \quad (5.41)$$

Furthermore, by using the following equation and Proposition B.41 in [1]:

$$\tilde{E}\left[\int_0^t \varphi(X_s)^2 \frac{1}{(1+\epsilon\tilde{T}_s)^2} \frac{1}{\epsilon^2} \left(\frac{\epsilon\tilde{T}_s}{1+\epsilon\tilde{T}_s}\right)^2 \|h(X_s)\|^2 ds\right]$$

$$\leq \frac{\|\varphi\|_\infty^2}{\epsilon^2} \tilde{E}\left[\int_0^t \|h(X_s)\|^2 ds\right]$$

$$= \frac{\|\varphi\|_\infty^2}{\epsilon^2} E\left[\int_0^t T_s \|h(X_s)\|^2 ds\right], \quad (5.42)$$

that $\lim_{\epsilon \to 0} \int_0^t \tilde{E}\left[\tilde{T}_t^\epsilon \frac{1}{1+\epsilon\tilde{T}_t} \varphi(X_s) h^T(X_s) | \mathcal{Y}\right] dY_s = \int_0^t \tilde{E}\left[\tilde{T}_t \varphi(X_s) h^T(X_s) | \mathcal{Y}\right] dY_s$.

Now let ϵ tend to 0. Writing λ for Lebesgue measure on $[0, \infty)$, the following hold:

$$\lim_{\epsilon \to 0} \tilde{T}_t^\epsilon = \tilde{T}_t$$

$$\lim_{\epsilon \to 0} \tilde{E}[\tilde{T}_t^\epsilon \varphi(X_t) | \mathcal{Y}] = \rho_t(\varphi) \quad \tilde{P} - a.s.$$

$$\lim_{\epsilon \to 0} \tilde{E}[\tilde{T}_t^\epsilon \mathscr{A}\varphi(X_t) | \mathcal{Y}] = \rho_t(\mathscr{A}\varphi) \quad \lambda \otimes \tilde{P} - a.s. \quad (5.43)$$

We further assume boundness of $\mathscr{A}\varphi$ the random variable $\|\mathscr{A}\varphi\|_\infty \tilde{E}[\tilde{T}_t | \mathcal{Y}]$, which can be seen in $L^1([0, t] \times \Omega; \lambda \otimes \tilde{P})$ since

$$\tilde{E}\left[\int_0^t \|\mathscr{A}\varphi\|_\infty \tilde{E}[\tilde{T}_s | \mathcal{Y}] ds\right] \leq \|\mathscr{A}\varphi\|_\infty \int_0^t \tilde{E}[\tilde{Z}_s] ds \leq \|\mathscr{A}\varphi\|_\infty t < \infty.$$

Consequently by the conditional form of the dominated convergence theorem as $\epsilon \to 0$,

$$\tilde{E}\left[\int_0^t \tilde{E}[\tilde{T}_s^\epsilon \mathscr{A}\varphi(X_s) | \mathcal{Y}] ds\right] \to \tilde{E}\left[\int_0^t \rho_s(\mathscr{A}\varphi) ds\right], \quad \tilde{P} - a.s.$$

Using the definition of ρ_t, we see that by Fubini's theorem

$$\int_0^t \tilde{E}[\tilde{T}_s^\epsilon \mathscr{A}\varphi(X_s) | \mathcal{Y}] ds \to \int_0^t \rho_s(\mathscr{A}\varphi) ds, \quad \tilde{P} - a.s.$$

Next we have that for almost every t,

$$\lim_{\epsilon \to 0} \epsilon \varphi(X_s)(\tilde{T}_s^\epsilon)^2 (1 + \epsilon \tilde{T}_s)^{-1} \|h(X_S)\|^2 = 0, \quad \tilde{P} - a.s.$$

and

$$\left| \epsilon \varphi(X_s)(\tilde{T}_s^\epsilon)^2 (1 + \epsilon \tilde{T}_s) \|h(X_s)\| \right|$$

$$= \left| \varphi(X_s) \tilde{T}_s \|h(X_s)\| \frac{\epsilon \tilde{T}_s}{1 + \epsilon \tilde{T}_s} (1 + \epsilon \tilde{T}_s)^{-2} \right|$$

$$\leq \|\varphi\|_\infty \tilde{T}_s \|h(X_s)\|^2. \tag{5.44}$$

The right-hand side of (5.44) is integrable over $[0, t] \times \Omega$ with respect to $\lambda \otimes \tilde{P}$ using condition 1:

$$\tilde{E}\left[\int_0^t \tilde{T}_s \|h(X_s)\|^2 ds \right] = E\left[\int_0^t \|h(X_s)\|^2 \right] < \infty.$$

Thus, we can use the conditional form of the dominated convergence theorem to obtain that

$$\lim_{\epsilon \to 0} \int_0^t \epsilon \tilde{E}\left[\varphi(X_s)(\tilde{T}_s^\epsilon)^2 (1 + \epsilon \tilde{T}_s)^{-1} \|h(X_s)\|^2 | \mathcal{Y} \right] ds = 0.$$

So the proof is complete. □

5.4 The Innovation Process Method

In this section, we will use another method to derive the evolution equation of conditional probability, which is based on a relevant process called innovation process.

Definition 5.2 Assume that $Y = \{Y_t : t \geq 0\}$ is the observation process of a filtering problem satisfying the evolution equation:

$$Y_t = Y_0 + \int_0^t h(X_s)ds + W_t. \tag{5.45}$$

5.4 The Innovation Process Method

π_s is the solution to the filtering problem, i.e., π_s is the conditional distribution of the state process X_s given $\mathcal{Y}_s$, and then the process $I = \{I_t : t \geq 0\}$ given by

$$I_t = Y_t - \int_0^t \pi_s(h) ds, \qquad (5.46)$$

is called the innovation process.

The innovation process defined above is in fact a Brownian motion.

Proposition 5.2 *The innovation process defined by (5.46) is a $\mathcal{Y}_t$-adapted Brownian motion.*

Proof It is obvious that the innovation process is $\mathcal{Y}_t$-adapted. It is a continuous martingale because

$$E[I_t|\mathcal{Y}_s] - I_s = E\left[\int_s^t (h(X_r) - \pi_r(h)) dr \,\Big|\, \mathcal{Y}_s\right]. \qquad (5.47)$$

Since

$$E[\pi_r(h)|\mathcal{Y}_s] = E[E[h_r(X_r)|\mathcal{Y}_r]|\mathcal{Y}_s] = E[h(X_r)|\mathcal{Y}_s], \qquad (5.48)$$

we have

$$E[I_t|\mathcal{Y}_s] = I_s. \qquad (5.49)$$

In order to show that the innovation process I_t is a Brownian motion, it suffices to calculate its cross variation. For any i, j, we have

$$\langle I^i, I^j \rangle_t = \langle W^i, W^j \rangle_t = t\delta_{ij}, \qquad (5.50)$$

and $\{I_t : t \geq 0\}$ is a Brownian motion by Lemma 3.3. $\square$

In order to introduce the innovation process method, we need the following version of the well-known martingale representation theorem. The details of the theorem can be found in monograph on stochastic analysis such as [6].

Proposition 5.3 (Martingale Representation Theorem) *Assume that the following conditions hold:*

$$E\left[\int_0^t \|h(X_s)\| ds\right] < \infty,$$
$$P\left(\int_0^t \|\pi_s(h)\|^2 ds < \infty\right) = 1. \qquad (5.51)$$

Then, every square integrable random variable η, which is $\mathcal{Y}_\infty = \sigma\left(\cup_{t\geq 0}\mathcal{Y}_t\right)$-measurable, has a representation of the form

$$\eta = E[\eta] + \int_0^\infty v_s^\top dI_s, \tag{5.52}$$

where $v = \{v_t : t \geq 0\}$ is a progressively measurable $\mathcal{Y}_t$-adapted process such that

$$E\left[\int_0^\infty \|v_s\|^2 ds\right] < \infty. \tag{5.53}$$

Proof Let us consider the Hilbert subspace of square-integrable random variables generated by $I = \{I_t : t \geq 0\}$:

$$\mathcal{I} = \left\{\int_0^\infty v_s^\top dI_s : E\left[\int_0^\infty \|v_s\|^2 ds\right] < \infty\right\}. \tag{5.54}$$

Also, let us denote by $\mathcal{I}^\perp$ the orthogonal complement of $\mathcal{I}$. Then, there is unique decomposition for every $\mathcal{Y}_\infty$-measurable square-integrable random variable η, which is given by

$$\eta = Z + \int_0^\infty v_s^\top dI_s, \tag{5.55}$$

and $Z \in \mathcal{I}^\perp$ is orthogonal to every elements in $\mathcal{I}$. It suffices to show that Z is constant, and therefore, $Z = E[\eta]$.

In fact, for every partition of the time interval $[0, t]$, given by $0 = s_0 < s_1 < \cdots < s_n = t$, and bounded functions $f_k : \mathbb{R}^d \to \mathbb{R}, k = 0, \cdots, n$, we have

$$E\left[(Z - E[Z]) \prod_{k=0}^n f_k(I_{s_k})\right] = 0, \tag{5.56}$$

according to the orthogonality of Z and $\mathcal{I}$. The desired result holds because of the Dynkin system theorem, the proof of which is left as the following exercise. □

Exercise 5.1 (Dynkin System Theorem) A collection $\mathcal{D}$ of subsets of a set Ω is called a Dynkin system:

1. If $\Omega \in \mathcal{D}$.
2. If $A, B \in \mathcal{D}$ and $B \subset A$, then $A/B \in \mathcal{D}$.
3. If $\{A_n\}_{n=1}^\infty \subset \mathcal{D}$ and $A_1 \subset A_2 \subset \cdots \subset A_n \subset \cdots$, then $\cup_{n=1}^\infty A_n \in \mathcal{D}$.

Let C be another collection of subset of Ω, which is closed under pairwise intersection. If $\mathcal{D}$ is a Dynkin system containing C, prove that $\mathcal{D}$ also contains the σ-field $\sigma(C)$ generated by C.

5.4 The Innovation Process Method

The following result is a direct corollary of the above martingale representation theorem.

Corollary 5.2 *Under the conditions in Proposition 5.3, every continuous square integrable martingale that is $\mathcal{Y}_t$-adapted has a representation*

$$\eta_t = \eta_0 + \int_0^t v_s^\top dI_s, \quad t \geq 0. \tag{5.57}$$

Proof For any $n \in \mathbb{N}$, by Proposition 5.3, since $\eta_n - \eta_0$ is $\mathcal{Y}_\infty$-measurable, there exists a progressively measurable $\mathcal{Y}_t$-adapted process $v^{(n)}$, such that

$$\eta_n - \eta_0 = \int_0^\infty (v_s^{(n)})^\top dI_s. \tag{5.58}$$

Since η_t is a martingale, by conditioning on $\mathcal{Y}_t$, $t \in [0, n]$, we have

$$\eta_t = \eta_0 + \int_0^t (v_s^{(n)})^\top dI_s. \tag{5.59}$$

In the meanwhile, processes $v^{(n)}$ and $v^{(m)}$ coincide with each other on the interval $[0, \min(m, n)]$. Therefore, the process v in Eq. (5.57) can be taken as

$$v_s = v_s^{(n)}, \tag{5.60}$$

for some $n \in \mathbb{N}$, $n > s$. □

With Corollary 5.2, we can derive the following Kushner-Stratonovich (K-S) equation, which is the evolution equation satisfied by the conditional distribution [7, 9].

Theorem 5.3 *If conditions in Proposition 5.3 are satisfied, then the conditional distribution of the state process X_t, $\pi_t(\varphi) \triangleq E[\varphi(X_t)|\mathcal{Y}_t]$, satisfies the following equation, which is also referred to as Kushner-Stratonovich equation:*

$$\pi_t(\varphi) = \pi_0(\varphi) + \int_0^t \pi_s(\mathscr{A}\varphi)ds + \int_0^t \left(\pi_s(\varphi h^\top) - \pi_s(h^\top)\pi_s(\varphi) \right)$$
$$\times (dY_s - \pi_s(h)ds), \tag{5.61}$$

for all $\varphi \in C_0^2(R^d)$.

Proof Define $N_t = \pi_t(\varphi) - \int_0^t \pi_s(\mathscr{A}\varphi)ds$. It is easy to show that N_t is a square integrable $\mathcal{Y}_t$ martingale. In light of Corollary 5.2, there exists a progressively measurable $\mathcal{Y}_t$-adapted process v, such that

$$N_t = \pi_0(\varphi) + \int_0^t v_s^\top dI_s, \tag{5.62}$$

that is,

$$\pi_t(\varphi) = \pi_0(\varphi) + \int_0^t \pi_s(\mathscr{A}\varphi)ds + \int_0^t v_s^\top dI_s. \tag{5.63}$$

Next, we need to show that the process v is of the form

$$v_t = \pi_t(\varphi h^\top) - \pi_t(\varphi)\pi_t(h). \tag{5.64}$$

In fact, Eq. (5.64) holds if we apply Itô's formula to both $\varphi(X_t)$ and $h(X_t)$ and then take conditional expectation with respect to $\mathscr{Y}_t$.

Therefore, the Kushner-Stratonovich equation (5.61) holds. □

Up to now, we have derived the Kushner-Stratonovich (K-S) equation from the innovation process approach. In the previous section, the K-S equation is deduced from the Zakai equation satisfied by the unnormalized conditional probability. In fact, the Zakai equation can also be deduced from this K-S equation. For this purpose, we need first introduce the following exponential martingale, $\{\widehat{Z}_t, t > 0\}$:

$$\widehat{Z}_t = \exp\left(\int_0^t \pi_s(h^\top)dY_s - \frac{1}{2}\int_0^t \|\pi_s(h)\|^2 ds\right). \tag{5.65}$$

Then, by Itô's formula, we have

$$d\left(\frac{1}{\widehat{Z}_t}\right) = -\frac{1}{\widehat{Z}_t}\pi_t(h^\top)dI_t, \tag{5.66}$$

and

$$\widehat{Z}_t = \widetilde{E}\left[\widetilde{Z}_t|\mathscr{Y}_t\right] = \rho_t(1). \tag{5.67}$$

According to Kallianpur-Striebel formula, the unnormalized conditional probability ρ_t satisfies the evolution equation

$$\begin{aligned}d\rho_t(\varphi) =& d\left(\widehat{Z}_t\pi_t(\varphi)\right) \\ =& \widehat{Z}_t d\pi_t(\varphi) + \pi_t(\varphi)d\widehat{Z}_t + d\langle\widehat{Z}, \pi.(\varphi)\rangle_t \\ =& \widehat{Z}_t d\pi_t(\varphi) + \pi_t(\varphi)\pi_t(h^\top)\widehat{Z}_t dY_t \\ &+ \pi_t(h^\top)\widehat{Z}_t(\pi_t(\varphi h^\top) - \pi_t(\varphi)\pi_t(h^\top))dt \\ =& \widehat{Z}_t\left(\pi_t(\mathscr{A}\varphi)dt + \pi_t(\varphi h^\top)dY_t\right) \\ =& \rho_t(\mathscr{A}\varphi)dt + \rho_t(\varphi h^\top)dY_t,\end{aligned} \tag{5.68}$$

which is just the Zakai equation.

5.5 The Density Function of Conditional Distribution

Zakai equations as well as Kushner-Stratonovich equations discussed in the previous sections are evolution equations for abstract measure-valued stochastic processes, ρ_t and π_t, and can only be analyzed with the help of test function φ. In fact, there exists a more convenient form of the Zakai equations and K-S equations. We will show that if the additional regularity of coefficients in the filtering system is satisfied, then conditional probability measures are absolutely continuous with respect to the Lebesgue measure, and the evolution equation of the density function can also be derived, which is more commonly used in practice.

The existence of a density function of the measures ρ_t and π_t is based on the following Lemma 5.3.

Lemma 5.3 *Let $\{\varphi_i\}_{i=1}^{\infty}$ be an orthonormal basis of $L^2(R^d)$ with the property that $\varphi_i \in C^b(R^d)$, for each $i \in \mathbb{N}$. Let $\mu \in \mathcal{M}(R^d)$ be a finite measure. If*

$$\sum_{i=1}^{\infty} \mu(\varphi_i)^2 \triangleq \sum_{i=1}^{\infty} \left(\int_{R^d} \varphi_i d\mu \right)^2 < \infty, \qquad (5.69)$$

then μ is absolutely continuous with respect to Lebesgue measure. Denote $f_\mu : R^d \to R$ is the density function of μ with respect to Lebesgue measure; then $f_\mu \in L^2(R^d)$

Proof Let $f_\mu : R^d \to R$ be defined as

$$f_\mu = \sum_{i=1}^{\infty} \mu(\varphi_i) \varphi_i. \qquad (5.70)$$

Because of the condition (5.69), we have $\|f_\mu\|_{L^2(R^d)} < \infty$, and thus $f_\mu \in L^2(R^d)$.

Consider the measure absolutely continuous with respect to Lebesgue measure with density f_μ. For each φ_i, $i \in \mathbb{N}$, we have

$$\int_{R^d} \varphi_i f_\mu dx = \sum_{j=1}^{\infty} \mu(\varphi_j) \int_{R^d} \varphi_i \varphi_j dx = \mu(\varphi_i). \qquad (5.71)$$

Since φ_i is a basis of $L^2(R^d)$, we have f_μ as the density function with respect to Lebesgue measure, and therefore, μ is absolutely continuous with respect to Lebesgue measure and the density function $f_\mu \in L^2(R^d)$. □

Theorem 5.4 *If π_0 is absolutely continuous with respect to Lebesgue measure with a density that is in $L^2(R^d)$ and coefficients in the filtering system are uniformly bounded, then almost surely, ρ_t has a density with respect to Lebesgue measure, and this density is square-integrable.*

Proof In light of Lemma 5.3, it suffices to show that the series

$$\sum_{i=1}^{\infty} \rho_t(\varphi_i). \tag{5.72}$$

has finite expectations under probability measure $\widetilde{P}$, where $\{\varphi_i\}_{i\geq 1}$ is an orthonormal basis of $L^2(R^d)$.

The idea of the proof is to apply regularization method to the conditional probability measure ρ_t and estimate the upper bound of the series (5.72) for the regularized measure.

The regularization procedure can be found in monographs on Sobolev spaces or functional analysis. Here, we only give a brief introduction to the whole procedure. Let $\{\psi_\epsilon\}_{\epsilon>0}$ be the collection of regularization kernels defined by

$$\psi_\epsilon(x) = (2\pi\epsilon)^{-\frac{d}{2}} \exp\left(-\frac{\|x\|^2}{2\epsilon}\right). \tag{5.73}$$

The regularized measure of ρ_t, $J_\epsilon \rho_t$, is defined to be the measure absolutely continuous to Lebesgue measure with density

$$J_\epsilon \rho_t(y) = \int_{R^d} \psi_\epsilon(x-y) \rho_t(dx). \tag{5.74}$$

From now on, without confusion of notations, we do not distinguish the probability measure, which is absolutely continuous with respect to Lebesgue measure, such as $J_\epsilon \rho_t$, and its density function, and denote both of them by $J_\epsilon \rho_t$.

Since ρ_t satisfies the Zakai equation (5.2), we have

$$J_\epsilon \rho_t(\varphi_i) = J_\epsilon \pi_0(\varphi_i) + \int_0^t \rho_s(\mathscr{A}(J_\epsilon \varphi_i)) ds + \sum_{j=1}^m \int_0^t \rho_s(h^j J_\epsilon \varphi_i) dY_s^j, \tag{5.75}$$

where

$$J_\epsilon \varphi(x) = \int_{R^d} \psi_\epsilon(x-y) \varphi(y) dy, \tag{5.76}$$

for any $\varphi \in L^2(R^d)$.

According to Itô's formula,

$$(J_\epsilon \rho_t(\varphi))^2 = (J_\epsilon \pi_0(\varphi))^2 + 2 \int_0^t J_\epsilon \rho_s(\varphi_i) \rho_s(\mathscr{A}(J_\epsilon(\varphi_i))) ds$$

$$+ 2 \sum_{j=1}^m \int_0^t J_\epsilon \rho_s(\varphi_i) \rho_s(h^j J_\epsilon \varphi_i) dY_s^j \tag{5.77}$$

$$+ \sum_{j=1}^m \int_0^t (\rho_s(h^j J_\epsilon \varphi_i))^2 ds.$$

5.5 The Density Function of Conditional Distribution

Since $\{\varphi_i\}_{i\geq 1}$ form an orthonormal basis of $L^2(R^d)$ and take summation over $i \geq 1$ for (5.77), we have

$$\widetilde{E}\left[\|J_\epsilon \rho_t\|^2\right] = \widetilde{E}\left[\sum_{i=1}^{\infty}(J_\epsilon \rho_t(\varphi_i))^2\right]$$

$$\leq \|J_\epsilon \pi_0\|^2 + 2\sum_{i=1}^{d}\int_0^t \widetilde{E}[\langle J_\epsilon \rho_s, \mathscr{A} J_\epsilon \rho_s\rangle]\,ds \qquad (5.78)$$

$$+ \sum_{j=1}^{m}\int_0^t \widetilde{E}\left[\|h^j J_\epsilon \rho_s\|^2\right]ds,$$

where the inequality is because of Fatou's lemma when we exchange the order of computing series and expectations and the stochastic differential term in (5.77) vanishes because of the martingale property.

Since coefficients in the filtering system are uniformly bound,

$$\widetilde{E}\left[\|J_\epsilon \rho_t\|^2\right] \leq \|J_\epsilon \pi_0\|^2 + c\int_0^t \widetilde{E}\left[\|J_\epsilon \rho_s\|^2\right]ds, \qquad (5.79)$$

where $0 < c < \infty$ is a constant that may only depend on coefficients of the filtering system.

Then, according to Gronwall's inequality, we have

$$\widetilde{E}\left[\|J_\epsilon \rho_t\|^2\right] \leq e^{ct}\|J_\epsilon \pi_0\|^2 \leq e^{ct}\|\pi_0\|^2, \qquad (5.80)$$

holds for all $\epsilon > 0$.

Therefore, by Fatou's lemma again, we have

$$\widetilde{E}\left[\sum_{i=1}^{\infty}(\rho_t(\varphi_i))^2\right] = \widetilde{E}\left[\lim_{\epsilon>0}\sum_{i=1}^{\infty}(J_\epsilon \rho_t(\varphi_i))^2\right]$$

$$\leq \liminf_{\epsilon>0}\widetilde{E}\left[\|J_\epsilon \rho_t\|^2\right] \leq e^{ct}\|\pi_0\|^2 < \infty, \qquad (5.81)$$

which is our desired result. □

If further regularity of the coefficients is also assumed, such that all kinds of Fubini's theorem and the integral-by-part formulae hold in our following discussion, then we can also derive the evolution equation satisfied by the density function of the unnormalized conditional probability measure, as is stated in the following theorem.

Theorem 5.5 *If we further assume that coefficients in the filtering equation are smooth enough, such that there exists a solution p_t to the stochastic partial differential equation*

$$p_t(x) = p_0(x) + \int_0^t \mathscr{A}^* p_s(x)ds + \int_0^t h^\top(x) p_s(x) dY_s. \tag{5.82}$$

Let ρ_t be the measure that is absolutely continuous with respect to the Lebesgue measure with density function p_t; then $\{\rho_t, t \geq 0\}$ satisfies the Zakai equation, that is, for any test function $\varphi \in C_0^2(R^d)$,

$$\rho_t(\varphi) = \pi_0(\varphi) + \int_0^t \rho_s(\mathscr{A}\varphi)ds + \int_0^t \rho_s(\varphi h^\top)dY_s, \tag{5.83}$$

where $\mathscr{A}^*$ is the adjoint operator of $\mathscr{A}$, with

$$\mathscr{A}\varphi = \sum_{i,j=1}^d a^{ij} \frac{\partial^2}{\partial x_i \partial x_j}\varphi + \sum_{i=1}^d f^i \frac{\partial \varphi}{\partial x_i}, \tag{5.84}$$

and

$$\mathscr{A}^*\varphi = \sum_{i,j=1}^d \frac{\partial^2}{\partial x_i \partial x_j}(a^{ij}\varphi) - \sum_{i=1}^d \frac{\partial}{\partial x_i}(f^i \varphi). \tag{5.85}$$

Proof Since we have assumed that coefficients in the filtering system are smooth enough, and the Fubini's theorem as well as the integral-by-part formula holds, then

$$\begin{aligned}\rho_t(\varphi) &= \int_{R^d} \varphi(x) p_t(x) dx \\ &= \int_{R^d} \varphi(x) \left(p_0(x) + \int_0^t \mathscr{A}^* p_s(x) ds + \int_0^t h^\top(x) p_s(x) dY_s \right) dx \\ &= \int_{R^d} \varphi(x) p_0(x) dx + \int_0^t \left(\int_{R^d} \varphi(x) \mathscr{A}^* p_s(x) dx \right) ds \\ &\quad + \int_0^t \left(\int_{R^d} \varphi(x) h^\top(x) p_s(x) dx \right) dY_s \\ &= \pi_0(\varphi) + \int_0^t \rho_s(\mathscr{A}\varphi) ds + \int_0^t \rho_s(\varphi h^\top) dY_s,\end{aligned} \tag{5.86}$$

which is the Zakai equation as desired.

We close this section by the evolution equation satisfied by the density function of π_t, which can also be derived with additional regularity of the system.

$$\pi_t(x) = \pi_0(x) + \int_0^t \mathscr{A}^* \pi_s(x) ds + \int_0^t \pi_s(x)(h^\top(x) - \pi_s(h^\top))(dY_s - \pi_s(h)ds). \tag{5.87}$$

5.6 Robust DMZ Equation

Since the deduction procedure is quite similar to the case of ρ_t, we would like to skip the proof here and reserve it for interested readers. □

5.6 Robust DMZ Equation

In this section, we will focus on a special change in the probability measure method. In the real application, the observation data is sometimes a path. However, in the filtering equation, the observation data is in the form of martingale integral, which is difficult to match with the orbital observation data. So the starting point of the robust DMZ equation in this chapter starts from rewriting the martingale integral.

Let us first recall the notation of change the probability methods,

$$\tilde{T}_t = \exp\left(\int_0^t h(X_t)^\top dY_s - \frac{1}{2}\int_0^t \|h(X_t)\|^2 dt\right). \tag{5.88}$$

For a single path $Y_t(\omega)$, integral $\int_0^t h(X_t)^\top dY_s$ is ill-defined, so we need to rewrite it by using the partial integral method of stochastic analysis, which is

$$\Theta(y.) := \exp(h^\top(X_t)y_t - \int_0^t y_s^\top dh(X_s) - \frac{1}{2}\int_0^t \|h(X_s)\|^2 ds). \tag{5.89}$$

So, the difficult part in the integral in the new representation (5.92) is

$$\int_0^t y_s^\top dh(X_s). \tag{5.90}$$

In the following, we will require that $s \to h(X_s)$ be a semimartingale that makes the (5.90) well-defined.

Let

$$h(X_s) = M_s^h + V_s^h, \quad s \geq 0.$$

be the Doob-Meyer decomposition of $h(X_s)$ with $V_s^h = (V_s^{h,i})_{i=1}^m$ the finite variation part of $h(X_s)$ and $M_s^h = (M_s^{h,i})_{i=1}^m$ assumed to be square integrable, similarly with (5.15) and (5.16).

Now, we can denote $Y(\cdot)$ to be an arbitrary element of the set $C([0,t], R^m)$, where $t \geq 0$ is arbitrary, and we fixed t throughout the section in order to fix the change measure martingale in (5.88). So in this case, $Y.(\omega) : [0,t] \to R^m$ is a continuous function. And $Y.$ is a random variable.

$$Y.(\omega) := \Omega \to C([0,t], R^m), Y.(\omega) = (Y_s(\omega), 0 \leq s \leq t). \tag{5.91}$$

Main Goal

For any test function φ, a bounded Borel-measurable function, the posterior expectation $\pi_t(\varphi)$ can be written as a function of the observation path. That is, there exists a bounded measurable function $U^\varphi : C([0, t], R^m) \to R$, which means the following equation holds:

$$\pi_t(\varphi) = U^\varphi(Y.) \quad P-a.s. \tag{5.92}$$

It is easy to find that U^φ is not unique. Any other function $\bar{U}^\varphi$ such that

$$P \circ Y^{-1}(U^\varphi \neq \bar{U}^\varphi) = 0, \tag{5.93}$$

where $P \circ Y^{-1}$ is the distribution of $Y.$ on the path space can replace U^φ in (5.92). In the following, we obtain a robust representation of the conditional expectation $\pi_t(\varphi)$ (following [1]). That is, we will show that there exists a continuous function $\hat{U}^\varphi$,

$$\pi_t(\varphi) = \hat{U}^\varphi(Y.) \quad P-a.s. \tag{5.94}$$

And similarly with DMZ equation and Kallianpur-Striebel formula,

$$\hat{U}^\varphi(Y.) = \frac{\hat{G}^\varphi(Y.)}{\hat{G}^1(Y.)}, \tag{5.95}$$

where $\hat{G}^\varphi(Y.) = \tilde{E}[\Theta(Y.)]$. So far, well-posedness of (5.89) only depends on (5.90). The following assumptions are required:

$$c^{fv} = \tilde{E}\left(\sum_{i=1}^{m} \int_0^t |dV_s^{h,i}|\right) < \infty, \tag{5.96}$$

$$c^m = \tilde{E}\left(\sum_{i=1}^{m} \int_0^t d\langle M^{h,i}\rangle_s\right) < \infty, \tag{5.97}$$

where $s \to M_s^h$ is the quadratic variation of $M_s^{h,i}$, for $i = 1, \cdots, m$, and $\int_0^t |dV_s^{h,i}|$ is the total variation of $V_s^{h,i}$ on $[0, t]$ for $i = 1, \cdots, m$.

The conciseness of (5.89) are as follows:

- $\Theta(Y.)$ is bounded with any bounded $Y.$.
- $\hat{G}^\varphi(Y.)$ is well-posed.

Lemma 5.4 *For any $R > 0$, there exists a positive constant B_R^Θ such that*

$$\sup_{\|Y.\|_\infty \leq R} \|\Theta(y.)\| \leq B_R. \tag{5.98}$$

5.6 Robust DMZ Equation

Proof In the following, for any arbitrary $Y(\cdot) \in C([0,t], R_m)$,

$$\Theta(Y_\cdot) = \exp\left(\int_0^t (\mathbf{R} + Y_s^T) dh(X_s) - \frac{1}{2}\int_0^t \|h(X_s)\|^2 ds\right)$$

$$\leq \exp\left(\int_0^t (\mathbf{R} + Y_s)^\top dV_s^h + \max_{\|\mathbf{r}\|_\infty \leq R}\int_0^t (\mathbf{r} + Y_s)^\top dM_s^h\right), \quad (5.99)$$

where $\mathbf{R}$ is a m dimensional vector for any component is R.

Furthermore, by using (5.96)

$$\tilde{E}\left[\exp\left(\int_0^t (\mathbf{R} + Y_s)^\top dV_s^h\right)\right] \leq \tilde{E}\left[\exp\left(2R\int_0^t |dVs^h|\right)\right] = e^{2Rc^{fv}},$$

and by using the Cauchy-Schwartz inequality

$$\tilde{E}\left[\exp\left(\max_{\|\mathbf{r}_s\|_\infty \leq R}\int_0^t (\mathbf{r}_s + Y_s)^\top dM_s^h\right)\right]$$

$$\leq \max_{\|\mathbf{r}_s\|_\infty \leq R} \tilde{E}\left[\exp\left(\int_0^t (2\mathbf{r}_s)^\top dM_s^h\right)\right] < \infty. \quad (5.100)$$

where the last inequality holds by using (5.96). $\square$

A different proof is given in [1].

Lemma 5.5 *For any $R > 0$, there exists a positive constant B_R such that*

$$\|\Theta(Y_\cdot^1) - \Theta(Y_\cdot^2)\|_{L^2(\Omega, \tilde{P}_t)} \leq B_R \|Y_\cdot^1 - Y_\cdot^2\|, \quad (5.101)$$

for any two paths $Y_\cdot^1$, $Y_\cdot^2$ such that $|Y_\cdot^1|, |Y_\cdot^2| \leq R$. In particular, (5.11) implies that g^1 is locally Lipschitz, more precisely

$$|G^1(Y_\cdot^1) - G^2(Y_\cdot^2)| \leq B_R \|Y_\cdot^1 - Y_\cdot^2\|.$$

for any two paths $Y_\cdot^1$, $Y_\cdot^2$ such that $|Y_\cdot^1|_\infty, |Y_\cdot^2|_\infty \leq R$.

Proof For two paths $Y_\cdot^1$ and $Y_\cdot^2$, let us denote by their difference by $Y_\cdot^{12} = Y_\cdot^1 - Y_\cdot^2$. Then,

$$|\Theta(Y_\cdot^1) - \Theta(Y_\cdot^2)| \leq (\Theta(Y_\cdot^1) + \Theta(Y_\cdot^2))\left|\int_0^t (Y_t^{12} - Y_s^{12}) dh(X_s)\right| \quad (5.102)$$

According to Lemma 3.4 and Cauchy-Schwartz inequality, we have

$$\|\Theta(Y_\cdot^1) - \Theta(Y_\cdot^2)\|_{L^2(\Omega,\tilde{P}_t)} \leq C_R \left\| \int_0^t (Y_t^{12} - Y_s^{12}) dh(X_s) \right\|_{L^2(\Omega,\tilde{P}_t)}, \quad (5.103)$$

where $C_R > 0$ is a constant depending on R, and the desired result follows from the Burkholder-Davis-Gundy inequality. □

Lemma 5.6 *The function G^φ is locally Lipschitz and locally bounded.*

Proof For a given $R > 0$, let $Y_\cdot^1$ and $Y_\cdot^2$ be two paths such that $|Y_\cdot^1|, |Y_\cdot^2| \leq R$. According to Cauchy-Schwartz inequality, we have

$$\tilde{E}\left[|\varphi(X_t)||\Theta(Y_\cdot^1) - \Theta(Y_\cdot^2)|\right] \leq \|\varphi(X_t)\|_{L^2(\Omega,\tilde{P}_t)} B_R \|Y_\cdot^1 - Y_\cdot^2\|. \quad (5.104)$$

Therefore, G^φ is locally Lipschitz. The local boundedness of G^φ follows directly from Lemma 5.4. □

Here we finish the proofs for well-defined robust representation. Now, we can introduce Clark's robustness result.

Theorem 5.6 (Clark) *The random variable $\hat{U}^\varphi(Y_\cdot)$ is a version of $\pi_t(\varphi)$, that is, $\pi_t(\varphi) = \hat{U}^\varphi(Y_\cdot)$, $\tilde{P}$-almost surely. Hence, $\hat{U}^\varphi(Y_\cdot)$ is the unique robust representation of $\pi_t(\varphi)$.*

Proof It suffices to prove that P-almost surely (or, equivalently, $\tilde{P}$-almost surely),

$$\rho_t(\varphi) = \hat{g}^\varphi(Y_\cdot) \quad \text{and} \quad \rho_t(1) = \hat{g}^1(Y_\cdot).$$

We need only prove the first identified as the second is just a special case obtained by setting $\varphi = 1$ in the first. From the definition of abstract conditional expectation, therefore, it suffices to show

$$\tilde{E}[\rho_t(\varphi)b(Y_\cdot)] = \tilde{E}[\hat{g}^\varphi b(Y_\cdot)], \quad (5.105)$$

where b is an arbitrary continuous bounded function $b : C([0, t] \times R^m) \to R$. Since X and Y are independent under $\tilde{P}$, it follows that the pair processes (X, Y) under $\tilde{P}$ and $(\hat{X}, Y)$ under $\tilde{P}$ have the same distribution. Hence, the left-hand side of (5.105) has the following representation:

$$\tilde{E}[\rho_t(\varphi)b(Y_\cdot)]$$
$$= \tilde{E}\left[\varphi(X_t)\exp(\int_0^t h(X_s)^T dY_s - \frac{1}{2}\int_0^t \|h(X_s)\|^2 ds)b(Y_\cdot)\right]$$
$$= \tilde{E}\left[\varphi(X_t)\exp(\int_0^t h(\hat{X}_s)^T dY_s - \frac{1}{2}\int_0^t \|h(\hat{X}_s)\|^2 ds)b(Y_\cdot)\right]$$
$$= \tilde{E}\left[\varphi(X_t)\exp(h(X_t)^T Y_t - \int_0^t Y_s^T dh(\hat{X}_s) - \frac{1}{2}\int_0^t \|h(\hat{X}_s)\|^2 ds)b(Y_\cdot)\right].$$

On the other hand, if the transformation (5.89) is well-defined, more details can be found in [2] and [3]. Hence, by Fubini's theorem, we can finish the proof. □

5.7 Exercises

1. Deduce the Kushner-Stratonovich equation from the DMZ equation and the Kallianpur-Striebel formula.
2. Deduce the DMZ equation from the Kushner-Stratonovich equation.
3. Consider the stochastic exponential Y given by

$$dY_t = Y_t(\mu \, dt + \sigma \, dW_t), \quad Y_0 = 1,$$

where μ and σ are constants. The Euler approximation of this SDE is

$$\Delta \hat{Y}_{n+1} = \hat{Y}_n(\mu \Delta t + \sigma \Delta \hat{W}_n), \quad \hat{Y}_0 = 1,$$

where Δt is the step size, $\Delta \hat{Y}_{n+1} = \hat{Y}_{n+1} - \hat{Y}_n$, and $\Delta \hat{W}_n = W_{(n+1)\Delta t} - W_{n\Delta t}$. Show that the local weak error is of second order, i.e.,

$$\limsup_{\Delta t \to 0} \left(\frac{E[f(Y_{\Delta t})] - E[f(\hat{Y}_1)]}{\Delta t^2} \right) < \infty$$

holds for any bounded smooth function f with bounded derivatives.
4. Let X and Y be real-valued processes solving SDEs:

$$dX_t = (a_0 + a_1 X_t) \, dt + b \, dW_t^1,$$

$$dY_t = (L_0 + L_1 X_t) \, dt + B \, dW_t^2,$$

where W^1 and W^2 are independent Wiener processes on $\mathbb{R}$, with normally distributed initial conditions (X_0, Y_0). Assume that $p_t \sim N(\hat{X}_t, \hat{\Sigma}_t)$ for some $\hat{X}_t$ and $\hat{\Sigma}_t$, which are the mean and variance of the posterior.
Show that:

$$d\hat{X}_t = (a_0 + a_1 \hat{X}_t) \, dt + \frac{\hat{\Sigma}_t L_1}{B^2} (dY_t - (L_0 + L_1 \hat{X}_t) \, dt),$$

$$\frac{d\hat{\Sigma}_t}{dt} = 2a_1 \hat{\Sigma}_t + b^2 - \frac{(\hat{\Sigma}_t L_1)^2}{B^2},$$

with initial conditions:

$$\hat{X}_0 = E[X_0 \mid Y_0], \quad \hat{\Sigma}_0 = \text{Var}(X_0 \mid Y_0).$$

5. Let $(W_t)_{t\in[0,T]}$ be a standard Brownian motion, X an independent random variable with finite exponential moments, and $Y_t = tX + \sqrt{s}W_t$, $t \in [0,T]$. Define an equivalent probability measure $\tilde{P}$ on $\mathcal{F}_T$ such that Y becomes a martingale and independent of X. Show that the unnormalized filter is given by

$$r_t(A) = \int_A \exp\left(\frac{s^2}{2}xY_t - \frac{1}{2}s^2x^2t\right)\mu(dx).$$

6. Let $(W_t)_{t\in[0,T]}$ be a standard Brownian motion, X an independent random variable with finite exponential moments, and $Y_t = tX + \sqrt{s}W_t$, $t \in [0,T]$. Suppose in addition that X is normally distributed. Then p_t is normally distributed by part. Calculate the mean $\hat{X}_t$ and covariance.

7. Consider the following model for population growth with noisy observations:

$$dX_t = rX_t\,dt, \quad dY_t = X_t\,dt + m\,dW_t,$$

with $X_0 \sim N(b, a^2)$ and $Y_0 = 0$ for some constants $r, m, b, a > 0$.

Calculate $\lim_{t\to\infty}\hat{\Sigma}_t$. How is the asymptotic precision of the filter affected by the growth rate r?

8. Consider the following model for population growth with noisy observations:

$$dX_t = rX_t\,dt, \quad dY_t = X_t\,dt + m\,dW_t,$$

with $X_0 \sim N(b, a^2)$ and $Y_0 = 0$ for some constants $r, m, b, a > 0$. Implement the Kalman-Bucy filter for the model. In order to test your implementation, approximate a path of $(X_t, Y_t)_{t\in[0,1]}$ using the Euler-Maruyama scheme. Use your implementation of the Kalman-Bucy filter to recover the signal from the observation. Use the following parameters: $r = 0.5, m = 1, b = 1, a = 0.5$. What do you observe?

9. X satisfies

$$dX_t = \mu_t\,dt + s_t\,dW_t$$

for some predictable processes μ, s and Brownian motion W.

(a) Show that the process

$$Y_t = \exp\left(X_t - X_0 - \frac{1}{2}\langle X, X\rangle_t\right)$$

is a solution of the SDE

$$dY_t = Y_t\,dX_t, \quad Y_0 = 1.$$

(b) Show that Y is a local martingale if $\mu = 0$.
(c) Show that Y is a martingale if $\mu = 0$ and s is bounded.

References

1. A. Bain and D. Crisan. *Fundamentals of stochastic filtering*. Springer-Verlag, New York, 2009.
2. J. M. C. Clark. Conditions for one to one correspondence between an observation process and its innovation. *Technical Report, Centre for Computing and Automation*, 1969.
3. J. M. C. Clark. The design of robust approximations to the stochastic differential equations of nonlinear filtering. *Communication Systems and Random Process Theory*, 25:721–734, 1978.
4. T. Duncan. Probability densities for diffusion processes with applications to nonlinear filtering theory. (Ph. D. thesis), Stanford University, 1967.
5. G. Kallianpur and C. Striebel. Estimation of stochastic systems: arbitrary system process with additive white noise observation errors. *The Annals of Mathematical Statistics*, 39(3):785–801, 1968.
6. I. Karatzas and S. E. Shreve, *Brownian Motion and Stochastic Calculus*. Springer, New York, 1991.
7. H. J. Kushner. Dynamical equations for optimal nonlinear filtering. *Journal of Differential Equations*, 3:179–190, 1967.
8. R. Mortensen. Optimal control of continuous-time stochastic systems. (Ph. D. thesis), University of California at Berkeley, 1966.
9. R. L. Stratonovich. On the theory of optimal nonlinear filtration of random functions. *Theory of Probability and Its Applications*, 4:223–225, 1959.
10. M. Zakai. On the optimal filtering of diffusion processes. *Zeitschrift für Wahrscheinlichkeitstheorie und Verwandte Gebiete*, 11(3):230–243, 1969.

Chapter 6
Estimation Algebra

In this chapter, we will introduce the finite-dimensional filter estimation algebra technique. Since the proposal of the linear Kalman-Bucy filter, addressing nonlinear filter problems has emerged as a significant and widely discussed area of research. A key question that arises is how to assess the efficiency of different nonlinear filter solutions. Estimation algebra, as both a geometric and an algebraic technique, serves as a powerful tool to address this challenge. By utilizing estimation algebra, we can develop finite-dimensional nonlinear filters that are governed by a finite set of statistical quantities, thus enabling systematic control over their behavior. Importantly, this approach facilitates the classification of various nonlinear systems based on these statistics, paving the way for practical applications in analyzing nonlinear control systems such as observability and controllability. This represents a groundbreaking integration of geometric and algebraic methods into the realm of nonlinear filter theory.

6.1 Introduction

Ever since the technique of the Kalman-Bucy filter was popularized, there has been an intense interest in finding new classes of finite-dimensional recursive filters. In the 1960s and early 1970s, the basic approach to nonlinear filtering theory was via the "Innovation methods" originally proposed by Kailath and subsequently rigorously developed by Fujisaki et al. [14] in 1972.

As pointed out by Mitter [18], the difficulty with this approach is that the Innovation process is not, in general, explicitly computable except in the well-known Kalman-Bucy filter. In the late 1970s, Brockett and Clark [2], Brockett [3], and Mitter [18] proposed the idea of using estimation algebras to construct a finite-dimensional nonlinear filter. The Lie algebra approach has several advantages. First, it takes into account of geometrical aspects of the situation. Second, it explains convincingly why it is easy to find exact recursive filters for linear dynamical systems, while it is very difficult for some filters like the cubic sensor system described in the work of Hazewinkel et al. [11]. The third, and perhaps the most important, as long as the estimation algebra is finite dimensional, not only can the finite-dimensional recursive filter be constructed explicitly, but also the constructed filter is universal in the sense of [15]. Moreover, the number of sufficient statistics in the Lie algebra method, which requires computing the conditional probability density, is linear in n, where n is the dimension of the state space. However, even in the case of linear filtering with non-Gaussian initial condition, the number of sufficient statistics needed in Makowski's method [16] or Haussman and Pardoux's method [12] is a quadratic polynomial.

In his talk at the International Congress of Mathematics in 1983, Brockett proposed the problem of classifying finite-dimensional estimation algebras (FDEA). Since then, the concept of estimation algebra has been proven to be an invaluable tool in the study of nonlinear filtering problems. Nevertheless, the structure and classification of finite-dimensional estimation algebras were studied in detail only in the early 1990s by series of works by Yau and his collaborators [4–7, 23, 25, 28, 29]. In [27], the concept of Wong's Ω-matrix was introduced and played an important role in algebraic structure. The program of classifying finite-dimensional estimation algebras of maximal rank was begun in 1990 by Yau et al. There are four crucial steps here.

Step 1 In 1990, Yau first observed that Wong's Ω-matrix plays an important role. As the first crucial step, he classifies all finite-dimensional estimation algebras of maximal rank if Wong's matrix has entries in constant coefficients. His result was announced in 1990 [30] and the detail of the proof was published in 1994 [28]. In 1991 paper [31], Chiou and Yau formally introduced the concept of finite-dimensional estimation algebra of maximal rank and gave classification when the state space dimension n is at most 2. Their results were published in 1994 [5].

Step 2 The second crucial step was due to Chen and Yau in 1996 [6]. They developed quadratic structure theory for finite-dimensional estimation algebra. They laid down all the ingredients that are needed to give classification of finite-dimensional estimation algebras of maximal rank. In particular, they introduced the notion of quadratic rank k. In this way, Wong's Ω-matrix is divided into three parts: (1) $\omega_{ij}, 1 \leq i, j \leq k$; (2) $\omega_{ij}, k+1 \leq i, j \leq n$; and (3) $\omega_{ij}, 1 \leq i \leq k, k+1 \leq j \leq n$, or $k+1 \leq i \leq n, 1 \leq j \leq k$. In [6], Chen and Yau proved that part (1) $\omega_{ij}, 1 \leq i, j \leq k$, is a matrix with constant coefficients.

Step 3 In their 1997 published paper [8], Chen, Yau, and Leung proved the weak Hessian matrix nondecomposition theorem for $n \leq 4$. As a result, the part (2) $\omega_{ij}, k+1 \leq i, j \leq n$, is a matrix with constant coefficients. In their 1997 paper [25], Wu, Yau, and Hu proved the weak Hessian matrix nondecomposition theorem for general n. Thus, part (2) $\omega_{ij}, k+1 \leq i, j \leq n$, is also a matrix with constant coefficients for arbitrary n.

6.1 Introduction

Step 4 This final step was also done in 1997. Yau and Hu [29] use full power of the quadratic structure theory developed by Chen and Yau [6] to prove that the part (3) ω_{ij}, $1 \leq i \leq k, k+1 \leq j \leq n$, or $k+1 \leq i \leq n, 1 \leq j \leq k$, are matrix with the constant coefficients.

The above four steps complete the classification issue. Therefore, Yau and his coworkers have proved the following theorem and completely determine the structure of estimation algebra.

Theorem 6.1 (Classification Theorem) *Suppose that the state space of the filtering system is of dimension n. If E is the finite-dimensional estimation algebra with maximal rank, then E is a real vector space of dimension $2n+2$ with basis given by $1, x_1, \cdots, x_n, D_1, \cdots, D_n, L_0$, where $D_i, i = 1, \cdots, n$, and L_0 will be defined in the following section.*

In fact, Mitter and Levine have conjectured the following.

Mitter Conjecture Let E be a finite-dimensional estimation algebra. If ϕ is a function in E, then ϕ is a polynomial of degree at most 1.

Levine Conjecture Let E be a finite-dimensional estimation algebra. The differential operators in E have orders at most 2.

Direct corollary of the work of Yau and his coworkers is that they have proved Mitter conjecture and Levine conjecture in estimation algebra with maximal rank.

The following corollary is a direct immediate consequence of the above classification theorem.

Corollary 6.1 (Sufficient Statistics) *Suppose that the state space of the filtering system is of dimension n. If E is the finite-dimensional estimation algebra with maximal rank, then the number of sufficient statistics in order to compute the conditional density by Lie algebraic methods is linear in n.*

Since the 2000s, Yau and his team have been studying the classification of nonmaximal rank estimation algebra, which is a quite important and difficult problem. General classification of nonmaximal rank case is still an open problem. However, Yau and his coworkers have made profound contributions in this field. In 2006, Wu and Yau [26] have finished the complete classification of FDEA with state dimension 2 and linear rank 1. In [26], Wu and Yau developed many powerful tools to solve the structure of Wong's Ω-matrix in nonmaximal rank estimation algebra. In 2017, Shi et al. [20] construct a new class of finite-dimensional filters with state dimension 3 and linear rank 1, in which Wong's Ω-matrix is not necessary to be a constant matrix. Shi et al. give some sufficient condition that can make estimation algebra be finite dimensional. In 2018, Shi and Yau [21, 22] study the structure of finite-dimensional estimation algebra with state dimension 3 and linear rank 2, and they can prove Wong's Ω-matrix has linear structure and Mitter conjecture holds in this case. In their work, they develop many techniques to construct infinite operator sequence in the estimation algebra. Recently, Dong et al. [10] construct a new class of finite-dimensional filtering systems with state space dimension 4 and linear rank 1, in which entries of Wong's Ω-matrix can be polynomials of any degree.

The following is the layout of this chapter: In Sect. 6.2, we introduce some basic concepts and notations of estimation algebra. In Sect. 6.3, the algebraic classification of finite-dimensional filter will be given which includes the complete series of results of maximal rank case. In Sect. 6.4, we introduce the well-known Wei-Norman approach which is a useful tool to construct recursive finite-dimensional filters. In Sect. 6.5, we mainly introduce the important structure results of non-maximal rank case especially under low dimension. In Sect. 6.6, we will introduce construction of novel class of finite-dimensional filters which include the nonlinear filter besides Kalman, Benes, and Yau filters.

6.2 Basic Concepts and Preliminaries

The filtering problem considered here is based on the continuous signal observation model:

$$\begin{cases} dx(t) = f(x(t))dt + g(x(t))dv(t), \ x(0) = x_0, \\ dy(t) = h(x(t))dt + dw(t), \ y(0) = 0. \end{cases} \quad (6.1)$$

Here x, v, y, w are respectively R^n, R^p, R^m, R^m-valued processes, and v and w have components that are independent, standard Brownian motions. We assume that $n = p$ and f, h are C^∞-function, and g is an orthogonal matrix. We refer to $x(t)$ as the state of the system at time t and to $y(t)$ as the observation at time t.

Let $\rho(t, x)$ denote the conditional probability density of the state given the observation $\{y(s) : 0 \leq s \leq t\}$. It is well-known that $\rho(t, x)$ is given by normalizing $\sigma(t, x)$, which satisfies the following Duncan-Mortensen-Zakai (DMZ) equation:

$$\begin{cases} d\sigma(t, x) = L_0 \sigma(t, x)dx + \sum_{i=1}^m L_i \sigma(t, x)dy_i(t), \\ \sigma(0, x) = \sigma_0, \end{cases} \quad (6.2)$$

where

$$L_0 = \frac{1}{2} \sum_{i=1}^n \frac{\partial^2}{\partial x_i^2} - \sum_{i=1}^n f_i \frac{\partial}{\partial x_i} - \sum_{i=1}^n \frac{\partial f_i}{\partial x_i} - \frac{1}{2} \sum_{i=1}^m h_i^2, \quad (6.3)$$

and, for $i = 1, \cdots, m$, L_i is the zero degree differential operator of multiplication by h_i. The term σ_0 is the initial probability density.

DMZ equation is a stochastic partial differential equation. In real applications, we are interested in constructing robust state estimators from observed sample paths with some property of robustness. Davis [9] studies this problem and proposed some

6.2 Basic Concepts and Preliminaries

robust algorithms. In our case, his basic idea reduces to defining a new unnormalized density:

$$u(t, x) = \exp\left(-\sum_{i=1}^{m} h_i(x) y_i(t)\right) \sigma(t, x). \tag{6.4}$$

Davis reduced the DMZ equation to the following time-varying partial differential equation, which is called the robust DMZ equation:

$$\begin{cases} \frac{\partial u}{\partial t}(t, x) = L_0 u(t, x) + \sum_{i=1}^{m} y_i(t)[L_0, L_i] u(t, x) \\ \qquad\qquad + \frac{1}{2} \sum_{i,j=1}^{m} y_i(t) y_j(t)[[L_0, L_i], L_j] u(t, x), \\ u(0, x) = \sigma_0. \end{cases} \tag{6.5}$$

Let

$$D_i := \frac{\partial}{\partial x_i} - f_i, \quad 1 \leq i \leq n,$$

$$\eta := \sum_{i=1}^{n} \frac{\partial f_i}{\partial x_i} + \sum_{i=1}^{n} f_i^2 + \sum_{i=1}^{m} h_i^2, \tag{6.6}$$

and then we can obtain a more compact form of L_0:

$$L_0 = \frac{1}{2} \left(\sum_{i=1}^{n} D_i^2 - \eta \right). \tag{6.7}$$

This can be verified by direct computation which has been set in the exercise.

Definition 6.1 (Lie Bracket) If X and Y are differential operators, the Lie bracket of X and Y, $[X, Y]$, is defined by $[X, Y]\phi = X(Y\phi) - Y(X\phi)$ for any C^∞ function ϕ.

Definition 6.2 (Lie Algebra) A vector space $\mathfrak{g}$ with the Lie bracket operation $\mathfrak{g} \times \mathfrak{g} \to \mathfrak{g}$ denoted by $(x, y) \longmapsto [x, y]$ is called a Lie algebra if the following axioms are satisfied:

(1) The Lie bracket operation is bilinear;
(2) $[x, x] = 0$ for all $x \in \mathfrak{g}$;
(3) $[x, [y, z]] + [y, [z, x]] + [z, [x, y]] = 0$, $x, y, z \in \mathfrak{g}$.

Definition 6.3 (Estimation Algebra) The estimation algebra E of a filtering system (6.1) is defined to be the Lie algebra generated by $\{L_0, L_1, \cdots, L_m\}$, i.e., $E = \langle L_0, h_1, \cdots, h_m \rangle_{L.A.}$.

Definition 6.4 An estimation algebra E is said to be of maximal rank if, for any $1 \leq i \leq n$, there exists a constant c_i, such that $x_i + c_i \in E$.

Definition 6.5 (Wong Ω-Matrix) Wong's Ω-matrix is the matrix $\Omega = (\omega_{ij})$, where

$$\omega_{ij} = \frac{\partial f_j}{\partial x_i} - \frac{\partial f_i}{\partial x_j}, \quad \forall 1 \leq i, j \leq n. \tag{6.8}$$

Obviously, $\omega_{ij} = -\omega_{ji}$, i.e., Ω is an antisymmetric matrix. Furthermore, Ω satisfies the following cyclic conditions:

$$\frac{\partial \omega_{jk}}{\partial x_i} + \frac{\partial \omega_{ki}}{\partial x_j} + \frac{\partial \omega_{ij}}{\partial x_k} = 0, \quad \forall 1 \leq i, j, k \leq n. \tag{6.9}$$

Definition 6.6 Let U be the vector space of differential operators in the following form:

$$A = \sum_{(i_1, i_2, \cdots, i_n) \in I_A} a_{i_1, i_2, \cdots, i_n} D_1^{i_1} D_2^{i_2} \cdots D_n^{i_n}, \tag{6.10}$$

where nonzero functions $a_{i_1, i_2, \cdots, i_n} \in C^\infty(R^n)$ and $I_A \subseteq N^n$ are the finite set of A. For $i = (i_1, i_2, \cdots, i_n) \in N^n$, denote $|i| := \sum_{k=1}^n i_k$. The order of A is defined by $ord(A) := \max_i |i|$.

Definition 6.7 Let U_k be the subspace of U consisting of the elements with order less than or equal to k, where $k \geq 0$. In particular, $U_0 := C^\infty(R^n)$.

Definition 6.8 (Finite-Dimensional Filter) A filter system is called finite dimensional if its corresponding estimation algebra is finite dimensional.

Remark 6.1 The finite dimensionality of E can be measured in terms of the finite order of elements in E. If E is finite dimensional, then the orders of its elements will have an upper bound. In particular, if there exists a sequence of elements $A_j \in E$ such that the orders of A_j's are strictly increasing, E is not finite dimensional. This basic idea provides us an efficient way to detect whether a given filter is finite dimensional. It is also useful for us in studying the structure of FDEA.

Basic Notations Related to Lie Bracket Let $A, B \in E$ and V is a subspace of E. Then we define an equivalence relation $A = B$, mod V if $A - B \in V$. We define adjoint map $Ad : E \times E \to E$ by $Ad_A B = [A, B]$ and $Ad_A^k B = [A, Ad_A^{k-1} B]$. Euler operator is $E_S := \sum_{l \in S} x_l \frac{\partial}{\partial x_l}$, where S is an index subset of $\{1, 2, \cdots, n\}$. Estimation algebra is an operator algebra. The following calculation rule is useful in exploring the algebraic structure.

Simple Example Operator $A = D_1^2 + x_1 D_2 + x_2^2$ can be written as $A = D_1^2$, mod U_1 in short, where U_1 denotes the collection of first-order differential operators in E.

6.2 Basic Concepts and Preliminaries

Lemma 6.1 *Let E be a finite-dimensional estimation algebra and let the D_i's be defined as (6.6). If $l > 0$ and*

$$A = \sum_{|(i_1,i_2,\cdots,i_n)|=l+1} a_{i_1,i_2,\cdots,i_n} D_1^{i_1} D_2^{i_2} \cdots D_n^{i_n}, \quad \mod U_l, \tag{6.11}$$

is in E, then $a_{i_1,i_2,\cdots,i_n}$'s are polynomials.

Proof If not all coefficients $a_{i_1,i_2,\cdots,i_n}$ are polynomials, there exists a variable x_k such that $\frac{\partial^s a_{j_1,\cdots,j_n}}{x_k^s} \neq 0$, $\forall s \geq 0$. Consider the index set $(j_1, \cdots, j_n)$ of all $a_{j_1,\cdots,j_n}$ and assume $a_{j_1,\cdots,j_n}$ is transcendental in x_1 and j_1 is the largest among first indices whose coefficient functions are transcendental in x_1. Let

$$A_1 = [L_0, A]. \tag{6.12}$$

The coefficient function of $D_1^{j_1+1} D_2^{j_2} \cdots D_n^{j_n}$ in A_1 is $\frac{\partial a_{j_1,\cdots,j_n}}{\partial x_1} + \frac{\partial a_{j_1+1,j_2-1,\cdots,j_n}}{\partial x_2} + \frac{\partial a_{j_1+1,j_2,\cdots,j_n-1}}{\partial x_n}$. By the assumption j_1 is the largest first index, it can be deduced that terms $\frac{\partial a_{j_1+1,j_2-1,\cdots,j_n}}{\partial x_2}, \ldots, \frac{\partial a_{j_1+1,j_2,\cdots,j_n-1}}{\partial x_n}$ are all polynomials in x_1. So coefficient function of $D_1^{j_1+1} D_2^{j_2} \cdots D_n^{j_n}$ is transcendental in x_1.

Similarly, let $|(i_1, \ldots, i_n)| = l+2$, $i_1 \geq j_1 + 2$ and consider coefficient functions of $D_1^{i_1} \cdots D_n^{i_n}$ in A_1. By assumption on a's, these functions are all polynomials in x_1.

Thus, from A to A_1, the differential orders increase by 1, while the transcendental structure on x_1 remains unchanged. By keeping this process, $A_{k+1} = [L_0, A_k]$ has order of $l + k + 2 \to \infty$. $\square$

In the following, several elementary results are listed.

Lemma 6.2 *Let E be an estimation algebra for the filtering problem (6.1). $\Omega = (\omega_{ij})$ is defined as in Definition 6.5. Assume $X, Y, Z \in E$ and $g, h \in C^\infty(\mathbb{R}^n)$. Then:*

(1) $[XY, Z] = X[Y, Z] + [X, Z]Y$;
(2) $[gD_i, h] = g\frac{\partial h}{\partial x_i}$;
(3) $[gD_i, hD_j] = gh\omega_{ji} + g\frac{\partial h}{\partial x_i} D_j - h\frac{\partial g}{\partial x_j} D_i$;
(4) $[gD_i^2, h] = 2g\frac{\partial h}{\partial x_i} D_i + g\frac{\partial^2 h}{\partial x_i^2}$;
(5) $[D_i^2, hD_j] = 2\frac{\partial h}{\partial x_i} D_i D_j + 2h\omega_{ji} D_i + \frac{\partial^2 h}{\partial x_i^2} D_j + h\frac{\partial \omega_{ji}}{\partial x_i}$;

Proof

(1)

$$\phi = XYZ\phi - ZXY\phi$$
$$= XYZ\phi - XZY\phi + XZY\phi - ZXY\phi \tag{6.13}$$
$$= X[Y, Z]\phi + [X, Z]Y\phi$$

(2)
$$[gD_i, h] = [g, h]D_i + g[D_i, h] = g\frac{\partial h}{\partial x_i} \quad (6.14)$$

(3)
$$= g[D_i, hD_j] + [g, hD_j]D_i$$
$$= -gh\omega_{ij} + g\frac{\partial h}{\partial x_i}D_j - h\frac{\partial g}{\partial x_j}D_i \quad (6.15)$$

(4)
$$= g[D_i^2, h] + [g, h]D_i^2$$
$$= g\frac{\partial^2 h}{\partial x_i^2} + 2g\frac{\partial h}{\partial x_i}D_i \quad (6.16)$$

(5)
$$= -[hD_j, D_i^2]$$
$$= 2h\omega_{ji}D_i + h\frac{\partial \omega_{ji}}{\partial x_i} + (\frac{\partial^2 h}{\partial x_i^2} + 2\frac{\partial h}{\partial x_i}D_i)D_j \quad (6.17)$$

□

Lemma 6.3

(1) $[L_0, x_j + c_j] = D_j$, $1 \leq j \leq n$;
(2) $[D_i, x_j + c_j] = \delta_{ij}$, $1 \leq i, j \leq n$;
(3) $[D_i, D_j] = \omega_{ji}$, $1 \leq i, j \leq n$;
(4) $Y_j := [L_0, D_j] = \sum_{i=1}^{n}\left(\omega_{ji}D_i + \frac{1}{2}\frac{\partial \omega_{ji}}{\partial x_i}\right) + \frac{1}{2}\frac{\partial \eta}{\partial x_j}$, $1 \leq j \leq n$;
(5) $[Y_j, \omega_{kl}] = \sum_{i=1}^{n} \omega_{ji}\frac{\partial \omega_{kl}}{\partial x_i}$, $1 \leq j, k, l \leq n$;
(6) $[Y_j, D_k] = \sum_{i=1}^{n}\left(\omega_{ji}\omega_{ki} - \frac{\partial \omega_{ji}}{\partial x_k}D_i\right) - \frac{1}{2}\sum_{i=1}^{n}\frac{\partial^2 \omega_{ij}}{\partial x_k \partial x_i} - \frac{1}{2}\frac{\partial^2 \eta}{\partial x_k \partial x_j}$.

These results can be obtained by applying Lemma 6.2 which has been left to readers as exercises.

The following theorem is the first result that allows us to understand what kind of functions can be contained in FDEA.

Theorem 6.2 (Ocone [19]) *Let E be finite-dimensional estimation algebra. If a function ξ is in E, then ξ is a polynomial of degree no more than 2.*

Proof Let $Ad_{L_0}(\xi) = [L_0, \xi]$ and $Ad_{L_0}^k(\xi) = [L_0, Ad_{L_0}^{k-1}(\xi)]$. Then it is easy to see that

6.2 Basic Concepts and Preliminaries

$$Ad_{L_0}^k(\xi) = \sum_{i_1,\cdots,i_k=1}^{n} \frac{\partial^k \xi}{\partial x_{i_1} \cdots \partial x_{i_k}} D_{i_1} \cdots D_{i_k}, \quad \mod U_{k-1} \in E. \tag{6.18}$$

Since $Ad_{L_0}^k(\xi)$ is in E for all k, the finite dimensionality of E implies that $\frac{\partial^k \xi}{\partial x_{i_1} \cdots \partial x_{i_k}} = 0$ for $1 \le x_{i_1} \cdots \partial x_{i_k} \le n$, if k is large enough. It follows that ξ is a polynomial.

Observe that $\xi \in E$ implies $\sum_{i=1}^{n} \left(\frac{\partial \xi}{\partial x_i}\right)^2 = [Ad_{L_0}(\xi), \xi] \in E$. The facts that ξ is a polynomial and E is finite dimensional imply ξ is a polynomial of degree at most 2. □

Next we shall prove a very useful theorem in terms of PDEs appeared in estimation algebra.

Theorem 6.3 *Let $F(x_1, \cdots, x_n)$ be a C^∞ function on R^n. Suppose that there exists a path $c : R \to R^n$ and $\delta > 0$ such that $\lim_{t \to \infty} \|c(t)\| = \infty$ and $\lim_{t \to \infty} \sup_{B_\delta(c(t))} F = -\infty$, where $B_\delta(c(t)) = \{x \in R^n : \|x - c(t)\| < \delta\}$ and sup denotes the supremum value. Then there are no C^∞ functions $f_1, f_2, \cdots, f_n$ on R^n satisfying the following equation:*

$$\sum_{i=1}^{n} \frac{\partial f_i}{\partial x_i} + \sum_{i=1}^{n} f_i^2 = F. \tag{6.19}$$

Proof Let $\psi \in C_0^\infty$ be any C^∞ function with compact support. Multiplying (6.19) with ψ^2 and integrating the equation over R^n, we get

$$\int_{R^n} (\nabla \cdot f)\psi^2 + \int_{R^n} \psi^2(f \cdot f) = \int_{R^n} F\psi^2, \tag{6.20}$$

where $f = (f_1, \cdots, f_n)$ and $\nabla \cdot f = \sum_{i=1}^{n} \frac{\partial f_i}{\partial x_i}$. In view of divergence theorem, we have

$$\int_{R^n} F\psi^2 = -\int_{R^n} 2\psi \nabla \psi \cdot f + \int_{R^n} \psi^2(f \cdot f). \tag{6.21}$$

By Schwartz inequality, we have

$$\int_{R^n} 2\psi \nabla \psi \cdot f = \int_{R^n} 2\nabla \psi \cdot (\psi f)$$

$$\le 2 \left(\int_{R^n} |\nabla \psi|^2\right)^{\frac{1}{2}} \left(\int_{R^n} \psi^2(f \cdot f)\right)^{\frac{1}{2}} \tag{6.22}$$

$$\le \int_{R^n} |\nabla \psi|^2 + \int_{R^n} \psi^2(f \cdot f).$$

Then we have

$$\int_{R^n} F\psi^2 \geq -\int_{R^n} |\nabla \psi|^2. \tag{6.23}$$

Therefore, we get

$$\int_{R^n} F\psi^2 + \int_{R^n} |\nabla \psi|^2 \geq 0, \tag{6.24}$$

for all $\psi \in C_0^\infty$. Take any nonzero C^∞ function θ with compact support in the ball $B_\delta(0)$ of radius δ. Define ψ to be θ followed by a translation by $c(t)$. Observe that $\int_{R^n} |\nabla \psi|^2$ is an integral of fixed function over R^n and independent of the translation selected. $\int_{R^n} F\psi^2 \to -\infty$ as $t \to \infty$ by our assumption. This leads to a contradiction to (6.24). □

6.3 Algebraic Classification of Finite-Dimensional Filter

6.3.1 *Maximal Rank Classification: Structures of Quadratic Forms*

We shall first recall the theory of quadratic forms in estimation algebras developed by Chen and Yau [5].

Let Q be the space of quadratic forms in n variables, that is, real vector space spanned by $x_i x_j$, $1 \leq i, j \leq n$. Let $M_n(R)$ be the set of $n \times n$ matrices. In the following, we give the definition of quadratic rank which describes the quadratic polynomials contained in EA.

Definition 6.9 For any quadratic form $p \in Q$, there exists a symmetric matrix A such that $p(x) = x^T A x$, where $x = (x_1, \cdots, x_n)^T$. The rank of the quadratic form p is denoted by $r(p)$ and is defined to be the rank of the matrix A.

Definition 6.10 (Quadratic Rank) A fundamental quadratic form of the estimation algebra E is an element $p_0 \in E \cap Q$ with the greatest positive rank, that is, $r(p_0) \geq r(p)$ for any $p \in E \cap Q$. The maximal rank of quadratic forms in E is defined to be $k = r(p_0)$ and is called the quadratic rank of estimation algebra.

After an orthogonal transformation on variable x, p_0 can be written as

$$p_0 = c_1 x_1^2 + \cdots + c_k x_k^2, \quad c_i \neq 0, \quad 0 \leq k \leq n. \tag{6.25}$$

From $p_0(x)$, we can construct a sequence of quadratic forms in $E \cap Q$ as follows:

6.3 Algebraic Classification of Finite-Dimensional Filter

$$\begin{cases} q_0(x) = p_0(x) \\ q_j(x) = [[L_0, q_{j-1}], q_0] = \sum_{i=1}^{k} 4^j c_i^{j+1} x_i^2. \end{cases} \tag{6.26}$$

In view of the invertibility of the Vandermonde matrix, we can assume that

$$p_0(x) = x_1^2 + x_2^2 + \cdots + x_k^2 \in E. \tag{6.27}$$

In the following lemma, we show that any quadratic polynomial in the estimation algebra only depends on the variables $x_1, \cdots, x_k$.

Lemma 6.4 *If p is a quadratic form in E, then p is independent of x_j for $j > k$, where $k = r(p_0)$. In other words,*

$$\frac{\partial p}{\partial x_j} = 0$$

for $k + 1 \leq j \leq n$.

Proof Suppose on the contrary that $\frac{\partial p}{\partial x_j} \neq 0$ for some $j > k$. Let A be a symmetric matrix such that $p = X^T A X$. A can be written as

$$A = \begin{pmatrix} A_1 & A_2 \\ A_2^T & A_4 \end{pmatrix}, \tag{6.28}$$

where $A_1 \in R^{k \times k}$, $A_4 \in R^{(n-k) \times (n-k)}$ are symmetric. There is a $k \times k$ orthogonal matrix S_1 and an $(n-k) \times (n-k)$ orthogonal matrix S_2 such that $S_1^T A_1 S_1$ and $S_2^T A_4 S_2$ are diagonal matrices. So we can assume that A_1 and A_4 are diagonal matrices. By condition, $\frac{\partial p}{\partial x_j} \neq 0$ for some $j > k$ implies that $A_2 \neq 0$ or $A_4 \neq 0$. Since

$$r(\lambda p_0 + p) = rank \begin{pmatrix} \lambda I + A_1 & A_2 \\ A_2^T & A_4 \end{pmatrix}, \tag{6.29}$$

if we choose λ large enough, it is easy to prove that

$$r(\lambda p_0 + p) > k. \tag{6.30}$$

This contradicts the greatest positive rank assumption of p_0. □

After we found the quadratic polynomial with the greatest quadratic rank, what we are still interested in is polynomial with least quadratic rank in E. Let $p_1 \in E \cap Q$ be an element with least positive rank, that is, $0 < r(p_1) \leq r(q)$ for any nonzero $q \in E \cap Q$. After an orthogonal transformation that fixes $x_{k+1}, \cdots, x_n$ variables then take the Vandermonde matrix procedure repeatedly as Eq. (6.26), we can assume without loss of generality,

$$p_1 = \sum_{i=1}^{k_1} x_i^2 \in E, \quad 1 \le k_1 \le k. \tag{6.31}$$

Note that the orthogonal transformation on $x_1, \cdots, x_n$ leaves p_0 invariant. In summary, we deduce that $p_0 = \sum_{i=1}^{k} x_i^2$ has the greatest positive rank and $p_1 = \sum_{i=1}^{k_1} x_i^2$ has the least positive rank. Define

$$S_1 = \{1, 2, \cdots, k_1\} \subseteq S = \{1, \cdots, k\}, \tag{6.32}$$

and $Q_1 =$ real vector space spanned by $\{x_i x_j : k_1 + 1 \le i \le j \le k\} \subseteq Q$.

If $k_1 < k$, then $Q_1 \cap E$ is a nontrivial space. In a similar procedure as above, there exists

$$p_2 = \sum_{i=k_1+1}^{k_2} x_i^2 \in E \cap Q_1, \tag{6.33}$$

with the least positive rank in $E \cap Q_1$. By induction, we construct a series of tuples $\{S_i, Q_i, p_i\}$ shown as below:

$$S_i = \{k_{i-1} + 1, \cdots, k_i\}, k_0 = 0, k_i \le k, \tag{6.34}$$

and $Q_i =$ real vector space spanned by $\{x_l x_j : k_i + 1 \le l \le j \le k\} \subseteq Q$:

$$p_i = \sum_{i=k_{i-1}+1}^{k_i} x_i^2 = \sum_{j \in S_i} x_j^2 \in E \cap Q_{i-1}, i > 0, \tag{6.35}$$

and p_i has the least positive rank in $E \cap Q_{i-1}$ for $i > 0$.

In the following, after we get the series of fundamental quadratic polynomials $\{p_i\}$, those can be utilized to describe the structure of any functions in E. Corresponding results are shown as follows.

Lemma 6.5 *If* $p \in E \cap Q$, *then*

$$p(0, \cdots, 0, x_{k_{i-1}+1}, \cdots, x_{k_i}, 0, \cdots, 0) = \lambda p_i, \text{ for } i > 0, \text{ where } \lambda \text{ is a real constant.} \tag{6.36}$$

Proof In view of Lemma 6.2 and the fact that $[L_0, p_i] \in E, [L_0, p_0 - p_i] \in E$, we have

$$\sum_{j \in S_i} x_j D_j \in E, \quad \sum_{j \in S-S_i} x_j D_j \in E. \tag{6.37}$$

Hence,

6.3 Algebraic Classification of Finite-Dimensional Filter

$$\left[\sum_{j \in S_i} x_j D_j, p\right] - \left[\sum_{j \in S-S_i} x_j D_j, \left[\sum_{j \in S_i} x_j D_j, p\right]\right] \quad (6.38)$$

$$= 2p(0, \cdots, 0, x_{k_{i-1}+1}, \cdots, x_{k_i}, 0, \cdots, 0) \in E.$$

Because p_i has the least positive rank for quadratic polynomials in $x_{k_{i-1}+1}, \cdots, x_{k_i}$, the matrix corresponding to $p(0, \cdots, 0, x_{k_{i-1}+1}, \cdots, x_{k_i}, 0, \cdots, 0)$ has $(k_i - k_{i-1})$ same eigenvalues. Then we deduce there is a λ such that

$$p(0, \cdots, 0, x_{k_{i-1}+1}, \cdots, x_{k_i}, 0, \cdots, 0) = \lambda p_i. \quad (6.39)$$

That finishes the whole proof. □

Lemma 6.6 *If* $p \in E \cap Q$, *then*

$$p(x_1, \cdots, x_{k_{i-1}}, 0, \cdots, 0, x_{k_i+1}, \cdots, x_k) \in E, \text{ for } i > 0. \quad (6.40)$$

Proof

$$p(x_1, \cdots, x_{k_{i-1}}, 0, \cdots, 0, x_{k_i+1}, \cdots, x_k)$$

$$= p - \left[\sum_{j \in S-S_i} x_j D_j, \left[\sum_{j \in S_i} x_j D_j, p\right]\right] \quad (6.41)$$

$$- p(0, \cdots, 0, x_{k_{i-1}+1}, \cdots, x_{k_i}, 0, \cdots, 0) \in E.$$

This lemma follows immediately from the above formula. □

The following lemma is a corollary of Lemma 6.5.

Lemma 6.7 *Let* $p = \sum_{i \in S_{l_1}} \sum_{j \in S_{l_2}} 2a_{ij} x_i x_j \in E$, *where* $a_{ij} \in R$ *and* $l_1 < l_2$. *Let* $X_i = (x_{k_{i-1}+1}, \cdots, x_{k_i})^T$ *be a* $(k_i - k_{i-1})$*-vector. Under this notation, p can be written as*

$$p = (X_{l_1}^T, X_{l_2}^T) \begin{pmatrix} 0 & A \\ A^T & 0 \end{pmatrix} \begin{pmatrix} X_{l_1} \\ X_{l_2} \end{pmatrix}. \quad (6.42)$$

Then $|S_{l_1}| = |S_{l_2}|$ *and* $A = bT$, *where b is a constant and T is an orthogonal matrix.*

Proof Direct calculations imply

$$[[L_0, p], p] = 4 \sum_{i,m \in S_{l_1}} \sum_{j \in S_{l_2}} a_{ij} a_{mj} x_i x_m + 4 \sum_{i \in S_{l_1}} \sum_{j,l \in S_{l_2}} a_{ij} a_{il} x_j x_l \in E. \quad (6.43)$$

Lemma 6.5 can be applied here and leads to the following:

$$\sum_{i,m \in S_{l_1}} \sum_{j \in S_{l_2}} a_{ij} a_{mj} x_i x_m = \lambda_1 p_{l_1}$$
$$\sum_{j,l \in S_{l_2}} \sum_{i \in S_{l_1}} a_{ij} a_{il} x_j x_l = \lambda_2 p_{l_2}$$
(6.44)

The above two equations imply that rows of A are orthogonal so are columns. Since the row rank is the same as the column rank for any matrix, A must be a square matrix which implies $|S_{l_1}| = |S_{l_2}|$. As the column vectors have the same length, A is a constant multiple of an orthogonal matrix. □

If E is a finite-dimensional estimation algebra with maximal rank, then by Ocone's theorem, $\omega_{ij} \in E$ is a polynomial of degree at most 2 for all $1 \leq i, j \leq n$. Let $\omega_{ij}^{(2)}, \omega_{ij}^{(1)}$ be the homogeneous part of degree 2 and 1 of ω_{ij}, respectively. Then we have the following lemma.

Lemma 6.8 *Suppose that E is a finite-dimensional estimation algebra of maximal rank. Then:*

(i) $\omega_{ij}^{(2)}$ *depends only on* $x_1, \cdots, x_k$, *for* $i \leq k$ *or* $j \leq k$;

(ii) $\omega_{ij}^{(2)} = 0$, *for* $k + 1 \leq i, j \leq n$;

(iii) $\frac{\partial \omega_{ij}^{(2)}}{\partial x_l} + \frac{\partial \omega_{jl}^{(2)}}{\partial x_i} + \frac{\partial \omega_{li}^{(2)}}{\partial x_j} = 0, \quad \forall 1 \leq i, j, l \leq n$;

(iv) $\frac{\partial \omega_{ij}^{(1)}}{\partial x_l} + \frac{\partial \omega_{jl}^{(1)}}{\partial x_i} + \frac{\partial \omega_{li}^{(1)}}{\partial x_j} = 0, \quad \forall 1 \leq i, j, l \leq n$.

Proof Since E is finite dimensional of maximal rank and $\omega_{ij} \in E$, it follows that $\omega_{ij}^{(2)} \in E$. Hence, $\omega_{ij}^{(2)} \in E$ depends only on $x_1, \cdots, x_k$ by Lemma 6.4. The cyclic conditions of part (iii) and part (iv) of this lemma follow from the corresponding cyclic conditions:

$$\frac{\partial \omega_{ij}}{\partial x_l} + \frac{\partial \omega_{jl}}{\partial x_i} + \frac{\partial \omega_{li}}{\partial x_j} = 0, \quad \forall 1 \leq i, j, l \leq n. \tag{6.45}$$

Let $k + 1 \leq i, j \leq n$, and $1 \leq l \leq k$. Then (iii) gives $\frac{\partial \omega_{ij}^{(2)}}{\partial x_l} = 0$. It follows that $\omega_{ij}^{(2)} = 0$ for $k + 1 \leq i, j \leq n$. □

The following three theorems about the Euler operator are due to Yau and Rasoulian [32].

Theorem 6.4 *Let $E_l = \sum_{j=1}^{l} x_j \frac{\partial}{\partial x_j}$ be an Euler operator in $x_1, \cdots, x_l$ variables. Suppose that m is an integer and ξ is a C^∞ function on R^n such that $E_l(\xi) + m\xi$ is a polynomial of degree k, k a positive integer, in $x_1, \cdots, x_l$ variables with coefficients in C^∞ function of $x_{l+1}, \cdots, x_n$ variables.*

6.3 Algebraic Classification of Finite-Dimensional Filter   183

(i) If $k + m \geq 0$, then ξ is a polynomial of degree k in $x_1, \cdots, x_l$ variables with coefficients in C^∞ functions of $x_{l+1}, \cdots, x_n$.

(ii) If $k + m < 0$, then ξ is a polynomial of degree at most $-m$ in $x_1, \cdots, x_l$ variables with coefficients in C^∞ functions of $x_{l+1}, \cdots, x_n$.

Proof First, let $k + m \geq 0$. Also let $D := (\frac{\partial}{\partial x_1})^{\alpha_1} \cdots (\frac{\partial}{\partial x_l})^{\alpha_l}, \alpha_1 + \ldots + \alpha_l = k + 1$ be a differential operator. By assumption that $E_l(\xi) + m\xi$ is a polynomial of degree k in $x_1, \ldots, x_l$, then we have $D[E_l(\xi) + m\xi] = 0$. It can be directly calculated that

$$D[E_l(\xi) + m\xi] = E_l(D\xi) + (\alpha_1 + \ldots + \alpha_l + m)D\xi. \tag{6.46}$$

So $E_l(D\xi) + (k + 1 + m)D\xi = 0$. Observe

$$E_l[x_1^{k+1+m} D\xi] = x_1^{k+1+m}[E_l(D\xi) + (k + 1 + m)D\xi] = 0. \tag{6.47}$$

Denote $\phi = x_1^{k+1+m} D\xi$. Because $k + 1 + m > 0$, we have

$$\phi(x_1, \ldots, x_l, \ldots, x_n) - \phi(\epsilon x_1, \ldots, \epsilon x_l, \ldots, x_n)$$

$$= \int_\epsilon^1 \frac{1}{t}(E_l\phi)(tx_1, \ldots, tx_l, x_{l+1}, \ldots, x_n)dt$$

$$= 0. \tag{6.48}$$

For $\epsilon > 0$. Now, let $\epsilon \to 0$ and we get $\phi = 0$ which implies that $D\xi = 0$. In other words, ξ is a polynomial of degree at most k in $x_1, \ldots, x_l$ with coefficients in smooth function $x_{l+1}, \ldots, x_n$. Next, we proceed to prove that ξ is a polynomial of degree at most k. The main idea is to use induction on k and using the same method as above. □

Theorem 6.5 *Let $E_S := \sum_{l \in S} x_l \frac{\partial}{\partial x_l}$ be an Euler operator, where S is a subset of index $\{1, 2, \cdots, n\}$. $P_k(x)$ denotes the set of polynomials of degree no more than k in variable $x_1, \cdots, x_n$. Assume $\zeta \in C^\infty(\mathbb{R}^n)$ and m is a positive constant. If $E_S(\zeta) + m\zeta \in P_k(x)$, then $\zeta \in P_k(x)$.*

Proof For simplicity of expression, we let $S = \{1, 2, \cdots, l\} \subset \{1, 2, \cdots, n\}$. Our proof includes two parts.

Step 1. ζ is a polynomial in variables $x_1, \cdots, x_l$ of degree k with smooth coefficients of $x_{l+1}, \cdots, x_n$.

Next we define multi-index $\beta = (\beta_1, \cdots, \beta_l)$ with $|\beta| = k + 1$ and differential operator $D := (\frac{\partial}{\partial x_1})^{\beta_1} \cdots (\frac{\partial}{\partial x_l})^{\beta_l}$. The only thing we need to do is to show that $D(\zeta) = 0$.

It is obvious that $D[E_S(\zeta) + m\zeta] = 0$. Next we need to simplify term $DE_S(\zeta)$ by exchanging order of operators D and E_S. First, we have following rules by basic computation:

$$\left(\frac{\partial}{\partial x_j}\right)^p E_S = E_S \left(\frac{\partial}{\partial x_j}\right)^p + p\left(\frac{\partial}{\partial x_j}\right)^p, \quad \text{for } j \in S \text{ and } p \in \mathbb{Z}_+. \tag{6.49}$$

By using the above relations and from $D[E_S(\zeta) + m\zeta] = 0$, we get

$$\begin{aligned} 0 &= D[E_S(\zeta) + m\zeta] \\ &= \left(\frac{\partial}{\partial x_{s_1}}\right)^{\beta_1} \cdots \left(\frac{\partial}{\partial x_{s_l}}\right)^{\beta_l} E_S(\zeta) + mD\zeta \\ &= E_S(D\zeta) + (|\beta| + m)D\zeta. \end{aligned} \tag{6.50}$$

Next, we notice following operator rules:

$$\begin{aligned} E_S(x_j)^p &= x_j^p E_S + p x_j^p, \quad \text{for } j \in S \text{ and } p \in \mathbb{Z}_+ \\ &= x_j^p (E_S + p). \end{aligned} \tag{6.51}$$

Then by using (6.51), we get

$$0 = x_1^{k+m+1}[E_S(D\zeta) + (k+m+1)D\zeta] = E_S(x_1^{k+m+1}D\zeta). \tag{6.52}$$

Next, we define $\phi(x) = x_1^{k+m+1} D\zeta$ and our goal is to prove $\phi \equiv 0$. It will derive $D\zeta = 0$ directly. Notice

$$\begin{aligned} 0 &= \int_\varepsilon^1 \frac{1}{t} E_S(\phi(tx_1, \cdots, tx_l, x_{l+1}, \cdots, x_n))dt \\ &= \int_\varepsilon^1 \frac{d\phi}{dt}(tx_1, \cdots, tx_l, x_{l+1}, \cdots, x_n)dt \\ &= \phi(x) - \phi(\varepsilon x_1, \cdots, \varepsilon x_l, x_{l+1}, \cdots, x_n), \end{aligned} \tag{6.53}$$

which yields that $\phi(x) = \phi(0, \cdots, 0, x_{l+1} \cdots x_n) = 0$ by letting $\varepsilon \to 0$.

Step 2. ζ is a polynomial of degree at most k. In the following, we shall assume

$$\zeta = \sum_{0 \le |\alpha| \le k} a_\alpha(x_{l+1}, \cdots, x_n) x_1^{\alpha_1} \cdots x_l^{\alpha_l}, \quad a_\alpha(x_{l+1}, \cdots, x_n) \in C^\infty(\mathbb{R}^{n-l}) \tag{6.54}$$

and

$$E_S(\zeta) + m\zeta = \sum_{0 \le |\alpha| \le k} b_\alpha(x_{l+1}, \cdots, x_n) x_1^{\alpha_1} \cdots x_l^{\alpha_l}, \tag{6.55}$$

6.3 Algebraic Classification of Finite-Dimensional Filter

where $b_\alpha(x_{l+1}, \cdots, x_n)$ is a polynomial of degree at most $k - |\alpha|$. By (6.54) to (6.55), $a_\alpha(x_{l+1}, \cdots, x_n)$ can be shown as a polynomial with degree at most $k - |\alpha|$ by comparing coefficients on both sides. □

Theorem 6.6 *Let $E_k = \sum_{j=1}^{k} x_j \frac{\partial}{\partial x_j}$ be an Euler operator in $x_1, \cdots, x_k$ variables. Suppose m is a positive integer and ξ is a C^∞ function on R^n such that $E_k(\xi) + m\xi$ is a polynomial of degree r in $x_1, \cdots, x_n$ variables. Then ξ is a polynomial of degree r in $x_1, \cdots, x_n$ variables.*

Proof Due to $m > 0$, then Theorem 6.4 (i) holds. We have

$$\xi = \sum_{0 \leq |\alpha| \leq r} a_\alpha(x_{k+1}, \cdots, x_n) x_1^{\alpha_1} \cdots x_k^{\alpha_k}, \qquad (6.56)$$

where $\alpha = (\alpha_1, \cdots, \alpha_k)$ and $|\alpha| = \alpha_1 + \cdots + \alpha_k$ and $a_\alpha(x_{k+1}, \cdots, x_n)$ is C^∞.

Next we only need to prove $a_\alpha(x_{k+1}, \cdots, x_n)$'s are polynomials. We calculate

$$E_k(\xi) + m\xi = \sum_{0 < |\alpha| \leq r} (|\alpha| + m) a_\alpha(x_{k+1}, \cdots, x_n) x_1^{\alpha_1} \cdots x_k^{\alpha_k}$$

$$+ m a_0(x_{k+1}, \cdots, x_n) \qquad (6.57)$$

$$= \sum_{0 < |\alpha| \leq r} p_\alpha(x_{k+1}, \cdots, x_n) x_1^{\alpha_1} \cdots x_k^{\alpha_k},$$

where $p_\alpha(x_{k+1}, \cdots, x_n)$'s are polynomials in $x_{k+1}, \cdots, x_n$. Now, looking at both sides, we conclude that $(|\alpha| + m) a_\alpha = p_\alpha$, for all $\alpha = (\alpha_1, \cdots, \alpha_k)$, $0 < |\alpha| \leq r$; in other words, all a_α, $0 < |\alpha| \leq r$ are polynomials and also $a_0 = \frac{p_0}{m}$ is a polynomial, and hence ξ is a polynomial. □

Remark 6.2 Theorem 6.6 is false if $m = 0$. It is possible that $E_k(\xi)$ is a polynomial of degree r in $x_1, \cdots, x_n$ variables, but ξ is not a degree r polynomial in $x_1, \cdots, x_n$. For example, we can simply take ξ to be any degree r polynomial in $x_1, \cdots, x_n$ plus a C^∞ function in $x_{k+1}, \cdots, x_n$.

Theorem 6.7 *Let $E_k = \sum_{j=1}^{k} x_j \frac{\partial}{\partial x_j}$ be an Euler operator in $x_1, \cdots, x_k$ variables. Suppose that ξ is a C^∞ function on R^n such that $E_k(\xi)$ is a polynomial of degree r in $x_1, \cdots, x_n$ variables. Then $\xi = P_r(x_1, \cdots, x_n) + a(x_{k+1}, \cdots, x_n)$ where $P_r(x_1, \cdots, x_n)$ is a polynomial of degree r and $a(x_{k+1}, \cdots, x_n)$ is a C^∞ function in $x_{k+1}, \cdots, x_n$.*

Proof In view of Theorem 6.4,

$$\xi = \sum_{0 \leq |\alpha| \leq r} a_\alpha(x_{k+1}, \cdots, x_n) x_1^{\alpha_1} \cdots x_k^{\alpha_k}, \qquad (6.58)$$

where $\alpha = (\alpha_1, \cdots, \alpha_k)$ and $|\alpha| = \alpha_1 + \cdots + \alpha_k$ and $a_\alpha(x_{k+1}, \cdots, x_n)$ is C^∞. Then $E_k(\xi) = \sum_{0<|\alpha|\leq r} |\alpha| a_\alpha(x_{k+1}, \cdots, x_n) x_1^{\alpha_1} \cdots x_k^{\alpha_k}$, which is a polynomial of degree r in $x_1, \cdots, x_n$. Therefore, $a_\alpha(x_{k+1}, \cdots, x_n)$ for $|\alpha| \geq 1$, are polynomials. Theorem 6.7 follows immediately. □

Lemma 6.9 *Let E be a finite-dimensional estimation algebra of maximal rank. Let k be the quadratic rank of E. For $1 \leq i, j \leq n$, ω_{ij}, and $\alpha_i = \sum_{j=1}^{k} x_j \omega_{ij} \in E$ are polynomials of degree 2 in $x_1, \cdots, x_n$ variables. Furthermore, we have the following relationships:*

(i) $E_k(\omega_{ij}) + 2\omega_{ij} = \frac{\partial \alpha_i}{\partial x_j} - \frac{\partial \alpha_j}{\partial x_i}, \forall 1 \leq i, j \leq k;$

(ii) $E_k(\omega_{ij}) + \omega_{ij} = \frac{\partial \alpha_i}{\partial x_j} - \frac{\partial \alpha_j}{\partial x_i}, \forall 1 \leq i \leq k, k+1 \leq j \leq n;$

(iii) $E_k(\omega_{ij}) + \omega_{ij} = \frac{\partial \alpha_i}{\partial x_j} - \frac{\partial \alpha_j}{\partial x_i}, \forall 1 \leq j \leq k, k+1 \leq i \leq n;$

(iv) $E_k(\omega_{ij}) = \frac{\partial \alpha_i}{\partial x_j} - \frac{\partial \alpha_j}{\partial x_i}, \forall k+1 \leq i, j \leq n.$

Proof First, we have $\omega_{ij} \in E$ and $\alpha_i = \frac{1}{2}[[L_0, D_j], p_0] \in E$ where p_0 is defined by (6.27). By Ocone's theorem, ω_{ij} and α_i are polynomials of degree 2 in $x_1, \cdots, x_n$. Relationships (i)–(iv) follow immediately from the definition of $E_k(\omega_{ij})$ and α_i. For example, we give the proof of (i) here:

$$\frac{\partial \alpha_i}{\partial x_j} = \sum_{l=1}^{k} \frac{\partial (x_l \omega_{il})}{\partial x_j} = \omega_{ij} + \sum_{l=1}^{k} x_l \frac{\partial \omega_{il}}{\partial x_j}$$

$$\frac{\partial \alpha_j}{\partial x_i} = \sum_{l=1}^{k} \frac{\partial (x_l \omega_{jl})}{\partial x_i} = \omega_{ji} + \sum_{l=1}^{k} x_l \frac{\partial \omega_{jl}}{\partial x_i} \quad (6.59)$$

$$\frac{\partial \alpha_j}{\partial x_i} - \frac{\partial \alpha_i}{\partial x_j} = 2\omega_{ji} + \sum_{l=1}^{k} x_l \frac{\partial \omega_{ji}}{\partial x_l} = 2\omega_{ji} + E_k(\omega_{ji}).$$

□

Corollary 6.2 *Suppose that E is a finite-dimensional estimation algebra of maximal rank. Then entries of Wong's Ω matrix are polynomials of degree at most one, i.e.,*

$$\omega_{ij} = P_1(x_1, x_2, \cdots, x_n) \; for \; 1 \leq i, j \leq n. \quad (6.60)$$

Proof This follows from Theorems 6.6, 6.7 and Lemmas 6.9, 6.4. □

In the following, the linear structure of Ω matrix will be derived by applying results of Lemma 6.4 and cyclic condition. Based on this linear property, we shall proceed to prove the left top corner in Ω in fact is a constant matrix. The illustration includes two steps in which first ω_{ij} for $i \in S_p, j \in S_q, p \neq q$ will be proved as constant and then ω_{ij} for $i, j \in S_l$ will be shown constant.

6.3 Algebraic Classification of Finite-Dimensional Filter

Lemma 6.10 ([33]) *Suppose that E is a finite-dimensional estimation algebra of maximal rank. Then,*

$$\Omega = (\omega_{ij}) = \left(\begin{array}{c|c} P_1(x_1, x_2, \cdots, x_k) & P_1(x_1, x_2, \cdots, x_k) \\ P_1(x_1, x_2, \cdots, x_k) & P_1(x_{k+1}, \cdots, x_n) \end{array} \right), \tag{6.61}$$

i.e.,

(i) ω_{ij}'s are polynomials of degree 1 in $x_1, \cdots, x_k$ for $1 \leq i \leq k$ or $1 \leq j \leq k$.
(ii) ω_{ij}'s are polynomials of degree 1 in $x_{k+1}, \cdots, x_n$ for $k+1 \leq i, j \leq n$.

Proof Since $\alpha_i = \sum_{j=1}^{k} x_j \omega_{ij}$ is a quadratic polynomial in E and by Lemma 6.4, it cannot depend on $x_{k+1}, \cdots, x_n$ for $1 \leq i \leq n$. Thus, (i) follows immediately. If $k+1 \leq i, j \leq n$, by using the cyclic relationship

$$\frac{\partial \omega_{ij}}{\partial x_l} + \frac{\partial \omega_{jl}}{\partial x_i} + \frac{\partial \omega_{li}}{\partial x_j} = 0, \quad \forall 1 \leq i, j, l \leq n. \tag{6.62}$$

we have $\frac{\partial \omega_{ij}}{\partial x_l} = 0$ for $1 \leq l \leq k$. This means that ω_{ij}'s are independent of $x_1, \cdots, x_k$ for $k+1 \leq i, j \leq n$. □

Lemma 6.11 ([33]) *Suppose that E is a finite-dimensional estimation algebra of maximal rank. With the same notation in Lemma 6.9, if*

$$\sum_{i \in S_l} x_i \alpha_i = 0, \tag{6.63}$$

where α_i's are homogeneous polynomials of degree 2 in E, then $\alpha_i = 0$ for all $i \in S_l$.

Proof Let $X_i = (x_{k_{i-1}+1}, \cdots, x_{k_i})^T$ and $X = (x_1, x_2, \cdots, x_k)^T$. Without loss of generality, we assume that $l = 1$. Let $X^T = (X_1^T, \bar{X}_1^T)$ where $\bar{X}_1$ is the complementing variable of X_1 in X. Write

$$\alpha_i(X) = \alpha_i(X_1, 0) + \alpha_i(0, \bar{X}_1) + [\alpha_i - \alpha_i(X_1, 0) - \alpha_i(0, \bar{X}_1)]. \tag{6.64}$$

Hence, (6.63) is still true if we replace α_i in (6.63) by one of the three terms on the right-hand side of (6.64). We can see immediately that

$$\alpha_i(0, \bar{X}_1) = 0 \quad \forall i \in S_i. \tag{6.65}$$

By Lemma 6.5, we have

$$\alpha_i(X_1, 0) = \lambda_i p_1. \tag{6.66}$$

So the corresponding Eq. (6.63) for $\alpha_i(X_1, 0)$ gives

$$\sum_{i \in S_l} x_i \lambda_i p_1 = 0. \tag{6.67}$$

It follows that $\lambda_i = 0$, that is,

$$\alpha_i(X_1, 0) = 0 \quad \forall i \in S_1. \tag{6.68}$$

Therefore, $\alpha_i - \alpha_i(X_1, 0) - \alpha_i(0, \bar{X}_1) = \sum_{l \geq 2} X_1^T B_{il} X_l \in E$, where $B_{il} \in R^{|S_1| \times |S_l|}$. By Lemma 6.6, we can deduce $X_1^T B_{il} X_l \in E$ for $l \geq 2$. By Lemma 6.7, we have $|S_1| = |S_l|$, $l \geq 2$ and $X_1^T B_{il} X_l = 2X_1^T R_{il} X_l$, where R_{il} is a constant multiple of an orthogonal matrix. Therefore, the corresponding equation of (6.63) for $\alpha_i - \alpha_i(X_1, 0) - \alpha_i(0, \bar{X}_1)$ gives

$$\sum_{l \geq 2} X_1^T \left(\sum_{i \in S_1} 2x_i R_{il} \right) X_l = \sum_{i \in S_1} x_i \sum_{l \geq 2} 2X_1^T R_{il} X_l = 0. \tag{6.69}$$

This implies

$$X_1^T \left(\sum_{i \in S_1} 2x_i R_{il} \right) = 0 \quad \forall l \geq 2. \tag{6.70}$$

Fix $i_0 \in S_1$ and let $x_{i_0} = 1$ and $x_i = 0$ for $i \neq i_0$. Then we obtain

$$(0, \cdots, 0, 1, 0, \cdots, 0) R_{i_0 l} = 0 \quad l \geq 2. \tag{6.71}$$

Since $R_{i_0 l}$ is a constant multiple of an orthogonal matrix, we see that $R_{i_0 l} = 0$, $\forall l \geq 2$. This is true for all $i_0 \in S_1$. Thus,

$$\alpha_i - \alpha_i(X_1, 0) - \alpha_i(0, \bar{X}_1) = 0. \tag{6.72}$$

So we have proved $\alpha_i = 0$. □

Theorem 6.8 ([33]) *Suppose that E is a finite-dimensional estimation algebra of maximal rank. With the same notation in Lemma 6.9, if $p \neq q$ and $i \in S_p$, $j \in S_q$, then ω_{ij} is a constant.*

Proof Recall that Lemma 6.5, and we have $\sum_{i \in S_p} x_i D_i$ and $\sum_{j \in S_q} x_j D_j$ in E. Hence,

$$\sum_{i \in S_p} \sum_{j \in S_q} x_i x_j \omega_{ij}^{(1)} = \sum_{i \in S_p} x_i \left(\sum_{j \in S_q} x_j \omega_{ij}^{(1)} \right) = \sum_{j \in S_q} x_j \left(\sum_{i \in S_p} x_i \omega_{ij}^{(1)} \right) = 0. \tag{6.73}$$

6.3 Algebraic Classification of Finite-Dimensional Filter

Hence, $\omega_{ij}^{(1)}$ depends on x_m, where $m \in S_p \cup S_q$ for $i \in S_p$ and $j \in S_q$. Since E is of maximal rank, $D_j \in E$ for any j. In particular, we have

$$-\left[\sum_{i \in S_p} x_i D_i, D_j\right] = \sum_{i \in S_p} x_i \omega_{ij} \in E \quad for \ j \in S_q,$$

$$\left[\sum_{j \in S_q} x_j D_j, D_i\right] = \sum_{j \in S_q} x_j \omega_{ij} \in E \quad for \ i \in S_p. \tag{6.74}$$

Then we have

$$\sum_{i \in S_p} x_i \omega_{ij}^{(1)} \in E \quad for \ j \in S_q,$$

$$\sum_{j \in S_q} x_j \omega_{ij}^{(1)} \in E \quad for \ i \in S_p. \tag{6.75}$$

By Lemma 6.11, we obtain

$$\sum_{i \in S_p} x_i \omega_{ij}^{(1)} = 0 \quad for \ j \in S_q,$$

$$\sum_{j \in S_q} x_j \omega_{ij}^{(1)} = 0 \quad for \ i \in S_p. \tag{6.76}$$

The first equation says that, for $i \in S_p$, $j \in S_q$, $\omega_{ij}^{(1)}$ does not depend on the variable x_m for $m \in S_q$. The second equation says that, for $i \in S_p$, $j \in S_q$, $\omega_{ij}^{(1)}$ does not depend on the variable x_m for $m \in S_p$. Hence, $\omega_{ij}^{(1)} = 0$. □

Theorem 6.9 ([33]) *Suppose that E is a finite-dimensional estimation algebra of maximal rank. With the same notation in Lemma 6.9, if $i, j \in S_l$, then ω_{ij} is a constant.*

Proof Without loss of generality, we shall assume that $l = 1$. For $1 \leq i \leq k_1$, $\alpha_i = \frac{1}{2}[[L_0, D_i], p_0] = \sum_{j=1}^{k} x_j \omega_{ij} \in E$. Since E is of maximal rank, then $\alpha_i^{(2)} = \sum_{j=1}^{k} x_j \omega_{ij}^{(1)} \in E$. By Lemma 6.5, we have

$$\alpha_i^{(2)}(x_1, \cdots, x_{k_1}, 0, \cdots, 0) = \sum_{j=1}^{k_1} x_j \omega_{ij}^{(1)}(x_1, \cdots, x_{k_1}, 0, \cdots, 0) = \lambda \sum_{i=1}^{k_1} x_i^2 \in E. \tag{6.77}$$

Since ω_{ij} is a degree 1 polynomial in $x_1, \cdots, x_k$ for $1 \leq i, j \leq k$, we can write

$$\omega_{ij}^{(1)} = \sum_{l=1}^{k} A_l(i, j) x_l. \tag{6.78}$$

Then we substitute it into the previous equation:

$$\sum_{j,l=1}^{k_1} A_l(i, j) x_j x_l = \lambda \sum_{i=1}^{k_1} x_i^2. \tag{6.79}$$

Next, we discuss the case of $1 \leq i, j, l \leq k_1$. The above equation implies

$$A_l(i, j) = 0 \ for \ 1 \leq j \neq l \leq k_1, 1 \leq i \leq k_1. \tag{6.80}$$

For $j = l$ case, if $i \neq l$, by Eq. (6.80), we have

$$A_l(i, l) = -A_l(l, i) = 0. \tag{6.81}$$

For $j = l$ case, if $i = l$, then $A_l(l, l) = 0$ holds obviously due to antisymmetry of Wong's matrix. In the view of Eqs. (6.80) and (6.81), we have

$$A_l(i, j) = 0 \ for \ 1 \leq i, j, l \leq k_1. \tag{6.82}$$

Observe that $A_l(i, j) = \frac{\partial \omega_{ij}^{(1)}}{\partial x_l}$. Therefore, (iv) of Lemma 6.8 implies

$$A_l(i, j) + A_j(l, i) + A_i(j, l) = 0 \ for \ 1 \leq i, j \leq k_1, k_1 + 1 \leq l \leq k. \tag{6.83}$$

Since $A_j(l, i) = \frac{\partial \omega_{li}^{(1)}}{\partial x_j} = 0$ and $A_i(j, i) = \frac{\partial \omega_{jl}^{(1)}}{\partial x_i} = 0$ by Theorem 6.8, we have

$$A_l(i, j) = 0 \ for \ 1 \leq i, j \leq k_1, k_1 + 1 \leq l \leq k. \tag{6.84}$$

Therefore, we have shown that $\omega_{ij}^{(1)} = 0$ for $1 \leq i, j \leq k_1$. □

Theorem 6.10 *Suppose that E is a finite-dimensional estimation algebra of maximal rank. Then,*

$$\Omega = (\omega_{ij}) = \left(\begin{array}{c|c} Constants & P_1(x_1, x_2, \cdots, x_k) \\ \hline P_1(x_1, x_2, \cdots, x_k) & P_1(x_{k+1}, \cdots, x_n) \end{array} \right), \tag{6.85}$$

i.e.,

(i) ω_{ij} is a constant for $1 \leq i, j \leq k$;

(ii) ω_{ij} is a polynomial of degree 1 in $x_1, \cdots, x_k$ for $1 \leq i \leq k, k+1 \leq j \leq n$ or $1 \leq j \leq k, k+1 \leq i \leq n$;
(iii) ω_{ij} is a polynomial of degree 1 in $x_{k+1}, \cdots, x_n$ for $k+1 \leq i, j \leq n$.

Proof This is an immediate consequence of Lemma 6.10, Theorems 6.8, and 6.9.
□

6.3.2 Maximal Rank Classification: Hessian Matrix Nondecomposition Theorem

In this section, we are going to prove that ω_{ij} is a constant for $k+1 \leq i, j \leq n$. We shall see that this statement follows from the Hessian matrix nondecomposition theorem which is a general theorem and has nothing to do with estimation algebras. The Hessian matrix nondecomposition theorem was first proved by Yau et al. [34].

Lemma 6.12 *Suppose that E is a finite-dimensional estimation algebra of maximal rank. Then:*

(i) $\sum_{l=1}^{n} \omega_{jl}\omega_{il} - \frac{1}{2}\frac{\partial^2 \eta}{\partial x_j \partial x_i} \in E$ for any $1 \leq i, j \leq n$;
(ii) η is a polynomial of degree 4.

Proof (i) follows from (vi) of Lemma 6.3 and Theorem 6.10. From (i) and Theorem 6.10, $\frac{\partial^2 \eta}{\partial x_i \partial x_j}$ is a degree 2 polynomial for all $1 \leq i, j \leq n$. Therefore, η is a polynomial of degree 4.
□

Lemma 6.13 *Suppose that E is a finite-dimensional estimation algebra of maximal rank. Let k be the quadratic rank. Let $\eta = \eta_4(x_{k+1}, \cdots, x_n) +$ polynomial of degree 3 in $x_{k+1}, \cdots, x_n$ with coefficients degree at most four polynomials in $x_1, \cdots, x_k$. Then for any $k+1 \leq i, j \leq n$,*

$$\sum_{l=k+1}^{n} \omega_{jl}^{(1)}\omega_{il}^{(1)} = \frac{1}{2}\frac{\partial^2 \eta_4}{\partial x_j \partial x_i}, \quad (6.86)$$

where $\eta_4 = \eta_4(x_{k+1}, \cdots, x_n)$ is a homogeneous polynomials of degree 4 in $x_{k+1}, \cdots, x_n$.

Proof From Lemma 6.12 and Theorem 6.10, we know that for $k+1 \leq i, j \leq n$, $\sum_{l=k+1}^{n} \omega_{jl}^{(1)}\omega_{il}^{(1)} - \frac{1}{2}\frac{\partial^2 \eta_4}{\partial x_j \partial x_i}$ is the homogeneous polynomial of degree 2 part of $\sum_{l=1}^{n} \omega_{jl}\omega_{il} - \frac{1}{2}\frac{\partial^2 \eta}{\partial x_j \partial x_i}$ in $x_{k+1}, \cdots, x_n$ variables. The result follows immediately from Lemma 6.4.
□

The following notations and lemma were used and observed by Chen et al. [8]. Define

$$\Delta := (\omega_{il}^{(1)}), \ k+1 \leq i, l \leq n$$

$$= \sum_{j=k+1}^{n} A_j x_j, \qquad (6.87)$$

where Δ is an $(n-k) \times (n-k)$ antisymmetric matrix and $A_j = A_j(p,q), k+1 \leq p, q \leq n$, are $(n-k) \times (n-k)$ antisymmetric matrices with constant coefficients. The anti-symmetry of Δ and A_j follows directly from that of Ω.

Lemma 6.14 *Suppose that E is a finite-dimensional estimation algebra of maximal rank. With the notations as above, then*

(i) $\Delta \Delta^T = \frac{1}{2} H(\eta_4)$, where $H(\eta_4) = \left(\frac{\partial^2 \eta_4}{\partial x_i \partial x_j} \right)$, $k+1 \leq i, j \leq n$, is the Hessian matrix of $\eta_4 = \eta_4(x_{k+1}, \cdots, x_n)$.
(ii) $A_i(j,l) + A_l(i,j) + A_j(l,i) = 0$.

Proof (i) follows from Lemma 6.13, while (ii) is a consequence of Lemma 6.8 (iv). □

The following Hessian matrix non-decomposition theorem is a general mathematical theorem that has independent interest besides nonlinear filtering theory. For a $(n-k) \times (n-k)$ matrix with $n-k$ less than or equal to 4, the theorem was proved in Chen et al. [8].

Theorem 6.11 *Let $\Delta = \sum_{j=k+1}^{n} A_j x_j$ be an $(n-k) \times (n-k)$ antisymmetric matrix, where $A_j = (A_j(p,q)), k+1 \leq p, q \leq n$, is an antisymmetric matrix with constant coefficients. Suppose*

$$A_i(j,l) + A_l(i,j) + A_j(l,i) = 0, \ \forall k+1 \leq i, j, k \leq n. \qquad (6.88)$$

Let $\eta_4 = \eta_4(x_{k+1}, \cdots, x_n)$ be a homogeneous polynomial of degree 4 in $x_{k+1}, \cdots, x_n$. Let $H(\eta_4) = \left(\frac{\partial^2 \eta_4}{\partial x_i \partial x_j} \right)$, $k+1 \leq i, j \leq n$, be the Hessian matrix of η_4. If $\Delta \Delta^T = \frac{1}{2} H(\eta_4)$, then $\Delta \equiv 0$, i.e., $A_j = 0$ for all $k+1 \leq j \leq n$.

Proof Let $\Delta = (\beta_{ij})$. Then $\frac{\partial \beta_{ij}}{\partial x_l} = A_l(i,j)$. Observe that from $\Delta \Delta^T = \frac{1}{2} H(\eta_4)$, we have

6.3 Algebraic Classification of Finite-Dimensional Filter

$$\frac{\partial^2}{\partial x_j^2}\left(\frac{\partial^2 \eta_4}{\partial x_i^2}\right) = \frac{\partial^2}{\partial x_i^2}\left(\frac{\partial^2 \eta_4}{\partial x_j^2}\right) = \frac{\partial^2}{\partial x_i \partial x_j}\left(\frac{\partial^2 \eta_4}{\partial x_i \partial x_j}\right)$$

$$\Longrightarrow \frac{\partial^2}{\partial x_j^2}\left[\sum_{l=k+1}^{n} \beta_{il}^2\right] = \frac{\partial^2}{\partial x_i^2}\left[\sum_{l=k+1}^{n} \beta_{jl}^2\right] = \frac{\partial^2}{\partial x_i \partial x_j}\left[\sum_{l=k+1}^{n} \beta_{il}\beta_{jl}\right] \quad (6.89)$$

$$\Longrightarrow \sum_{l=k+1}^{n} [A_j(i,l)]^2 = \sum_{l=k+1}^{n} [A_i(j,l)]^2 = \frac{1}{2}\sum_{l=k+1}^{n} [A_i(i,l)A_j(j,l)$$
$$+ A_j(i,l)A_i(j,l)]$$

$$\frac{\partial^2}{\partial x_i \partial x_j}\left(\frac{\partial^2 \eta_4}{\partial x_j^2}\right) = \frac{\partial^2}{\partial x_j^2}\left(\frac{\partial^2 \eta_4}{\partial x_i \partial x_j}\right)$$

$$\Longrightarrow \frac{\partial^2}{\partial x_i \partial x_j}\left[\sum_{l=k+1}^{n} \beta_{jl}^2\right] = \frac{\partial^2}{\partial x_j^2}\left[\sum_{l=k+1}^{n} \beta_{il}\beta_{jl}\right] \quad (6.90)$$

$$\sum_{l=k+1}^{n} A_j(j,l)A_i(j,l) = \sum_{l=k+1}^{n} A_j(j,l)A_j(i,l)$$

$$\Longrightarrow \sum_{l=k+1}^{n} A_j(j,l)[A_i(j,l) + A_j(l,i)] = 0 \quad \text{(by anti-symmetry of } A\text{)} \quad (6.91)$$

$$\Longrightarrow \sum_{l=k+1}^{n} A_j(j,l)A_l(j,i) = 0 \quad \text{(by cyclic condition)}$$

$$\frac{\partial^2}{\partial x_p \partial x_q}\left(\frac{\partial^2 \eta_4}{\partial x_j^2}\right) = \frac{\partial^2}{\partial x_j^2}\left(\frac{\partial^2 \eta_4}{\partial x_p \partial x_q}\right)$$

$$\Longrightarrow \frac{\partial^2}{\partial x_p \partial x_q}\left[\sum_{l=k+1}^{n} (\beta_{jl})^2\right] = \frac{\partial^2}{\partial x_j^2}\left[\sum_{l=k+1}^{n} \beta_{pl}\beta_{ql}\right] \quad (6.92)$$

$$\Longrightarrow \sum_{l=k+1}^{n} A_p(j,l)A_q(j,l) = \sum_{l=k+1}^{n} A_j(p,l)A_j(q,l).$$

Observe that $(A_j^2)^T = (A_j A_j)^T = A_j^T A_j^T = (-A_j)(-A_j) = A_j^2$. Denote $A_j^2(p,q)$ the (p,q)-entry of A_j^2 matrix. Then for any $k+1 \le j \le n$,

$$\sum_{l=k+1}^{n} [A_j^2(j,l)]^2 = \sum_{l=k+1}^{n} A_j^2(j,l) A_j^2(l,j)$$

$$= \sum_{l=k+1}^{n} \left[\sum_{q=k+1}^{n} A_j(j,q) A_q(j,l) \right] \left[\sum_{p=k+1}^{n} A_j(j,p) A_p(j,l) \right]$$

$$= 0. \quad \text{by (6.92)} \tag{6.93}$$

In particular, we have $A_j^2(j,j) = 0$ for all $k+1 \le j \le n$:

$$\sum_{l=k+1}^{n} [A_j(j,l)]^2 = - \sum_{l=k+1}^{n} A_j(j,l) A_j(l,j) = -A_j^2(j,j) = 0. \tag{6.94}$$

$$\implies A_j(j,l) = 0 \quad \text{for all } k+1 \le j, l \le n.$$

Now (6.89) becomes, for any $k+1 \le i, j \le n$,

$$\sum_{l=k+1}^{n} [A_j(i,l)]^2 = \sum_{l=k+1}^{n} [A_i(j,l)]^2 = \frac{1}{2} \sum_{l=k+1}^{n} A_j(i,l) A_i(j,l)$$

$$\le \frac{1}{4} \sum_{l=k+1}^{n} [A_j(i,l)]^2 + \frac{1}{4} \sum_{l=k+1}^{n} [A_i(j,l)]^2. \tag{6.95}$$

Then we have

$$\frac{3}{4} \sum_{l=k+1}^{n} [A_i(j,l)]^2 \le \frac{1}{4} \sum_{l=k+1}^{n} [A_j(i,l)]^2 = \frac{1}{4} \sum_{l=k+1}^{n} [A_i(j,l)]^2$$

$$\implies \sum_{l=k+1}^{n} [A_i(j,l)]^2 = 0. \tag{6.96}$$

$$\implies A_i(j,l) = 0 \quad \text{for all } k+1 \le i, j, l \le n.$$

$$\implies \Delta = 0.$$

$\square$

Theorem 6.12 *Suppose that E is a finite-dimensional estimation algebra of maximal rank. Then,*

$$\Omega = (\omega_{ij}) = \left(\frac{\text{Constants}}{P_1(x_1, x_2, \cdots, x_k)} \middle| \frac{P_1(x_1, x_2, \cdots, x_k)}{\text{Constants}} \right), \tag{6.97}$$

i.e.,

(i) ω_{ij} is a constant for $1 \le i, j \le k$ or $k+1 \le i, j \le n$;
(ii) ω_{ij} is a polynomial of degree 1 in $x_1, \cdots, x_k$ for $1 \le i \le k, k+1 \le j \le n$ or $1 \le j \le k, k+1 \le i \le n$.

Proof This follows from Theorem 6.10, 6.11 directly. □

6.3.3 Maximal Rank Classification: Complete Classification Theorem

In this section, the main goal that we shall finish is to prove the remaining part of the Ω is still constant, i.e., two non-diagonal parts $1 \le i \le k, k+1 \le j \le n$ and $k+1 \le i \le n, 1 \le j \le k$. Since the similar technique of Lie algebra computations will be utilized, the details of the proof of the lemmas and propositions below will be omitted which can be found in Yau et al. [33]. The following propositions and lemmas will facilitate the proof of our classification theorem.

Theorem 6.13 *Let E be an estimation algebra of the filtering system (6.1) whose Ω-matrix has constant entries.*

(i) *If η is a polynomial of degree at most 2 and $h_1, \cdots, h_m$ are affine in x, then E is finite dimensional and has a basis consisting of $E_0 = L_0, E_1, \cdots, E_p, E_{p+1}, \cdots, E_q, 1$ (for some $p < p$). The differential operators $E_1, \cdots, E_p$ have the form*

$$\sum_{j=1}^{n} \alpha_{ij} D_j + \beta_j, \quad 1 \le i \le p, \tag{6.98}$$

where α_{ij}'s are constants and β_i's are affine in x, and the differential operators $E_{p+1}, \cdots, E_q$ are affine in x. Moreover, the quadratic part of $\eta - \sum_{i=1}^{m} h_i^2$ is positive semidefinite.

(ii) *Conversely, if E is finite dimensional, then $h_1, \cdots, h_m$ are affine in x, i.e., the observation matrix $H = [\nabla h_1, \cdots, \nabla h_m]$ is a constant matrix. Furthermore, if the observation matrix has rank n, then η is a polynomial of degree at most 2 and E is of dimension $2n+2$ with a basis given by $1, x_1, \cdots, x_n, D_1, \cdots, D_n, L_0$.*

Lemma 6.15 *Let E be a finite-dimensional estimation algebra with maximal rank. Let k be the quadratic rank of E. Then, $\frac{\partial \omega_{il}}{\partial x_j} = \frac{\partial \omega_{jl}}{\partial x_i}$ for all $k+1 \le l \le n$ and $1 < i, j \le k$.*

Proposition 6.1 *If $x_{k_{p-1}+1}^2 + \cdots + x_{k_p}^2$ is a basic quadratic form in E and $\frac{\partial \omega_{jl}}{\partial x_i} = 0$ for all $k+1 \le l \le n, k_{p-1}+1 \le i, j \le k_p$, and $i \ne j$, then $\frac{\partial \omega_{il}}{\partial x_i} = 0$ for all $k_{p-1}+1 \le i \le k_p$.*

Lemma 6.16 Let $x_{k_{r-1}+1}^2 + \cdots + x_{k_r}^2$ and $x_{k_{s-1}+1}^2 + \cdots + x_{k_s}^2$ be the basic forms in E, where $k_{r-1} < k_r \leq k_{s-1} \leq k_s$. Let $\xi_{ij} = \sum_{l=k+1}^{n} \left(\frac{\partial \omega_{jl}}{\partial x_i}\right) D_l$. Suppose $\sum_{j=k_{s-1}+1}^{k_s} \xi_{pj}\xi_{qj} = 0$ for all $k_{r-1}+1 \leq p, q \leq k_r, p \neq q$. Then $\frac{\partial \omega_{jl}}{\partial x_i} = 0$ for all $k+1 \leq l \leq n, k_{r-1}+1 \leq i \leq k_r$ and $k_{s-1}+1 \leq j \leq k_s$.

Lemma 6.17 Let $x_{k_{r-1}+1}^2 + \cdots + x_{k_r}^2$ and $x_{k_{s-1}+1}^2 + \cdots + x_{k_s}^2$ be the basic forms in E, where $k_{r-1} < k_r \leq k_{s-1} \leq ks$. Let $\xi_{ij} = \sum_{l=k+1}^{n}\left(\frac{\partial \omega_{jl}}{\partial x_i}\right) D_l$. Then $\sum_{j=k_{s-1}+1}^{k_s} \xi_{pj}\xi_{qj} = 0$ for all $k_{r-1}+1 \leq p, q \leq k_r, p \neq q$ if and only if $\sum_{j=k_{s-1}}^{k_s} a_{jl_1}^p a_{jl_2}^q = 0$ for all $k+1 \leq l_1, l_2 \leq n, k_{r-1}+1 \leq p, q \leq k_r, p \neq q$, where $a_{jl_1}^p = \frac{\partial \omega_{jl_1}}{\partial x_p}$.

Lemma 6.18 Let $x_{k_{r-1}+1}^2 + \cdots + x_{k_r}^2$ and $x_{k_{s-1}+1}^2 + \cdots + x_{k_s}^2$ be the basic forms in E, where $k_{r-1} < k_r \leq k_{s-1} \leq k_s$. Assume that $Q_l = \sum_{i=k_{r-1}+1}^{k_r} \sum_{j=k_{s-1}+1}^{k_s} a_{jl}^i x_i x_j \in E$ for all $k+1 \leq l \leq n$, where $a_{jl}^i = \frac{\partial \omega_{jl}}{\partial x_i}$. Then $\sum_{j=k_{s-1}+1}^{k_s} a_{jl_1}^p a_{jl_2}^q$ for all $k+1 \leq l_1, l_2 \leq n, k_{r-1}+1 \leq p, q \leq k_r$.

Proposition 6.2 Let $x_{k_{r-1}+1}^2 + \cdots + x_{k_r}^2$ and $x_{k_{s-1}+1}^2 + \cdots + x_{k_s}^2$ be the basic forms in E, where $k_{r-1} < k_r \leq k_{s-1} \leq k_s$. Then $\frac{\partial \omega_{jl}}{\partial x_i} = 0$ for all $k+1 \leq l \leq n, k_{r-1}+1 \leq i \leq k_r$ and $k_{s-1}+1 \leq j \leq k_s$.

Proposition 6.3 Let $x_{k_{r-1}+1}^2 + \cdots + x_{k_r}^2$ be the basic forms in E. Then $\frac{\partial \omega_{jl}}{\partial x_i} = 0$ for all $k+1 \leq l \leq n, k_{r-1}+1 \leq i, j \leq k_r$ and $i \neq j$.

Theorem 6.14 Suppose that E is a finite-dimensional estimation algebra of maximal rank. Then $\Omega = (\omega_{ij})$ is a matrix with constant coefficients.

Proof This follows from Propositions 6.1–6.3. □

Theorem 6.15 $\frac{\partial f_j}{\partial x_i} - \frac{\partial f_i}{\partial x_j} = c_{ij}$ are constants for all i and j if and only if

$$(f_1, \cdots, f_n) = (l_1, \cdots, l_n) + \left(\frac{\partial \psi}{\partial x_1}, \cdots, \frac{\partial \psi}{\partial x_n}\right), \quad (6.99)$$

where $l_1, \cdots, l_n$ are polynomials of degree 1 and ψ is a C^∞ function.

Proof Sufficiency is easy to obtain by applying direct calculus computation. Our goal aims to solve necessity. Let $b_{ij} = -\frac{1}{2}c_{ij}$. Then we have $b_{ji} - b_{ij} = c_{ij}$. Let $l_i(x) = \sum_{j=1}^{n} b_{ij}x_j$. In the following, we find that the following two exterior derivatives of differential forms $\sum_{i=1}^{n} f_i dx_i$ and $\sum_{i=1}^{n} l_i dx_i$ are the same, i.e.,

$$d(\sum_{i=1}^{n} f_i dx_i) = d(\sum_{i=1}^{n} l_i dx_i). \quad (6.100)$$

6.3 Algebraic Classification of Finite-Dimensional Filter

Equivalently,

$$d(\sum_{i=1}^{n} f_i dx_i - \sum_{i=1}^{n} l_i dx_i) = 0. \tag{6.101}$$

By Poincare lemma, every d-closed differential form on R^n are d-exact; there exists a smooth function ψ such that

$$\sum_{i=1}^{n} f_i dx_i - \sum_{i=1}^{n} l_i dx_i = d\psi = \sum_{i=1}^{n} \frac{\partial \psi}{\partial x_i} dx_i. \tag{6.102}$$

Results follows immediately. □

In the following, after we get the constant structure of Ω matrix, estimation algebra of maximal rank can be proved to be a linear vector space with a specific basis.

Theorem 6.16 (Complete Classification) *Let E be the finite-dimensional estimation algebra with maximal rank and $\omega_{ij} = \frac{\partial f_j}{\partial x_i} - \frac{\partial f_i}{\partial x_j} = c_{ij}$. Then E is a real vector space of dimension $2n + 2$ with basis given by $1, x_1, \cdots, x_n, D_1, \cdots, D_n, L_0$ and η is a quadratic polynomial.*

Proof With the condition of maximal rank, without loss of generality, there is $x_i + c_i$ in E for $i = 1, \cdots, n$. Then

$$[L_0, x_i + c_i] = D_i \in E$$
$$[D_i, x_i + c_i] = \delta_{ij} \in E \tag{6.103}$$
$$[L_0, D_i] = \sum_{i=1}^{n} c_{ij} D_j + \frac{1}{2} \frac{\partial \eta}{\partial x_i} \in E.$$

Then $\frac{\partial \eta}{\partial x_i} \in E$ for all $i = 1, \cdots, n$. If η is a quadratic polynomial, then we easily see that E is a finite-dimensional real vector space spanned by basis $1, x_1, \cdots, x_n, D_1, \cdots, D_n, L_0$. To finish the whole proof, the only thing that remains to do is to prove that η is a polynomial of degree at most 2. To this end, it can be observed that $\frac{\partial \eta}{\partial x_i}$ are polynomials of degree at most 2 because of Ocone's lemma. It follows that η is a polynomial of degree at most 3. If homogeneous degree 3 part of η is nonzero, then clearly there exists a straight line $c(t)$ satisfying $\lim_{t \to \infty} \eta(c(t)) = \infty$. Then

$$\lim_{t \to \infty} (\eta - \sum_{i=1}^{m} h_i^2)(c(t)) = -\infty. \tag{6.104}$$

Recall that nonexistence property is satisfied by underdetermined PDE and Theorem 6.3. We get a contradiction. □

6.4 Wei-Norman Approach

Constructing a robust finite-dimensional filter for (6.1) is defined to find a smooth manifold M, complete C^∞ vector fields μ_i on M, C^∞ function ν on $M \times R \times R^n$, and ω_i's on R^m such that $u(t, x)$ in robust DMZ equation can be represented in the following form:

$$\begin{cases} \dfrac{dz}{dt}(t) = \sum_{i=1}^{k} \mu_i(z(t))\omega_i(y(t)), \; z(0) \in M, \\ u(t,x) = \nu(z(t), t, x). \end{cases} \quad (6.105)$$

Following Chaleyat et al. [15], we say that system (6.1) has a robust universal finite-dimensional filter if, for each initial probability density σ_0, there exists a z_0 such that (6.105) holds if $z(0) = z_0$ and μ_i, ω_i are independent of σ_0.

The method of Wei and Norman [24] of using Lie algebraic ideas to solve time-varying linear differential equations is roughly as follows. Consider the following equation:

$$\frac{d}{dt} X(t) = A(t)X(t) = \sum_{i=1}^{m} a_i(t) A_i X(t), \quad X(0) = X_0, \quad (6.106)$$

where X and A_i's are $n \times n$ matrices and a_i's are scalar-valued functions. Let $B_1, \cdots, B_l$ be a basis of the Lie algebra generated by $A_1, \cdots, A_m$. Then the Wei-Norman theorem states that, locally in t, $X(t)$ has a representation of the following form:

$$X(t) = e^{b_1(t) B_1} \cdots e^{b_l(t) B_l} X_0, \quad (6.107)$$

where the b_i's satisfy an ordinary differential equation of the following form:

$$\frac{db_i}{dt} = c_i(b_1, \cdots, b_l), \quad b_i(0) = 0, \; 1 \le i \le l. \quad (6.108)$$

The functions c_i, $1 \le i \le l$ in the above equation are determined by the structure constraints of the Lie algebra (generated by the A_i's) relative to the basis $\{B_1, \cdots, B_l\}$.

6.4 Wei-Norman Approach

The extension of Wei and Norman's approach to the nonlinear filtering problem is much more complicated. Instead of an ordinary differential equation, we have to solve the robust DMZ equation, which is a time-varying partial differential equation.

Suppose that the Wei-Norman theory is applied to solve partial differential equations of the following form:

$$\frac{\partial u}{\partial t} = a_1 A_1 u + \cdots + a_m A_m u, \tag{6.109}$$

where the A_i, $1 \leq i \leq m$, are linear partial differential operators in $x_1, \cdots, x_n$, and the a_i, $1 \leq i \leq m$, are given functions of time t.

We shall assume that the Lie algebra generated by the operators $A_1, \cdots, A_m$ in (6.109) is finite dimensional. By setting, if necessary, some of the $a_i(t)$ equal to zero, and by combining other $a_j(t)$ in case of linear dependence among the operators on the r.h.s. of (6.109), without loss of generality, we can assume $A_1, \cdots, A_m$ consist a basis of Lie algebra and satisfy

$$[A_i, A_j] = \sum_{k}^{m} \gamma_{ij}^k A_k, \quad 1 \leq i, j \leq m, \tag{6.110}$$

for suitable real constants γ_{ij}^k, $1 \leq i, j, k \leq m$.

The central idea of Wei-Norman theory is to try for a solution of the following form:

$$u(t, x) = e^{g_1(t)A_1} \cdots e^{g_m(t)A_m} \psi, \tag{6.111}$$

where the g_i, $1 \leq i \leq m$, are undetermined functions of time. The next step is to insert such representation (6.111) to (6.109) and derive the ordinary differential equation satisfied by g_i, $1 \leq i \leq m$. We shall give a sketch of the procedure of obtaining evolution equation of g_i, $1 \leq i \leq m$. First,

$$\begin{aligned}\frac{\partial u}{\partial t} &= \dot{g}_1 A_1 e^{g_1(t)A_1} \cdots e^{g_m(t)A_m} \psi + e^{g_1(t)A_1} \dot{g}_2 A_2 \cdots e^{g_m(t)A_m} \psi + \cdots \\ &+ e^{g_1(t)A_1} \cdots e^{g_{m-1}(t)A_{m-1}} \dot{g}_m A_m e^{g_m(t)A_m} \psi.\end{aligned} \tag{6.112}$$

Now for $i = 2, \cdots, n$, insert a term

$$e^{-g_{i-1}A_{i-1}} \cdots e^{-g_1 A_1} e^{g_1 A_1} \cdots e^{g_{i-1}A_{i-1}}, \tag{6.113}$$

just behind $\dot{g}_i A_i$ in the i-th term of (6.112). Then use the adjoint representation formula

$$e^A B e^{-A} = B + [A, B] + \frac{1}{2!}[A, [A, B]] + \frac{1}{3!}[A, [A, [A, B]]] + \cdots \tag{6.114}$$

and (6.110) repeatedly; the right-hand side of (6.112) can be expanded as linear combination of $\{A_1 u, A_2 u, \cdots, A_m u\}$. The coefficients will involve combinations of derivatives of $g_i(t)$. For example, the second term of RHS of (6.112) shall be

$$e^{g_1(t)A_1} \dot{g}_2 A_2 e^{-g_1(t)A_1} e^{g_1(t)A_1} \cdots e^{g_m(t)A_m} \psi$$

$$= e^{g_1(t)A_1} \dot{g}_2 A_2 e^{-g_1(t)A_1} u \tag{6.115}$$

$$= (\dot{g}_2 A_2 + g_1 \dot{g}_2 [A_1, A_2] + \frac{1}{2} g_1^2 \dot{g}_2 [A_1, [A_1, A_2]] + \cdots) u.$$

This summation will be finite because such Lie algebra is assumed to be finite dimensional.

By matching both sides of (6.109), systems of ODEs for the $g_1, \cdots, g_m$ will be obtained. These systems of ODEs are always solvable for small time. However, they may not be solvable for all time, meaning that finite escape time phenomena may occur. Fortunately, Theorem 6.16 will allow us to prove the following theorem which shows concretely how to construct robust finite-dimensional filters from finite-dimensional estimation algebra. Since the estimation algebra is solvable, the corresponding systems of ODEs are solvable for all $t \geq 0$.

Theorem 6.17 *Let E be a finite-dimensional estimation algebra in system (6.1) satisfying $\frac{\partial f_j}{\partial x_i} - \frac{\partial f_i}{\partial x_j} = c_{ij}$ where c_{ij} are constants for all $1 \leq i, j \leq n$. Suppose E is finite dimensional, then $h_1, \cdots, h_m$ are affine. Suppose further that $m \geq n$ and the observation matrix has full rank, then $\eta = \sum_{i,j=1}^{n} a_{ij} x_i x_j + \sum_{i=1}^{n} b_i x_i + d$ where a_{ij}, b_i and d are constants for all $1 \leq i, j \leq n$ and the robust DMZ equation has a solution for all $t \geq 0$ of the following form:*

$$u(t,x) = e^{T(t)} e^{R^n(t)x_n} \cdots e^{r_1(t)x_1} e^{s_n(t)D_n} \cdots e^{s_1(t)D_1} e^{tL_0} \sigma_0, \tag{6.116}$$

where $T(t), r_1(t), \cdots, R^n(t), s_1(t), \cdots, s_n(t)$ satisfy the following ordinary differential equations:
For $1 \leq i \leq n$

$$\frac{ds_i(t)}{dt} = r_i(t) + \sum_{j=1}^{n} s_j(t) c_{ji} + \sum_{k=1}^{m} h_{ki} y_k(t), \tag{6.117}$$

where $h_k = \sum_{j=1}^{n} h_{kj} x_j + e_k$, $1 \leq k \leq m$; h_{kj} and e_k are constants.
For $1 \leq j \leq n$

$$\frac{dr_j(t)}{dt} = \frac{1}{2} \sum_{i=1}^{n} s_i(t)(a_{ij} + a_{ji}), \tag{6.118}$$

and

$$\begin{aligned}\frac{dT(t)}{dt} =& -\frac{1}{2}\sum_{i=1}^{n} r_i^2(t) - \frac{1}{2}\sum_{i=1}^{n} s_i^2(t)\left(\sum_{j=1}^{n} c_{ij}^2 - a_{ii}\right) \\ & + \sum_{1\leq i<k\leq n} s_i(t)s_k(t) \times \left(\sum_{j=1}^{n} c_{ij}c_{jk} + \frac{1}{2}(a_{ik}+a_{ki})\right) \\ & + \sum_{i=1}^{n} r_i(t) - \sum_{j=2}^{n}\sum_{i=1}^{j} s_j(t)c_{ij} + \frac{1}{2}\sum_{i=1}^{n} s_i(t)b_i \\ & + \frac{1}{2}\sum_{i,j=1}^{m} y_i(t)y_j(t)\left(\sum_{k=1}^{n} h_{ik}h_{jk}\right) - \sum_{i,j=1}^{n} s_i(t)s_j(t)c_{ij}.\end{aligned} \quad (6.119)$$

It follows that a universal finite-dimensional filter exists for the system (6.1).

Proof The detail can be found in Yau [28]. The basic idea of proof is to submit Expression (6.116) to the robust DMZ equation. By comparing the coefficients on both sides, dynamical evolution of coefficient functions $T(t), r_i(t), s_j(t)$ will be proved to satisfy ordinary differential equations. The detail is left as an exercise.
□

6.5 Classification with Nonmaximal Rank

6.5.1 State Dimension 2

The most basic and important situation in nonmaximal rank estimation algebra is that state dimension equal to 2. In Wu and Yau [26], general considerations and approaches toward the classification of finite-dimensional estimation algebra are proposed. Some structural results are obtained. The properties of Euler operators and the solution to an underdetermined PDE are extended and compared to [28, 32]. These tools and techniques are applied to the study of finite-dimensional estimation algebras with state dimension 2 to obtain a complete classification result. It is shown that the dimension of finite-dimensional estimation algebra is no more than 6. Moreover, the Mitter conjecture and the Levine conjecture hold for the case of state dimension 2.

Unlike estimation algebras of maximal rank, it is not clear whether x_i's are elements of a general estimation algebra E. To simplify the situation, the concept of linear rank is introduced and some basic properties of linear rank are proved. In order to explore the structure of the function elements in E, quadratic rank is extended. In what follows, for a polynomial ϕ, $\phi^{(k)}$ denotes its homogeneous degree k part. The following theorem is from [26], which is the basics of linear rank.

Theorem 6.18 *Let* $E = \langle L_0, h_1, \cdots, h_m \rangle_{L.A.}$ *and* $\bar{E} = \langle 1, L_0, h_1, \cdots, h_m \rangle_{L.A.}$. *Then E is finite dimensional if and only if $\bar{E}$ is finite dimensional.*

Proof Since $E \subset \bar{E}$, it suffices to show $\dim \bar{E} - \dim E \leq 1$ if $\dim E < \infty$. Start from the generators of $\bar{E}$ to construct increasing subsets $A_k \subset \bar{E}$, as follows:

- $A_0 = \{1, L_0, h_1, \cdots, h_m\}$;
- for $k \geq 1$, $A_k = \{a_1 B_1 + a_2 B_2 + a_3 [B_3, B_4] : a_i \in R, B_j \in A_{k-1}, i = 1, 2, 3, j = 1, 2, 3, 4\}$;

Clearly, $A_k \uparrow \bar{E}$ as $k \to \infty$. Now, assume that $\dim E < \infty$. Let vector space $E^* = E + 1 := \{aX + b : X \in E, a, b \in R\}$. It is easy to show by induction that $A_k \subset E^*$ for $k \geq 0$. Thus,

$$\bar{E} = \cup_{k \geq 0} A_k \subset E^*, \tag{6.120}$$

which means

$$\dim \bar{E} \leq \dim E^* \leq \dim E + 1 < \infty. \tag{6.121}$$

$\square$

Thus, the estimation algebra E of a filtering problem being finite dimensional is equivalent to the finite dimensionality of $\langle E, 1 \rangle_{L.A.}$. In the discussion of the finite dimensionality of E, 1 can always be assumed an element of E.

Under the assumption that $1 \in E$, any degree 1 polynomial in $E \implies$ its homogeneous degree 1 part is in E.

Definition 6.11 (Wu and Yau [26]) Let $L(E) \subset E$ be the vector space consisting of all the homogeneous degree 1 polynomials in E. Then the linear rank of estimation algebra E is defined by $r := \dim L(E)$.

Note that an estimation algebra is associated with a filtering system and is coordinate-dependent. The recognition of the structurally equivalent estimation algebras is very important in the classification problem. Since an estimation algebra is essentially a Lie algebra, the definitions of homomorphism and isomorphism of estimation algebras follow from those of Lie algebras. It is shown in [1, 4], and [17] that orthogonal variable transformation and affine transformations extend to estimation algebra isomorphisms.

Let estimation algebra E possess linear rank r. Clearly, $r \leq n$ and there exist r independent linear functions (the basis of $L(E)$) $l_1(x), \cdots, l_r(x)$ such that

$$(l_1(x), \cdots, l_r(x))^T = Ax, \tag{6.122}$$

where A is an $r \times n$ matrix with rank r. From the singular value decomposition theorem, there exist orthogonal matrices $U \in R^{r \times r}$, $V \in R^{n \times n}$ such that

6.5 Classification with Nonmaximal Rank

$$A = U[D \ 0]V^T, \tag{6.123}$$

where $D = diag(d_1, \cdots, d_r)$, with $d_1, \cdots, d_r \neq 0$ the singular values of A. Thus,

$$(l_1(x), \cdots, l_r(x))^T = U[D \ 0]V^T x. \tag{6.124}$$

After an orthogonal variable transformation $y = V^T x$, it is easy to see that for $1 \leq i \leq r$, y_i is a linear combination of $l_j(x) (1 \leq j \leq r)$. Therefore, $y_i (1 \leq i \leq r)$ are independent linear functions in E, i.e., $\{y_i, 1 \leq i \leq r\}$ is the basis of $L(E)$.

Hence, by an orthogonal variable transformation, if necessary, a linear function l is in E if and only if

$$l(x) \in L(E) := span\{x_1, \cdots, x_r\}. \tag{6.125}$$

Quadratic rank has been introduced in Definition 6.10. However, in order to deal with estimation algebra with nonmaximal rank, more properties of quadratic rank have been explored by Wu and Yau. More detailed results can be found in [26].

Lemma 6.19 *Let E be an estimation algebra with linear rank r. Then for any $\phi \in E$, the quadratic part of ϕ is*

$$\phi^{(2)} = x^T \begin{pmatrix} A_1 & 0 \\ 0 & A_2 \end{pmatrix} x, \tag{6.126}$$

where A_1, A_2 are symmetric matrices with dimensions $r \times r$ and $(n-r) \times (n-r)$.

Proof $\phi^{(2)}$ is a degree 2 homogeneous polynomial. Therefore, it can be written as $\phi^{(2)} = x^T A x$ for a symmetric matrix $A = (a_{ij})$ of dimension $n \times n$.

Since linear rank of E is r, $x_i \in E$ if and only if $1 \leq i \leq r$. For $1 \leq i \leq r$, $[L_0, x_i] = D_i \in E$, and

$$[D_i, \phi] = \frac{\partial \phi}{\partial x_i} = 2 \sum_{j=1}^{n} a_{ij} x_j, \text{ mod } R \in E. \tag{6.127}$$

Then $a_{ij} = 0$ for $r+1 \leq j \leq n, 1 \leq i \leq r$. By the symmetry of A, the lemma follows. □

Now, consider an estimation algebra E with linear rank r and quadratic rank k. By the definition of the quadratic rank of E, there exists $p_0 \in E$ such that the quadratic rank of $p_0 = k$. By Lemma 6.19,

$$p_0^{(2)} = x^T \begin{pmatrix} A_1 & 0 \\ 0 & A_2 \end{pmatrix} x. \tag{6.128}$$

Let $k_1 = rank(A_1)$, $k_2 = rank(A_2)$. Then $k = k_1+k_2, 0 \le k_1 \le r, 0 \le k_2 \le n-r$. Since A_1 and A_2 are real symmetric, there are orthogonal matrices U_1 and U_2 such that

$$A_1 = U_1 \begin{pmatrix} D_1 & 0 \\ 0 & 0 \end{pmatrix} U_1^T, \qquad (6.129)$$

and

$$A_2 = U_2 \begin{pmatrix} 0 & 0 \\ 0 & D_2 \end{pmatrix} U_2^T, \qquad (6.130)$$

where D_1, D_2 are nonsingular diagonal matrices with dimensions $k_1 \times k_1$ and $k_2 \times k_2$. By taking the orthogonal variable transformation $T = \begin{pmatrix} U_1 & 0 \\ 0 & U_2 \end{pmatrix}$, $p_0^{(2)} = \sum_{i=1}^{k_1} d_i x_i^2 + \sum_{i=n-k_2+1}^{n} d_i x_i^2$, where $d_i \ne 0$. Moreover, by an affine variable transformation, if necessary,

$$p_0 = \sum_{i=1}^{k_1} d_i x_i^2 + \sum_{i=n-k_2+1}^{n} d_i x_i^2 + \sum_{i=k_1+1}^{n-k_2} c_i x_i + c_0 \in E$$

$$\Longrightarrow [[L_0, p_0], p_0] = \sum_{i=1}^{k_1} 4d_i^2 x_i^2 + \sum_{i=n-k_2+1}^{n} 4d_i^2 x_i^2 + \sum_{i=k_1+1}^{n-k_2} c_i^2 \in E$$

$$\Longrightarrow q_0 := \sum_{i=1}^{k_1} d_i^2 x_i^2 + \sum_{i=n-k_2+1}^{n} d_i^2 x_i^2 \in E$$

$$\Longrightarrow q_j := [[L_0, q_{j-1}], q_{j-1}] = \sum_{i=1}^{k_1} 4^j d_i^{2j+2} x_i^2 + \sum_{i=n-k_2+1}^{n} 4^j d_i^{2j+2} x_i^2 \in E, \ j \ge 1. \qquad (6.131)$$

If no d_i^2's are equal, then the coefficient matrix of q_j for x_i^2 forms a Vandermonde matrix. By the invertibility of the Vandermonde matrix, x_i^2 can be represented as a linear combination of q_j's, and therefore $x_i^2 \in E$. Hence,

$$\sum_{i=1}^{k_1} x_i^2 + \sum_{i=n-k_2+1}^{n} x_i^2 \in E. \qquad (6.132)$$

If some d_i^2's are equal, for example, d_i^2's equal for $l_1 \le i \le l_2$ in p_0, they can be grouped to be solved as one variable. Instead of individual $x_i^2 \in E$ for $l_1 \le i \le l_2$, $x_{l_1}^2 + \cdots + x_{l_2}^2$ is obtained as a group under the above Vandermonde argument.

6.5 Classification with Nonmaximal Rank

In any case, (6.132) can always be constructed as long as the quadratic rank of E is k. An important observation is that both orthogonal variable transformation so used and affine variable transformation do not change the basis of $L(E)$. In summary, we have the following theorem, which describes the structure of linear and quadratic functions in estimation algebra.

Theorem 6.19 *Let E be a finite-dimensional estimation algebra with linear rank r and quadratic rank k.*

There exists $p_0 = \sum_{i=1}^{k_1} x_i^2 + \sum_{i=n-k_2+1}^{n} x_i^2 \in E$, where $k_1 + k_2 = k$, $k_1 \leq r$ and $k_2 \leq n - r$.

If $\phi \in E$ is a degree 1 polynomial, then ϕ is independent of $x_{r+1}, \cdots, x_n$.

If $\phi \in E$ is a degree 2 polynomial, then $\phi^{(2)}$ is independent of $x_{k_1+1}, \cdots, x_{n-k_2}$.

The theory of Euler operator is largely extended to more general case by Wu and Yau. The proof details can be found in [26]. Relative theorems are listed below. Notations of Euler operator are the same as before.

Theorem 6.20 (Theorem 3.13 [26]) *Let m be a constant integer and $\xi \in C^\infty(R^n)$ such that $E_l(\xi) + m\xi$ is a polynomial of degree $k(\geq 0)$ in $x_1, \cdots, x_l$ variables with coefficients in C^∞-functions of $x_{l+1}, \cdots, x_n$ variables.*

If $m + k + 1 > 0$, ξ is a polynomial of degree k in $x_1, \cdots, x_l$ variables with coefficients in C^∞-functions of $x_{l+1}, \cdots, x_n$ variables.

If $m + k + 1 \leq 0$, ξ is a polynomial of degree k or degree $-m (\geq k + 1 > 0)$ in $x_1, \cdots, x_l$ variables with coefficients in C^∞-functions of $x_{l+1}, \cdots, x_n$ variables.

Theorem 6.21 (Theorem 3.15 [26]) *Let m be a constant integer and $\xi \in C^\infty(R^n)$ such that $E_l(\xi) + m\xi \in P_k(x_1, \cdots, x_n)$, a polynomial of degree $k \geq 0$ in $x_1, \cdots, x_n$ variables.*

If $m > 0$, $\xi \in P_k(x_1, \cdots, x_n)$.

If $m = 0$, $\xi \in P_k(x_1, \cdots, x_n) + a(x_{l+1}, \cdots, x_n)$, where $a(x_{l+1}, \cdots, x_n)$ is a C^∞-function in $x_{l+1}, \cdots, x_n$ variables.

The theory of underdetermined partial differential equation is much helpful for classification problem of estimation algebra. Notice that the filtering system (6.1) is completely parameterized by the pair, (f, h). It follows by Eq. (6.6) that the underdetermined partial differential equation

$$\sum_{i=1}^{n} \frac{\partial f_i}{\partial x_i} + \sum_{i=1}^{n} f_i^2 = F \qquad (6.133)$$

provides a complete characterization of the realization set of such systems. Therefore, it is of primary interest to investigate the solution and solution properties for this class of equations.

In (6.133), $f_1, \cdots, f_n$ and F are C^∞-functions on R^n. F is given and $f_1, \cdots, f_n$ are treated as unknown. Although there is only one equation with n unknowns, (6.133) may not have solutions. In the work of Wu and Yau, they extend the result of Yau in 1994 for underdetermined partial differential equation.

Following some theorems are developed by Wu and Yau [26] based on the previous result Theorem 6.3.

Theorem 6.22 (Theorem 3.18 [26]) *Let $F(x_1, \cdots, x_n)$ be a degree $d \geq 1$ polynomial on R^n. The homogeneous degree d part of F is denoted by $F_d = \sum_{|i|=d} a_i x_1^{i_1} \cdots x_n^{i_n}$, where $i = (i_1, \cdots, i_n)$. If there exist n numbers $b_1, \cdots, b_n$ such that $F_d(b_1, \cdots, b_n) < 0$, there are no C^∞-functions $f_1, \cdots, f_n$ on R^n satisfying Eq. (6.133).*

Theorem 6.23 (Theorem 3.21 [26]) *Let d and $r \leq n$ be two positive integers and*

$$F(x_1, \cdots, x_n) = \sum_{|i| \leq d} a_i(x_{r+1}, \cdots, x_n) x_1^{i_1} \cdots x_r^{i_r}, \quad (6.134)$$

where $i = (i_1, \cdots, i_r)$, and where a_i's are C^∞-functions in $x_{r+1}, \cdots, x_n$ variables. The homogeneous degree d part in $x_1, \cdots, x_r$ variables of F is denoted by $F_d = \sum_{|i|=d} a_i(x_{r+1}, \cdots, x_n) x_1^{i_1} \cdots x_r^{i_r}$. If there exist n numbers $b_1, \cdots, b_n$ such that $F_d(b_1, \cdots, b_n) < 0$, there are no C^∞-functions $f_1, \cdots, f_n$ on R^n satisfying Eq. (6.133).

In the following, we will introduce classification result of estimation algebras with state dimension 2. First, the state dimension is assumed to be $n = 2$, i.e., there are two state variables x_1, x_2. Some definitions of differential operators and estimation algebra have been shown in previous sections. Wong's Ω-matrix is a 2×2 antisymmetric matrix. Therefore, only $\omega_{12} = -\omega_{21}$ is unknown. In the following, we prove the linear structure of the Ω-matrix, i.e., ω_{12} is a degree 1 polynomial in variables x_1, x_2.

Theorem 6.24 *Suppose* dim $E < \infty$ *and* $Y = p(x)D_2$, mod $U_0 \in E$. *Then p is a polynomial in x_1, x_2 of degree at most 1.*

Proof By Theorem 6.1, p is a polynomial in x_1, x_2. Let $l = \deg p$ and $p^{(l)}$ be the homogeneous degree l part of p. Then

$$p^{(l)} = \sum_{i=0}^{l} a_i x_1^{l-i} x_2^i = \sum_{i=s}^{t} a_i x_1^{l-i} x_2^i, \quad (6.135)$$

where a_i's are constants, $0 \leq s \leq t \leq l$, $a_s \neq 0$, $a_t \neq 0$, and $a_i = 0$ for $i < s$ or $i > t$. Let $Y_k = Ad_{L_0}^k Y$ for $k \geq 0$. By induction,

$$Y_k = Ad_{L_0}^k Y = \sum_{j=0}^{k} C_k^j \frac{\partial^k p}{\partial x_1^{k-j} \partial x_2^j} D_1^{k-j} D_2^{j+1}, \text{ mod } U_k, \quad (6.136)$$

where C_k^j's are binomial numbers. In particular,

6.5 Classification with Nonmaximal Rank

$$Y_l = \sum_{j=0}^{l} C_l^j (l-j)! j! a_j D_1^{l-j} D_2^{j+1} = l! \sum_{j=s}^{t} a_j D_1^{l-j} D_2^{j+1}, \text{ mod } U_l, \quad (6.137)$$

and

$$Y_{l-1} = \sum_{j=0}^{l-1} C_{l-1}^j \frac{\partial^{l-1} p}{\partial x_1^{l-j-1} \partial x_2^j} D_1^{l-j-1} D_2^{j+1}$$

$$= \sum_{j=0}^{l-1} (l-1)! \left((l-j) a_j x_1 + (j+1) a_{j+1} x_2 + c_j \right) D_1^{l-j-1} D_2^{j+1}, \text{ mod } U_{l-1}, \quad (6.138)$$

where c_j's are constants from the $(l-1)$-th partial derivatives of the homogeneous degree $l-1$ part of p.

Now, depending on whether or not $s = 0$, and in the case when $s = 0$ whether a_1, a_2 are 0, there are four cases for which similar constructions of sequences with different calculations will show that l must be less than 2 if E is finite dimensional.

(i) Case 1. $s \neq 0$.

$$A_0 := Y_{l-1} \quad (6.139)$$

$$= (d_0 x_2 + (l-1)! c_{s-1}) D_1^{l-s} D_2^s$$

$$+ \text{ terms with lower order in } D_1, \text{ mod } U_{l-1},$$

$$A_1 := Y_l$$

$$= d_1 D_1^{l-s} D_2^{s+1} + \text{ terms with lower order in } D_1, \text{ mod } U_l,$$

$$A_2 := [A_1, A_0]$$

$$= (s+1) d_1 d_0 D_1^{2(l-s)} D_2^{2s} + \text{ terms with lower order in } D_1, \text{ mod } U_{2l-1},$$

$$\vdots$$

$$A_{r+1} := [A_r, A_0]$$

$$= d_{r+1} D_1^{(r+1)(l-s)} D_2^{(r+1)s-r+1}$$

$$+ \text{ terms with lower order in } D_1, \text{ mod } U_{(r+1)l-r},$$

where $d_0 = (l-1)! s a_s \neq 0$, $d_1 = l! a_s \neq 0$, and $d_{r+1} = (rs - r + 2) d_r d_0 \neq 0$ for $r \geq 1$ by induction. The orders of A_{r+1}'s are $(r+1)l - r + 1 \to \infty$ unless $l < 2$.

(ii) Case 2. $s = 0$ and $a_1 \neq 0$.

$$A_0 := Y_{l-1} \tag{6.140}$$
$$= (l!a_0 x_1 + d_0 x_2 + (l-1)!c_0) D_1^{l-1} D_2 + \text{terms with lower order in } D_1,$$
$$\text{mod } U_{l-1},$$
$$A_1 := Y_l$$
$$= d_1 D_1^l D_2 + \text{terms with lower order in } D_1, \text{ mod } U_l,$$
$$A_2 := [A_1, A_0]$$
$$= d_1 d_0 D_1^{2l-1} D_2 + \text{terms with lower order in } D_1,$$
$$\text{mod } U_{2l-1},$$
$$\vdots$$
$$A_{r+1} := [A_r, A_0]$$
$$= d_{r+1} D_1^{(r+1)l-r} D_2 + \text{terms with lower order in } D_1, \text{ mod } U_{(r+1)l-r},$$

where $d_0 = (l-1)!a_1 \neq 0, d_1 = l!a_0 \neq 0$ and $d_{r+1} = d_r d_0 \neq 0$ for $r \geq 1$. The orders of A_{r+1}'s are $(r+1)l - r + 1 \to \infty$ unless $l < 2$.

(iii) Case 3. $s = 0$ and $a_1 = a_2 = 0$.

$$A_0 := Y_{l-1} = (d_0 x_1 + (l-1)!c_0) D_1^{l-1} D_2 + (l-1)!c_2 D_1^{l-2} D_2^2 \tag{6.141}$$
$$+ \text{degree 1 coeff. terms with lower order in } D_1, \text{ mod } U_{l-1},$$
$$A_1 := Y_l = d_1 D_1^l D_2$$
$$+ \text{constant coeff. terms with lower order in } D_1, \text{ mod } U_l,$$
$$A_2 := [A_1, A_0] = l d_1 d_0 D_1^{2l-2} D_2^2$$
$$+ \text{constant coeff. terms with lower order in } D_1, \text{ mod } U_{2l-1},$$
$$\vdots$$
$$A_{r+1} := [A_r, A_0] = d_{r+1} D_1^{(r+1)l-2r} D_2^{r+1}$$
$$+ \text{constant coeff. terms with lower order in } D_1, \text{ mod } U_{(r+1)l-r},$$

(iv) Case 4. $s = 0$ and $a_1 = 0$, but $a_2 \neq 0$.

$$Y_l = d_1 D_1^l D_2 + d_3 D_1^{l-2} D_2^3 + \text{terms with lower order in } D_1, \text{ mod } U_l, \tag{6.142}$$

where $d_1 = l!a_0 \neq 0$ and $d_3 = l!a_2 \neq 0$. If $l \geq 2$, consider $Z = [Y_l, pD_2]$ and $A_0 = Ad_{L_0}^{l-2} Z$:

6.5 Classification with Nonmaximal Rank

$$Z = [Y_l, pD_2]$$
$$= d_1 \frac{\partial p}{\partial x_2} D_1^l D_2 + l d_1 \frac{\partial p}{\partial x_1} D_1^{l-1} D_2^2$$
$$+ \text{terms with lower order in } D_1, \text{ mod } U_l,$$

$$A_0 = Ad_{L_0}^{l-2} Z$$
$$= d_1 \frac{\partial^{l-1} p}{\partial x_1^{l-2} \partial x_2} D_1^{2l-2} D_2 + \text{terms with lower order in } D_1, \text{ mod } U_{2l-2},$$
(6.143)

Since $p^{(l)} = a_0 x_1^l + a_2 x_1^{l-2} x_2^2 + \text{terms with lower degree in } x_1$, $\frac{\partial^{l-1} p}{\partial x_1^{l-2} \partial x_2} = 2(l-2)! a_2 x_2 + c_2$, where c_2 is $(l-2)!$ multiplied by the coefficient of $x_1^{l-2} x_2$ in p. Hence,

$$A_0 = (e_0 x_2 + d_1 c_2) D_1^{2l-2} D_2 + \text{terms with lower order in } D_1, \text{ mod } U_{2l-2},$$

$$A_1 = Y_l$$
$$= d_1 D_1^l D_2 + d_3 D_1^{l-2} D_2^3 + \text{terms with lower order in } D_1, \text{ mod } U_l,$$

$$A_2 = [A_1, A_0]$$
$$= d_1 e_0 D_1^{3l-2} D_2 + \text{terms with lower order in } D_1, \text{ mod } U_{3l-2},$$

$$\vdots$$

$$A_{r+1} = [A_r, A_0]$$
$$= e_{r+1} D_1^{(2r+1)l-2r} D_2 + \text{terms with lower order in } D_1, \text{ mod } U_{(2r+1)l-2r},$$
(6.144)

where $e_0 = 2(l-2)! a_2 d_1 \neq 0$, $e_1 = d_1 \neq 0$ and $e_{r+1} = e_r e_0 \neq 0$. The orders of A_{r+1}'s are $(2r+1)l - 2r + 1 \to \infty$ unless $l < 2$. □

Theorem 6.25 *Suppose* $\dim E < \infty$. $x_1 + c \in E \implies \omega_{12}$ *is a polynomial of degree at most 1.*

Proof Clearly, $L_0, x_1, D_1 \in E$.

$$H_0 := [L_0, D_1] = \omega_{12} D_2 + \frac{1}{2} \frac{\partial \omega_{12}}{\partial x_2} + \frac{1}{2} \frac{\partial \eta}{\partial x_1} \in E. \tag{6.145}$$

By Theorem 6.24, ω_{12} is a polynomial of degree at most 1. □

Remark 6.3 Theorem 6.24 shows that if the linear rank of estimation algebra is equal to 1, then entry of Wong's Ω-matrix is a degree 1 polynomial.

In the following, we prove some powerful tools to try to prove Mitter conjecture.

Theorem 6.26 *Suppose* $\dim E < \infty$. $x_1^2 + c \in E \implies \omega_{21}$ *is a constant.*

Proof

$$K_0 = \left[L_0, \frac{1}{2}(x_1^2 + c)\right] - \frac{1}{2} = x_1 D_1 \in E,$$

$$K_1 = [L_0, K_0] = D_1^2 - \alpha_2 D_2 + \frac{1}{2}E_1(\eta) - \frac{1}{2}\frac{\partial \alpha_2}{\partial x_2} \in E,$$

$$K_2 = [K_1, K_0] = 2D_1^2 + E_1(\alpha_2)D_2 + \alpha_2^2 - \frac{1}{2}E_1^2(\eta) + \frac{1}{2}E_1\left(\frac{\partial \alpha_2}{\partial x_2}\right) \in E$$

$$Z_0 = K_2 - 2K_1 = (E_1(\alpha_2) + 2\alpha_2)D_2 + \gamma(x) \in E,$$

(6.146)

where $\alpha_2 = x_1 \omega_{21}$, $E_1(\cdot) = x_1 \frac{\partial}{\partial x_1}$ and $\gamma(x) = \alpha_2^2 - E_1(\eta) - \frac{1}{2}E_1^2(\eta) + \frac{\partial \alpha_2}{\partial x_2} + \frac{1}{2}E_1\left(\frac{\partial \alpha_2}{\partial x_2}\right)$. By Theorem 6.24, $E_1(\alpha_2) + 2\alpha_2$ is a polynomial of at most 1, and so is $\alpha_2 = x_1 \omega_{21}$ by Theorem 6.21. Since $\omega_{21} \in C^\infty(R^2)$, ω_{21} must be a constant. $\square$

Lemma 6.20 *Suppose* $K_0 = x_1 D_1 + x_2 D_2 \in E$ *and* $Y = \sum_{i=0}^k b_i(x) D_1^{k-i} D_2^i$, *mod* $U_{k-1} \in E$. *Let* $b_i^{(r)}$ *denote the homogeneous degree* r *part of* b_i *for* $0 \leq i \leq k$. *Then* $\sum_{i=0}^k b_i^{(r)} D_1^{k-i} D_2^i$, *mod* $U_{k-1} \in E$ *for* $r \geq 0$.

Proof Let $E(\cdot) = \sum_{i=1}^2 x_i \frac{\partial}{\partial x_i}$:

$$\left[K_0, \sum_{i=0}^k b_i(x) D_1^{k-i} D_2^i, \text{mod } U_{k-1}\right] = \sum_{i=0}^k (E(b_i) - kb_i) D_1^{k-i} D_2^i, \text{mod } U_k - 1.$$

(6.147)

Let $l = \max_{0 \leq i \leq k} \deg b_i$. Use the above equation to construct a sequence of elements:

$$Z_r = \sum_{j=0}^{l-r} c_{rj} \sum_{i=0}^k b_i^{(j)} D_1^{k-i} D_2^j \in E, \text{ mod } U_{k-1},$$ (6.148)

in E as follows:

(1) $Z_0 = y$, i.e., $c_{0j} = 1, 1 \leq j \leq l$;
(2) $Z_{r+1} = (l-r-k)Z_r - [K_0, Z_r] \implies c_{r+1,j} = (l-r-j)c_{rj}, 0 \leq j \leq l-r-1$.

Note that in (6.148), $c_{rj} \neq 0$ for $r \leq l$. Starting from Z_l, one can solve $\sum_{i=0}^k b_i^{(r)} D_1^{k-1} D_2^i$, mod U_{k-1} successively. This will lead to

$$\sum_{i=0}^k b_i^{(r)} D_1^{k-1} D_2^i, \text{ mod } U_{k-1} \in E.$$ (6.149)

$\square$

6.5 Classification with Nonmaximal Rank

Lemma 6.21 *Suppose* $\dim E < \infty$. *Let* $Y = \sum_{i=i_0}^{i_k} b_i(x) D_1^{k-i} D_2^i$, $\mod U_{k-1} \in E$ *be a differential operator with the highest order k in E (obviously $k \geq 2$), where $0 \leq i_0 \leq i_k \leq k$. Then*

$$\frac{\partial b_{i_0}}{\partial x_1} = \frac{\partial b_{i_k}}{\partial x_2} = 0, \quad \frac{\partial b_i}{\partial x_1} + \frac{\partial b_{i-1}}{\partial x_2} = 0, \text{ for } i_0 + 1 \leq i \leq i_k. \tag{6.150}$$

Proof

$$[L_0, Y] = \sum_{i=i_0}^{i_k} \frac{\partial b_i}{\partial x_1} D_1^{k+1-i} D_2^i + \sum_{i=i_0}^{i_k} \frac{\partial b_i}{\partial x_2} D_1^{k-i} D_2^{i+1}$$

$$= \frac{\partial b_{i_0}}{\partial x_1} D_1^{k+1} + \sum_{i=i_0+1}^{i_k} \left(\frac{\partial b_i}{\partial x_1} + \frac{\partial b_{i-1}}{\partial x_2} \right) D_1^{k+1-i} D_2^i + \frac{\partial b_{i_k}}{\partial x_2} D_2^{k+1}, \mod U_k. \tag{6.151}$$

Since elements in E have the highest differential order k, all the coefficient functions of the order $k+1$ terms in the above equation must be 0. □

Lemma 6.22 *Suppose* $\dim E < \infty$ *and* $Y = p_1 D_1 + p_2 D_2$, $\mod U_0 \in E$, *where* $p_1 = \sum_{j=1}^{l} a_j x_1^{l-j} x_2^j$ *and* $p_2 = -\sum_{j=1}^{l} a_j x_1^{l+1-j} x_2^{j-1}$, a_j*'s are constants and* $a_j \neq 0$ *for some j. Further we assume $K_0 = x_1 D_1 + x_2 D_2 \in E$. Then $l \leq 2$.*

Remark 6.4 The proof depends on Lemma 6.20. Detailed proof can be found in Wu and Yau [26].

Next by applying Lemma 6.22, we will obtain the following important structure of estimation algebra.

Theorem 6.27 *Suppose* $\dim E < \infty$. $x_1^2 + x_2^2 + c \in E \implies \omega_{12}$ *is a polynomial of degree at most 1.*

Proof

$$K_0 = \left[L_0, \frac{1}{2}(x_1^2 + x_2^2 + c) \right] - 1 = x_1 D_1 + x_2 D_2 \in E \tag{6.152}$$

$$K_1 = [L_0, K_0] = \sum_{i=1}^{2} D_i^2 - \sum_{i=1}^{2} \alpha_i D_i + \frac{1}{2} E(\eta) - \frac{1}{2} \sum_{i=1}^{2} \frac{\partial \alpha_i}{\partial x_i} \in E \tag{6.153}$$

$$K_2 = 2L_0 - K_1 = \sum_{i=1}^{2} \alpha_i D_i + \frac{1}{2} \sum_{i=1}^{2} \frac{\partial \alpha_i}{\partial x_i} - \eta - \frac{1}{2} E(\eta) \in E, \tag{6.154}$$

where $\alpha_1 = x_2 \omega_{12}, \alpha_2 = x_1 \omega_{21}, E(\cdot) = \sum_{i=1}^{2} x_i \frac{\partial}{\partial x_i}$. Due to dim $E < \infty$ and (6.154), α_i is a polynomial. Therefore, ω_{12} must be a polynomial as well.

Let $\alpha^{(r)}$ be homogeneous degree r part of α_i. By Lemma 6.20,

$$\sum_{i=1}^{2} \alpha_i^{(r)} D_i, \mod U_0 \in E. \tag{6.155}$$

On the other hand,

$$\sum_{i=1}^{2} x_i \alpha_i = \sum_{i,j=1}^{2} x_i x_j \omega_{ij} = 0 \Longrightarrow x_1 \alpha_1^{(r)} + x_2 \alpha_2^{(r)} = 0. \tag{6.156}$$

By Lemma 6.22, we obtain $r \leq 2$. Therefore, ω_{12} is a polynomial of degree at most 1. □

Theorems 6.26 and 6.27 are significant for the following two results. First, we can prove that all observation terms are polynomials at most of degree 1. Furthermore, we can deduce that Mitter conjecture holds.

Theorem 6.28 *Suppose* dim $E < \infty$. h_i's *are polynomials at most of degree 1.*

Proof Without loss of generality, h_1 is assumed to be a polynomial of degree 2. By orthogonal and affine transformations, h_1 is either $ax_1^2 + cx_2 + d$ or $ax_1^2 + bx_2^2 + d$, where $a, b \neq 0$.

(i) If $h_1 = ax_1^2 + cx_2 + d$, then $[[L_0, h_1], h_1] = 4a^2 x_1^2 + c^2 \in E$. By Theorem 6.26, ω_{21} is a constant. Hence, E is not finite dimensional according to Theorem 6.13 (ii).

(ii) If $h_1 = ax_1^2 + bx_2^2 + d$, then $[[L_0, h_1], h_1] = 4a^2 x_1^2 + 4b^2 x_2^2 \in E$. If $a \neq b$, by technique of Vandermonde determinant, we can deduce both $x_1^2 \in E$ and $x_2^2 \in E$ hold. From the discussion (i), E is not finite dimensional. Hence, $h_1 = ax_1^2 + ax_2^2 + c$. By Theorem 6.27, ω_{12} is a polynomial of degree at most 1. Moreover, ω_{12} must be a nondegenerate degree 1 polynomial by Theorem 6.13, i.e., $\omega_{12} = c_1 x_1 + c_2 x_2 + c_0$, where $c_1 \neq 0$ or $c_2 \neq 0$.

From $K_0, K_2 \in E$ in Theorem 6.27, let $Z_0 = K_2$ and

$$Z_{r+1} = (1-r)Z_r - [K_0, Z_r] \text{ for } r = 0, 1. \tag{6.157}$$

We have

6.5 Classification with Nonmaximal Rank

$$Z_0 = \sum_{i=1}^{2} \alpha_i D_i + q_0 \in E,$$

$$Z_1 = c_0 x_2 D_1 - c_0 x_1 D_2 + q_0 - E(q_0) - \sum_{i=1}^{2} \alpha_i^2 \in E, \tag{6.158}$$

$$Z_2 = -c_0 x_2 \alpha_1 + c_0 x_1 \alpha_2 + E\left(\sum_{i=1}^{2} \alpha_i^2\right) - E(q_0 - E(q_0)) \in E,$$

where $q_0 = \frac{1}{2}\sum_{i=1}^{2} \frac{\partial \alpha_i}{\partial x_i} - \eta - \frac{1}{2} E(\eta)$. Since Z_2 must be a polynomial of degree less than or equal to 2, regardless of whether c_0 is 0 or not, then $q_0 - E(q_0)$ or $E(q_0 - E(q_0))$ is a degree 4 polynomial by Theorem 6.21. So η is a degree 4 polynomial by applying Theorems 6.20 and 6.21. Moreover, the homogeneous degree 4 part of η is

$$\eta^{(4)} = c\left(\sum_{i=1}^{2} \alpha_i^2\right)^{(4)} = c(x_1^2 + x_2^2)(c_1 x_1 + c_2 x_2)^2. \tag{6.159}$$

Hence, homogeneous degree 4 part of $\eta - \sum_{i=1}^{m} h_i^2$ is

$$c(x_1^2 + x_2^2)(c_1 x_1 + c_2 x_2)^2 - a^2(x_1^2 + x_2^2)^2 - \sum_{i>1}^{m} \left(h_i^{(2)}\right)^2. \tag{6.160}$$

By taking $x_1 = -c_2$ and $x_2 = c_1$, the above expression results in a negative number. By Theorem 6.22, there are no smooth solutions in f_i. Contradiction!

In summary of case (i) and (ii), h_i's are polynomials at most of degree 1. □

Theorem 6.29 (Mitter Conjecture) *Suppose* $\dim E < \infty$. *If* $\phi \in E$, ϕ *is a polynomial of degree at most 1.*

Proof

(i) If E has linear rank 2, E is of maximal rank, and therefore no degree 2 polynomials are in E by classification of maximal rank estimation algebra of Theorem 6.16.
(ii) If E has linear rank 0, then h_i's must be constants. E is finite dimensional and $E \subset \langle L_0, 1 \rangle_{L.A.}$. Any function element in E must be a constant.
(iii) Let E have linear rank 1, i.e., there exists a function $d_i x_1 + e_i \in E \Longrightarrow D_1 \in E$.

From the proof of Theorem 6.28, by orthogonal and affine transformations, a degree 2 polynomial $\phi \in E$ only has the form $ax_1^2 + ax_2^2 + d$, where $a \neq 0$. Again by Theorem 6.28, $\omega_{12} = c_1 x_1 + c_2 x_2 + c_0$, where $c_1 \neq 0$ or $c_2 \neq 0$ (otherwise ω_{12} is a constant). Thus,

$$[L_0, D_1] = \omega_{12} D_2 + \phi \in E$$

$$[\omega_{12} D_2 + \phi, ax_1^2 + ax_2^2 + d] = 2ax_2 \omega_{12} = 2ac_1 x_1 x_2 + 2ac_2 x_2^2 + 2ac_0 x_2 \in E. \tag{6.161}$$

If $c_1 \neq 0$, $[D_1, 2ac_1 x_1 x_2 + 2ac_2 x_2^2 + 2ac_0 x_2] = 2ac_1 x_2$ is dependent on x_2. This contradicts the linear rank of E being 1. If $c_1 = 0$ and $c2 \neq 0$, from $2ac_2 x_2 + 2ac_0 x_2$ and $ax_1^2 + ax_2^2 + d$, are in E, one has that $x_1^2 + c_4 x_2 + c_5 \in E$. From the discussion (i) in the proof of Theorem 6.28, E is not finite dimensional. $\square$

In the next discussion, E is assumed to be finite dimensional and has linear rank 1. As is shown in the previous theorem (Mitter conjecture), the structure of E is very clear when E's linear rank is 0 or 2. Therefore, in the following, we mainly focus on the case that linear rank is 1.

Assume $x_1 \in E$. Then if a function $p \in E$, p is a degree 1 polynomial in x_1 since Mitter conjecture holds. By Theorem 6.25, we assume $\omega_{12} = c_1 x_1 + c_2 x_2 + c_0$. It will be shown that ω_{12} must be a constant in this section.

By assumptions, $L_0, x_1, D_1, 1 \in E$. Moreover, the following elements are in E:

$$H_0 = [L_0, D_1] = \omega_{12} D_2 + \frac{1}{2} \frac{\partial \eta}{\partial x_1} + \frac{1}{2} c_2, \tag{6.162}$$

$$H_1 = [D_1, H_0] = c_1 D_2 - \omega_{12}^2 + \frac{1}{2} \frac{\partial^2 \eta}{\partial x_1^2}, \tag{6.163}$$

$$H_2 = [D_1, H_1] = -3c_1 \omega_{12} + \frac{1}{2} \frac{\partial^3 \eta}{\partial x_1^3}, \tag{6.164}$$

$$H_3 = [H_1, H_0] - c_2 H_1$$
$$= 3c_2 \omega_{12}^2 + \frac{1}{2} c_1 \frac{\partial^2 \eta}{\partial x_1 \partial x_2} - \frac{1}{2} c_2 \frac{\partial^2 \eta}{\partial x_1^2} - \frac{1}{2} \omega_{12} \frac{\partial^3 \eta}{\partial x_1^2 \partial x_2}, \tag{6.165}$$

$$X_0 = [L_0, H_0]$$
$$= c_1 D_1 D_2 + c_2 D_2^2 + \left(-\omega_{12}^2 + \frac{1}{2} \frac{\partial^2 \eta}{\partial x_1^2} \right) D_1 + \frac{1}{2} \frac{\partial^2 \eta}{\partial x_1 \partial x_2} D_2$$
$$+ \left(-\frac{1}{2} c_1 \omega_{12} + \frac{1}{2} \omega_{12} \frac{\partial \eta}{\partial x_2} + \frac{1}{4} \frac{\partial^3 \eta}{\partial x_1^3} + \frac{1}{4} \frac{\partial^3 \eta}{\partial x_1 \partial x_2^2} \right). \tag{6.166}$$

6.5 Classification with Nonmaximal Rank

Lemma 6.23 η is a polynomial in x_1 of degree at most 4 with coefficients being functions of x_2. Let $\eta = \sum_{i=0}^{4} a_i x_1^i$, where a_i's are functions of x_2. Then:

(i) a_4 is a constant;
(ii) $a_3 = c_1 c_2 x_2 + constant$;
(iii) $(c_1^2 - 4a_4)c_2 = 0$;
(iv) $3c_2^3 x_2^2 + 6c_0 c_2^2 x_2 - c_2 a_2 + \frac{1}{2}c_1 a_1' - c_0 a_2' - c_2 a_2' x_2 = constant$.

Proof Since H_2 is a function in E, H_2 is independent of x_2 and a degree 1 polynomial in x_1. Therefore, η is a polynomial in x1 of degree at most 4 with coefficient functions in x_2. Let $\eta = \sum_{i=0}^{4} a_i x_1^i$, where a_i's are functions of x_2.
$H_2 \in E \Longrightarrow a_4$ is a constant, and $a_3 = c_1 c_2 x_2 + constant$.
(iii) and (iv) follow from $H_3 \in E$ by substituting $\eta = \sum_{i=0}^{4} a_i x_1^i$ into (6.165). □

Now, from (6.166), $X_0 \in E$. Next we calculate

$$X = [L_0, X_0]$$
$$= \left(-3c_1 \omega_{12} + \frac{1}{2}\frac{\partial^3 \eta}{\partial x_1^3}\right) D_1^2 + \left(-4c_2 \omega_{12} + \frac{\partial^3 \eta}{\partial x_1^2 \partial x_2}\right) D_1 D_2 \quad (6.167)$$
$$+ \left(c_1 \omega_{12} + \frac{1}{2}\frac{\partial^3 \eta}{\partial x_1 \partial x_2^2}\right) D_2^2$$
$$= q_0 D_1^2 + q_1 D_1 D_2 + q_2 D_2^2, \text{ mod } U_1 \in E,$$

where $q_0 = -3c_1 \omega_{12} + \frac{1}{2}\frac{\partial^3 \eta}{\partial x_1^3}$, $q_1 = -4c_2 \omega_{12} + \frac{\partial^3 \eta}{\partial x_1^2 \partial x_2}$, $q_2 = c_1 \omega_{12} + \frac{1}{2}\frac{\partial^3 \eta}{\partial x_1 \partial x_2^2}$. By Lemma 6.23 (i) and (ii),

$$\frac{\partial q_0}{\partial x_1} = 12a_4 - 3c_1^2, \quad \frac{\partial q_0}{\partial x_2} = 0,$$

$$\frac{\partial q_1}{\partial x_1} = 2c_1 c_2, \quad \frac{\partial q_1}{\partial x_2} = -4c_2^2 + 2a_2'', \quad (6.168)$$

$$\frac{\partial q_2}{\partial x_1} = c_1^2 + a_2'', \quad \frac{\partial q_2}{\partial x_2} = c_1 c_2 + \frac{1}{2}a_1''' + a_2''' x_1.$$

Lemma 6.24 $a_4 = \frac{1}{4}c_1^2$.

Proof If $a_4 \neq \frac{1}{4}c_1^2$, i.e., $r := \frac{\partial q_0}{\partial x_1} = 12a_4 - 3c_1^2 \neq 0$, then from (6.167) and L_0, one can construct a sequence of elements in E whose orders strictly increase as follows:

$$Y_1 = [L_0, X] = rD_1^3 + \sum_{i=1}^{3} p_{1i} D_1^{3-i} D_2^i, \text{ mod } U_2,$$

$$Y_2 = [Y_1, X] = [rD_1^3 + p_{11} D_1^2 D_2 + \cdots, q_0 D_1^2 + q_1 D_1 D_2 + q_2 D_2^2]$$

$$= \left(3r \frac{\partial q_0}{\partial x_1} + p_{11} \frac{\partial q_0}{\partial x_2}\right) D_1^4 + \sum_{i=1}^{4} p_{2i} D_1^{4-i} D_2^i \tag{6.169}$$

$$= 3r^2 D_1^4 + \sum_{i=1}^{4} p_{2i} D_1^{4-i} D_2^i, \text{ mod } U_3,$$

...

Assume that $Y_k = r_k D_1^{k+2} + \sum_{i=1}^{k+2} p_{ki} D_1^{k+2-i} D_2^i$, mod U_{k+1} with $r_k \neq 0$:

$$Y_{k+1} = r_{k+1} D_1^{k+3} + \sum_{i=1}^{k+3} p_{(k+1)i} D_1^{k+3-i} D_2^i, \text{ mod } U_{k+2}, \tag{6.170}$$

where $r_{k+1} = (k+2) r r_k \neq 0$. □

The following theorem plays a significant role in constant structure of Ω-matrix.

Theorem 6.30 *If $c_1 c_2 \neq 0$, then E is not finite dimensional.*

Proof Assume that $c_1 c_2 \neq 0$.

(i) From Lemma 6.24, q_0 is a constant.
(ii) Let $r_1 := \frac{\partial q_1}{\partial x_1} = 2c_1 c_2$, and then we calculate

$$\begin{aligned} Z &= [L_0, X] = [L_0, q_0 D_1^2 + q_1 D_1 D_2 + q_2 D_2^2 \text{ mod } U_1] \\ &= r_1 D_1^2 D_2 + \text{terms with lower order in} D_1, \text{ mod } U_2 \in E. \end{aligned} \tag{6.171}$$

Suppose q_1 is a polynomial of degree k in x_2, i.e., $\frac{\partial^k q_1}{\partial x_2^k}$ is a nonzero constant.
If $k \geq 1$,

$$\begin{aligned} A_1 &= [Z, X] \\ &= r_1 \frac{\partial q_1}{\partial x_2} D_1^3 D_2 + \text{terms with lower order in} D_1, \text{ mod } U_3 \in E. \end{aligned}$$

6.5 Classification with Nonmaximal Rank

$$A_2 = [Z, A_1]$$

$$= r_1^2 \frac{\partial^2 q_1}{\partial x_2^2} D_1^5 D_2 + \text{terms with lower order in } D_1, \text{ mod } U_5 \in E.$$

$$\vdots$$

$$A_k = [Z, A_{k-1}]$$

$$= r_1^k \frac{\partial^k q_1}{\partial x_2^k} D_1^{2k+1} D_2 + \text{terms with lower order in } D_1, \text{ mod } U_{2k+1} \in E. \tag{6.172}$$

Now one can repeat the above process by letting $Z = A_k$ to construct a sequence of elements in E whose orders strictly increase.

Hence, k must be zero, which means $\frac{\partial q_1}{\partial x_2} = 0 \Longrightarrow a_2'' = 2c_2^2$.

(iii) By taking twice the derivative with respect to x_2 in Lemma 6.23 (iv) and substituting $a_2'' = 2c_2^2$, one has $c_1 a_1''' = 0 \Longrightarrow a_1''' = 0$.

Thus, $X = q_0 D_1^2 + q_1 D_1 D_2 + q_2 D_2^2$, mod $U_1 \in E$, with

$$\frac{\partial q_0}{\partial x_1} = \frac{\partial q_0}{\partial x_2} = 0,$$

$$\frac{\partial q_1}{\partial x_1} = 2c_1 c_2 = r_1 \neq 0, \quad \frac{\partial q_1}{\partial x_2} = 0, \tag{6.173}$$

$$\frac{\partial q_2}{\partial x_1} = c_1^2 + 2c_2^2 = r_2 > 0, \quad \frac{\partial q_2}{\partial x_2} = c_1 c_2 = \frac{1}{2} r_1,$$

$$Y_0 = [L_0, X] = p_{01} D_1^2 D_2 + p_{02} D_1 D_2^2 + p_{03} D_2^3, \text{ mod } U_2 \in E,$$

where $p_{01} = r_1$, $p_{02} = r_2$, $p_{03} = \frac{1}{2} r_1$. For $l \geq 0$, suppose $Y_l = p_{l1} D_1^2 D_2^{l+1} + p_{l2} D_1 D_2^{l+2} + p_{l3} D_2^{l+3} \in E$, mod U_{l+2}, where p_{l1}, p_{l2}, p_{l3} are constants:

$$Y_{l+1} = [Y_l, X]$$

$$= \frac{l+5}{2} r_1 p_{l1} D_1^2 D_2^{l+2} + \left(2 r_2 p_{l1} + \frac{l+4}{2} r_1 p_{l2}\right) D_1 D_2^{l+3}$$

$$+ \left(r_2 p_{l2} + \frac{l+3}{2} r_1 p_{l3}\right) D_2^{l+4}$$

$$= p_{(l+1)1} D_1^2 D_2^{l+2} + p_{(l+1)2} D_1 D_2^{l+3} + p_{(l+1)3} D_2^{l+4}, \text{ mod } U_{l+3} \in E. \tag{6.174}$$

By comparing the coefficient, we obtain $p_{(l+1)1} = \frac{l+5}{2} r_1 p_{l1}$ is a constant. Since $p_{01} = r_1 \neq 0$, $p_{l1} \neq 0$ for any l. Thus, Y_l is a differential operator of degree $l + 3$. Contradiction!

□

Remark 6.5 This theorem shows that if E is a finite-dimensional estimation algebra, then at least one of c_1, c_2 is 0. In the following, we just need to discuss two cases (i) $c_1 = 0, c_2 \neq 0$ and (ii) $c_1 \neq 0, c_2 = 0$. If we find contradiction in both (i) and (ii), then we can deduce if dim $E < \infty$, then $c_1 = c_2 = 0$, i.e., Ω-matrix has constant entries.

Lemma 6.25 *If $c_1 = 0, c_2 \neq 0$, we have (i) $a_4 = 0$; (ii) $a_3 = 0$; (iii) $a_2 = \omega_{12}^2 + $ constant; and (iv) $a_1''' = 0$.*

Proof (i) follows from Lemma 6.24.

By Lemma 6.23 (ii), a_3 is a constant. Since a_3 is the coefficient of an odd order term in η with the highest order in $x1$, $a_3 = 0$ by (6.19).

(iii) follows from (6.163). H_1 is a function in E when $c_1 = 0$.

Now, q_0, q_1 are actually constants, while $\frac{\partial q_2}{\partial x_1} = a_2'' = 2c_2^2$ and $\frac{\partial q_2}{\partial x_2} = \frac{1}{2} a_1'''$. Let a_1'' be a degree k polynomial in x_2, i.e., $\frac{d^{k+2} a_1}{dx_2^{k+2}}$ is a nonzero constant. If $k > 1$,

$$Z_0 = Ad_{L_0}^k X = \frac{1}{2} \frac{d^{k+2} a_1}{dx_2^{k+2}} D_2^{k+2}, \text{ mod } U_{k+1},$$

$$Y_0 = [Z_0, X] = (k+2) \frac{1}{4} \frac{d^{k+2} a_1}{dx_2^{k+2}} \frac{d^3 a_1}{dx_2^3} D_2^{k+3}, \text{ mod } U_{k+2}.$$

(6.175)

By letting $Z_l = Ad_{L_0}^{k-1} Y_{l-1}$ and $Y_l = [Z_l, X]$ for $l \geq 1$, one can construct a sequence of elements in E with strictly increasing orders.

If $k = 1$, consider $Z_0 = [L_0, X] = \frac{\partial q_2}{\partial x_1} D_1 D_2^2 + \frac{\partial q_2}{\partial x_2} D_2^3$, mod U_2 and $Z_{l+1} = [Z_l, X]$ for $l \geq 0$. Assume

$$Z_l = p_{l0} D_1 D_2^{l+2} + p_{l1} D_2^{l+3}, \text{ mod } U_{l+2},$$

(6.176)

where $p_{00} = \frac{\partial q_2}{\partial x_1} = 2c_2^2$ and $p_{01} = \frac{\partial q_2}{\partial x_2} = \frac{1}{2} \frac{d^3 a_1}{dx_2^3}$. Then

$$Z_{l+1} = [Z_l, X] = (l+2) p_{l0} \frac{\partial q_2}{\partial x_2} D_1 D_2^{l+3}$$
$$+ \left(p_{l0} \frac{\partial q_2}{\partial x_1} + (l+3) p_{l1} \frac{\partial q_2}{\partial x_2} \right) D_2^{l+4}, \text{ mod } U_{l+3}.$$

(6.177)

Therefore,

6.5 Classification with Nonmaximal Rank

$$p_{(l+1)0} = (l+2)p_{l0}\frac{\partial q_2}{\partial x_2} \implies p_{l0} = (l+1)!\frac{\partial q_2}{\partial x_1}\left(\frac{\partial q_2}{\partial x_2}\right)^l \neq 0. \tag{6.178}$$

Z_l's have strictly increasing orders. Contradiction! Hence, k must be zero $\implies a_1''' = 0$. □

Theorem 6.31 *If $c_1 = 0$, $c_2 \neq 0$, E is not finite dimensional.*

Proof In the proof of this theorem, $r_1, r_2, \cdots$ are used to denote constants. The exact values of these r_i's are not important. Some r_i's may be used repeatedly to denote different constants.

By substituting Lemma 6.25 (i)–(iv) into (6.166), we obtain

$$\begin{cases}
Z_0 = X_0 - r_1 D_1 - r_2 x_1 - r_3 \\
\quad = c_2 D_2^2 + \frac{1}{2}\frac{\partial^2 \eta}{\partial x_1 \partial x_2} D_2 + \frac{1}{2}\omega_{12}\frac{\partial \eta}{\partial x_2}, \\
Z_1 = [L_0, Z_0] - r_5 D_1 \\
\quad = \frac{1}{2}\frac{\partial^3 \eta}{\partial x_1 \partial x_2^2} D_2^2 + \left(\frac{3}{2}c_2\frac{\partial \eta}{\partial x_2} + \frac{1}{2}\omega_{12}\frac{\partial^2 \eta}{\partial x_2^2}\right) D_2, \bmod U_0 \\
Z_2 = [L_0, Z_1] = a_2'' D_1 D_2^2 + \left(\frac{3}{2}c_2\frac{\partial^2 \eta}{\partial x_1 \partial x_2} - \frac{1}{2}\omega_{12}\frac{\partial^3 \eta}{\partial x_1 \partial x_2^2}\right) D_1 D_2 \\
\quad + \left(2c_2\frac{\partial^2 \eta}{\partial x_2^2} + \frac{1}{2}\omega_{12}\frac{\partial^3 \eta}{\partial x_2^3}\right) D_2^2, \bmod U_1 \\
Z_3 = [L_0, Z_2] = p_1 D_1 D_2^2 + p_2 D_2^3, \bmod U_2.
\end{cases} \tag{6.179}$$

where $p_1 = 3c_2\frac{\partial^3 \eta}{\partial x_1 \partial x_2^2} = 3c_2(a_1'' + 2a_2'' x_1)$ is degree 1 in x_1 and independent of x_2 with $\frac{\partial p_1}{\partial x_1} = 6c_2 a_2'' = 12c_2^3 \neq 0$, and $p_2 = a_2'' \omega_{12} + \frac{5}{2}c_2\frac{\partial^3 \eta}{\partial x_2^3} + \frac{1}{2}\omega_{12}\frac{\partial^4 \eta}{\partial x_2^4}$ is a function of x_2 and independent of x_1. Therefore, p_2 must be a polynomial in x_2; otherwise, E is not finite dimensional.

Next, we have

$$Y_0 = [Z_0, Z_3] = 2c_2\frac{\partial p_2}{\partial x_2} D_2^4, \bmod U_3 \in E,$$

$$Y_1 = [Z_0, Y_0] = (2c_2)^2\frac{\partial^2 p_2}{\partial x_2^2} D_2^5, \bmod U_4 \in E, \tag{6.180}$$

$$\cdots$$

$$Y_l = Ad_{Z_0}^{l+1} Z_3 \in E.$$

If deg $p_2 = k \geq 1$, one can construct an infinite sequence of elements in E whose orders strictly increase as follows:

$$W_0 = Ad_{Z_0}^k Z_3 = d_0 D_2^{e_0}, \text{ mod } U_{e_0-1}, \quad d_0 = (2c_2)^k \frac{\partial^k p_2}{\partial x_2^k} \neq 0, e_0 = k+3,$$

$$W_{l+1} = Ad_{Z_0}^{k-1}[W_l, Z_3] = d_{l+1} D_2^{e_{l+1}}, \text{ mod } U_{e_{l+1}-1},$$
(6.181)

where $d_{l+1} = e_l d_l (2c_2^{k-1}) \frac{\partial^k p_2}{\partial x_2^k} \neq 0$ and $e_{l+1} = e_l + k + 1 > e_l$. Therefore, p_2 is a constant.

Now, p_1 is a degree 1 polynomial in x_1, and p_2 is a constant:

$$V_0 = [L_0, Z_3] = \frac{\partial p_1}{\partial x_1} D_1^2 D_2^2, \text{ mod } U_3,$$

$$V_{l+1} = [V_l, Z_3] = 2^{l+1} \left(\frac{\partial p_1}{\partial x_1}\right)^{l+2} D_1^2 D_2^{2(l+2)}, \text{ mod } U_{2l+5}.$$
(6.182)

Again, V_l's have increasing orders. Hence, E is not finite dimensional. □

Lemma 6.26 *If $c_1 \neq 0, c_2 = 0$, we have (i) $a_4 = \frac{1}{4}c_1^2$; (ii) a_3 is a constant; (iii) $a_2'' = 0$; and (iv) $a_1'' = 0$.*

In a similar way, we can discuss a case that $c_1 \neq 0, c_2 = 0$. We can explore the structure of η in detail and then deduce a contradiction.

Lemma 6.27 (Lemma 4.16 [26]) *If $c_1 \neq 0, c_2 = 0$, then η is a degree 4 polynomial in x_1, x_2 with its principal part $\eta^{(4)} = \frac{1}{4}c_1^2 x_1^4$.*

Theorem 6.32 (Theorem 4.17 [26]) *If $c_1 \neq 0, c_2 = 0$, then E is not finite dimensional.*

Finally, let us show the constant structure of Ω.

Theorem 6.33 *If E has linear rank 1 and is finite dimensional, then the Ω-matrix has constant entries.*

Proof By Theorems 6.30–6.32, we can obtain that if $\dim E < \infty$, $c_1 = c_2 = 0$. It shows that Ω-matrix has constant entries. □

Finally, we can finish complete classification of estimation algebra with state dimension 2. First, assume E has linear rank 1. By Theorem 6.29, h_i's must be degree at most 1 polynomials in x_1. By Theorem 6.33, ω_{12} is a constant.

(i) If $\omega_{12} = 0$, $[L_0, D_1] = \frac{1}{2}\frac{\partial \eta}{\partial x_1} \in E$. Thus, η must be a degree 2 polynomial in x_1 plus a C^∞-function in x_2. $E = \{L_0, x_1, D_1, 1\}$. For example, $f_1 = x_1, f_2 = \sin x_2, h_1 = x_1, \eta = 2x_1^2 + 1 + \cos x_2 + \sin^2 x_2$.
(ii) If $\omega_{12} \neq 0$,

6.5 Classification with Nonmaximal Rank

$$A_1 := [L_0, D_1] = \omega_{12} D_2 + \frac{1}{2}\frac{\partial \eta}{\partial x_1} \in E, \tag{6.183}$$

$$A_2 := [D_1, A_1] = -\omega_{12}^2 + \frac{1}{2}\frac{\partial^2 \eta}{\partial x_1^2} \in E, \tag{6.184}$$

$$A_3 := [L_0, A_1]$$
$$= \left(-\omega_{12}^2 + \frac{1}{2}\frac{\partial^2 \eta}{\partial x_1^2}\right) D_1 + \frac{1}{2}\frac{\partial^2 \eta}{\partial x_1 \partial x_2} D_2$$
$$+ \frac{1}{2}\omega_{12}\frac{\partial \eta}{\partial x_2} + \frac{1}{4}\frac{\partial^3 \eta}{\partial x_1^3} + \frac{1}{4}\frac{\partial^3 \eta}{\partial x_1 \partial x_2^2} \in E. \tag{6.185}$$

By (6.184), $\eta = d_0 x_1^3 + d_1 x_1^2 + e_2(x_2) x_1 + e_3(x_2)$. By Theorem 6.23, $d_0 = 0$. By (6.185) and $D_1 \in E$,

$$\frac{1}{2}\frac{\partial^2 \eta}{\partial x_1 \partial x_2} D_2 + \frac{1}{2}\omega_{12}\frac{\partial \eta}{\partial x_2} + \frac{1}{4}\frac{\partial^3 \eta}{\partial x_1 \partial x_2^2} \in E. \tag{6.186}$$

By Theorem 6.24, $\frac{\partial^2 \eta}{\partial x_1 \partial x_2} = e_2'(x_2)$ is a polynomial of degree at most 1. Thus, e_2 is a polynomial of degree at most 2 in x_2. By substituting e_2 and e_3 into η and removing D_1, x_1, and 1 from (6.183) and (6.186), one has

$$\bar{A}_1 = \omega_{12} D_2 + \frac{1}{2} e_2 \in E,$$
$$\bar{A}_3 = \frac{1}{2} e_2' D_2 + \frac{1}{2}\omega_{12}(e_2' x_1 + e_3') \in E. \tag{6.187}$$

If e_2 is a degree 2 polynomial in x_2, then $[\bar{A}_1, \bar{A}_3] - \frac{1}{2} e_2'' \bar{A}_1 = \frac{1}{2}\omega_{12}^2 e_3'' - \frac{1}{4}(e_2 e_2'' + e_2' e_2') + \frac{1}{2}\omega_{12}^2 e_2'' x_1 \in E$. Since the degree 2 term of $e_2 e_2'' + e_2' e_2'$ will never be zero, $e_3(x_2)$ must be a degree 4 polynomial. Now, consider

$$B_1 = [L_0, \bar{A}_3] = \frac{1}{2} e_2'' D_2^2 + \frac{1}{2}\omega_{12}(e_2'' x_1 + e_3'') D_2 + \frac{1}{4} e_2'(e_2' x_1 + e_3')$$
$$+ \frac{1}{4}\omega_{12} e_3'' \in E,$$

$$B_2 = [L_0, B_1] = \frac{1}{2}\omega_{12} e_3''' D_2^2 - \frac{1}{2}\omega_{12} e_2'' D_2 D_1, \mod U_1 \in E. \tag{6.188}$$

Therefore, $e_3''' = constant$; otherwise, we can construct an infinite sequence by using L_0 and B_2. Contradiction! Hence, e_2 must be a degree 1 polynomial.

Consider $\omega_{12}\bar{A}_3 - \frac{1}{2}e'_2\bar{A}_1 = \frac{1}{2}\omega_{12}^2 e'_2 x_1 + \frac{1}{2}\omega_{12}^2 e'_3 - \frac{1}{4}e'_2 e_2 \in E \Longrightarrow \frac{1}{2}\omega_{12}^2 e'_3 - \frac{1}{4}e'_2 e_2$ is independent of x_2 and e_3 must be a degree 2 polynomial. Hence, η is a degree 2 polynomial in x_1 and x_2. Therefore, E is of dimension 5 and $E = \{L_0, x_1, D_1, D_2 + cx_2, 1\}$.

For example, $f_1 = 5x_1 - 3x_2$, $f_2 = 4x_2$, $h_1 = x_1$. Then $\omega_{12} = 3$ and $\eta = 26x_1^2 - 30x_1 x_2 + 25x_2^2 + 9$. It is easy to show that $E = \{L_0, x_1, D_1, D_2 - 5x_2, 1\}$.

If E has linear rank 0, h_i's must be constants, and $E = \{L_0\}$ or $E = \{L_0, 1\}$.

If E has linear rank 2, E is of maximal rank. The Ω-matrix must have constant entries and $E = \{L_0, x_1, x_2, D_1, D_2, 1\}$ by classification result of maximal rank. In summary, we have the following.

Theorem 6.34 (Complete Classification Result) *Let state dimension $n = 2$. If E is finite dimensional, then*

(1) if h_i's are constants, $E = \{L_0\}$ or $E = \{L_0, 1\}$.
(2) otherwise, Ω-matrix has constant entries. h_i's must be affine in x_1 and x_2. E has dimension of either 4, 5, or 6.

Moreover, from the above discussion, it is easy to see that if E is finite dimensional, it has only elements with order less than or equal to 2. Thus, the Levine conjecture holds for the finite-dimensional estimation algebras with state dimension 2.

Remark 6.6 Finally, we finish complete classification of estimation algebra with state dimension 2. The important step is that Wu and Yau [26] can prove that Wong's Ω-matrix has constant entries. Especially, this is difficult for the case of quadratic rank equal 0. Based on results of constant structure of Wong's Ω-matrix, it can be show η function is a degree 2 polynomial in x_1, x_2. This will determine the explicit basis of nonmaximal rank estimation algebra and will finish complete classification.

6.5.2 State Dimension 3

In this section, we study the structure of finite-dimensional estimation algebras with state dimension 3 and rank 2 arising from a nonlinear filtering system by using the theories of the Euler operator and underdetermined partial differential equations.

6.5.2.1 Linear Structure of Wong's Ω-Matrix

In this subsection, the structure of Wong's Ω-matrix is shown to be linear, i.e., the entries of the Ω-matrix are polynomials of degree ≤ 1. The main reference used in this subsection is the work of Shi and Yau [21]. The fundamental strategy we use to prove these results is to show that if they were not true, then infinite sequences could be constructed in the finite-dimensional estimation algebra.

6.5 Classification with Nonmaximal Rank

Assumption: In this section, we consider state dimension $n = 3$ estimation algebra E of system (6.1) with linear rank 2, also $\dim(E) < \infty$. Without loss of generality, we assume there exist constants c_i, $1 \leq i \leq 2$, such that $x_i + c_i \in E$, $1 \leq i \leq 2$ and for any constant c, $x_3 + c \notin E$.

The following notations are used in this section:

$P_k(x_{i_1}, \ldots, x_{i_m})$ denotes the space consisting of polynomials of degree at most k in $x_{i_1}, \ldots, x_{i_m}$, and $pol_k(x_{i_1}, \ldots, x_{i_m})$ denotes a polynomial in $P_k(x_{i_1}, \ldots, x_{i_m})$.

We give the following elementary lemma.

Lemma 6.28

$$[L_0, x_i + c_i] = D_i \in E, \ 1 \leq i \leq 2, \tag{6.189}$$

$$[D_2, D_1] = \omega_{12} \in E, [D_1, x_1 + c_1] = 1 \in E \tag{6.190}$$

$$Y_1 := [L_0, D_1] = \omega_{12} D_2 + \omega_{13} D_3 + \frac{1}{2}\frac{\partial \omega_{12}}{\partial x_2} + \frac{1}{2}\frac{\partial \omega_{13}}{\partial x_3} + \frac{1}{2}\frac{\partial \eta}{\partial x_1} \tag{6.191}$$

$$= \omega_{12} D_2 + \omega_{13} D_3, \ mod \ U_0 \in E \tag{6.192}$$

$$Y_2 := [L_0, D_2] = \omega_{21} D_1 + \omega_{23} D_3 + \frac{1}{2}\frac{\partial \omega_{21}}{\partial x_1} + \frac{1}{2}\frac{\partial \omega_{23}}{\partial x_3} + \frac{1}{2}\frac{\partial \eta}{\partial x_2} \tag{6.193}$$

$$= \omega_{21} D_1 + \omega_{23} D_3, \ mod \ U_0 \in E. \tag{6.194}$$

So $P_1(x_1, x_2) \subset E$.

Lemma 6.29 *For any function $\phi \in E$, ϕ does not contain $x_1 x_3$, $x_2 x_3$ terms.*

Proof By Ocone's Theorem 6.2, every function in estimation algebra E is a polynomial of degree at most 2. Since $P_1(x_1, x_2) \subset E$, without loss of generality, assume ϕ in E be

$$\phi = ax_1^2 + bx_2^2 + cx_3^2 + dx_1 x_2 + ex_1 x_3 + fx_2 x_3 + gx_3, \tag{6.195}$$

where a, b, c, d, e, f, g are constants:

$$[D_1, \phi] = 2ax_1 + dx_2 + ex_3 \in E$$
$$[D_2, \phi] = 2bx_2 + dx_1 + fx_3 \in E. \tag{6.196}$$

So $ex_3, fx_3 \in E$. By assumption $x_3 \notin E$; hence, $e = f = 0$. □

Theorem 6.35 *ω_{12} is a degree no more than 1 polynomial of x_1, x_2.*

Proof Step [1]: We prove that the degree 2 part of ω_{12} can only be $const \cdot x_3^2$, where const means a constant.

From Theorem 6.1 and equations $Y_1, Y_2, \omega_{12}, \omega_{13}, \omega_{23}$ are polynomials. From Lemma 6.29, we may assume any $\phi \in E$ is of the following form:

$$\phi = ax_1^2 + bx_2^2 + cx_3^2 + dx_1x_2 + gx_3, \tag{6.197}$$

where a, b, c, d are constants. Consider

$$Z := [L_0, \phi]$$
$$= (2ax_1 + dx_2)D_1 + (2bx_2 + dx_1)D_2 + (2cx_3 + g)D_3 + a + b + c \in E$$
$$[D_1, Z] = 2aD_1 + dD_2 + (2bx_2 + dx_1)\omega_{21} + (2cx_3 + g)\omega_{31} \in E$$
$$[D_2, Z] = dD_1 + 2bD_2 + (2ax_1 + dx_2)\omega_{12} + (2cx_3 + g)\omega_{32} \in E. \tag{6.198}$$

Since $D_1 \in E, D_2 \in E$, we have

$$(2bx_2 + dx_1)\omega_{21} + (2cx_3 + g)\omega_{31} \in E \tag{6.199}$$

$$(2ax_1 + dx_2)\omega_{12} + (2cx_3 + g)\omega_{32} \in E. \tag{6.200}$$

Case (1): There exists $\phi \in E$ in which a, b, d are not all 0. Note in Theorem 6.1 and Lemma 6.29 that any function in E is a degree no more than 2 polynomial and does not contain x_1x_3, x_2x_3 terms.

Case (1.1): If $c \neq 0$, then ω_{13}, ω_{23} are degree at most 2 polynomials. If $a \neq 0$, then from Eq. (6.199), the degree 2 part of ω_{12} cannot contain x_1^2, x_1x_2, x_2^2 terms; that is, the degree 2 part of ω_{12} can only be $const \cdot x_3^2$. If $b \neq 0$ or $d \neq 0$, we can easily find the same conclusion holds.

Case (1.2): If $c = g = 0$, then from Eqs. (6.199) and (6.200), we can easily find that the degree 2 part of ω_{12} can only be $const \cdot x_3^2$.

Case (1.3): If $c = 0, g \neq 0$, then

$$Z = [L_0, \phi] = (2ax_1 + dx_2)D_1 + (2bx_2 + dx_1)D_2 + gD_3 + a + b \in E,$$
$$[Z, \phi] = (4a^2 + d^2)x_1^2 + (4b^2 + d^2)x_2^2 + 4(a+b)dx_1x_2 + g^2 \in E,$$
$$\implies \psi := \hat{a}x_1^2 + \hat{b}x_2^2 + \hat{d}x_1x_2 \in E, \tag{6.201}$$

where $\hat{a} = 4a^2 + d^2, \hat{b} = 4b^2 + d^2, \hat{d} = 4(a+b)d$ are not all 0. By the above case (1.2), the degree 2 part of ω_{12} can only be $const \cdot x_3^2$, where const means constant (hereinafter).

Case (2): For any $\phi \in E, a = b = d = 0$. In this case, since $\omega_{12} \in E$, the degree 2 part of ω_{12} can only be $const \cdot x_3^2$. From case (1) and case (2), we can assume $\omega_{12} = \frac{1}{2}kx_3^2 + gx_3 + mx_1 + nx_2 + l \in E$.

Step [2]: In this step, we prove that ω_{12} is a degree at most 1 polynomial in x_1, x_2. If $k = 0$, then $g = 0$ by assumption, and the conclusion holds. If $k \neq 0$, without loss of generality, assume $k = 1$; then

6.5 Classification with Nonmaximal Rank

$$\omega_{12} \in E \implies \frac{1}{2}x_3^2 + gx_3 \in E \tag{6.202}$$

$$[L_0, \omega_{12}] = mD_1 + nD_2 + (x_3 + g)D_3 + \frac{1}{2} \in E \implies (x_3 + g)D_3 \in E \tag{6.203}$$

$$[D_1, (x_3 + g)D_3] = (x_3 + g)\omega_{31} \in E \tag{6.204}$$

$$[D_2, (x_3 + g)D_3] = (x_3 + g)\omega_{32} \in E. \tag{6.205}$$

Therefore, ω_{13}, ω_{23} are degree at most 1 polynomials of x_3.

(i) If ω_{13}, ω_{23} are both degree 1, without loss of generality, assume $\omega_{13} = x_3 + \alpha, \omega_{32} = x_3 + \beta$, where α, β are constants. From (6.204),

$$(x_3 + g)(x_3 + \alpha) = x_3^2 + (g + \alpha)x_3 + g\alpha \in E \implies x_3^2 + (g + \alpha)x_3 \in E, \tag{6.206}$$

combining this with (6.202) $\implies (g - \alpha)x_3 \in E \implies \alpha = g$. Similarly, $\beta = g$. So $\omega_{31} = \omega_{32} = x_3 + g$. Recall that

$$\begin{aligned} Y_1 &= \omega_{12}D_2 + \omega_{13}D_3, \bmod U_0 = \omega_{12}D_2 - (x_3 + g)D_3, \bmod U_0 \in E \\ Y_2 &= \omega_{21}D_1 + \omega_{23}D_3, \bmod U_0 = \omega_{21}D_1 - (x_3 + g)D_3, \bmod U_0 \in E. \end{aligned} \tag{6.207}$$

Combining Y_1, Y_2 with (6.203), we have

$$\omega_{12}D_2, \bmod U_0 \in E, \omega_{12}D_1, \bmod U_0 \in E. \tag{6.208}$$

(ii) Only one of ω_{13}, ω_{23} is degree 1. Without loss of generality, assume $\omega_{13} = x_3 + \alpha, \omega_{32} = \beta$. The proof in (i) shows that $\omega_{12}D_2, \bmod U_0 \in E$. From (6.205), $\omega_{32} = \beta = 0$. From Y_2, we have $\omega_{12}D_1, \bmod U_0 \in E$. Therefore, (6.208) holds.

(iii) If ω_{13}, ω_{23} are all constants, from the proof of (ii), we can see that (6.208) also holds. Namely, (6.208) always holds. Consider

$$\begin{aligned} N_0 &= [(x_3 + g)D_3, \omega_{12}D_2, \bmod U_0] = (x_3 + g)^2 D_2, \bmod U_0 \in E \\ M_1 &= [L_0, N_0] = 2(x_3 + g)D_2 D_3, \bmod U_1 \in E \\ N_1 &= [M_1, N_0] = 2^2(x_3 + g)^2 D_2^2, \bmod U_1 \in E \\ &\cdots \\ M_n &= [L_0, N_{n-1}] = 2^{2n-1}(x_3 + g)D_2^n D_3, \bmod U_n \in E \\ N_n &= [M_n, N_0] = 2^{2n}(x_3 + g)^2 D_2^{n+1}, \bmod U_1 \in E. \end{aligned} \tag{6.209}$$

Continuing this procedure, we can gain an infinite sequence in E which contradicts with the finite dimensionality of E. Hence, ω_{12} must be a degree 1 polynomial of x_1, x_2. □

The following several lemmas are very important tools to solve the linear structure of ω_{13}, ω_{23}.

Lemma 6.30 *Suppose that*

$$K := cD_3^{n+1} + (2ax_1 + dx_2 + e)D_1D_3^n + (2bx_2 + dx_1 + f)D_2D_3^n$$
$$+ \cdots, \bmod U_n \in E$$
$$A := (2ax_1 + dx_2 + e)D_3^l + \cdots, \bmod U_{l-1} \in E \quad (6.210)$$
$$B := (2bx_2 + dx_1 + f)D_3^l + \cdots, \bmod U_{l-1} \in E,$$

where a, b, c, d, e, f are constants, $n \geq 1, l \geq 1$. The $(\cdots)$ part means terms with the highest order but lower order in D_3. Then $a = b = d = 0$.

Proof If

$$\det \begin{pmatrix} 2a & d \\ d & 2b \end{pmatrix} = 4ab - d^2 \neq 0, \quad (6.211)$$

then a, d are not all zero, and from A and B we have

$$\begin{cases} C_{11} := (x_1 + \tilde{c}_1)D_3^l + \cdots, \bmod U_{l-1} \in E \\ C_{12} := (x_2 + \tilde{c}_2)D_3^l + \cdots, \bmod U_{l-1} \in E \\ B_{21} := [K, C_{11}] = (2ax_1 + dx_2 + e)D_3^{l+n} + \cdots, \bmod U_{l+n-1} \in E \\ B_{22} := [K, C_{12}] = (dx_1 + 2bx_2 + f)D_3^{l+n} + \cdots, \bmod U_{l+n-1} \in E. \end{cases}$$
(6.212)

For the same reason, we have

$$\begin{cases} C_{21} := (x_1 + \tilde{c}_1)D_3^{l+n} + \cdots, \bmod U_{l+n-1} \in E \\ C_{22} := (x_2 + \tilde{c}_2)D_3^{l+n} + \cdots, \bmod U_{l+n-1} \in E. \end{cases} \quad (6.213)$$

Continuing this procedure, we can gain an infinite sequence in E, a contradiction! Hence, $d^2 = 4ab$.

Suppose $a \neq 0$, and let $d = k_1 \cdot 2a$, where $k_1 = \frac{d}{2a}$; then $2b = k_1 \cdot d$.
If $a + b \neq 0$, then

$$K = cD_3^{n+1} + (2ax_1 + dx_2 + e)D_1D_3^n$$
$$+ (k_1(2ax_1 + dx_2 + e) + c')D_2D_3^n + \cdots, \bmod U_n \in E, \quad (6.214)$$

where $c' = f - k_1 \cdot e$.

6.5 Classification with Nonmaximal Rank

$$T_1 = [K, A] = (2(a+b)(2ax_1 + dx_2 + e) + dc')D_3^{l+n} + \cdots, \mod U_{l+n-1} \in E$$

$\ldots$

$$T_n = [K, T_{n-1}]$$
$$= (2(a+b))^{n-1}(2(a+b)(2ax_1 + dx_2 + e) + dc')D_3^{l+kn} + \cdots,$$
$$\mod U_{l+kn-1} \in E.$$

$\ldots$

(6.215)

Continuing this procedure, we can gain an infinite sequence T_n in E, a contradiction! Hence, $a + b = 0$; thus, a, b have the opposite sign. However, this contradicts with $d^2 = 4ab$. So $a = 0$, and therefore $d = 0$. Similarly, $b = 0$. □

Lemma 6.31 *Since ω_{13} is a polynomial of x_1, x_2, x_3, we may assume that*

$$\omega_{13} = a_l x_3^l + \cdots + a_1 x_3 + a_0, \tag{6.216}$$

where $a_i, 0 \leq i \leq l$ are polynomials of x_1, x_2, $a_l \neq 0$. If $l \geq 1$, then $a_l \in P_1(x_1, x_2)$.

Proof First, we calculate

$$Ad_{L_0} Y_1 = \frac{\partial \omega_{13}}{\partial x_3} D_3^2 + \frac{\partial \omega_{13}}{\partial x_1} D_1 D_3 + \frac{\partial \omega_{13}}{\partial x_2} D_2 D_3 + \cdots, \mod U_1,$$

$$Ad_{L_0}^2 Y_1 = \frac{\partial^2 \omega_{13}}{\partial x_3^2} D_3^3 + 2\frac{\partial^2 \omega_{13}}{\partial x_1 \partial x_3} D_1 D_3^2 + 2\frac{\partial^2 \omega_{13}}{\partial x_2 \partial x_3} D_2 D_3^2 + \cdots, \mod U_2,$$

$\ldots$

$$Ad_{L_0}^l Y_1 = \frac{\partial^l \omega_{13}}{\partial x_3^l} D_3^{l+1} + l\frac{\partial^l \omega_{13}}{\partial x_1 \partial x_3^{l-1}} D_1 D_3^l + l\frac{\partial^l \omega_{13}}{\partial x_2 \partial x_3^{l-1}} D_2 D_3^l + \cdots, \mod U_l,$$

$$Ad_{L_0}^{l+1} Y_1 = (l+1)\frac{\partial^{l+1} \omega_{13}}{\partial x_1 \partial x_3^l} D_1 D_3^{l+1} + (l+1)\frac{\partial^{l+1} \omega_{13}}{\partial x_2 \partial x_3^l} D_2 D_3^{l+1} + \cdots, \mod U_{l+1},$$

(6.217)

where $(\ldots)$ in the above equations means terms with the highest order but lower order in D_3.

Define

$$M_1 = \frac{1}{l!} Ad_{L_0}^l Y_1 = a_l D_3^{l+1} + \cdots, \mod U_l,$$

$$M_2 = \frac{1}{(l+1)!} Ad_{L_0}^{l+1} Y_1 - \frac{\partial a_l}{\partial x_1} D_1 D_3^{l+1} - \frac{\partial a_l}{\partial x_2} D_2 D_3^{l+1} + \cdots, \mod U_{l+1}.$$

(6.218)

Suppose $\deg(a_l) = k \geq 2$, where $\deg(a_l)$ means the degree of the polynomial a_l. Assume that the homogeneous degree k part of a_l is

$$a_l^{(k)} = b_0 x_1^k + b_1 x_1^{k-1} x_2 + \cdots + b_k x_2^k, \tag{6.219}$$

where $b_0, b_1, \ldots, b_k$ are not all zero constants:

$$\begin{aligned}
A &:= Ad_{D_1}^{k-i-2}(Ad_{D_2}^i M_1) \\
&= (\frac{1}{2} i!(k-i)! b_i x_1^2 + (i+1)!(k-i-1)! b_{i+1} x_1 x_2 \\
&\quad + \frac{1}{2}(i+2)!(k-i-2)! b_{i+2} x_2^2 \\
&\quad + pol_1(x_1, x_2)) D_3^{l+1} + \cdots, \text{ mod } U_l \\
&:= p(x_1, x_2) D_3^{l+1} + \cdots, \text{ mod } U_l,
\end{aligned} \tag{6.220}$$

where

$$\begin{aligned}
p(x_1, x_2) &= \frac{1}{2} i!(k-i)! b_i x_1^2 + (i+1)!(k-i-1)! b_{i+1} x_1 x_2 \\
&\quad + \frac{1}{2}(i+2)!(k-i-2)! b_{i+2} x_2^2 + pol_1(x_1, x_2) \\
&:= a x_1^2 + b x_2^2 + d x_1 x_2 + pol_1(x_1, x_2),
\end{aligned} \tag{6.221}$$

with $a = \frac{1}{2} i!(k-i)! b_i$, $b = \frac{1}{2}(i+2)!(k-i-2)! b_{i+2}$, $d = (i+1)!(k-i-1)! b_{i+1}$, $i = 0, 1, \cdots, k-2$. Consider

$$\begin{aligned}
A_2 &:= Ad_{D_1}^{k-i-2}(Ad_{D_2}^i M_2) \\
&= (2ax_1 + dx_2 + c_1) D_1 D_3^{l+1} + (2bx_2 + dx_1 + c_2) D_2 D_3^{l+1} + \cdots, \\
&\text{mod } U_{l+1} \in E,
\end{aligned} \tag{6.222}$$

where c_1, c_2 are constants. Consider

$$\begin{cases} B := [D_1, A_1] = (2ax_1 + dx_2 + c_1) D_3^{l+1} + \cdots, \text{ mod } U_l \in E \\ C := [D_2, A_1] = (dx_1 + 2bx_2 + c_2) D_3^{l+1} + \cdots, \text{ mod } U_l \in E. \end{cases} \tag{6.223}$$

Note $A_2, B, C \in E$ satisfy the assumption of Lemma 6.30, and we have $a = b = d = 0$. That is, $b_i = b_{i+1} = b_{i+1} = 0$, $0 \le i \le k-2$, which contradict with that $b_i, i = 0, 1, \cdots, k$ are not all zero. So we have proved that a_l must be a polynomial of x_1, x_2 with degree no more than 1. □

Lemma 6.32 *Suppose*

$$\omega_{13} = \alpha_k x_1^k + \cdots + \alpha_1 x_1 + \alpha_0, k \ge 1, \alpha_k \ne 0, \tag{6.224}$$

6.5 Classification with Nonmaximal Rank

where α_i, $0 \leq i \leq k$ are polynomials of x_2, x_3. Then $\alpha_k \in P_1(x_2, x_3)$.

Proof First, we have

$$Ad_{D_1}^k Y_1 = const \cdot D_2 + \frac{\partial^k \omega_{13}}{\partial x_1^k} D_3, \text{ mod } U_0 \in E.$$

$$\implies \frac{\partial^k \omega_{13}}{\partial x_1^k} D_3, \text{ mod } U_0 = k! \alpha_k D_3, \text{ mod } U_0 \in E. \tag{6.225}$$

$$M_0 := \alpha_k D_3, \text{ mod } U_0 \in E,$$

$$M_1 := [L_0, M_0] = \frac{\partial \alpha_k}{\partial x_2} D_2 D_3 + \frac{\partial \alpha}{\partial x_3} D_3^2, \text{ mod } U_1 \in E.$$

We first prove that when α_k is a degree 2 polynomial of x_2, x_3, there exists a contradiction in Part (I). When the degree of α_k is higher than 2, we will reduce it to degree 2 case in Part (II). Therefore, α_k must be degree less than 2 polynomial of x_2, x_3. Detailed proof can be found in [21]. □

Lemma 6.33 Suppose that $\omega_{13} = a_l x_3^l + \cdots + a_1 x_3 + a_0$, $(l \geq 1)$, $a_l \neq 0$, where a_i, $0 \leq i \leq l$ are polynomials of x_1, x_2. Then $l < 2$.

Proof Assume that $\omega_{13} = \alpha_k x_1^k + \cdots + \alpha_1 x_1 + \alpha_0$, $k \geq 0$, $\alpha_k \neq 0$. Suppose $l \geq 2$.
Part (I): $k \geq 1$ case. We claim that $l < 2$.

By Lemma 6.32, we can assume that $\alpha_k = ax_2 + bx_3 + c$, where a, b, c are not all zero constants. From Lemma 6.31, a_l is a degree ≤ 1 polynomial of x_1, x_2; assume $a_l = c_1 x_1 + c_2 x_2 + c_0$, where c_0, c_1, c_2 are not all zero. Note that ω_{12} is a degree ≤ 1 polynomial of x_1, x_2; we have

$$\frac{1}{k!} Ad_{D_1}^k Y_1 = const \cdot D_2 + \frac{1}{k!} \frac{\partial^k \omega_{13}}{\partial x_1^k} D_3, \text{ mod } U_0 \implies Z := \alpha_k D_3, \text{ mod } U_0 \in E$$

$$T_1 := [L_0, Z] = aD_2 D_3 + bD_3^2, \text{ mod } U_1 \in E$$

$$T_2 := [T_1, Z] = (a^2 + 2b^2) D_3^2 + ab D_2 D_3, \text{ mod } U_1 \in E$$

$$T_2 - bT_1 = (a^2 + b^2) D_3^2, \text{ mod } U_1 \in E. \tag{6.226}$$

Step [1]. We claim that when a, b are not all zero, then $l < 2$ holds.
If a, b are not all zero, then $K_0 := D_3^2$, mod $U_1 \in E$. Consider

$$Ad_{K_0}^l Y_1 = 2^l \ l! a_l D_3^{l+1}, \text{ mod } U_1 \in E \implies a_l D_3^{l+1}, \text{ mod } U_1 \in E. \tag{6.227}$$

If $c \neq 0$, then $[D_1, a_l D_3^{l+1}] = c_1 D_3^{l+1}$, mod $U_l \in E$. If $c_2 \neq 0$, then $[D_2, a_l D_3^{l+1}] = c_2 D_3^{l+1}$, mod $U_l \in E$. If $c_1 = c_2 = 0$, then $c_0 \neq 0$, and we have $c_0 D_3^{l+1}$, mod $U_l \in E$. In both cases, we have $K_1 := D_3^{l+1}$, mod $U_l \in E$. Consider

$$Ad^l_{K_1}Y_1 = (l+1)^l l! a_l D_3^{l^2+1}, \text{ mod } U_{l^2} \in E \iff a_l D_3^{l^2+1}, \text{ mod } U_{l^2} \in E. \tag{6.228}$$

Similarly, we have $K_2 := D_3^{l^2+1}$, mod $U_{l^2} \in E$. Repeat the above procedure, and we can get an infinite sequence $\{K_n\} \in E$, a contradiction. Hence, $l < 2$ holds.

Step [2]. We claim that when $a = b = 0$, $l < 2$ holds.

If $a = b = 0$, then $\alpha_k = c \neq 0$. Without loss of generality, we can assume $\alpha_k = 1$. Then $Z = D_3$, mod $U_0 \in E$. Consider

$$M_0 := \frac{1}{(l-1)!} Ad_Z^{(l-1)} Y_1 = (la_l x_3 + a_{l-1}) D_3, \text{ mod } U_0 \in E$$

$$N_0 := \frac{1}{l!} Ad_Z^l Y_1 = a_l D_3, \text{ mod } U_0 \in E \tag{6.229}$$

$$[L_0, N_0] = c_1 D_1 D_3 + c_2 D_2 D_3, \text{ mod } U_1 \in E$$

$$[[L_0, N_0], N_0] = (c_1^2 + c_2^2) D_3^2, \text{ mod } U_1 \in E.$$

Step [2.a]. If c_1, c_2 are not all zero, then D_3^2, mod $U_1 \in E$. Just like the proof in Step [1], we have $l < 2$ in this case.

Step [2.b]. If $c_1 = c_2 = 0$, without loss of generality, we may assume that $a_l = 1$. Then $M_0 = (lx_3 + a_{l-1}) D_3$, mod $U_0 \in E$. Since a_{l-1} is a polynomial of x_1, x_2, suppose $\deg(a_{l-1}) = r$.

(*). $r \geq 2$ case. We may assume that the homogeneous degree r part of a_{l-1} is

$$a_{l-1}^{(r)} = b_s x_1^{r-s} x_2^s + \cdots + b_t x_1^{r-t} x_2^t, \tag{6.230}$$

where $b_s, \ldots, b_t$ are constants and $0 \leq s \leq t \leq r$, $b_s \neq 0$, $b_t \neq 0$. Consider

$$R := Ad_{D_1}^{r-s-1}(Ad_{D_2}^s M_0) = (ex_1 + fx_2 + const) D_3, \text{ mod } U_0 \in E,$$

$$T := [L_0, R] = e D_1 D_3 + f D_2 D_3, \text{ mod } U_1 \in E, \tag{6.231}$$

$$[T, R] = (e^2 + f^2) D_3^2, \text{ mod } U_1 \in E,$$

where $e = s!(r-s)! b_s \neq 0$, $f = (s+1)!(r-s-1)! b_{s+1}$. From the proof in Step [1], we have $l < 2$ in this case.

(**). $r \leq 1$ cases. Assume $a_{l-1} = k_1 x_1 + k_2 x_2 + k_0$, where k_1, k_2, k_0 are constants. Consider

$$M_1 = [L_0, M_0], \quad M_2 = [M_1, M_0]$$
$$M_2 - l M_1 \in E \implies D_3^2, \text{ mod } U_1 \in E. \tag{6.232}$$

From the proof in Step [1], we have $l < 2$ in this case.

By Steps [1] and [2], we have proved that for $k \geq 1$ case, $l < 2$ holds.

Part (II): $k = 0$ case. In this case, ω_{13} is a polynomial of x_2, x_3.

6.5 Classification with Nonmaximal Rank

(1) Suppose the degree of ω_{13} with respect to x_2 is at least 1 since x_2 plays the same role as x_1 in ω_{13}; by the argument of Part (I), we have $l < 2$ also holds.

(2) Suppose ω_{13} is a polynomial of x_3 and is irrelevant with x_1, x_2 variables. Then $a_l, \cdots, a_0$ are constants. Without loss of generality, assume $a_l = 1$. Consider $A_1 = \frac{1}{l!} Ad_{L_0}^{l-1} Y_1$, $A_2 = [L_0, A_1]$, $A_n = [A_{n-1}, A_1]$, and we obtain an infinite sequence in E, a contradiction. Hence, $l < 2$ holds.

From Part (I) and Part (II), we have proved $l < 2$. □

Lemma 6.34 *By Lemma 6.33, $\omega_{13} = a_1 x_3 + a_0$, where $a_1 \in P_1(x_1, x_2)$. Then $a_0 \in P_2(x_1, x_2)$.*

Proof Assume $a_1 = k_1 x_1 + k_2 x_2 + k_0$, where k_0, k_1, k_2 are constants. Suppose $\deg(a_0) = k \geq 2$ and homogeneous degree k part of a_0 is $a_0^{(k)} = b_0 x_1^k + b_1 x_1^{k-1} x_2 + \cdots + b_k x_2^k$, where $b_0, \ldots, b_k$ are not all zero constants. Consider

$$Ad_{L_0} Y_1 = a_1 D_3^2 + \left(k_1 x_3 + \frac{\partial a_0}{\partial x_1}\right) D_1 D_3 + \left(k_2 x_3 + \frac{\partial a_0}{\partial x_2}\right) D_2 D_3 + \cdots, \text{ mod } U_1 \in E. \quad (6.233)$$

For $i = 0, \cdots, k-2$, denote

$$p_i(x_1, x_2) = \frac{\partial^{k-2} a_0}{\partial x_1^{k-i-2} \partial x_2^i} = ax_1^2 + bx_2^2 + dx_1 x_2 + pol_1(x_1, x_2), \quad (6.234)$$

where $a = \frac{1}{2}(k-i)! i! b_i$, $b = \frac{1}{2}(k-i-2)!(i+2)! b_{i+2}$, $d = (k-i-1)!(i+1)! b_{i+1}$. Consider

$$K := Ad_{D_2}^i (Ad_{D_1}^{k-i-2} (Ad_{L_0} Y_1))$$
$$= const \cdot D_3^2 + (2ax_1 + dx_2 + e) D_1 D_3 + (2bx_2 + dx_1 + f) D_2 D_3 + \cdots,$$
$$\text{mod } U_1 \in E,$$

$$B := Ad_{D_1}^{k-i-1} Ad_{D_2}^i Y_1 = (2ax_1 + dx_2 + e) D_3, \text{ mod } U_0 \in E,$$

$$C := Ad_{D_1}^{k-i-2} Ad_{D_2}^{i+1} Y_1 = (2bx_2 + dx_1 + f) D_3, \text{ mod } U_0 \in E,$$
$$(6.235)$$

where e, f are constants and $K, B.C$ satisfy the assumption of Lemma 6.30, so $a = b = d = 0$. That is, $b_i = b_{i+1} = b_{i+2} = 0, i = 0, \cdots, k-2$, a contradiction. Therefore, $a_0 \in P_2(x_1, x_2)$. □

Theorem 6.36 *ω_{13}, ω_{23} are degree at most 1 polynomials of x_1, x_2, x_3.*

Proof By (6.34), we may assume that $\omega_{13} = a_1 x_3 + a_0$, $a_1 = c_1 x_1 + c_2 x_2 + c$, $a_0^{(2)} = ax_1^2 + bx_2^2 + dx_1 x_2$, where c_0, c_1, c_2, a, b, d are constants. Consider

$$[D_1, Y_1] \Longrightarrow G_1 := (c_1 x_3 + 2ax_1 + dx_2 + e)D_3, \text{ mod } U_0 \in E,$$
$$[D_2, Y_1] \Longrightarrow G_2 := (c_2 x_3 + 2bx_2 + dx_1 + f)D_3, \text{ mod } U_0 \in E, \quad (6.236)$$

where e, f are constants.

Step [1]. We claim that a_1 is a constant, that is, $c_1 = c_2 = 0$.

If c_1, c_2 are not all zero, without loss of generality, we can assume that $c1 \neq 0$. Then

$$T := \frac{c_2}{c_1} G_1 - G_2 = \left(\left(\frac{2ac_2}{c_1} - d \right) x_1 + \left(\frac{dc_2}{c_1} - 2b \right) x_2 + const \right) D_3, \text{ mod } U_0$$

$$= (\alpha_1 x_1 + \alpha_2 x_2 + const) D_3, \text{ mod } U_0 \in E$$

$$[L_0, T] = \alpha_1 D_1 D_3 + \alpha_2 D_2 D_3, \text{ mod } U_1 \in E$$

$$[[L_0, T], T] = (\alpha_1^2 + \alpha_2^2) D_3^2, \text{ mod } U_1 \in E, \quad (6.237)$$

where $\alpha_1 = \frac{2ac_2}{c_1} - d$ and $\alpha_2 = \frac{dc_2}{c_1} - 2b$.

If α_1, α_2 are not all zero, denote $A_1 = D_3^2$, mod $U_1 \in E$ and calculate

$$[A_1, Y_1] \Longrightarrow a_1 D_3^2, \text{ mod } U_1 \in E$$
$$[L_0, a_1 D_3^2, \text{ mod } U_1] = c_1 D_1 D_3^2 + c_2 D_2 D_3^2, \text{ mod } U_2 \in E \quad (6.238)$$
$$[[L_0, a_1 D_3^2, \text{ mod } U_1], a_1 D_3^2, \text{ mod } U_1] \Longrightarrow D_3^4, \text{ mod } U_3 \in E.$$

Continue this process and we obtain a contradiction. Hence, $\alpha_1 = \alpha_2 = 0$, which lead to $2ac_2 = dc_1, 2bc_1 = dc_2$. Using $2ac_2 = dc_1$ and recalling $c_1 \neq 0$, we have

$$N_0 = c_1 G_1, F = [L_0, N_0], H = [F, N_0]$$
$$H - c_1^2 F \in E \Longrightarrow B_1 := D_3^2, \text{ mod } U_1 \in E. \quad (6.239)$$

Replacing A_1 with B_1 in (6.238) and repeating the procedure, we can get an infinite sequence $\{Bn\}$ in E, a contradiction. Hence, a_1 is a constant.

Step [2]. We claim that $a_0 \in P_1(x_1, x_2)$. Now G_1, G_2 become

$$G_1 = (2ax_1 + dx_2 + e)D_3, \text{ mod } U_0 \in E$$
$$G_2 = (2bx_2 + dx_1 + f)D_3, \text{ mod } U_0 \in E. \quad (6.240)$$

Consider $K := Ad_{L_0} Y_1$ and we notice $K, G_1, G_2 \in E$ satisfy the assumption of Lemma 6.30. hence, $a = b = d = 0$. That is, a_0 is a degree 1 polynomial of x_1, x_2.

By Steps [1] and [2], we have proved ω_{13} is a degree 1 polynomial of x_1, x_2, x_3. We can similarly prove that ω_{23} is a degree at most 1 polynomial of x_1, x_2, x_3. □

6.5.2.2 Mitter Conjecture

In this section, we will give some structure results about estimation algebra E, and finally we prove Mitter conjecture holds for finite-dimensional estimation algebra with state dimension 3 and linear rank 2. We mainly refer to the results in Shi and Yau [22] and Yau [28].

By using the same assumption of the previous section, we first do the following basic calculations:

$$[L_0, x_1] = D_1 \in E, [L_0, x_2] = D_2 \in E,$$
$$[D_2, D_1] = \omega_{12}, [D_1, x_1] = 1 \in E. \tag{6.241}$$

We denote $E_0 = span\{1, x_1, x_2, D_1, D_2\} \subseteq E$. In Sect. 6.5.2.1, Wong's matrix is shown to have linear structure, and we assume that

$$\omega_{12} = k_1 x_1 + k_2 x_2 + k_0,$$
$$\omega_{13} = \alpha_1 x_1 + \alpha_2 x_2 + \alpha_3 x_3 + \alpha_0, \tag{6.242}$$
$$\omega_{23} = \beta_1 x_1 + \beta_2 x_2 + \beta_3 x_3 + \beta_0,$$

where k_i, α_i, β_i's are constants. Since ω_{ij} is linear, we have

$$Y_1 := [L_0, D_1], \text{mod } E_0 = \omega_{12} D_2 + \omega_{13} D_3 + \frac{1}{2}\frac{\partial \eta}{\partial x_1} \in E,$$
$$Y_2 := [L_0, D_2], \text{mod } E_0 = \omega_{21} D_1 + \omega_{23} D_3 + \frac{1}{2}\frac{\partial \eta}{\partial x_2} \in E. \tag{6.243}$$

In the following lemma, we discuss about the type of quadratic polynomials in estimation algebra E.

Lemma 6.35 *If a degree 2 polynomial is in the estimation algebra E of the system (2), then we only need to consider the following cases:*

Case (A): For any function $\phi \in E$, $\psi^{(2)} = cst \cdot x_3^2$.
Case (B):
Case (B.1): There exists $\phi = x_1^2 + x_2^2 + x_3^2 \in E$;
Case (B.2): There exists $\phi = x_1^2 + x_3^2 \in E$.
Case (C):
Case (C.1): There exists $\phi = x_1^2 + x_2^2 \in E$.
Case (C.2): For any function $\phi \in E$, $\phi^{(2)} = cst \cdot x_1^2$.

where cst represents an arbitrary constant number.

Proof Since there exists a degree 2 polynomial in estimation algebra, quadratic rank $1 \leq k \leq 3$. By Theorem 3.7 in Wu and Yau [26], we have the cases stated in the lemma. □

Lemma 6.36 *Suppose there exists $\phi = ax_1^2 + bx_2^2 + cx_3^2 \in E$; here a, b, c are constants. If $c \neq 0$, then $\omega_{13} = \alpha_3 x_3 + \alpha_0$, $\omega_{23} = \beta_3 x_3 + \beta_0$.*

Proof If $c \neq 0$, then

$$\frac{1}{2}[D_1, [L_0, \phi]] = bx_2\omega_{21} + cx_3\omega_{31} \in E,$$

$$\frac{1}{2}[D_2, [L_0, \phi]] = ax_1\omega_{12} + cx_3\omega_{32} \in E, \quad (6.244)$$

by Lemma 6.29, $\omega_{13}, \omega_{23} \in P_1(x_3)$. □

Lemma 6.37

(1) *If $x_1^2 + x_2^2 \in E$ or $x_1 x_2 \in E$, then $x_1^2 + x_2^2 \in E$, and ω_{12} is a constant;*
(2) *If $x_1^2 \in E$, then $k_2 = 0$. If $x_2^2 \in E$, then $k_1 = 0$.*

Proof If $\phi = x_1^2 + x_2^2 \in E$,

$$A_1 := [D_1, \frac{1}{2}[L_0, \phi]], \bmod E_0 = k_1 x_1 x_2 + k_2 x_2^2 \in E,$$

$$A_2 := [D_2, \frac{1}{2}[L_0, \phi]], \bmod E_0 = k_1 x_1^2 + k_2 x_1 x_2 \in E. \quad (6.245)$$

If $k_1 \neq 0$, then

$$A_3 := A_2 - \frac{k_2}{k_1} A_1 = k_1 x_1^2 - \frac{k_2^2}{k_1} x_2^2 \in E,$$

$$\frac{k_1}{k_1^2 + k_2^2}(k_1 \phi - A_3) = x_2^2 \in E, \phi - x_2^2 = x_1^2 \in E, \quad (6.246)$$

$$\frac{1}{4}[[L_0, x_1^2], [L_0, x_2^2]] = x_1 x_2 \cdot \omega_{21} \in E,$$

and then ω_{12} is a constant, contradiction. Hence, $k_1 = 0$. Similarly, $k_2 = 0$. Then ω_{12} is a constant. If $x_1 x_2 \in E$, $[[L_0, x_1 x_2], x_1 x_2] = x_1^2 + x_2^2$, the proof follows as above.

(2) Consider $[D_2, \frac{1}{2}[L_0, x_1^2]] = k_1 x_1^2 + k_2 x_1 x_2 + k_0 x_1 \implies k_2 x_1 x_2 \in E$. If $k_2 \neq 0$, then $x_1 x_2 \in E \implies \omega_{12} = const$, a contradiction. Hence, $k_2 = 0$. □

In the following three lemmas, we will prove Wong's matrix is a constant matrix in all cases in Lemma 6.35.

6.5 Classification with Nonmaximal Rank

Lemma 6.38 *For case (A) in Lemma 6.35, Wong's matrix is a constant matrix; moreover, $\omega_{13} = \omega_{23} = 0$.*

Proof For case (A) in Lemma 6.35, $\phi = x_3^2 \in E$. By Lemma 6.36, $\omega_{13} = \alpha_3 x_3 + \alpha_0$, $\omega_{23} = \beta_3 x_3 + \beta_0$. Define $Z := \frac{1}{2}[L_0, \phi]$, mod $E_0 = x_3 D_3$, mod $E_0 \in E$:

$$[D_1, Z] = -\alpha_3 x_3^2 - \alpha_0 x_3 \in E \Longrightarrow \alpha_0 = 0. \tag{6.247}$$

Similarly, we get $\beta_0 = 0$. Now $\omega_{13} = \alpha_3 x_3$, $\omega_{23} = \beta_3 x_3$. If $\alpha_3 \neq 0$ or $\beta_3 \neq 0$, consider

$$\begin{cases} Ad_{L_0} Z = D_3^2 - \alpha_3 x_3^2 D_1 - \beta_3 x_3^2 D_2, \text{ mod } U_0 \in E, \\ [Z, Ad_{L_0} Z] = -2D_3^2 - 2\alpha_3 x_3^2 D_1 - 2\beta_3 x_3^2 D_2, \text{ mod } U_0 \in E, \\ 2 Ad_{L_0} Z - [Z, Ad_{L_0} Z] = 4D_3^2, \text{ mod } U_0 \in E \Longrightarrow D_3^2, \text{ mod } U_0 \in E, \\ T := \alpha_3 x_3^2 D_1 + \beta_3 x_3^2 D_2, \text{ mod } U_0 \in E, \\ T_0 := \frac{1}{4}[K, T] = \alpha_3 x_3 D_1 D_3 + \beta_3 x_3 D_2 D_3, \text{ mod } U_1 \in E, \\ T_1 := \frac{1}{2}[K, T_0] = \alpha_3 D_1 D_3^2 + \beta_3 D_2 D_3^2, \text{ mod } U_2 \in E, \\ T_{k+1} := \frac{1}{2}[T_k, T_0] = \alpha_3^{k+1} D_1^{k+1} D_3^2 + \beta_3^{k+1} D_2^{k+1} D_3^2 + \cdots, \text{ mod } U_k \in E, \end{cases} \tag{6.248}$$

we can see $\{T_k\}$ is an infinite sequence in E, a contradiction. Hence, $\alpha_3 = \beta_3 = 0$; then $\omega_{13} = \omega_{23} = 0$. Define $M = L_0 - K = \frac{1}{2}(D_1^2 + D_2^2 - \eta_1) \in E$; then

$$[D_1, [M, D_1]] - k_1 D_2 = \frac{1}{2}\frac{\partial^2 \eta_1}{\partial x_1^2} - \omega_{12}^2 \in E,$$

$$[D_2, [M, D_2]] + k_2 D_1 = \frac{1}{2}\frac{\partial^2 \eta_1}{\partial x_2^2} - \omega_{12}^2 \in E, \tag{6.249}$$

$$[D_2, [M, D_2]] - k_2 D_2 = \frac{1}{2}\frac{\partial^2 \eta_1}{\partial x_1 \partial x_2} \in E.$$

From the above equations, we can easily deduce that η_1 is a polynomial of degree at most 4 in x_1, x_2. Assume that

$$\eta_1 = a_{40} x_1^4 + a_{31} x_1^3 x_2 + a_{22} x_1^2 x_2^2 + a_{13} x_1 x_2^3 + a_{04} x_2^4 \\ + \text{degree 4 polynomial with respect to } x_1, x_2, \tag{6.250}$$

where $a_{ij} \in C^\infty(x_3)$. Hence,

$$E \ni \omega_{12}^2 - \frac{1}{2}\frac{\partial^2 \eta_1}{\partial x_1^2} = (k_1^2 - 6a_{40})x_1^2 + (2k_1k_2 - 3a_{31})x_1x_2$$

$$+ (k_2^2 - a_{22})x_2^2 + \text{polynomial of degree 1 with respect to } x_1, x_2,$$

$$E \ni \omega_{12}^2 - \frac{1}{2}\frac{\partial^2 \eta_1}{\partial x_2^2} = (k_1^2 - a_{22})x_1^2 + (2k_1k_2 - 3a_{13})x_1x_2 \qquad (6.251)$$

$$+ (k_2^2 - 6a_{04})x_2^2 + \text{polynomial of degree 1 with respect to } x_1, x_2,$$

$$E \ni \frac{1}{2}\frac{\partial^2 \eta_1}{\partial x_1 \partial x_2} = 3a_{31}x_1^2 + 4a_{22}x_1x_2 + 3a_{13}x_2^2$$

$$+ \text{polynomial of degree 1 with respect to } x_1, x_2.$$

Note for Case (A), the degree 2 part of any function in E is $cst \cdot x_3^2$. Hence, we have $k_1 = k_2 = a_{22} = 0$, which means ω_{12} is constant. □

Lemma 6.39 *For Case (B) in Lemma 6.35, we have $\omega_{13} = \omega_{23} = 0$ and Wong's matrix is a constant matrix.*

Proof Recall Lemma 6.36, and we have $\omega_{13} = \alpha_3 x_3 + \alpha_0$, $\omega_{23} = \beta_3 x_3 + \beta_0$.
Part (I): For Case (B.1), there exists $\phi := x_1^2 + x_2^2 + x_3^2 \in E$. Consider

$$Z := \frac{1}{2}[L_0, \phi], \text{ mod } E_0 = x_1 D_1 + x_2 D_2 + x_3 D_3 \in E,$$

$$K_1 := [L_0, Z] - 2L_0 = \sum_{i=1}^{3} \gamma_i D_i, \text{ mod } U_0 \in E, \qquad (6.252)$$

$$N_0 := [D_1, Ad_{L_0} K_1] = \alpha_3 D_3^2 + \cdots *, \text{ mod } U_1 \in E,$$

$$H_0 := [D_2, Ad_{L_0} K_1] = \beta_3 D_3^2 + \cdots *, \text{ mod } U_1 \in E,$$

where $\gamma_i = \sum_{j \neq i} x_j \omega_{ji}$ and $\cdots *$ means terms with the highest order but lower order in D_3 and do not contain $D_1 D_3$, $D_2 D_3$ terms.

Step [1]. In this step, we will show $\alpha_3 = \beta_3 = 0$, that is, ω_{13}, ω_{23} are constants. If $\alpha_3 \neq 0$, then

$$N_1 := \left[Ad_{L_0} K_1, \frac{1}{2\alpha_3} N_0 \right] = \alpha_3 D_1 D_3^2 + \beta_3 D_2 D_3^2$$

$$+ \cdots, \text{ mod } U_2 \in E,$$

$$M_0 := \frac{1}{\alpha_3^2 + \beta_3^2}[N_1, K_1] = (x_3 + cst) D_3^2 + \cdots, \text{ mod } U_2 \in E, \qquad (6.253)$$

$$M_1 := [L_0, M_0] = D_3^4 + \cdots, \text{ mod } U_3 \in E,$$

$$M_{n+1} := \frac{1}{2(n+1)}[M_n, M_0] = D_3^{2(n+2)} + \cdots, \text{ mod } U_{2n+3} \in E,$$

6.5 Classification with Nonmaximal Rank

we can get an infinite sequence $\{M_n\}$ in E, contradiction. Hence, $\alpha_3 = 0$. Similarly, we can prove $\beta_3 = 0$ starting from H_0. Therefore, ω_{12}, ω_{23} are constants.

Step [2]. In this step, we will prove ω_{12} is a constant and $\alpha_0 = \beta_0 = 0$. Consider

$$B_1 := [Z, D_1] - D_1 = k_1 x_1 x_2 + k_2 x_2^2 + \alpha_0 x_3 \in E, \tag{6.254}$$
$$B_2 := [D_2, Z] - D_2 = k_1 x_1^2 + k_2 x_1 x_2 - \beta_0 x_3 \in E.$$

If $k_1 \neq 0, k_2 = 0$, then $B_1 = k_1 x_1 x_2 + \alpha_0 x_3 \in E$.

$$[[L_0, B_1], B_1] - \alpha_0 = k_1^2 (x_1^2 + x_2^2) \in E, \tag{6.255}$$

by Lemma 6.37(3), ω_{12} is a constant, a contradiction. Similarly, one can show that the case $k_2 \neq 0, k_1 = 0$ cannot occur. If $k_1 \cdot k_2 \neq 0$, now consider

$$B_3 := B_2 - \frac{k_2}{k_1} B_1 + \frac{k_2^2}{k_1} \phi = ex_1^2 + fx_3^2 + gx_3^2 \in E, \tag{6.256}$$
$$B_4 := \frac{1}{4}[[L_0, B_3], B_3] - \frac{g^2}{4} = e^2 x_1^2 + f^2 x_3^2 + fgx_3 \in E,$$

where $e \neq f$. Then from B_3, B_4, we have $x_1^2 \in E$. However, by Lemma 6.37(2), $k_2 = 0$, a contradiction.

Hence, $k_1 = k_2 = 0$, that is, ω_{12} is a constant. From B_1, B_2 in Step [2], $\alpha_0 = \beta_0 = 0$, so $\omega_{13} = \omega_{23} = 0$.

Part (II). For Case (B.2), there exists $\phi = x_1^2 + x_3^2 \in E$. Consider

$$Z := \frac{1}{2}[L_0, \phi] - 1 = x_1 D_1 + x_3 D_3 \in E,$$
$$[D_1, Z] - D_1 = x_3 \cdot \omega_{31} \in E \tag{6.257}$$
$$[Z, x_3 \omega_{13}] = 2x_3 \omega_{13} - \alpha_0 x_3 \in E \Longrightarrow \alpha_3 x_3^2 \in E$$
$$\alpha_0 x_3 \in E \Longrightarrow \alpha_0.$$

If $\alpha_3 \neq 0$, we have $x_3^2 \in E$; then $x_1^2 = \phi - x_3^2 \in E$:

$$[[L_0, x_1^2], [L_0, x_3^2]] = 4x_1 x_3 \omega_{31}. \tag{6.258}$$

Then ω_{31} is a constant, a contradiction. Hence, $\alpha_3 = 0$ and $\omega_{13} = 0$. Now we have $Y_1 = \omega_{12} D_2$, mod U_0:

$$[Y_1, [D_2, Z]] = k_2 x_1 \omega_{12} \in E. \tag{6.259}$$

If $k_2 \neq 0$, then $x_1\omega_{12} \in E$:

$$[Z, x_1\omega_{12}] \Longrightarrow x_1 x_2 \in E \tag{6.260}$$

By Lemma 6.37(1), ω_{12} is a constant, a contradiction; then $k_2 = 0$.

Now $[D_2, Z] = k_1 x_1^2 - \beta_3 x_3^2 + k_0 x_1 - \beta_0 x_3 \in E$. Combining with ϕ, we have $(k_1 + \beta_3)x_3^2 + \beta_0 x_3 \in E$. $[Z, (k_1 + \beta_3)x_3^2 + \beta_0 x_3] \in E \Longrightarrow \beta_0 = 0$.

If $k_1 \beta_3 \neq 0$, define

$$\begin{cases} M_0 := \frac{1}{4} Ad_{L_0}^2 Z \\ M_1 := [L_0, M_0] \\ M_{n+1} = \frac{1}{2}[M_n, M_0] = k_1^{n+1} D_1^2 D_2^{n+1} - \beta_3^{n+1} D_2^{n+1} D_3^2, \text{ mod } U_{n+2} \in E; \end{cases} \tag{6.261}$$

then we can get an infinite sequence $\{M_n\}$ in E, a contradiction. Hence, $k_1 = \beta_3 = 0$, that is, ω_{12} is constant and $\omega_{23} = 0$. □

Lemma 6.40 *For case (C) in Lemma 6.35, Wong's matrix is a constant matrix.*

Proof Part (I). For case (C.2), $\phi = x_1^2 \in E$. Note Lemma 6.37(2), and we have $\omega_{12} = k_1 x_1 + k_0$. Consider

$$Z_1 := \frac{1}{2}[L_0, \phi], \text{ mod } E_0 = x_1 D_1 \in E$$

$$T = Ad_{L_0} Z_1 - [Z_1, Ad_{L_0} Z_1] = 3D_1^2 - k_1 x_1^2 D_2 - \alpha_1 x_1^2 D_3, \text{ mod } U_0 \in E$$

$$K := \frac{1}{12}(2T + [T, Z_1]) = D_1^2, \text{ mod } U_0 \in E$$

$$R := 3K - T = k_1 x_1^2 D_2 + \alpha_1 x_1^2 D_3, \text{ mod } U_0 \in E. \tag{6.262}$$

Step [1]: In this step, we will show that ω_{12}, ω_{13} is a constant. Define

$$N_0 := \frac{1}{4}[K, R], \ N_1 = \frac{1}{2}[K, N_0], \ N_{k+1} = \frac{1}{2}[N_k, N_0]. \tag{6.263}$$

We can find $\{N_k\}$ is an infinite sequence in E, a contradiction. Hence, $k_1 = \alpha_1 = 0$; then ω_{12} is constant. If $\alpha_3 \neq 0$, consider

$$\begin{cases} M_0 := 2K + [Z_1, Ad_{L_0} Z_1], \ K_1 := [L_0, [D_1, M_0]] \\ T_1 = \frac{1}{\alpha_2^2 + \alpha_3^2}([K_1, [D_1, M_0]] - \alpha_3 K_1) = D_3^2, \text{ mod } U_1 \in E \\ R_1 = \frac{1}{2\alpha_3}[T_1, M_0], \ R^n = [T_{n-1}, M_0], \ T_n = Ad_{R_{n-1}}^2 L_0 = t_n D_3^{2^n}, \\ \quad \text{mod } U_{2^n - 1} \in E, t_n \neq 0. \end{cases} \tag{6.264}$$

6.5 Classification with Nonmaximal Rank

Then $\{T_n\}$ is an infinite sequence in E, a contradiction. Hence, $\alpha_3 = 0$. If $\alpha_2 \neq 0$, define $A_n = c(n) Ad_{L_0}^n M_0$, where $c(n) \neq 0$ is a constant. $\{A_n\}$ is an infinite sequence in E, a contradiction. Hence, $\alpha_2 = 0$ and then ω_{13} is constant.

Step [2]. In this step, we prove ω_{23} is constant. Recall Y_1, Y_2; now we have

$$G_1 := \omega_{13} D_3 + \frac{1}{2} \frac{\partial \eta}{\partial x_1} \in E$$

$$G_2 := \omega_{23} D_3 + \frac{1}{2} \frac{\partial \eta}{\partial x_2} \in E \qquad (6.265)$$

$$[D_1, G_2] - [D_2, G_1] = \beta_1 D_3 \in E.$$

If $\beta_1 \neq 0$, then $D_3 \in E, [D_3, D_2] = \omega_{23} \in E \Longrightarrow \beta_3 = 0$. From Step [1], ω_{12}, ω_{13} are constants; then $G_1 \in E \Longrightarrow \frac{\partial \eta}{\partial x_1} \in E$. Then we assume $\eta = a_3 x_1^3 + a_2 x_1^2 + a_2 x_1^2 + a_0, a_i \in C^\infty(x_2, x_3)$. Now $\frac{\partial \eta}{\partial x_1} = 3a_3 x_1^2 + 2a_2 x_1 + a_1 \in E$; we have a_3, a_2 a constant and $a_1 \in P_1(x_2)$.

$$L_1 := [L_0, D_3], \mod E_0 = \omega_{32} D_2 + \frac{1}{2} \frac{\partial a_0}{\partial x_3} \in E$$

$$[G_2, L_1] - \beta_2 G_2 = -\omega_{23}^2 - \frac{\beta_2}{2} \left(\frac{\partial a_0}{\partial x_2} + \frac{\partial a_1}{\partial x_2} x_1 \right) \qquad (6.266)$$

$$+ \frac{\omega_{23}}{2} \left(\frac{\partial^2 a_0}{\partial x_2^2} + \frac{\partial^2 a_0}{\partial x_3^2} \right),$$

which is a degree 3 polynomial in x_1, a contradiction. Hence, $\beta_1 = 0$. Now $\omega_{23} = \beta_2 x_2 + \beta_3 x_3 + \beta_0$. Define $\tilde{L}_0 = L_0 - K = \frac{1}{2}(D_2^2 + D_3^2 - \tilde{\eta}) \in E$ and consider the subalgebra $\tilde{E} := <\tilde{L}_0, D_2, x_2, 1>_{L.A.}$. $\tilde{E}$ can be considered as the estimation algebra of a filtering system with 2-D state and rank 1, which is exactly the case in Wu and Yau [26]; thus, $\beta_2 = \beta_3 = 0$.

Part (II). For case (C.1), there exists $\phi = x_1^2 + x_2^2 \in E$. From Lemma 6.37(1), $\omega_{12} = k_0$. Consider

$$Z_2 := \frac{1}{2}[L_0, \phi], \mod E_0 = x_1 D_1 + x_2 D_2, \mod E_0 \in E$$

$$L_2 := Ad_{L_0} Z_2 - 2L_0, \mod E_0$$

$$\zeta := \frac{1}{2} E_2(\eta) + \eta \qquad (6.267)$$

$$L_3 := [Z_2, L_2], \mod E_0$$

$$= p(x_1, x_2, x_3) D_3, \mod U_0 \in E,$$

where $p(x_1, x_2, x_3) = 4\alpha_1 x_1^2 + 4\beta_2 x_2^2 + 4(\alpha_2 + \beta_1)x_1x_2 + 3\alpha_3 x_1 x_3 + 3\beta_3 x_2 x_3 + 3\alpha_0 x_1 + 2\beta_0 x_2$. By Corollary 3.11 in Shi and Yau [21], we have $p(x_1, x_2, x_3)$ that must be a degree at most 1 polynomial, that is, $\alpha_1 = \alpha_3 = \beta_2 = \beta_3 = (\alpha_2+\beta_1) = 0$. Now

$$L_3 = E_2^2(\zeta) - E_2(\zeta) - 4(\alpha_0 x_1 + \beta_0 x_2)^2 \in E. \qquad (6.268)$$

Then due to the theory of Euler operator, η is a degree at most 2 polynomial with respect to x_1, x_2. Suppose $\eta = a_2 x_1^2 + a_1 x_1 x_2 + a_0 x_2^2 + b_1 x_1 + b_2 x_2 + b_0, a_i, b_i \in C^\infty(x_3)$. Consider

$$\begin{aligned}
&[D_1, Y_1], \mod E_0 = a_2 - \alpha_2^2 x_2^2 \in E \\
&[D_2, Y_2], \mod E_0 = a_0 - \beta_1^2 x_2^2 \in E \\
&[Z_2, a_2 - \alpha_2^2 x_2^2] = -2\alpha_2^2 x_2^2 \in E \\
&[Z_2, a_0 - \beta_1^2 x_2^2] = -2\beta_1^2 x_2^2 \in E.
\end{aligned} \qquad (6.269)$$

If $\alpha_2 \neq 0$ or $\beta_1 \neq 0$, then $x_1^2, x_2^2 \in E$. By Step [1] of Part (I), we have ω_{13}, ω_{23} are constants, a contradiction. Hence, $\alpha_2 = \beta_1 = 0$, which means ω_{13}, ω_{23} are constants. □

Theorem 6.37 *If there exists a degree 2 polynomial in E, then Wong's matrix must be a constant matrix and h_i, $1 \leq i \leq m$ are affine in x.*

Proof If a degree 2 polynomial ϕ is in E, then by Lemma 6.35, we only need to consider cases (A), (B), and (C). The conclusion comes from Lemmas 6.38–6.40 and Theorem 3 in Yau [28]. □

The following two lemmas describe the structure of η in the case of Lemma 6.35 and are essential for proving Mitter conjecture. Detailed proof can be found in Shi and Yau [22].

Lemma 6.41 *Under the assumption of Theorem 6.37, if $T = cD_3 + \phi(x_3) \in E$, where $c \neq 0$ is a constant, $\phi(x_3) \in C^\infty(x_3)$, then $\phi(x_3) \in P_2(x_3)$.*

Lemma 6.42 *Under the assumption of Theorem 6.37,*

(i) *For case (A) and (B): $\eta \in P_2(x_1, x_2, x_3)$;*
(ii) *For case (C): $\eta = P_2(x_1, x_2) + \psi(x_3)$, where $\psi(x_3) \in C^\infty(x_3)$; moreover, if $\omega_{13} \neq 0$ or $\omega_{23} \neq 0$, then $\eta \in P_2(x_1, x_2, x_3)$.*

Up to now, under the assumption of Lemma 6.35, we have proved that Wong's matrix has a constant structure. η is a degree 2 polynomial in variables x_1, x_2 for all cases of Lemma 6.35. Finally, by using the tool proposed by Yau [28], Mitter conjecture can be proved below.

6.6 Novel Finite-Dimensional Filter

Theorem 6.38 *The Mitter conjecture holds for state dimension 3, linear rank 2 case, that is, any function in estimation algebra E is affine in x.*

Proof If a degree 2 polynomial is in estimation algebra E, then by Lemma 6.35, we only need to consider cases (A), (B), and (C).

For cases (A) and (B), from Lemma 6.42, $\eta \in P_2(x_1, x_2, x_3)$. However, by Theorem 5(i) in Yau [28], any function in E is degree at most 1; thus, cases (A) and (B) are impossible.

For case (C), if $\omega_{13} \neq 0$ or $\omega_{23} \neq 0$, then $\eta \in P_2(x_1, x_2, x_3)$; by Theorem 5(i) in Yau [28], it is impossible. If $\omega_{13} = \omega_{23} = 0$, note h_i's are degree at most 1 polynomials of x_1, x_2 and $\eta = P_2(x_1, x_2) + \psi(x_3)$ by Lemma 6.42. Then

$$E = < L_0, h_1, \cdots, h_m >_{L.A.} \subset < L_0, D_1, D_2, x_1, x_2, 1 >_{L.A.} . \qquad (6.270)$$

It can be easily checked that the latter is finite dimensional and it does not contain degree 2 polynomials; then E cannot contain degree 2 polynomials, a contradiction. Therefore, the estimation algebra E only contains linear functions. □

6.6 Novel Finite-Dimensional Filter

In this section, we mainly focus on the construction of finite-dimensional filters with nonmaximal rank. In the previous section, we introduce finite-dimensional estimation algebra with maximal rank and proved that Wong's matrix is a constant matrix. However, in FDEA with nonmaximal rank, this is not true. We will show there exist new classes of finite-dimensional filter, in which Wong's matrix is not necessary to be a constant matrix.

6.6.1 State Dimension 3

First, we consider the finite-dimensional estimation algebra E corresponding to (6.1) with state dimension $n = 3$ and linear rank $r = 1$. This result can be found in Shi [20]. Without loss of generality, we may assume that $x_1 \in E$, $x_2, x_3 \notin E$.

It is easy to see that

$$[L_0, x_1] = D_1 \in E, [D_1, x_1] = 1 \in E,$$

$$[L_0, D_1] = \omega_{12} D_2 + \omega_{13} D_3 + \frac{1}{2}\frac{\partial \omega_{12}}{\partial x_2} \qquad (6.271)$$

$$+ \frac{1}{2}\frac{\partial \omega_{13}}{\partial x_3} + \frac{1}{2}\frac{\partial \eta}{\partial x_1} \in E.$$

If we impose the following conditions:

(I) $\omega_{12} = \omega_{13} = 0$,
(II) $\eta = P_2(x_1) + \phi(x_2, x_3)$,

where $P_2(x_1)$ denotes degree at most 2 and $\phi(x_2, x_3)$ is a C^∞ function of x_2, x_3. Then from (6.271), it is easy to see that the estimation algebra E is finite dimensional with basis given by $\{1, x_1, D_1, L_0\}$.

Next we construct a class of nonlinear filtering systems which satisfy conditions (I) and (II). By condition (II),

$$\eta = \sum_{i=1}^{3}\left(f_i^2 + \frac{\partial f_i}{\partial x_i}\right) + \sum_{i=1}^{m} h_i^2, \tag{6.272}$$

is a polynomial of degree at most 2 with respect to x_1, then we may assume that $f_i, 1 \le i \le 3$ are polynomials of degree at most 1 with respect to x_1, i.e., we assume that for $1 \le i \le 3$,

$$f_i = a_i(x_2, x_3)x_1 + \phi_i(x_2, x_3), \tag{6.273}$$

where $a_i(x_2, x_3)$ and $\phi_i(x_2, x_3)$ are C^∞ function of x_2, x_3. By condition (I), we have

$$\begin{aligned}\omega_{12} &= \frac{\partial f_2}{\partial x_1} - \frac{\partial f_1}{\partial x_2} = a_2 - \left(\frac{\partial a_1}{\partial x_2}x_1 + \frac{\partial \phi_1}{\partial x_2}\right) = 0, \\ \omega_{13} &= \frac{\partial f_3}{\partial x_1} - \frac{\partial f_1}{\partial x_3} = a_3 - \left(\frac{\partial a_1}{\partial x_3}x_1 + \frac{\partial \phi_1}{\partial x_3}\right) = 0.\end{aligned} \tag{6.274}$$

Hence, we have

$$\begin{cases} \frac{\partial a_1}{\partial x_2} = 0, \; \frac{\partial a_1}{\partial x_3} = 0 \\ \frac{\partial \phi_1}{\partial x_2} = a_2, \; \frac{\partial \phi_1}{\partial x_3} = a_3. \end{cases} \tag{6.275}$$

From (6.275), a_1 must be a constant. Now

$$\begin{aligned}\eta &= \sum_{i=1}^{3}\left(f_i^2 + \frac{\partial f_i}{\partial x_i}\right) + \sum_{i=1}^{m} h_i^2 \\ &= \left(\sum_{i=1}^{3} a_i^2\right) x_1^2 + \left(2\sum_{i=1}^{3} a_i \phi_i + \frac{\partial a_2}{\partial x_2} + \frac{\partial a_3}{\partial x_3}\right) x_1 \\ &\quad + \sum_{i=1}^{3} \phi_i^2 + a_1 + \frac{\partial \phi_2}{\partial x_2} + \frac{\partial \phi_3}{\partial x_3} + \sum_{i=1}^{m} h_i^2.\end{aligned} \tag{6.276}$$

6.6 Novel Finite-Dimensional Filter

Since the estimation algebra has linear rank 1, we assume that h_i, $1 \leq i \leq m$ are degree 1 polynomials of x_1. Now condition (II) implies

(II.1) $\sum_{i=1}^{3} a_i^2(x_2, x_3)$ is a constant;
(II.2) $2\sum_{i=1}^{3} a_i \phi_i + \frac{\partial a_2}{\partial x_2} + \frac{\partial a_3}{\partial x_3}$ is a constant.

To summarize, in order to satisfy conditions (I) and (II), it suffices to satisfy the following conditions:

(i) $f_i = a_i x_1 + \phi(x_2, x_3)$, $1 \leq i \leq 3$,
(ii) $\frac{\partial \phi_1}{\partial x_2} = a_2$, $\frac{\partial \phi_1}{\partial x_3} = a_3$,
(iii) a_1 is a constant, $\sum_{i=1}^{3} a_i^2$ is a positive constant,
(iv) $2\sum_{i=1}^{3} a_i \phi_i + \frac{\partial a_2}{\partial x_2} + \frac{\partial a_3}{\partial x_3}$ is a constant,
(v) h_i's are degree at most 1 polynomials of x_1, and then the estimation algebra E is finite dimensional with basis $\{L_0, D_1, x_1, 1\}$.

Example Now we give a nonlinear filtering system example that satisfies conditions (i)–(v). We let all the a_i's be constants, e.g., if we take $a_1 = 1, a_2 = 1, a_3 = -1$, then condition (iii) is satisfied and from (ii) we can see that ϕ_1 is degree at most 1 polynomial of x_2, x_3. Thus, we can take $\phi_1 = x_2 - x_3$. Now condition (iv) that says

$$\phi_1 + \phi_2 - \phi_3 = \phi_2 - \phi_3 + x_2 - x_3 \tag{6.277}$$

is a constant can be easily satisfied. For example, if we take $\phi_2 = x_2^2 + x_3^2 + x_3 - x_2 + 1$, $\phi_3 = x_2^2 + x_3^2$, then condition (iv) is satisfied. Condition (v) is easily satisfied by letting the observation term $h(x) = x_1$. Now the Ω-matrix is given by

$$\begin{pmatrix} 0 & 0 & 0 \\ 0 & 0 & 2x_2 - 2x_3 - 1 \\ 0 & 2x_3 - 2x_2 + 1 & 0 \end{pmatrix} \tag{6.278}$$

and $\eta = 4x_1^2 + 2x_1 + \gamma(x_2, x_3)$. Then the estimation algebra corresponding to this class of nonlinear filtering systems is finite dimensional with basis $\{L_0, D_1, x_1, 1\}$. More importantly, the entries of Ω-matrix are not necessary to be constant.

Next, we use the structure results to derive finite-dimensional filters for the robust DMZ equation by the Wei-Norman approach.

In real applications, we are interested in considering robust state estimator from observed sample paths with some properties of robustness. Davis [9] considered this problem and proposed some robust algorithms. In our case, his basic idea reduced to define a new unnormalized density

$$u(t, x) = \exp\left(-\sum_{i=1}^{m} h_i(x) y_i(t)\right) \sigma(t, x). \tag{6.279}$$

Then $u(t, x)$ satisfies the following robust DMZ equations:

$$\begin{cases} \frac{\partial u}{\partial t}(t, x) = & L_0 u(t, x) + \sum_{i=1}^{m} y_i [L_0, L_i] u(t, x) \\ & + \frac{1}{2} \sum_{i,j=1}^{m} y_i(t) y_j(t) [[L_0, L_i], L_j] u(t, x), \\ u(0, x) = & \sigma_0(x). \end{cases} \quad (6.280)$$

The objective of constructing a robust finite-dimensional filter to (6.280) is equivalent to finding a smooth manifold M and complete C^∞ vector fields μ_i on M and C^∞ functions v on $M \times R \times R^n$ and ω_i's on R^m, such that $u(t, x)$ can be represented in the following form:

$$\begin{cases} \frac{dz(t)}{dt} = & \sum_{i=1}^{k} \mu_i(z(t)) \omega_i(y(t)), \quad z(0) \in M, \\ u(t, x) = & v(z(t), t, x). \end{cases} \quad (6.281)$$

Following [15], we say that system (6.1) has a robust universal finite-dimensional filter if for each initial probability density σ_0, there exists a z_0, such that (6.281) holds if $z(0) = z_0$, and μ_i, ω_i are independent of σ_0.

The following theorem gives the solution of the above robust-DMZ equation by the basis of the corresponding estimation algebra in terms of ordinary differential equations. The detailed calculations can be found in Shi [20].

Theorem 6.39 *If the nonlinear filtering system (6.1) satisfies conditions (i)–(v), then we can assume* $\eta = a_2 x_1^2 + a_1 x_1 + a_0(x_2, x_3)$, $h_i = c_{i1} x_1 + c_{i0}$, $1 \le i \le m$, *where* c_{i1}, c_{i0}, a_2, a_1 *are constants and* $a_0(x_2, x_3)$ *is a* C^∞ *function of* x_2, x_3. *Then the robust DMZ equation (6.280) has a solution for all t of the following form:*

$$u(t, x) = e^{r_0(t)} e^{r_1(t) x_1} e^{r_2(t) D_1} e^{t L_0} \sigma_0, \quad (6.282)$$

where r_i's *satisfy the following ordinary differential equations for all* $t \ge 0$:

$$\begin{aligned} \dot{r}_1(t) &= a_2 r_2(t), \\ \dot{r}_2(t) &= r_1(t) + \sum_{i=1}^{m} c_{i1} y_i(t), \\ \dot{r}_3(t) &= \frac{r_1(t)^2}{2} + \frac{a_2}{2} r_2(t)^2 + \sum_{i=1}^{m} c_{i1} y_i(t) r_1(t) \\ &\quad + \frac{1}{2} a_1 r_2(t) + \frac{1}{2} \sum_{i,j=1}^{m} c_{i1} c_{j1} y_i(t) y_j(t), \end{aligned} \quad (6.283)$$

where the initial condition is given $r_0(0) = r_1(0) = r_2(0) = 0$.

6.6 Novel Finite-Dimensional Filter

Proof As described previously, the estimation algebra E satisfies conditions (I) and (II) with basis of $\{L_0, D_1, x_1, 1\}$. By differentiating $u(t, x)$, we have

$$\begin{aligned}
\frac{\partial u}{\partial t} &= e^{r_0(t)} e^{r_1(t) x_1} e^{r_2(t) D_1} L_0 e^{t L_0} \sigma_0 \\
&\quad + \dot{r}_2(t) e^{r_0(t)} e^{r_1(t) x_1} D_1 e^{r_2(t) D_1} e^{t L_0} \sigma_0 \\
&\quad + (\dot{r}_0(t) + \dot{r}_1(t) x_1) u(t, x) \\
&= A + B + (\dot{r}_0(t) + \dot{r}_1(t) x_1) u(t, x),
\end{aligned} \qquad (6.284)$$

where we denote

$$\begin{aligned}
A &:= e^{r_0(t)} e^{r_1(t) x_1} e^{r_2(t) D_1} L_0 e^{t L_0} \sigma_0 \\
B &:= \dot{r}_2(t) e^{r_0(t)} e^{r_1(t) x_1} D_1 e^{r_2(t) D_1} e^{t L_0} \sigma_0.
\end{aligned} \qquad (6.285)$$

Recall the classical Baker-Campbell-Hausdorff-type relation, i.e.,

$$\begin{aligned}
e^{r(t) E_i} E_k e^{s(t) E_j} &= \left(E_k + r(t)[E_i, E_k] + \frac{r(t)^2}{2!}[E_i, [E_i, E_k]] + \cdots \right) \\
&\quad \times e^{r(t) E_i} e^{s(t) E_j},
\end{aligned} \qquad (6.286)$$

where E_i, E_k, E_j are elements of a Lie algebra. The following calculations basically come from (6.286); we have

$$A := e^{r_0(t)} e^{r_1(t) x_1} \left(L_0 + r_2(t)[D_1, L_0] + \frac{r_2(t)^2}{2}[D_1, [D_1, L_0]] + \cdots \right)$$

$$e^{r_2(t) D_1} e^{t L_0} \sigma_0 = e^{r_0(t)} e^{r_1(t) x} L_0 e^{r_2(t) D_1} e^{t L_0} \sigma_0$$

$$- \left(r_2(t)(a_2 x_1 + \frac{1}{2} a_1) + \frac{r_2(t)^2}{2} a_2 \right) u(t, x), \qquad (6.287)$$

and

$$e^{r_0(t)} e^{r_1(t) x} L_0 e^{r_2(t) D_1} e^{t L_0} \sigma_0 = $$
$$e^{r_0(t)} \left(L_0 - r_1(t) D_1 + \frac{r_1(t)^2}{2} \right) e^{r_1(t) x} e^{r_2(t) D_1} e^{t L_0} \sigma_0, \qquad (6.288)$$

and

$$\begin{aligned}
e^{r_0(t)} L_0 e^{r_1(t) x} e^{r_2(t) D_1} e^{t L_0} \sigma_0 &= L_0 u(t, x) \\
e^{r_0(t)} r_1(t) D_1 e^{r_1(t) x} e^{r_2(t) D_1} e^{t L_0} \sigma_0 &= r_1(t) D_1 u(t, x).
\end{aligned} \qquad (6.289)$$

Putting (6.288) and (6.289) into (6.287), we have

$$A = L_0 u(t, x) - r_1(t) D_1 u(t, x)$$
$$+ \left(\frac{r_1(t)^2}{2} - \frac{r_2(t)^2}{2} a_2 - r_2(t)(a_2 x_1 + \frac{1}{2} a_1) \right) u(t, x). \quad (6.290)$$

Similarly, we have

$$B = \dot{r}_2(t) D_1 u(t, x) - \dot{r}_2(t) r_1(t) u(t, x). \quad (6.291)$$

Then put simplified A, B into (6.284); we have

$$\frac{\partial u}{\partial t} = L_0 u(t, x) + (\dot{r}_2(t) - r_1(t)) D_1 u(t, x)$$
$$+ \left(\frac{r_1^2}{2} - r_2 \left(a_2 x_1 + \frac{1}{2} a_1 \right) - \frac{r_2^2}{2} a_2 + \dot{r}_0(t) + \dot{r}_1(t) x_1 - \dot{r}_2(t) r_1(t) \right) u(t, x).$$
$$(6.292)$$

Note that L_i is the zero degree differential operator of multiplication by h_i; then robust DMZ equation becomes

$$\frac{\partial u}{\partial t} = L_0 u(t, x) + \left(\sum_{i=1}^{m} c_{i1} y_i(t) \right) D_1 u(t, x) + \left(\frac{1}{2} \sum_{i,j=1}^{m} c_{i1} c_{j1} y_i(t) y_j(t) \right) u(t, x).$$
$$(6.293)$$

Comparing (6.292) and (6.293), we have

$$\dot{r}_2(t) - r_1(t) = \sum_{i=1}^{m} c_{i1} y_i(t) \quad (6.294)$$

and

$$\frac{r_1^2}{2} - r_2 \left(a_2 x_1 + \frac{1}{2} a_1 \right) - \frac{r_2^2}{2} a_2 + \dot{r}_0(t) + \dot{r}_1(t) x_1 - \dot{r}_2(t) r_1(t)$$
$$= \frac{1}{2} \sum_{i,j=1}^{m} c_{i1} c_{j1} y_i(t) y_j(t). \quad (6.295)$$

From (6.294) and (6.295), we have

$$\dot{r}_1(t) = a_2 r_2(t), \quad (6.296)$$

$$\dot{r}_2(t) = r_1(t) + \sum_{i=1}^{m} c_{i1} y_i(t) \quad (6.297)$$

$$\dot{r}_3(t) = \frac{r_1(t)^2}{2} + \frac{a_2}{2}r_2(t)^2 + \sum_{i=1}^{m} c_{i1}y_i(t)r_1(t)$$
$$+ \frac{1}{2}a_1 r_2(t) + \frac{1}{2}\sum_{i,j=1}^{m} c_{i1}c_{j1}y_i(t)y_j(t). \tag{6.298}$$

Considering $u(0, x) = \sigma_0$, the initial condition can be given $r_0(0) = r_1(0) = r_2(0) = 0$.

Let $r_i(t), 0 \leq i \leq 2$ play the role of $z(t)$ in (6.281); then it is easy to check that (6.282) are of the form (6.281), i.e., a universal finite-dimensional filter exists for finite-dimensional filter with state dimension 3 and linear rank 1. □

Therefore, for the filters with finite-dimensional estimation algebra with state dimension 3 and linear rank 1, universal finite-dimensional filter exists.

Later, Dong et al. [10] use similar methods to study construction of finite-dimensional estimation algebra with state dimension 4 and linear rank 1 and further obtain a new class of nonlinear finite-dimensional filters. They prove that there is a class of polynomial finite-dimensional system in state dimension 4 with linear rank 1, but the entries in Wong's matrix are polynomials of degree 2 or higher.

6.6.2 Arbitrary State Dimension

For the problem to construct FDF on arbitrary dimension n, Jiao and Yau [13] have made a great progress and successfully gave the procedure of construction of novel filters. In this chapter, by applying Wong's theorem [27], we construct a new class of finite-dimensional filters with arbitrary state space dimension n and linear rank $n - 2$. Importantly, we show that in the new class of nonlinear filtering systems, the entries of Wong's Ω-matrix are not necessary to be constants or polynomials and can be C^∞ functions. This result largely extends the types of finite-dimensional filter which already include Kalman-Bucy filter and Yau filter.

In the following, we sketch the main idea and procedures of construction. First, we construct state evolutionary stochastic differential equations and calculate corresponding Ω-matrix. Inspired by Wong's theorem, we define corresponding observation equations. In order to construct filters that satisfy all assumptions of Wong's theorem, we make an orthogonal transformation for H_i to obtain a new class of finite-dimensional filters with any state dimension n and linear rank $n - 2$. Matrix H_i is defined as follows. More significantly, we can prove in such constructed filters, entries of Wong's Ω-matrix are not necessarily to be constants or polynomials and can be C^∞ functions.

We consider the following filtering system:

$$\begin{cases} dx(t) = f(x(t))dt + Gdw(t), & x(0) \in R^n, \\ dy(t) = Hx(t)dt + dv(t), & y(0) \in R^m, \end{cases} \tag{6.299}$$

where x, w, y, v are, respectively, R^n, R^n, R^m, R^m-valued processes. $x(t)$ represents the state of system and $y(t)$ represents observation. $w(t), v(t)$ are independent standard Brownian motions and independent of the initial conditions $x(0), y(0)$. We assume that f is a vector-valued C^∞ function and G, H are constant matrices. Besides, G is an orthogonal matrix. Next we denote $H = (H_1, H_2, \cdots, H_m)^T$, where $H_i = (H_{i1}, \cdots, H_{in})^T, 1 \leq i \leq m$.

Next we introduce Wong's theorem about finite-dimensional estimation algebra.

Theorem 6.40 (Wong [27]) *Let U denote the associative algebra of n by n matrix-valued functions generated by $\{\Omega C, J_\eta C, I\}$, where I stands for the identity matrix and $C = GG^T$. If $H_i^T C\Gamma$ is a constant vector for any $1 \leq i \leq m$ and any $\Gamma \in U$, then estimation algebra of system is finite dimensional and $\dim E \leq 2n + m + 2$.*

In order to construct a finite-dimensional filter, our main method is based on Theorem 6.40 to find proper f, G, H in (6.299).

In the following, we define

$$\begin{cases} f_1 &= x_1 + x_2 + \cdots + x_n + \gamma(x_1 + x_2 + \cdots + x_n), \\ f_2 &= x_1 + x_3 + \cdots + x_n, \\ f_3 &= x_1 + x_2 + x_4 + \cdots + x_n, \\ \cdots \\ f_n &= x_1 + x_2 + \cdots + x_{n-1}, \end{cases} \quad (6.300)$$

where γ is a C^∞ function with a bounded, nonzero first derivative. Let G be the identity matrix; then the corresponding state evolutionary equation is

$$\begin{cases} dx_1 = (x_1 + x_2 + \cdots + x_n + \gamma(x_1 + x_2 + \cdots + x_n))dt + dw_1, \\ dx_2 = (x_1 + x_3 + \cdots + x_n)dt + dw_2, \\ dx_3 = (x_1 + x_2 + x_4 + \cdots + x_n)dt + dw_3, \\ \cdots \\ dx_n = (x_1 + x_2 + \cdots + x_{n-1})dt + dw_n, \end{cases} \quad (6.301)$$

where $w_i, 1 \leq i \leq n$ are independent standard Brownian motions.

Wong's Ω-matrix can be calculated:

$$\Omega = \begin{pmatrix} 0 & -1 & -1 & \cdots & -1 \\ 1 & 0 & 0 & \cdots & 0 \\ 1 & 0 & 0 & \cdots & 0 \\ \vdots & \vdots & \vdots & \ddots & \vdots \\ 1 & 0 & 0 & \cdots & 0 \end{pmatrix} \cdot \gamma'(x_1 + x_2 + \cdots + x_n). \quad (6.302)$$

Then we can calculate eigenvalues and corresponding eigenvectors of Ω^T:

6.6 Novel Finite-Dimensional Filter

$$\lambda_1 = \lambda_2 = \cdots = \lambda_{n-2} = 0, \ \lambda_{n-1} = \sqrt{n-1}\gamma' i, \ \lambda_n = -\sqrt{n-1}\gamma' i,$$

$$H_1 = \begin{pmatrix} 0 \\ 1 \\ -1 \\ 0 \\ \vdots \\ 0 \end{pmatrix}, \ H_2 = \begin{pmatrix} 0 \\ 1 \\ 0 \\ -1 \\ \vdots \\ 0 \end{pmatrix}, \cdots, H_{n-2} = \begin{pmatrix} 0 \\ 1 \\ 0 \\ 0 \\ \vdots \\ -1 \end{pmatrix},$$

$$H_{n-1} = \begin{pmatrix} \sqrt{n-1}i \\ 1 \\ 1 \\ 1 \\ \vdots \\ 1 \end{pmatrix}, \ H_n = \begin{pmatrix} -\sqrt{n-1}i \\ 1 \\ 1 \\ 1 \\ \vdots \\ 1 \end{pmatrix}. \tag{6.303}$$

Considering selection of H is restricted to the real matrix, we only consider real eigenvectors $H_1, H_2, \cdots, H_{n-2}$, which are $n-2$ linearly independent eigenvectors corresponding to eigenvalue 0.

Define $H = (H_1, H_2, \cdots, H_{n-2})^T$ and we obtain observation equation:

$$\begin{cases} dy_1 = H_1^T x dt + dv_1 = (x_2 - x_3)dt + dv_1, \\ dy_2 = H_2^T x dt + dv_2 = (x_2 - x_4)dt + dv_2, \\ dy_3 = H_3^T x dt + dv_2 = (x_2 - x_5)dt + dv_3, \\ \cdots \\ dy_{n-2} = H_{n-2}^T x dt + dv_2 = (x_2 - x_n)dt + dv_{n-2}, \end{cases} \tag{6.304}$$

where v_i, $1 \leq i \leq n-2$ are independent standard Brownian motions.

By combining state evolutionary equation (6.301) and observation equation (6.304), we obtain the filtering system (F1):

$$(F1): \begin{cases} dx_1 = (x_1 + x_2 + \cdots + x_n + \gamma(x_1 + x_2 + \cdots + x_n))dt + dw_1, \\ dx_2 = (x_1 + x_3 + \cdots + x_n)dt + dw_2, \\ \cdots \\ dx_n = (x_1 + x_2 + \cdots + x_{n-1})dt + dw_n, \\ dy_1 = (x_2 - x_3)dt + dv_1, \\ dy_2 = (x_2 - x_4)dt + dv_2, \\ \cdots \\ dy_{n-2} = (x_2 - x_n)dt + dv_{n-2}. \end{cases}$$

(6.305)

In filtering system (F1), H_i, $1 \leq i \leq n-2$ is the real eigenvector of Ω^T but not J_η. In the following calculations, we aim to make an orthogonal transformation for H_i, $1 \leq i \leq n-2$ and obtain a filtering system satisfying assumptions of (6.40).

First, η can be calculated by definition:

$$\eta = \sum_{i=1}^{n} \frac{\partial f_i}{\partial x_i} + \sum_{i=1}^{n} f_i^2 + \sum_{i=1}^{n-2} h_i^2$$

$$= (1+\gamma') + \left(\sum_{i=1}^{n} x_i + \gamma\right)^2 + \sum_{k=2}^{n}\left(\sum_{i=1}^{n} x_i - x_k\right)^2 + \sum_{i=3}^{n}(x_2 - x_i)^2. \tag{6.306}$$

Gradient of η can be calculated:

$$\begin{cases} \eta_{x_1} = \delta_0 + 2\left[(n-1)x_1 + (n-2)\sum_{i=2}^{n} x_i\right], \\[2mm] \eta_{x_2} = \delta_0 + 2\left[(n-2)x_1 + 2(n-2)x_2 + (n-4)\sum_{i=3}^{n} x_i\right], \\[2mm] \eta_{x_k} = \delta_0 + 2\left[(n-2)x_1 + (n-4)x_2 + (n-3)\sum_{i=3}^{n} x_i + 2x_k\right], \quad 3 \leq k \leq n, \end{cases} \tag{6.307}$$

where $\delta_0 = \gamma'' + 2(x_1 + x_2 + \cdots + x_n + \gamma)(1+\gamma')$.

Then we can calculate Hessian matrix J_η:

$$J_\eta = \eta_0 \times \begin{pmatrix} 1 & \cdots & 1 \\ \vdots & \ddots & \vdots \\ 1 & \cdots & 1 \end{pmatrix}$$

$$+ 2 \begin{pmatrix} n-1 & n-2 & n-2 & \cdots & n-2 & n-2 \\ n-2 & 2(n-2) & n-4 & \cdots & n-4 & n-4 \\ n-2 & n-4 & n-1 & \cdots & n-3 & n-3 \\ \vdots & \vdots & \vdots & \ddots & \vdots & \vdots \\ n-2 & n-4 & n-3 & \cdots & n-3 & n-1 \end{pmatrix}, \tag{6.308}$$

where $\eta_0 = \gamma''' + 2(x_1 + \cdots + x_n + \gamma)\gamma'' + 2(1+\gamma')^2$. Next we calculate $J_\eta H_i$, $1 \leq i \leq n-2$:

$$\begin{cases} J_\eta H_1 = (0, 2n, -6, -2, \cdots, -2)^T = 6H_1 + 2H_2 + \cdots + 2H_{n-2}, \\[2mm] J_\eta H_2 = (0, 2n, -2, -6, \cdots, -2)^T = 2H_1 + 6H_2 + \cdots + 2H_{n-2}, \\[2mm] \cdots \\[2mm] J_\eta H_{n-2} = (0, 2n, -2, -2, \cdots, -6)^T = 2H_1 + 2H_2 + \cdots + 6H_{n-2}. \end{cases} \tag{6.309}$$

6.6 Novel Finite-Dimensional Filter

Equation (6.309) can be expressed as the matrix form:

$$J_\eta(H_1, H_2, \cdots, H_{n-2}) = (H_1, H_2, \cdots, H_{n-2})A, \tag{6.310}$$

where

$$A = \begin{pmatrix} 6 & 2 & 2 & \cdots & 2 \\ 2 & 6 & 2 & \cdots & 2 \\ 2 & 2 & 6 & \cdots & 2 \\ \vdots & \vdots & \vdots & \ddots & \vdots \\ 2 & 2 & 2 & \cdots & 6 \end{pmatrix}. \tag{6.311}$$

Since A is a real symmetric matrix, it can be orthogonally diagonalized. First, we calculate eigenvalues and eigenvectors of A. Due to $\det(A - \lambda I) = (2n - \lambda)(4 - \lambda)^{n-3}$, we get eigenvalues of A:

$$\lambda_1 = \lambda_2 = \cdots = \lambda_{n-3} = 4, \quad \lambda_{n-2} = 2n. \tag{6.312}$$

Eigenvectors corresponding to first $n - 3$ eigenvalues are given:

$$\alpha_1 = \begin{pmatrix} 1 \\ -1 \\ 0 \\ 0 \\ \vdots \\ 0 \end{pmatrix}, \alpha_2 = \begin{pmatrix} 1 \\ 0 \\ -1 \\ 0 \\ \vdots \\ 0 \end{pmatrix}, \cdots, \alpha_{n-3} = \begin{pmatrix} 1 \\ 0 \\ 0 \\ 0 \\ \vdots \\ -1 \end{pmatrix}. \tag{6.313}$$

By using Schmidt orthogonalization for $\{\alpha_1, \alpha_2, \cdots, \alpha_{n-3}\}$, normalized orthogonal eigenvectors can be obtained:

$$v_1 = \frac{1}{\sqrt{2}}\begin{pmatrix} 1 \\ -1 \\ 0 \\ 0 \\ \vdots \\ 0 \end{pmatrix}, v_2 = \sqrt{\frac{2}{3}}\begin{pmatrix} \frac{1}{2} \\ \frac{1}{2} \\ -1 \\ 0 \\ \vdots \\ 0 \end{pmatrix}, v_3 = \frac{\sqrt{3}}{2}\begin{pmatrix} \frac{1}{3} \\ \frac{1}{3} \\ \frac{1}{3} \\ -1 \\ \vdots \\ 0 \end{pmatrix}, \cdots, v_{n-3} = \sqrt{\frac{n-3}{n-2}}\begin{pmatrix} \frac{1}{n-3} \\ \frac{1}{n-3} \\ \frac{1}{n-3} \\ \vdots \\ \frac{1}{n-3} \\ -1 \end{pmatrix}. \tag{6.314}$$

The eigenvector corresponding to $\lambda_{n-2} = 2n$ is

$$v_{n-2} = \frac{1}{\sqrt{n-2}}\begin{pmatrix} 1 \\ 1 \\ \vdots \\ 1 \end{pmatrix}. \tag{6.315}$$

We define $P = (v_1, v_2, \cdots, v_{n-2})$ which is an orthogonal matrix; then we get the diagonal decomposition $A = P \Lambda P^T$, where $\Lambda = \text{diag}(4, 4, \cdots, 4, 2n)$. Thus, Eq. (6.310) becomes

$$J_\eta(H_1, H_2, \cdots, H_{n-2}) = (H_1, H_2, \cdots, H_{n-2}) P \Lambda P^T,$$
$$\implies J_\eta(H_1, H_2, \cdots, H_{n-2}) P = (H_1, H_2, \cdots, H_{n-2}) P \Lambda. \tag{6.316}$$

Next we define $(\tilde{H}_1, \tilde{H}_2, \cdots, \tilde{H}_{n-2}) = (H_1, H_2, \cdots, H_{n-2}) P$, which is given by

$$\tilde{H}_1 = \frac{1}{\sqrt{2}} \begin{pmatrix} 0 \\ 0 \\ -1 \\ 1 \\ 0 \\ \vdots \\ 0 \end{pmatrix}, \tilde{H}_2 = \sqrt{\frac{2}{3}} \begin{pmatrix} 0 \\ 0 \\ -\frac{1}{2} \\ -\frac{1}{2} \\ 1 \\ \vdots \\ 0 \end{pmatrix}, \cdots, \tag{6.317}$$

$$\tilde{H}_{n-3} = \sqrt{\frac{n-3}{n-2}} \begin{pmatrix} 0 \\ 0 \\ -\frac{1}{n-3} \\ -\frac{1}{n-3} \\ \vdots \\ -\frac{1}{n-3} \\ 1 \end{pmatrix}, \tilde{H}_{n-2} = \frac{1}{\sqrt{n-2}} \begin{pmatrix} 0 \\ n-2 \\ -1 \\ \vdots \\ -1 \end{pmatrix}.$$

Clearly, $\tilde{H}_i$, $1 \leq i \leq n-2$ is the eigenvector of J_η.

Define $\tilde{H} = (\tilde{H}_1, \tilde{H}_2, \cdots, \tilde{H}_{n-2})^T$ and keep f, G unchanged; then we consider the following filtering system (F2):

(F2):
$$dx_1 = (x_1 + x_2 + \cdots + x_n + \gamma(x_1 + x_2 + \cdots + x_n))dt + dw_1,$$
$$dx_2 = (x_1 + x_3 + \cdots + x_n)dt + dw_2,$$
$$\cdots$$
$$dx_n = (x_1 + x_2 + \cdots + x_{n-1})dt + dw_n,$$
$$dy_1 = \frac{1}{\sqrt{2}}(-x_3 + x_4)dt + dv_1,$$
$$dy_2 = \sqrt{\frac{2}{3}}\left(-\frac{1}{2}x_3 - \frac{1}{2}x_4 + x_5\right)dt + dv_2,$$
$$dy_{n-3} = \sqrt{\frac{n-3}{n-2}}\left(-\frac{1}{n-3}x_3 - \frac{1}{n-3}x_4 - \cdots - \frac{1}{n-3}x_{n-1} + x_n\right)dt$$
$$+ dv_{n-3},$$
$$dy_{n-2} = \frac{1}{\sqrt{n-2}}((n-2)x_2 - x_3 - \cdots - x_n)dt + dv_{n-2}.$$
$$\tag{6.318}$$

6.6 Novel Finite-Dimensional Filter

In the filtering system (F2), we denote

$$\tilde{f}_i = f_i, \ 1 \leq i \leq n,$$
$$\tilde{h}_i = \tilde{H}_i^T x, \ 1 \leq i \leq n,$$
$$\tilde{\Omega} = (\tilde{\omega}_{ij}), \ \tilde{\omega}_{ij} = \frac{\partial \tilde{f}_j}{\partial x_i} - \frac{\partial \tilde{f}_i}{\partial x_j}, \ 1 \leq i, j \leq n, \quad (6.319)$$
$$\tilde{\eta} = \sum_{i=1}^{n} \frac{\partial \tilde{f}_i}{\partial x_i} + \sum_{i=1}^{n} \tilde{f}_i^2 + \sum_{i=1}^{n-2} \tilde{h}_i^2.$$

Due to $\tilde{f}_i = f_i$, $1 \leq i \leq n$ and

$$\sum_{i=1}^{n-2} \tilde{h}_i^2 = \sum_{i=1}^{n-2} h_i^2, \quad (6.320)$$

then we have $\tilde{\eta} = \eta$. By Eq. (6.316), we obtain

$$J_{\tilde{\eta}}(\tilde{H}_1, \tilde{H}_2, \cdots, \tilde{H}_{n-2}) = (\tilde{H}_1, \tilde{H}_2, \cdots, \tilde{H}_{n-2}) \Lambda, \quad (6.321)$$

which means $\tilde{H}_i$ is the eigenvector of $J_{\tilde{\eta}}$. Due to $\Omega^T H_i = 0$, $1 \leq i \leq n-2$, $\tilde{\Omega} = \Omega$ and definition of $\tilde{H}_i$, then $\tilde{\Omega}^T \tilde{H}_i = 0$, $1 \leq i \leq n-2$, which means that $\tilde{H}_i$ is the eigenvector of $\tilde{\Omega}^T$. Therefore, $\tilde{H}_i$'s are $n-2$ linearly independent common real eigenvectors of $\tilde{\Omega}^T$ and $J_{\tilde{\eta}}$.

Let U be the associative algebra of n by n matrix-valued functions generated by $\{\tilde{\Omega}, J_{\tilde{\eta}}, I\}$, where I stands for the identity matrix. We will obtain the following result. For simplicity, we omit tilde notation.

Theorem 6.41 *Estimation algebra of filtering system (F2) is finite dimensional.*

Proof Let A_0 be linear space generated by $\{\Omega, J_\eta, I\}$. $A_1 = \{X \in U | X = X_1 + X_2 X_3, X_i \in A_0\}$. Recursively, we define $A_k = \{X \in U | X = X_1 + X_2 X_3, X_i \in A_{k-1}\}$ for $k \geq 1$. Notice that

$$U = \bigcup_{k \in Z_{\geq 0}} A_k. \quad (6.322)$$

In the following, we claim $H_i^T \Gamma = \text{const} \cdot H_i^T$ for $1 \leq i \leq m$ and any $\Gamma \in A_k, k \in Z_{\geq 0}$. We use induction to prove this. For $k = 0$, for any $\Gamma \in A_0$, $\Gamma = a_1 \Omega + a_2 J_\eta a_3 I, a_i \in \mathbb{R}$. We calculate

$$H_i^T \Gamma = a_2 H_i^T J_\eta + a_3 H_i^T = (a_2 \lambda_i + a_3) H_i^T. \quad (6.323)$$

Assume the statement holds for $k = p$. Then $\forall \Gamma \in A_{p+1}$, $\Gamma = X_1 + X_2 X_3$, where $X_i \in A_p$. By assumption of induction, we calculate

$$H_i^T \Gamma = H_i^T (X_1 + X_2 X_3) = const \cdot H_i^T. \tag{6.324}$$

It follows that $H_i^T \Gamma$ is a constant vector for $\forall 1 \leq i \leq m$ and $\forall \Gamma \in U$.

By (6.40), it leads to finite dimensionality of estimation algebra. □

In the following theorem, we show the structure of estimation algebra and Wong's Ω-matrix of filtering system (F2).

Theorem 6.42 *Nonlinear filtering system is given by*

$$\begin{cases} dx_1 = (x_1 + x_2 + \cdots + x_n + \gamma(x_1 + x_2 + \cdots + x_n))dt + dw_1, \\ dx_2 = (x_1 + x_3 + \cdots + x_n)dt + dw_2, \\ dx_3 = (x_1 + x_2 + x_4 + \cdots + x_n)dt + dw_3, \\ \cdots \\ dx_n = (x_1 + x_2 + \cdots + x_{n-1})dt + dw_n, \\ dy_1 = \frac{1}{\sqrt{2}}(-x_3 + x_4)dt + dv_1, \\ dy_2 = \sqrt{\frac{2}{3}}(-\frac{1}{2}x_3 - \frac{1}{2}x_4 + x_5)dt + dv_2, \\ \cdots \\ dy_{n-3} = \sqrt{\frac{n-3}{n-2}}(-\frac{1}{n-3}x_3 - \frac{1}{n-3}x_4 - \cdots - \frac{1}{n-3}x_{n-1} + x_n)dt + dv_{n-3}, \\ dy_{n-2} = \frac{1}{\sqrt{n-2}}((n-2)x_2 - x_3 - \cdots - x_n)dt + dv_{n-2}, \end{cases} \tag{6.325}$$

where γ is a C^∞ function with a bounded, nonzero first derivative. w, v are vector-valued independent standard Brownian motions and independent of the initial conditions. Then in this filtering system, entries of Wong's Ω-matrix are not necessarily to be constants or polynomials. Dimension of estimation algebra is $2n - 2$ and linear rank of estimation algebra is $n - 2$.

Proof For convenience, we omit tilde in Eq. (6.319) and denote

$$\begin{cases} f_1 = x_1 + x_2 + \cdots + x_n + \gamma(x_1 + x_2 + \cdots + x_n), \\ f_k = \sum_{i=1}^{n} x_i - x_k, \quad 2 \leq k \leq n, \\ h_k = \sqrt{\frac{k}{k+1}}\left(-\frac{1}{k}x_3 - \frac{1}{k}x_4 - \cdots - \frac{1}{k}x_{k+2} + x_{k+3}\right), \quad 1 \leq k \leq n-3, \\ h_{n-2} = \frac{1}{\sqrt{n-2}}((n-2)x_2 - x_3 - \cdots - x_n). \end{cases} \tag{6.326}$$

6.6 Novel Finite-Dimensional Filter

Wong's Ω-matrix is given by Eq. (6.302):

$$\Omega = \begin{pmatrix} 0 & -1 & -1 & \cdots & -1 \\ 1 & 0 & 0 & \cdots & 0 \\ 1 & 0 & 0 & \cdots & 0 \\ \vdots & \vdots & \vdots & \ddots & \vdots \\ 1 & 0 & 0 & \cdots & 0 \end{pmatrix} \cdot \gamma'(x_1 + x_2 + \cdots + x_n). \tag{6.327}$$

Due to arbitrariness of γ, entries of Ω are not necessarily to be constants or polynomials. Next we calculate elements in estimation algebra E of filtering system. By definition, $L_0, h_1, h_2, \cdots, h_{n-2} \in E$.

First, we calculate elements $[L_0, h_i]$, $1 \leq i \leq n-2$:

$$[L_0, h_1] = \frac{1}{2}\left[\sum_{i=1}^{n} D_i^2 - \eta, \frac{1}{\sqrt{2}}(-x_3 + x_4)\right] \tag{6.328}$$

$$= \frac{1}{\sqrt{2}}(-D_3 + D_4) \in E,$$

$$[L_0, h_2] = \frac{1}{2}\left[\sum_{i=1}^{n} D_i^2 - \eta, \sqrt{\frac{2}{3}}\left(-\frac{1}{2}x_3 - \frac{1}{2}x_4 + x_5\right)\right]$$

$$= \sqrt{\frac{2}{3}}\left(-\frac{1}{2}D_3 - \frac{1}{2}D_4 + D_5\right) \in E,$$

$$\cdots$$

$$[L_0, h_k] = \frac{1}{2}\left[\sum_{i=1}^{n} D_i^2 - \eta, \sqrt{\frac{k}{k+1}}\left(-\frac{1}{k}\sum_{i=3}^{k+2} x_i + x_{k+3}\right)\right]$$

$$= \sqrt{\frac{k}{k+1}}\left(-\frac{1}{k}\sum_{i=3}^{k+2} D_i + D_{k+3}\right) \in E, \quad 1 \leq k \leq n-3,$$

$$\cdots$$

$$[L_0, h_{n-3}] = \sqrt{\frac{n-3}{n-2}}\left(-\frac{1}{n-3}\sum_{i=3}^{n-1} D_i + D_n\right) \in E,$$

$$[L_0, h_{n-2}] = \frac{1}{2}\left[\sum_{i=1}^{n} D_i^2 - \eta, \frac{1}{\sqrt{n-2}}\left((n-2)x_2 - \sum_{i=3}^{n} x_i\right)\right]$$

$$= \frac{1}{\sqrt{n-2}}\left((n-2)D_2 - \sum_{i=3}^{n} D_i\right) \in E.$$

Next we can calculate Lie bracket of L_0 and above first-order differential operators obtained by Eq. (6.328):

$$[L_0, -D_3 + D_4] = \frac{1}{2}\left(-\frac{\partial \eta}{\partial x_3} + \frac{\partial \eta}{\partial x_4}\right) \in E, \tag{6.329}$$

$$\left[L_0, -\frac{1}{2}D_3 - \frac{1}{2}D_4 + D_5\right] = \frac{1}{2}\left(-\frac{1}{2}\frac{\partial \eta}{\partial x_3} - \frac{1}{2}\frac{\partial \eta}{\partial x_4} + \frac{\partial \eta}{\partial x_5}\right) \in E, \tag{6.330}$$

$$\left[L_0, -\frac{1}{k}\sum_{i=3}^{k+2} D_i + D_{k+3}\right] = \frac{1}{2}\left(-\frac{1}{k}\sum_{j=3}^{k+2} \frac{\partial \eta}{\partial x_j} + \frac{\partial \eta}{\partial x_{k+3}}\right) \in E, \quad 1 \leq k \leq n-3, \tag{6.331}$$

$$\left[L_0, -\frac{1}{n-3}\sum_{i=3}^{n-1} D_i + D_n\right] = \frac{1}{2}\left(-\frac{1}{n-3}\sum_{i=3}^{n-1} \frac{\partial \eta}{\partial x_i} + \frac{\partial \eta}{\partial x_n}\right) \in E,$$

$$\left[L_0, (n-2)D_2 - \sum_{i=3}^{n} D_i\right] = \frac{1}{2}\left((n-2)\frac{\partial \eta}{\partial x_2} - \sum_{i=3}^{n} \frac{\partial \eta}{\partial x_i}\right) \in E,$$

where η is given by Eq. (6.306). By using Eq. (6.307) and (6.329) can be simplified as

$$\left[L_0, -\frac{1}{k}\sum_{i=3}^{k+2} D_i + D_{k+3}\right] = 2\left(-\frac{1}{k}\sum_{j=3}^{k+2} x_j + x_{k+3}\right) \in E, \quad 1 \leq k \leq n-3,$$

and

$$\left[L_0, (n-2)D_2 - \sum_{i=3}^{n} D_i\right] = n\left((n-2)x_2 - \sum_{i=3}^{n} x_i\right) \in E \tag{6.332}$$

Due to

$$\left[\frac{1}{\sqrt{2}}(-D_3 + D_4), \frac{1}{\sqrt{2}}(-x_3 + x_4)\right] = 1 \in E, \tag{6.333}$$

all constants are in E. Finally, we calculate Lie bracket between first-order differential operators obtained by Eq. (6.328):

$$\left[-\frac{1}{k_1}\sum_{j=3}^{k_1+2} D_j + D_{k_1+3}, -\frac{1}{k_2}\sum_{j=3}^{k_2+2} D_j + D_{k_2+3}\right] = 0, \quad 1 \leq k_1, k_2 \leq n-3,$$

$$\left[-\frac{1}{k}\sum_{j=3}^{k+2} D_j + D_{k+3}, (n-2)D_2 - \sum_{i=3}^{n} D_i\right] = 0, \quad 1 \leq k \leq n-3. \tag{6.334}$$

6.7 Exercises

Therefore, estimation algebra E of the filtering system (6.325) is a $2n-2$ dimensional Lie algebra with basis given by

$$\{1, -x_3 + x_4, -\frac{1}{2}x_3 - \frac{1}{2}x_4 + x_5, \cdots, -\frac{1}{n-3}x_3 - \frac{1}{n-3}x_4 - \cdots$$

$$-\frac{1}{n-3}x_{n-1} + x_n,$$

$$(n-2)x_2 - x_3 - \cdots - x_n, -D_3 + D_4, -\frac{1}{2}D_3 - \frac{1}{2}D_4 + D_5, \cdots, \qquad (6.335)$$

$$-\frac{1}{n-3}D_3 - \frac{1}{n-3}D_4 - \cdots - \frac{1}{n-3}D_{n-1} + D_n,$$

$$(n-2)D_2 - D_3 - \cdots - D_n, L_0\}.$$

Clearly, linear rank of estimation algebra E is $n-2$. □

6.7 Exercises

1. Prove that L_0 has the following compact form:

$$L_0 = \frac{1}{2}\left(\sum_{i=1}^{n} D_i^2 - \eta\right). \qquad (6.336)$$

2. Prove the following bracket results:
   (1) $[L_0, x_i] = D_i$;
   (2) $[[L_0, \phi], \phi] = |\nabla \phi|^2 = \sum_{i=1}^{n}(\frac{\partial \phi}{\partial x_i})^2$;
   (3) $[L_0, D_j] = \sum_{i=1}^{n} \omega_{ji} D_i + \frac{1}{2}\frac{\partial \eta}{\partial x_j} + \frac{1}{2}\sum_{i=1}^{n} \frac{\partial \omega_{ji}}{\partial x_i}$;
   (4) $[L_0, x_j^2] = 2x_j D_j + 1$.

3. Prove that following bracket results:
   (1) $[L_0, x_j + c_j] = D_j$, $1 \leq j \leq n$;
   (2) $[D_i, x_j + c_j] = \delta_{ij}$, $1 \leq i, j \leq n$;
   (3) $[D_i, D_j] = \omega_{ji}$, $1 \leq i, j \leq n$;
   (4) $Y_j := [L_0, D_j] = \sum_{i=1}^{n}\left(\omega_{ji} D_i + \frac{1}{2}\frac{\partial \omega_{ji}}{\partial x_i}\right) + \frac{1}{2}\frac{\partial \eta}{\partial x_j}$, $1 \leq j \leq n$;
   (5) $[Y_j, \omega_{kl}] = \sum_{i=1}^{n} \omega_{ji} \frac{\partial \omega_{kl}}{\partial x_i}$, $1 \leq j, k, l \leq n$;
   (6) $[Y_j, D_k] = \sum_{i=1}^{n}\left(\omega_{ji}\omega_{ki} - \frac{\partial \omega_{ji}}{\partial x_k} D_i\right) - \frac{1}{2}\sum_{i=1}^{n}\frac{\partial^2 \omega_{ij}}{\partial x_k \partial x_i} - \frac{1}{2}\frac{\partial^2 \eta}{\partial x_k \partial x_j}$.

4. Let $m > 0$ be a positive integer and $\xi \in C^{\infty}(R^n)$. If $E_l(\xi) + m\xi = 0$, prove that $\xi = 0$.

5. Let $\xi \in C^\infty(R^n)$. If $E_l(\xi) = 0$, prove that ξ is a C^∞-function in $x_{l+1}, \cdots, x_n$ variables.

6. Suppose $i = (i_1, \cdots, i_n)$ and $|i| = \sum_{l=1}^n i_l \geq 2$. Prove that

$$gD_1^{i_1} \cdots D_n^{i_n} = gD_{k_1}^{i_{k_1}} \cdots D_{k_n}^{i_{k_n}}, \mod U_{|i|-2}, \quad (6.337)$$

where g is a C^∞-function of $x_1, \cdots, x_n$ and $k = (k_1, \cdots, k_n)$ is a permutation of $(1, 2, \cdots, n)$.

7. Suppose E is a nonmaximal rank estimation algebra. Its associated polynomial with greatest quadratic rank is denoted as $p_0 = \sum_{i \in S} x_i^2$ where index set $S := \{1, \cdots, k_1, n - k_2 + 1, \cdots, n\}$. If $p \in E$ is a quadratic function, prove that homogeneous quadratic part $p^{(2)}(x)$ is independent of x_j for $j \notin S$, i.e., $\frac{\partial p^{(2)}(x)}{\partial x_j} = 0$ for $j = k_1 + 1, \cdots, n - k_2$.

8. Prove the cyclical identity is satisfied by Ω, i.e.,

$$\frac{\partial \omega_{ij}}{\partial x_k} + \frac{\partial \omega_{jk}}{\partial x_i} + \frac{\partial \omega_{ki}}{\partial x_j} = 0 \quad (6.338)$$

9. Prove that $\frac{\partial f_j}{\partial x_i} - \frac{\partial f_i}{\partial x_j} = c_{ij} + D_{ij}^T x$, where $D_{ij} \in R^n$, $x = (x_1, x_2, \cdots, x_n)^T$ for all i, j if and only if

$$(f_1, \cdots, f_n) = (l_1, \cdots, l_n) + \left(\frac{\partial \psi}{\partial x_1}, \cdots, \frac{\partial \psi}{\partial x_n}\right), \quad (6.339)$$

where $l_1, \cdots, l_n$ are degree at most 2 polynomials and ψ is a C^∞ function.

References

1. R. W. Brockett. Classification and equivalence in estimation theory *18th IEEE Conference on Decision and Control including the Symposium on Adaptive Processes*, 2:172–175, 1979.
2. R. W. Brockett and J. M. C. Clark. The geometry of the conditional density function. *Analysis and Optimization of Stochastic Systems*. 1980.
3. R. W. Brockett. *Nonlinear systems and nonlinear estimation theory*. Springer, Dordrecht, 1981.
4. W. L. Chiou and S. S.-T. Yau. Finite-dimensional filters with nonlinear drift. II: Brockett's problem on classification of finite-dimensional estimation algebras. *SIAM Journal on Control and Optimization*, 32(1):297–310, 1994.
5. J. Chen and S. S.-T. Yau. Finite-dimensional filters with nonlinear drift, VI: Linear structure of Ω. *Mathematics of Control, Signals and Systems*, 9(4):370–385, 1996.
6. J. Chen and S. S.-T. Yau. Finite-dimensional filters with nonlinear drift VII: Mitter conjecture and structure of η. *SIAM Journal on Control and Optimization*, 35(4):1116–1131, 1997.
7. J. Chen, S. S.-T. Yau and C. W. Leung. Finite-dimensional filters with nonlinear drift IV: Classification of finite-dimensional estimation algebras of maximal rank with state–space dimension 3. *SIAM Journal on Control and Optimization*, 34(1):179–198, 1996.

8. J. Chen, S. S.-T. Yau and C. W. Leung. Finite-dimensional filters with nonlinear drift VIII: Classification of finite-dimensional estimation algebras of maximal rank with state-space dimension 4. *SIAM Journal on Control and Optimization*, 35(4):1132–1141, 1997.
9. M. H. A. Davis. On a multiplicative functional transformation arising in nonlinear filtering theory. *Zeitschrift für Wahrscheinlichkeitstheorie und verwandte Gebiete*, 54(2):125–139, 1980.
10. W. Dong, X. Chen and S. S.-T. Yau. The novel classes of finite dimensional filters with non-maximal rank estimation algebra on state dimension four and rank of one. *International Journal of Control*, 94(5):1156–1165, 2021.
11. M. Hazewinkel, S. I. Marcus and H. J. Sussmann. Nonexistence of finite-dimensional filters for conditional statistics of the cubic sensor problem. *Systems & control letters*, 3(6):331–340, 1983.
12. U. G. Haussmann and E. Pardoux. A conditionally almost linear filtering problem with non-Gaussian initial condition. *Stochastics: An International Journal of Probability and Stochastic Processes*, 23(2):241–275, 1988.
13. X. Jiao and S. S.-T. Yau. New classes of finite dimensional filters with nonmaximal rank estimation algebra on state dimension n and linear rank n-2. *SIAM Journal on Control and Optimization*, 58(6):3413–3427, 2020.
14. M. Fujisaki, G. Kallianpur and H. Kunita. Stochastic differential equations for the non linear filtering problem. *Osaka Journal of Mathematics*, 9(1):19–40, 1972.
15. M. C. Maurel and D. Michel. Des résultats de non existence de filtre de dimension finie. *Stochastics: An International Journal of Probability and Stochastic Processes*, 13(1–2):83–102, 1984.
16. A. M. Makowski. Filtering formulae for partially observed linear systems with non-Gaussian initial conditions. *Stochastics: An International Journal of Probability and Stochastic Processes*, 16(1–2):1–24, 1986.
17. S. I. Marcus. Algebraic and geometric methods in nonlinear filtering. *SIAM Journal on Control and Optimization*, 22(6):817–844, 1984.
18. S. K. Mitter. On the analogy between mathematical problems of non-linear filtering and quantum physics. *MIT Cambridge Lab For Information and decision systems*, 1980.
19. D. Ocone. Topics in nonlinear filtering theory. *Massachusetts Institute of Technology*, 1980.
20. J. Shi, X. Chen, W. Dong and S. S.-T. Yau. New classes of finite dimensional filters with non-maximal rank. *IEEE Control Systems Letters*, 1(2):233–237, 2017.
21. J. Shi and S. S.-T. Yau. Finite dimensional estimation algebras with state dimension 3 and rank 2, I: linear structure of Wong matrix. *SIAM Journal on Control and Optimization*, 55(6):4227–4246, 2017.
22. J. Shi and S. S.-T. Yau. Finite dimensional estimation algebras with state dimension 3 and rank 2, Mitter conjecture. *International Journal of Control*, 93(9):2177–2186, 2020.
23. L. F. Tam, W. S. Wong and S. S.-T. Yau. On a necessary and sufficient condition for finite dimensionality of estimation algebras. *SIAM Journal on Control and Optimization*, 28(1):173–185, 1990.
24. J. Wei and E. Norman. On global representations of the solutions of linear differential equations as a product of exponentials. *Proceedings of the American Mathematical Society*, 15(2):327–334, 1964.
25. X. Wu, S. S.-T. Yau and G.-Q. Hu. Finite-dimensional filters with nonlinear drift. XII: Linear and constant structure of Wong-matrix. *Stochastic Theory and Control*, 507–518, 2002.
26. X. Wu and S. S.-T. Yau. Classification of estimation algebras with state dimension 2. *SIAM Journal on Control and Optimization*, 45(3):1039–1073, 2006.
27. W. S. Wong. On a new class of finite dimensional estimation algebras. *Systems & Control Letters*, 9(1): 79–83, 1987.
28. S. S.-T. Yau. Finite-dimensional filters with nonlinear drift. I: A class of filters including both Kalman-Bucy and Benes filters. *Journal of Mathematical Systems, Estimation, and Control*, 4:181–203, 1994.

29. S. S.-T. Yau, G. Q. Hu and W. L. Chiou. Finite dimensional filters with nonlinear drift XIV: Classification of finite-dimensional estimation algebras of maximal rank with arbitrary state space dimension and Mitter conjecture. Preprint, 2000.
30. S. S.-T. Yau. Recent results on nonlinear filtering: New class of finite dimensional filters *29th IEEE Conference on Decision and Control*. 1990.
31. S. S.-T. Yau and W. L. Chiou. Recent results on classification of finite dimensional estimation algebras: dimension of state space ≤ 2 *Proceedings of the 30th IEEE Conference on Decision and Control*, 2758–2760, 1991.
32. S.S.-T. Yau and A. Rasoulian. Classification of four-dimensional estimation algebras. *IEEE Transactions on Automatic Control*, 44(12):2312–2318, 1999.
33. S. S.-T. Yau. Complete classification of finite-dimensional estimation algebras of maximal rank. *International Journal of Control*, 76(7): 657–677, 2003.
34. S. S.-T. Yau, X. Wu and W. S. Wong. Hessian matrix non-decomposition theorem. *Mathematical Research Letters*, 6(5):663–673, 1999.

Part III
Numerical Algorithms

Chapter 7
Yau-Yau Algorithm

In this chapter, we will present a numerical algorithm to solve general nonlinear filtering problems based on the DMZ equations introduced in Chap. 5. This algorithm, which is called the Yau-Yau algorithm, is named after Shing-Tung Yau and Stephen S.-T. Yau, who proposed this algorithm in the beginning of this century [4, 5]. Generally speaking, the idea of this algorithm is to separate the task of solving nonlinear filtering problems into two phases, and the procedure of solving partial differential equations (PDEs), which is computationally expensive, can be dealt with off-line. We will start with the formulation of the Yau-Yau algorithm and present a rigorous convergence analysis. Later on, its applications in time-variant filtering systems will be discussed and some suitable numerical methods of solving PDEs will also be summarized.

7.1 Introduction

In this chapter, we will consider the following filtering problem:

$$\begin{cases} dX_t = f(X_t)dt + dV_t, \\ dY_t = h(X_t)dt + dW_t, \end{cases} \quad (7.1)$$

in which X, V, Y, W are, respectively, R^n-, R^n-, R^m-, and R^m-valued processes and V and W are independent standard Brownian motion with corresponding dimensions.

We further assume that f and h are C^∞ smooth functions and $\rho(t, x)$ denotes conditional probability density of state x_t given the observations $\{Y_s : 0 \leq s \leq t\}$.

We have derived before that $\rho(t, x)$ is given by normalizing a function, $\sigma(t, x)$, which satisfies the following DMZ equation:

$$d\sigma(t,x) = L_0\sigma(t,x)dt + \sum_{i=1}^{n} h_i(x)\sigma(t,x)dY_t^i, \quad \sigma(0,x) = \sigma_0(x), \tag{7.2}$$

where

$$L_0 = \frac{1}{2}\sum_{i=1}^{n}\frac{\partial^2}{\partial x_i^2} - \sum_{i=1}^{n} f_i\frac{\partial}{\partial x_i} - \sum_{i=1}^{n}\frac{\partial f_i}{\partial x_i} - \frac{1}{2}\sum_{i=1}^{m} h_i^2. \tag{7.3}$$

As we mentioned before, we can define a new unnormalized density by

$$u(t,x) = \exp\left(-\sum_{i=1}^{m} h_i(x)Y_t^i\right)\sigma(t,x), \tag{7.4}$$

and $u(t,x)$ defined above satisfies the following robust DMZ equation:

$$\begin{cases} \dfrac{\partial u}{\partial t}(t,x) = \dfrac{1}{2}\Delta u(t,x) + (-f(x) + \nabla K(t,x))\cdot\nabla u(t,x) \\ \qquad + \left(-\nabla\cdot f(x) - \dfrac{1}{2}|h(x)|^2 + \dfrac{1}{2}\Delta K(t,x)\right. \\ \qquad \left. - f(x)\cdot\nabla K(t,x) + \dfrac{1}{2}|\nabla K(t,x)|^2\right)u(t,x), \\ u(0,x) = \sigma_0(x), \end{cases} \tag{7.5}$$

where $K(t,x) = \sum_{j=1}^{m} Y_t^j h_j(x)$, $f = (f_1, f_2, \cdots, f_n)$, and $h = (h_1, h_2, \cdots, h_m)$.

Generally speaking, the robust DMZ equation (7.5) does not have a closed-form solution. The main idea of Yau-Yau's algorithm is to construct a good approximation to this solution through a time discretization procedure. In this chapter, we will first summarize the formulation of this algorithm in Sect. 7.2. Then, we will give a detailed proof of the convergence results of this algorithm in Sect. 7.3. Finally, the extension of this algorithm in time-dependence case will be introduced in Sect. 7.5.

7.2 The Formulation of Yau-Yau Algorithm

The motivation of Yau-Yau algorithm is the following observation, which establishes the connection between the robust DMZ equation in a small interval and a parabolic equation with coefficients independent of the observations.

Lemma 7.1 $\tilde{u}(t,x)$ *satisfies the following parabolic equation:*

7.2 The Formulation of Yau-Yau Algorithm

$$\frac{\partial \tilde{u}}{\partial t}(t,x) = \frac{1}{2}\Delta \tilde{u}(t,x) - \sum_{i=1}^{n} f_i(x)\frac{\partial \tilde{u}}{\partial x_i}(t,x) - \left(\sum_{i=1}^{n}\frac{\partial f_i}{\partial x_i}(x) + \frac{1}{2}\sum_{i=1}^{m} h_i^2(x)\right)\tilde{u}(t,x), \quad (7.6)$$

for $\tau_{l-1} \le t \le \tau_l$ if and only if

$$u(t,x) = exp\left(-\sum_{i=1}^{m} Y_i(\tau_{l-1})h_i(x)\right)\tilde{u}(t,x)$$

satisfies the robust DMZ equation with observation being frozen at $y(\tau_l)$:

$$\begin{cases} \frac{\partial u}{\partial t}(t,x) = \frac{1}{2}\Delta u(t,x) + (-f(x) + \nabla K(\tau_{l-1},x)) \cdot \nabla u(t,x) \\ \quad + \left(-\nabla \cdot f(x) - \frac{1}{2}|h(x)|^2 + \frac{1}{2}\Delta K(\tau_{l-1},x)\right. \\ \quad \left. - f(x) \cdot \nabla K(\tau_{l-1},x) + \frac{1}{2}|\nabla K(\tau_{l-1},x)|^2\right)u(t,x), \end{cases} \quad (7.7)$$

where the definition of $K(t,x)$ is the same as Eq. (7.5).

Proof This equivalence follows by direct computation. □

We will later show that the solution of the robust DMZ equation with frozen observations (7.7) is a good approximation to the solution of the original robust DMZ equation in the interval $[\tau_{l-1}, \tau_l]$, when the length of the interval is sufficiently small. With the above observation, we can propose the formulation of Yau-Yau algorithm.

Suppose that $\{y_s : 0 \le s \le \tau\}$ is a realization of the observation process $Y = \{Y_s : 0 \le s \le \tau\}$, and $u(t,x)$ is the solution of the robust DMZ equation (7.5) and we want to compute $u(\tau, x)$. Let $\mathcal{P}_k = \{0 = \tau_0 < \tau_1 < \cdots < \tau_k = \tau\}$ be a partition of $[0, \tau]$. Let $u_i(t, x)$ be a solution of the following partial differential equation for $\tau_{i-1} \le t \le \tau_i$:

$$\begin{cases} \frac{\partial u_i}{\partial t}(t,x) = \frac{1}{2}\Delta u_i(t,x) + (-f(x) + \nabla K(\tau_{i-1},x)) \cdot \nabla u_i(t,x) \\ \quad + \left(-\nabla \cdot f(x) - \frac{1}{2}|h(x)|^2 + \frac{1}{2}\Delta K(\tau_{i-1},x)\right. \\ \quad \left. - f(x) \cdot \nabla K(\tau_{l-1},x) + \frac{1}{2}|\nabla K(\tau_{i-1},x)|^2\right)u_i(t,x), \\ u_i(\tau_{i-1},x) = u_{i-1}(\tau_{i-1},x). \end{cases} \quad (7.8)$$

By Lemma 7.1, each $u_i(\tau_i, x)$ can be computed by $\tilde{u}_i(\tau_i, x)$, where for $\tau_{i-1} \le t \le \tau_i$, $\tilde{u}_i(t,x)$ satisfies the following equation:

$$\frac{\partial \widetilde{u}_i}{\partial t}(t,x) = \frac{1}{2}\Delta \widetilde{u}_i(t,x) - \sum_{i=1}^{n} f_i(x)\frac{\partial \widetilde{u}_i}{\partial x_i}(t,x)$$
$$- \left(\sum_{i=1}^{n}\frac{\partial f_i}{\partial x_i}(x) + \frac{1}{2}\sum_{i=1}^{m} h_i^2(x)\right)\widetilde{u}_i(t,x), \quad (7.9)$$

with initial value

$$\widetilde{u}_i(\tau_{i-1}, x) = \exp\left(\sum_{j=1}^{m}(y_{\tau_{i-1}}^j - y_{\tau_{i-2}}^j h_j(x))\right)\widetilde{u}_{i-1}(\tau_{i-1}, x), \quad i \geq 2, \quad (7.10)$$

and

$$\widetilde{u}_1(0, x) = \sigma_0(x)\exp\left(\sum_{j=1}^{m} y_j(\tau_0)h_j(x)\right). \quad (7.11)$$

Notice that Eq. (7.9) is a parabolic equation independent of observation $y(t)$. It can thus be computed off-line. The entire procedure of this algorithm to compute $u(\tau, x)$ can be summarized as follows:

1. Partition the interval $[0, \tau]$ by $0 = \tau_0 < \tau_1 < \cdots < \tau_k = \tau$.
2. At each time τ_i, use the value of $\widetilde{u}_{i-1}(\tau_{i-1}, x)$ to obtain the initial value (7.10) and solve the parabolic differential equation (7.9). We then get the value of $\widetilde{u}_i(\tau_i, x)$.
3. At time $\tau = \tau_k$, we get the value of $\widetilde{u}_k(\tau_k, x)$, and

$$u_k(\tau_k, x) = \exp\left(-\sum_{j=1}^{m} y_j(\tau_k)h_j(x)\right)\widetilde{u}_k(\tau_k, x)$$

serves to be an approximation to $u(\tau, x)$.

7.3 L^1-Convergence

In this section, we will show that under some mild conditions, the solution obtained from Yau-Yau algorithm introduced in Sect. 7.2 converges to the solution of robust DMZ equation in the L^1 sense. The main convergence result stems from another two estimation results. (1) The first one is that the solution to a global DMZ equation can be approximated by a DMZ equation on the ball B_r, which we will introduce later; (2) the second one is that as $k \to \infty$, the algorithm introduced in the previous section converges to the solution of DMZ equation in every bounded domain.

7.3 L^1-Convergence

We first introduce the following robust DMZ equation on the ball B_r:

$$\begin{cases} \dfrac{\partial u_r}{\partial t}(t,x) = \dfrac{1}{2}\Delta u_r(t,x) + (-f(x)+\nabla K(t,x))\cdot\nabla u_r(t,x) \\ \qquad + \left(-\nabla\cdot f(x) - \dfrac{1}{2}|h(x)|^2 + \dfrac{1}{2}\Delta K(t,x)\right. \\ \qquad \left. -f(x)\cdot\nabla K(t,x) + \dfrac{1}{2}|\nabla K(t,x)|^2\right)u_r(t,x), \\ u_r(t,x) = 0 \quad (t,x)\in[0,T]\times\partial B_r, \\ u_r(0,x) = \sigma_0(x). \end{cases} \quad (7.12)$$

The next theorem shows that almost all the density can be captured by a ball large enough, which serves to be a foundation to the proof of the first estimation result.

Hereafter in this chapter, for the simplification of the notations, we would like to use $\int_U F dx$ or simply $\int_U F$ to indicate the integral of function $F(x)$ with respect to the variable x inside the domain U, if there is no ambiguity in the contexts.

Theorem 7.1 *Consider the filtering model (7.1). For any $T > 0$, let u be a solution of the robust DMZ equation (7.5) in $[0, T] \times R^n$. Assume the following condition holds:*

$$-\frac{1}{2}|h|^2 - \frac{1}{2}\Delta K - f\cdot\nabla K + \frac{1}{2}|\nabla K|^2 + |f-\nabla K| \leq c_1, \quad \forall (t,x)\in[0,T]\times R^n, \quad (7.13)$$

where c_1 is a constant possibly depending on T. Then

$$\sup_{0\leq t\leq T}\int_{R^n} e^{\sqrt{1+|x|^2}} u(t,x)dx \leq e^{(c_1+\frac{n+1}{2})T}\int_{R^n} e^{\sqrt{1+|x|^2}} u(0,x)dx. \quad (7.14)$$

In particular,

$$\sup_{0\leq t\leq T}\int_{|x|\geq r} u(t,x)dx \leq e^{-\sqrt{1+r^2}} e^{(c_1+\frac{n+1}{2})T}\int_{R^n} e^{\sqrt{1+|x|^2}} u(0,x)dx. \quad (7.15)$$

Proof For the proof of Theorem 7.1, we first introduce a group of test function e^ϕ, where ϕ is a C^∞ function on R^d. Let u_r be the solution of (7.12), the DMZ equation on B_r; then, according to the equation satisfied by u_R and the integration-by-part formula, we have

$$\frac{d}{dt}\int_{B_r} e^\phi u_r = \frac{1}{2}\int_{B_r} e^\phi \Delta u_r + \int_{B_r} e^\phi(-f+\nabla K)\cdot\nabla u_r$$

$$+ \int_{B_r} \left(-\nabla \cdot f - \frac{1}{2}|h|^2 + \frac{1}{2}\Delta K - f \cdot \nabla K + \frac{1}{2}|\nabla K|^2 \right) e^\phi u_r$$

$$= \frac{1}{2} \int_{B_r} e^\phi u_r \left(\Delta \phi + |\nabla \phi|^2 \right) + \int_{B_r} e^\phi u_r \nabla \phi \cdot (f - \nabla K)$$

$$+ \int_{B_r} e^\phi u_r \left(-\frac{1}{2}|h|^2 - \frac{1}{2}\Delta K - f \cdot \nabla K + \frac{1}{2}|\nabla K|^2 \right)$$

$$- \frac{1}{2} \int_{\partial B_r} e^\phi u_r \nabla \phi \cdot v + \frac{1}{2} \int_{\partial B_r} e^\phi \frac{\partial u_r}{\partial v}$$

$$+ \int_{\partial B_r} e^\phi u_r (-f + \nabla K) \cdot v,$$

where v is the unit outward normal vector of B_r. Choose $\phi = \sqrt{1 + |x|^2}$; then

$$\frac{\partial \phi}{\partial x_i} = \frac{x_i}{\sqrt{1+|x|^2}}, \quad \frac{\partial^2 \phi}{\partial x_i^2} = \frac{1}{\sqrt{1+|x|^2}} - \frac{x_i^2}{(1+|x|^2)^{3/2}}.$$

Since $u|_{\partial B_r} = 0$ and $\frac{\partial u_r}{\partial v}|_{\partial B_r} \leq 0$, we have

$$\frac{d}{dt}\int_{B_r} e^\phi u_r \leq \int_{B_r} e^\phi u_r \left[-\frac{1}{2}|h|^2 - \frac{1}{2}\Delta K - f \cdot \nabla K + \frac{1}{2}|\nabla K|^2 \right.$$

$$\left. + \frac{1}{2}\Delta \phi + \frac{1}{2}|\nabla \phi|^2 + \nabla \phi \cdot (f - \nabla K) \right]$$

$$= \int_{B_r} e^{\phi_r} \left[-\frac{1}{2}|h|^2 - \frac{1}{2}\Delta K - f \cdot \nabla K + \frac{1}{2}|\nabla K|^2 + \frac{n}{2\sqrt{1+|x|^2}} \right.$$

$$\left. - \frac{|x|^2}{2(1+|x|^2)^{3/2}} + \frac{1}{2}\frac{|x|^2}{1+|x|^2} + \frac{x}{\sqrt{1+|x|^2}}(f - \nabla K) \right]$$

$$\leq \int_{B_r} e^\phi u_r \left[-\frac{1}{2}|h|^2 - \frac{1}{2}\Delta K - f \cdot \nabla K + \frac{1}{2}|\nabla K|^2 + \frac{n+1}{2} + |f - \nabla K| \right]$$

$$\leq \left(c_1 + \frac{n+1}{2} \right) \int_{B_r} e^\phi u_r.$$

By Gronwall's inequality, we have

$$\int_{B_r} e^\phi u_r(t,x) dx \leq e^{(c_1 + \frac{n+1}{2})t} \int_{B_r} e^\phi u(0,x) dx \quad t \in [0, T]. \tag{7.16}$$

Let r go to infinity and because t is arbitrarily chosen from $[0, T]$, we have proved

7.3 L^1-Convergence

$$\sup_{0\leq t\leq T}\int_{R^n}e^{\sqrt{1+|x|^2}}u(t,x)dx \leq e^{(c_1+\frac{n+1}{2})T}\int_{R^n}e^{\sqrt{1+|x|^2}}u(0,x)dx.$$

In particular,

$$\sup_{0\leq t\leq T}\int_{|x|\geq r}u(t,x)dx \leq e^{-\sqrt{1+r^2}}\sup_{0\leq t\leq T}\int_{|x|\geq r}e^{\sqrt{1+|x|^2}}u(t,x)dx$$

$$\leq e^{-\sqrt{1+r^2}}\sup_{0\leq t\leq T}\int_{R^n}e^{\sqrt{1+|x|^2}}u(t,x)dx$$

$$\leq e^{-\sqrt{1+r^2}}e^{(c_1+\frac{n+1}{2})T}\int_{R^n}e^{\sqrt{1+|x|^2}}u(0,x)dx.$$

□

Theorem 7.1 shows that the density outside the ball B_r can be arbitrarily small for r large enough. This gives us the opportunity to only analyze the DMZ equation on a sufficiently large B_r. The next theorem shows that the solution of the original DMZ equation can be approximated in L^1 sense by the solution of the local DMZ equation on the ball B_r.

Theorem 7.2 *Consider the filtering model (7.1). For any $T > 0$, let u be a solution of the robust DMZ equation (7.5) in $[0, T] \times R^n$. Assume that*

(1) Condition (7.13) is satisfied;
(2) For all $(t, x) \in [0, T] \times R^n$,

$$-\frac{1}{2}|h|^2 - \frac{1}{2}\Delta K - f\cdot\nabla K + \frac{1}{2}|\nabla K|^2 + 12 + 2n + 4|f - \nabla K| \leq c_2, \quad (7.17)$$

where c_2 is a constant possibly depending on T.
(3) For all $(t, x) \in [0, T] \times R^n$,

$$e^{-\sqrt{1+|x|^2}}[12 + 2n + 4|f - \nabla K|] \leq c_3. \quad (7.18)$$

Let $r \geq 1$ and u_r be the solution of the following DMZ equation on the ball B_r. Let $v = u - u_r$. Then $v \geq 0$ for all $(t, x) \in [0, T] \times B_r$ and

$$\int_{B_{\frac{r}{2}}}v(T,x) \leq \frac{2(e^{c^2 T}-1)}{c_2}c_3 e^{-\frac{9}{16}r}e^{(c_1+\frac{n+1}{2})T}\int_{R^n}e^{\sqrt{1+|x|^2}}u(0,x)dx. \quad (7.19)$$

Proof Since u is an unnormalized conditional probability density, we have $u \geq 0$, for all $(t,x) \in [0, T] \times R^n$. Thus, $v = u - u_r \geq 0$, for all $(t,x) \in [0, T] \times \partial B_r$. By the maximum principle of parabolic partial differential equation, we have $v \geq 0$ in $[0, T] \times B_r$.

For a test function ψ in C^∞, we can imitate the calculation in the proof of Theorem 7.1 and obtain

$$\frac{d}{dt}\int_{B_r}\psi v = \frac{1}{2}\int_{B_r}(\Delta\psi)v - \frac{1}{2}\int_{\partial B_r}v\frac{\partial\psi}{\partial v} + \frac{1}{2}\int_{\partial B_r}\psi\frac{\partial v}{\partial v}$$
$$- \int_{B_r}\nabla\psi\cdot(-f+\nabla K)v - \int_{B_r}\psi(-\nabla\cdot f + \Delta K)v$$
$$+ \int_{\partial B_r}\psi v(-f+\nabla K)\cdot v$$
$$+ \int_{B_r}\left(-\nabla\cdot f - \frac{1}{2}|h|^2 + \frac{1}{2}\Delta K - f\cdot\nabla K + \frac{1}{2}|\nabla K|^2\right)\psi v, \quad (7.20)$$

where again, v is the unit outward normal of ∂B_r.

Here, we choose ϕ to be a radial symmetric function such that $\phi|_{\partial B_r} = r$, $\nabla\phi|_{\partial B_r} = 0$, and ϕ is increasing with $|x|$. Then, in Eq. (7.20), we take

$$\psi(x) = e^{-\phi(x)} - e^{-r},$$

and $\psi|_{\partial B_r} = 0$, $\nabla\psi|_{\partial B_r} = 0$. Hence,

$$\frac{d}{dt}\int_{B_r}\psi v = \frac{1}{2}\int_{B_r}(\Delta\psi)v - \int_{B_r}\nabla\psi\cdot(-f+\nabla K)v$$
$$+ \int_{B_r}\left(-\frac{1}{2}|h|^2 - \frac{1}{2}\Delta K - f\cdot\nabla K + \frac{1}{2}|\nabla K|^2\right)\psi v$$
$$\leq \sup_{B_r}\left[-\frac{1}{2}\Delta\phi + \frac{1}{2}|\nabla\phi|^2 - \nabla\phi\cdot(f-\nabla K)\right.$$
$$\left. - \frac{1}{2}|h|^2 - \frac{1}{2}\Delta K - f\cdot\nabla K + \frac{1}{2}|\nabla K|^2\right]\cdot\int_{B_r}\psi v$$
$$+ e^{-r}\sup_{B_r}\left\{e^{-\sqrt{1+|x|^2}}\left[\frac{1}{2}(-\Delta\phi + |\nabla\phi|^2) - \nabla\phi\cdot(f-\nabla K)\right]\right\}\times\int_{B_r}e^{\sqrt{1+|x|^2}}v.$$

Choose

$$\phi(x) = r - r(1 - \frac{|x|^2}{r^2})^2,$$

and use Conditions (2) and (3); we have

$$\frac{d}{dt}\int_{B_r}\psi v \leq c_2\int_{B_r}\psi v + e^{-r}c_3\int_{B_r}e^{\sqrt{1+|x|^2}}u. \quad (7.21)$$

7.3 L^1-Convergence

From Theorem 7.1,

$$\frac{d}{dt}\int_{B_r} \psi v \leq c_2 \int_{B_r} \psi v + e^{-r} c_3 e^{(c_1+\frac{n+1}{2})T} \int_{R^n} e^{\sqrt{1+|x|^2}} u(0,x). \tag{7.22}$$

Here again, with Gronwall's inequality, we have

$$\int_{B_r} \psi v(T,x) \leq \frac{e^{c_2 T}-1}{c_2} c_3 e^{-r} e^{(c_1+\frac{n+1}{2})T} \int_{R^n} e^{\sqrt{1+|x|^2}} u(0,x). \tag{7.23}$$

Substitute $\psi(x) = e^{\frac{|x|^4}{r^3}-2\frac{|x|^2}{r}} - e^{-r}$ in Eq. (7.23); we have

$$\int_{B_r} \psi v(T,x) \geq \int_{B_{\frac{r}{2}}} \left[e^{\frac{|x|^4}{r^3}-2\frac{|x|^2}{r}} - e^{-r} \right] v(T,x)$$

$$\geq \left(e^{-\frac{7}{16}r} - e^{-r}\right) \int_{B_{\frac{r}{2}}} v(T,x) \geq \frac{1}{2} e^{-\frac{7}{16}r} \int_{B_{\frac{r}{2}}} v(T,x).$$

Combining this with inequality (7.23), we obtain

$$\int_{B_{\frac{r}{2}}} v(T,x)dx \leq \frac{e^{c_2 T}-1}{c_2} c_3 e^{-\frac{9}{16}r} e^{(c_1+\frac{n+1}{2})T} \int_{R^n} e^{\sqrt{1+|x|^2}} u(0,x). \tag{7.24}$$

□

Here, we have finished the first estimation leading to the main result of this section. And now, we can focus on the convergence result of the algorithm on a bounded domain in R^n.

Theorem 7.3 *Let U be a bounded domain in R^n. Let $F : [0,T] \times U \to R^n$ be a family of vector fields C^∞ in x and Holder continuous in t with exponent α and let $J : [0,T] \times U \to R$ be a C^∞ function in x and Hölder continuous in t with exponent α such that the following properties are satisfied:*

$$|\nabla \cdot F(t,x)| + 2|J(t,x)| + |F(t,x)| \leq c \tag{7.25}$$

$$|F(t,x) - F(\bar{t},x)| + |\nabla \cdot F(t,x) - \nabla \cdot F(\bar{t},x)| + |J(t,x) - J(\bar{t},x)| \leq c_1 |t-\bar{t}|^\alpha, \tag{7.26}$$

for all $(t,x), (\bar{t},x) \in [0,T] \times U$.

Let $u(t, x)$ be the solution on $[0, T] \times U$ of the equation:

$$\begin{cases} \dfrac{\partial u}{\partial t}(t, x) = \dfrac{1}{2}\Delta u(t, x) + F(t, x) \cdot \nabla u(t, x) + J(t, x)u(t, x), \\ u(0, x) = \sigma_0(x), \\ u(t, x)|_{\partial U} = 0. \end{cases} \quad (7.27)$$

For any $0 \leq \tau \leq T$, let $\mathcal{P} = \{0 = \tau_0 < \tau_1 < \cdots < \tau_k = \tau\}$ be a partition of $[0, \tau]$, where $\tau_i = \frac{i\tau}{k}$. Let $u_i(t, x)$ be the solution on $[\tau_{i-1}, \tau_i] \times U$ of the equation:

$$\begin{cases} \dfrac{\partial u_i}{\partial t}(t, x) = \dfrac{1}{2}\Delta u_i(t, x) + F(\tau_{i-1}, x) \cdot \nabla u_i(t, x) + J(\tau_{i-1}, x)u_i(t, x), \\ u_i(\tau_{i-1}, x) = u_{i-1}(\tau_{i-1}, x), \\ u_i(t, x)|_{\partial U} = 0, \end{cases} \quad (7.28)$$

where we use the convention that $u_0(t, x) = \sigma_0(x)$. Then

$$u(\tau, x) = \lim_{k \to \infty} u_k(\tau, x) \quad (7.29)$$

in L^1 sense on U and

$$\int_U |u(\tau, x) - u_k(\tau, x)|dx \leq \frac{C}{k^\alpha}. \quad (7.30)$$

Here C is a constant only depending on $T, U, \alpha, \sigma_0, F$ and J.

Proof Let $U_t^+ = \{x \in U : u(t, x) - u_i(t, x) \geq 0\}$, $t \in [\tau_{i-1}, \tau_i]$; we will first estimate the integration in (7.30) on U_t^+. As we did before, we need to compute the derivative of this integral, $\frac{d}{dt} \int_{U_t^+} (u - u_i)(t, x)dx$. Notice that the integral region is not independent of t; therefore, we need to treat the calculation more carefully.

Lemma 7.2 *Let U be a bounded domain in R^n and let $v : [0, T] \times \bar{U} \to R$ be a C^1 function. Assume that $v(t, x) = 0$, for $(t, x) \in [0, T] \times \partial U$. Let $U_t^+ = \{x \in U : v(t, x) \geq 0\}$; then*

$$\frac{d}{dt} \int_{U_t^+} v(t, x)dx = \int_{U_t^+} \frac{dv}{dt}(t, x)dx, \quad \text{for almost all } t \in [0, T]. \quad (7.31)$$

□

Proof of Lemma 7.2: we use the definition of derivatives:

7.3 L^1-Convergence

$$\frac{d}{dt}\int_{U_t^+} v(t,x)dx = \lim_{\Delta t \to 0} \frac{\int_{U_{t+\Delta t}^+} v(t+\Delta t,x)dx - \int_{U_t^+} v(t,x)dx}{\Delta t}$$

$$= \int_{U_t^+} \frac{dv}{dt}(t,x)dx + \lim_{\Delta t \to 0} \frac{v(t,\xi_1)\,\text{Vol}\left(U_{t+\Delta t}^+ - U_t^+\right)}{\Delta t}$$

$$- \lim_{\Delta t \to 0} \frac{v(t,\xi_2)\,\text{Vol}\left(U_t^+ - U_{t+\Delta t}^+\right)}{\Delta t},$$

where $\xi_1 \in U_{t+\Delta t}^+ - U_t^+$ and $\xi_2 \in U_t^+ - U_{t+\Delta t}^+$.

Since $v(t,x) = 0$, for $(t,x) \in [0,T] \times \partial U$, we have $\lim_{\Delta t \to 0} v(t,\xi_1) = \lim_{\Delta t \to 0} v(\xi_2) = 0$. We only need to show that

$$\lim_{\Delta t \to 0} \frac{\text{Vol}\left(U_t^+ - U_{t+\Delta t}^+\right)}{\Delta t} \quad \text{and} \quad \lim_{\Delta t \to 0} \frac{\text{Vol}\left(U_{t+\Delta t}^+ - U_t^+\right)}{\Delta t}$$

are bounded for almost all $t \in [0,T]$. To this end, we need to apply the famous coarea formula for Euclidean space, which reads as follows:

Let $w : A \to R$ be a Lipschitz function, where $A \subset R^n$ is measurable. Then

$$\int_A h(x)|\nabla w(x)|dx = \int_R \int_{w^{-1}(y)} h(x)\mathcal{H}^{n-1}(x)dy, \quad (7.32)$$

where $\mathcal{H}^{n-1}$ denotes the Hausdorff measure with respect to the Euclidean distance and $h : A \to R$ is a measurable function.

Let $A = U_t^+ - U_{t+\Delta t}^+ = \{x \in U_t^+, v(t+\Delta t, x) \leq 0\}$. Let L be the Lipschitz constant such that

$$|v(t,x) - v(t+\Delta t, x)| \leq L\Delta t, \ \forall x \in \bar{U}.$$

Since $v(t,x) \geq 0$ for $x \in A$, we have

$$v(t+\Delta t, x) \geq v(t,x) - L\Delta t \geq -L\Delta t, \ for\ x \in A.$$

Let $h(x) = \frac{1}{|\nabla v(t+\Delta t, x)|}$ and $w(x) = v(t+\Delta t, x)$ in the coarea formula (7.32). We have

$$\text{Vol}(U_t^+ - U_{t+\Delta t}^+) = \int_{-L\Delta t}^0 \int_{\{x \in A:\ v(t+\Delta t, x) = y\}} \frac{1}{|\nabla v(t+\Delta t, x)|}\mathcal{H}^{n-1}(x)dy.$$
(7.33)

Consider the map $\Phi : [0,T] \times \bar{U} \to [0,T] \times R$ given by $\Phi(t,x) = (t, v(t,x))$. By Sard's theorem, the set of critical values of Φ has Lebesgue measure zero. Therefore, for almost all t, Δt and y, $\nabla v(t+\Delta t, x) \neq 0$ for all $x \in \{x \in A :$

$v(t+\Delta t, x) = y\}$. Therefore, $\lim_{\Delta t \to 0} \frac{\text{Vol}(U_t^+ - U_{t+\Delta t}^+)}{\Delta t}$ is bounded for almost all t. Similarly, $\lim_{\Delta t \to 0} \frac{\text{Vol}(U_{t+\Delta t}^+ - U_t^+)}{\Delta t}$ is also bounded.

Back to the Proof of Theorem 7.3: With Lemma 7.2, we can calculate the derivative of $\int_{U_t^+} (u - u_i)(t, x) dx$:

$$\frac{d}{dt} \int_{U_t^+} (u - u_i)(t, x) dx \leq c \int_{U_t^+} (u - u_i) dx + c_1 (t - \tau_{i-1})^\alpha \int_{U_t^+} u \, dx$$

$$+ c_1 (t - \tau_{i-1})^\alpha \int_{U_t^+} |\nabla u| dx.$$

Next, we will estimate both $\int_U u$ and $\int_U |\nabla u|$. For the first integration, notice that

$$\frac{d}{dt} \int_U u(t, x) dx = \frac{1}{2} \int_U \Delta u \, dx + \int_U F(t, x) \cdot \nabla u(t, x) dx + \int_U J(t, x) u(t, x) dx$$

$$= \frac{1}{2} \int_{\partial U} \frac{\partial u}{\partial \nu} d\sigma - \int_U \text{div } F(t, x) u(t, x) dx + \int_U J(t, x) u(t, x) dx$$

$$\leq c \int_U u(t, x) dx.$$

Therefore, for all $0 \leq t \leq T$,

$$\int_U u(t, x) dx \leq e^{cT} \int_U u(0, x) dx. \tag{7.34}$$

For the second integration, we can instead estimate $\int_U |\nabla u|^2$, because U is bounded and

$$\int_U |\nabla u| dx \leq \sqrt{\text{Vol}(U)} \left[\int_U |\nabla u|^2 dx \right]^{\frac{1}{2}}.$$

Notice that

$$\frac{d}{dt} \int_U |\nabla u|^2(t, x) = \int_U 2 \nabla \frac{\partial u}{\partial t}(t, x) \cdot \nabla u(t, x)$$

$$\leq 2c^2 \int_U |\nabla u(t, x)|^2 + 2c^2 \int_U u^2(t, x),$$

and we need then, estimate the L^2 norm of $u(t, x)$.

$$\frac{d}{dt} \int_U u^2(t, x) = 2 \int_U u(t, x) \frac{\partial u}{\partial t}(t, x) \leq c \int_U u^2(t, x).$$

7.3 L^1-Convergence

Thus,
$$\int_U u^2(t,x) \leq e^{cT} \int_U u^2(0,x).$$

Back to the estimation of $\int_U |\nabla u|^2$, we have

$$\int_U |\nabla u(t,x)|^2 \leq 2c^2 T e^{2c^2 T} \int_U u^2(0,x) + e^{2c^2 T} \int_U |\nabla u(0,x)|^2. \quad (7.35)$$

Putting (7.34) and (7.35) back to the estimation of $\int_{U_t^+} (u - u_i)(t,x)$, we have

$$\frac{d}{dt} \int_{U_t^+} (u - u_i)(t,x) \leq c \int_{U_t^+} (u - u_i)(t,x) + c_2(t - \tau_{i-1})^\alpha. \quad (7.36)$$

This implies

$$\int_{U_t^+} (u - u_i)(t,x) \leq e^{c(t-\tau_{i-1})} \left(\int_{U_{\tau_{i-1}}^+} (u - u_{i-1})(\tau_{i-1}, x) + c_2 \frac{(t-\tau_{i-1})^{\alpha+1}}{\alpha+1} \right). \quad (7.37)$$

Similarly, we have

$$\int_{U_t^-} (u - u_i)(t,x) \leq e^{c(t-\tau_{i-1})} \left(\int_{U_{\tau_{i-1}}^-} (u_{i-1} - u)(\tau_{i-1}, x) + c_2 \frac{(t-\tau_{i-1})^{\alpha+1}}{\alpha+1} \right). \quad (7.38)$$

Consequently, we have

$$\int_{U_t^+} |u - u_i|(t,x) \leq e^{c(t-\tau_{i-1})} \left(\int_U |u - u_{i-1}|(\tau_{i-1}, x) + 2c_2 \frac{(t-\tau_{i-1})^{\alpha+1}}{\alpha+1} \right). \quad (7.39)$$

By induction,

$$\int_U |u - u_k|(\tau_k, x) \leq e^{cT} \int_U |u - u_0|(0,x) + \frac{2c_2}{\alpha+1} \frac{T^{\alpha+1}}{k^{\alpha+1}}$$
$$\times \left[e^{c\frac{T}{k}} + e^{c\frac{2T}{k}} + \cdots + e^{kc\frac{T}{k}} \right] \quad (7.40)$$
$$\leq \frac{2c_2}{\alpha+1} \frac{T^{\alpha+1} e^{cT}}{k^\alpha}.$$

$\square$

Up to now, we have finished the proof of the main results in this section. The main L^1 convergence result goes as follows. Firstly, the solution to the DMZ equation on R^n can be approximated by the solution to the DMZ equation on a large enough ball

B_r. Then, on the bounded domain B_r, the solution obtained by Yau-Yau algorithm converges to the solution of DMZ equation in L^1 sense.

7.4 Lower Bound Estimation of Density Function

In practical nonlinear filtering computation, it is important to know how much density remains within a given ball, because the solution to the robust DMZ equation is an unnormalized probability density and is easily vanishing. In this section, we shall provide such a lower estimation to the density remaining in a given ball. In particular, the solution u of the DMZ equation in R^n obtained by taking $\lim_{r \to \infty} u_r$, where u_r is the solution of the DMZ equation in the ball B_r, is a nontrivial solution.

Theorem 7.4 *Let u_r be the solution of the DMZ equation on B_r, (7.12). Assume that*

(1) $f(x)$ and $h(x)$ have at most polynomial growth;
(2) For any $0 \leq t \leq T$, there exist positive integer m and positive constants c', c'', and c_1, which are independent of r, such that for all $x \in R^n$,

$$\frac{m^2}{2}|x|^{2m-2} - \frac{m}{2}(m+n-2)|x|^{m-2} - m|x|^{m-2}x \cdot (f - \nabla K) \tag{7.41}$$
$$-\frac{1}{2}\Delta K - \frac{1}{2}|h|^2 - f \cdot \nabla K + \frac{1}{2}|\nabla K|^2 \geq -c',$$

$$\left| \frac{m^2}{2}|x|^{2m-2} - \frac{m(m+n-2)}{2} - m|x|^{m-2}(f - \nabla K) \cdot x \right| \tag{7.42}$$
$$\leq \frac{m(m+1)}{2}|x|^{2m-2} + c'',$$

$$-\frac{1}{2}|h|^2 - \frac{1}{2}\Delta K - \sum_{j=1}^{n} f_j \frac{\partial K}{\partial x_j} + \frac{1}{2}|\nabla K|^2 \leq c_1. \tag{7.43}$$

Then, for any $r_0 \leq r$,

$$\int_{B_{r_0}} (e^{-|x|^m} - e^{-r_0^m}) u_r(T, x) dx$$

$$\geq e^{-c'T} \int_{B_{r_0}} (e^{-|x|^m} - e^{-r_0^m}) \sigma_0(x) \tag{7.44}$$

$$+ \frac{e^{-r_0^m}}{c'} \left(\frac{m(m+1)}{2} r_0^{2m-2} + c'' \right) (1 - e^{c'T}) \int_{B_r} \sigma_0(x) dx.$$

7.4 Lower Bound Estimation of Density Function

In particular, the solution u of the robust DMZ equation on R^n has the estimate

$$\int_{R^n} e^{-|x|^m} u(T,x)dx \geq e^{-c'T} \int_{R^n} e^{-|x|^m} \sigma_0(x)dx. \tag{7.45}$$

Proof Just as we did in the proof of the previous theorems, the key point of this proof is also choosing a suitable test function and fulfilling the estimations. Here, we choose a test function $\phi(x) = e^{-\rho(x)} - e^{-\rho(r_0)}$, where ρ is an increasing function of $|x|$. Therefore, $\phi \geq 0$, for all $x \in B_{r_0}$, $\phi|_{\partial B_{r_0}} = 0$ and $\frac{\partial \phi}{\partial \nu}\big|_{\partial B_{r_0}} \leq 0$, where ν is the unit outward normal vector of ∂B_{r_0}.

Since u_r is the solution of the DMZ equation on B_r, we have

$$\frac{d}{dt}\int_{B_{r_0}}\phi u_r = \frac{1}{2}\int_{B_{r_0}}\phi\Delta u_r + \int_{B_{r_0}}\phi(-f+\nabla K)\cdot\nabla u_r$$

$$+ \int_{B_{r_0}}\left(-\nabla\cdot f - \frac{1}{2}|h|^2 + \frac{\Delta K}{2} - f\cdot\nabla K + \frac{1}{2}|\nabla K|^2\right)\phi u_r$$

$$\geq \frac{1}{2}\int_{B_{r_0}} u_r\Delta\phi + \int_{B_{r_0}} u_r\nabla\phi\cdot(f-\nabla K)$$

$$+ \int_{B_{r_0}}\left(-\frac{\Delta K}{2} - \frac{1}{2}|h|^2 - f\cdot\nabla K + \frac{1}{2}|\nabla k|^2\right)\phi u_r.$$

Notice that $\nabla\phi = e^{-\rho(x)}\nabla\rho$ and $\Delta\phi = e^{-\rho(x)}(|\nabla\rho|^2 - \Delta\rho)$. Let $r = |x|$; we have $\nabla\rho = \frac{\rho'(r)}{r}x$, and $\Delta\rho = \rho''(r) + \rho'(r)\frac{n-1}{r}$.

Therefore,

$$\frac{d}{dt}\int_{B_{r_0}}\phi u_r \geq \int_{B_{r_0}} u_r e^{-\rho(x)}\left[-\frac{\Delta\rho}{2} + \frac{|\nabla\rho|^2}{2} - \nabla\rho\cdot(f-\nabla K)\right]$$

$$+ \int_{B_{r_0}}\left(-\frac{\Delta K}{2} - \frac{1}{2}|h|^2 - f\cdot\nabla K + \frac{1}{2}|\nabla K|^2\right)\phi u_r$$

$$= e^{-\rho(r_0)}\int_{B_{r_0}}\left[\frac{\rho'^2}{2} - \frac{\rho''}{2} - \frac{n-1}{2r}\rho' - \rho'(f-\nabla K)\cdot\frac{x}{r}\right]u_r$$

$$+ \int_{B_{r_0}}\left[-\frac{\Delta\rho}{2} + \frac{|\nabla\rho|^2}{2} - \nabla\rho\cdot(f-\nabla K)\right. \tag{7.46}$$

$$\left.-\frac{\Delta K}{2} - \frac{1}{2}|h|^2 - f\cdot\nabla K + \frac{1}{2}|\nabla K|^2\right]\phi u_r.$$

We now take $\rho = |x|^m$; then

$$\Delta\rho = m(m+n-2)r^{m-2} \text{ and } |\nabla\rho|^2 = m^2 r^{2m-2}.$$

Since f and h are of polynomial growth, we can choose a positive integer m large enough such that

$$-\frac{\Delta\rho}{2} + \frac{|\nabla\rho|^2}{2} - \nabla\rho\cdot(f-\nabla K) - \frac{\Delta K}{2} - \frac{1}{2}|h|^2 - f\cdot\nabla K + \frac{1}{2}|\nabla K|^2$$
$$= -\frac{1}{2}m(m+n-2)r^{m-2} + \frac{1}{2}m^2 r^{2m-2} - \frac{\rho'(r)}{r}x\cdot(f-\nabla K)$$
$$- \frac{\Delta K}{2} - \frac{1}{2}|h|^2 - f\cdot\nabla K + \frac{1}{2}|\nabla K|^2 \geq -c'$$
(7.47)

holds for every $x \in R^n$, where c' is a positive constant independent of r and r_0.

In the meanwhile, for m large enough,

$$\left|\frac{\rho'^2}{2} - \frac{\rho''}{2} - \frac{n-1}{2r}\rho' - \rho'(f-\nabla K)\cdot\frac{x}{r}\right|$$
$$= \left|\frac{m^2 r^{2m-2}}{2} - \frac{m(m+n-2)}{2}r^{m-2} - mr^{m-2}(f-\nabla K)\cdot x\right|$$
$$\leq \frac{m(m+1)}{2}r^{2m-2} + c'',$$

where c'' is independent of r and r_0.

Therefore,

$$e^{-\rho(r_0)} \left| \int_{B_{r_0}} \left[\frac{\rho'^2}{2} - \frac{\rho''}{2} - \frac{n-1}{2r}\rho' - \rho'(f-\nabla K)\cdot\frac{x}{r}\right] u_r(t,x) \right|$$
$$\leq e^{-r_0^m}\left(\frac{m(m+1)}{2}r_0^{2m-2} + c''\right)\int_{B_{r_0}} u_r(t,x).$$

Also,

$$\frac{d}{dt}\int_{B_r} u_r(t,x) \leq c_1 \int_{B_r} u_r(t,x).$$

Hence,

$$\int_{B_r} u_r(t,x) \leq e^{c't}\int_{B_r}(0,x), \ 0 \leq t \leq T$$

$$e^{-\rho(r_0)}\left|\int_{B_{r_0}}\left[\frac{\rho'^2}{2}-\frac{\rho''}{2}-\frac{n-1}{2r}\rho'-\rho'(f-\nabla K)\cdot\frac{x}{r}\right]u_r(t,x)\right|$$
$$\le e^{c'T-r_0^m}\left(\frac{m(m+1)}{2}r_0^{2m-2}+c''\right)\int_{B_r}u_r(0,x). \tag{7.48}$$

We denote the right-hand side of (7.48) by $\epsilon(r_0)$, which goes to zero as $r_0\to\infty$.

In the view of (7.46), (7.47), we have

$$\frac{d}{dt}\int_{B_{r_0}}\phi u_r \ge -\epsilon(r_0)-c'\int_{B_{r_0}}\phi u_r,$$

and thus,

$$\int_{B_{r_0}}\phi u_r(T,x)\ge e^{-c'T}\int_{B_r}\phi u_r(0,x)+\epsilon(r_0)e^{-c'T}\frac{1-e^{c'T}}{c'}. \tag{7.49}$$

Take $r_0\to\infty$, and we have

$$\int_{R^n}e^{-|x|^m}u(T,x)\ge e^{-c'T}\int_{R^n}e^{-|x|^m}u(0,x). \tag{7.50}$$

□

7.5 Algorithm in Time-Variant Systems

In the previous sections of this chapter, we mainly deal with the time-invariant filtering systems. We introduced Yau-Yau algorithm for these systems and gave detailed proofs of the convergence results. In the meanwhile, the algorithm introduced in Sect. 7.2 can also be extended to more general time-variant nonlinear filtering systems.

In this section, we will focus our attention on the nonlinear filtering systems with explicit time dependence in the drift term, observation term, and the variance of the noises:

$$\begin{cases} dX_t = f(X_t,t)dt+G(X_t,t)dV_t, \\ dY_t = h(X_t,t)dt+dW_t, \end{cases} \tag{7.51}$$

where the noise terms dv_t and dw_t satisfy $E[dv_t dv_t^\top]=Q(t)dt$, $E[dw_t dw_t^\top]=S(t)dt$ and $Q(t)$, $S(t)$ are positive-definite matrices.

For this more general case, the DMZ equation becomes

$$d\sigma(x,t) = L\sigma(x,t)dt + \sigma(x,t)h^\top(x,t)S^{-1}(x,t)dy_t, \quad \sigma(x,0) = \sigma_0(x), \tag{7.52}$$

where $\sigma_0(x)$ is the probability density of the initial state x_0, and

$$L(\cdot) = \frac{1}{2}\sum_{i,j=1}^{n} \frac{\partial^2}{\partial x_i \partial x_j}[(GQG^\top)_{ij} \cdot] - \sum_{i=1}^{n} \frac{\partial}{\partial x_i}(f_i \cdot).$$

As in the time-invariant case, for a given trajectory $\{y_t : 0 \le t < \infty\}$ of the observation process Y_t, the new unnormalized probability density

$$u(x,t) = \exp[-h^\top(x,t)S^{-1}(t)y_t]\sigma(x,t)$$

satisfies the following robust DMZ equation:

$$\frac{\partial}{\partial t}u(x,t) = \frac{1}{2}D_w^2 u(x,t) + F(x,t) \cdot \nabla u(x,t) + J(x,t)u(x,t), \tag{7.53}$$

with initial value $u(x,0) = \sigma_0(x)$, where

$$D_w^2 = \sum_{i,j=1}^{n}(GQG^\top)_{ij}\frac{\partial^2}{\partial x_i \partial x_j},$$

$$F(x,t) = \left(\sum_{j=1}^{n}\frac{\partial}{\partial x_j}(GQG^\top)_{ij} + \sum_{j=1}^{n}(GQG^\top)_{ij}\frac{\partial K}{\partial x_j} - f_i\right)_{i=1}^{n},$$

$$J(x,t) = -\frac{\partial}{\partial t}(h^\top S^{-1})^\top y_t + \frac{1}{2}\sum_{i,j=1}^{n}\frac{\partial^2}{\partial x_i \partial x_j}(GQG^\top)_{ij}$$

$$+ \sum_{i,j=1}^{n}\frac{\partial}{\partial x_i}(GQG^\top)_{ij}\frac{\partial K}{\partial x_j} + \frac{1}{2}\sum_{i,j=1}^{n}(GQG^\top)_{ij}\left[\frac{\partial^2 K}{\partial x_i \partial x_j} + \frac{\partial K}{\partial x_i}\frac{\partial K}{\partial x_j}\right]$$

$$- \sum_{i=1}^{n}\frac{\partial f_i}{\partial x_i} - \sum_{i=1}^{n}\frac{\partial K}{\partial x_i}f_i - \frac{1}{2}h^\top S^{-1}h,$$

in which

$$K(x,t) = h^T(x,t)S^{-1}(t)y(t).$$

Similar to the time-invariant case, we first partition the time interval $[0, \tau]$ by $\mathcal{P}_k = \{0 = \tau_0 < \tau_1 < \cdots < \tau_k = \tau\}$ and approximate the solution of the robust DMZ

7.5 Algorithm in Time-Variant Systems

equation on the interval $[\tau_{i-1}, \tau_i]$ by $u_i(x, t)$, which satisfies the DMZ equation with coefficients frozen at τ_{i-1}:

$$\begin{cases} \dfrac{\partial u_i}{\partial t}(x,t) + \dfrac{\partial}{\partial t}(h^\top S^{-1})^\top y_{\tau_{i-1}} u_i(x,t) \\ \qquad = \exp(-h^\top S^{-1} y_{\tau_{i-1}}) \left[L - \dfrac{1}{2} h^\top S^{-1} h \right] (\exp(h^\top S^{-1} y_{\tau_{i-1}}) u_i(x,t)), \\ u_i(\tau_{i-1}, x) = u_{i-1}(x, \tau_{i-1}, x), \end{cases} \quad (7.54)$$

with the convenience that $u_0(x, 0) = \sigma_0(x)$.

The problem of solving Eq. (7.54) can also be converted into solving an observation-independent Kolmogorov forward equation, because of the following theorem.

Theorem 7.5 *For each $\tau_{i-1} \le t < \tau_i$, $i = 1, 2, \cdots, k$, $u_i(x, t)$ satisfies Eq. (7.54) if and only if*

$$\tilde{u}_i(x, t) = \exp[h^\top(x, t) S^{-1}(t) y_{\tau_{i-1}}] u_i(x, t)$$

satisfies the following Kolmogorov forward equation:

$$\frac{\partial \tilde{u}_i}{\partial t}(x, t) = \left(L - \frac{1}{2} h^\top S^{-1} h \right) \tilde{u}_i(x, t). \quad (7.55)$$

Proof The result of Theorem 7.5 is obtained from direct calculations. □

The algorithm for this time-variant system is the same as that for time-invariant systems. At interval $[\tau_{i-1}, \tau_i]$, we first solve the Kolmogorov forward equation (7.55) with initial value:

$$\tilde{u}_i(x, \tau_{i-1}) = \exp[h^\top(x, \tau_{i-1}) S^{-1}(\tau_{i-1})(y_{\tau_{i-1}} - y_{\tau_{i-2}})] \tilde{u}_{i-1}(x, \tau_{i-1}).$$

Then, we can update the initial value of (7.55) with the new observation y_{τ_i} at τ_i and calculate $\tilde{u}_{i+1}(x, t)$ recursively.

Similar convergence results also hold for the algorithm in time-variant system. For the convenience of notations, we also use u_r to denote the solution to the robust DMZ equation restricted on B_r with zero Dirichlet boundary conditions. Besides, we denote

$$N(x, t) \triangleq -\frac{\partial}{\partial t}(h^\top S^{-1}) y_t - \frac{1}{2} D_w^2 K + \frac{1}{2} D_w K \cdot \nabla K - f \cdot \nabla K - \frac{1}{2}(h^\top S^{-1} h), \quad (7.56)$$

where

$$D_w(\cdot) = \left[\sum_{j=1}^{n}(GQG^\top)_{ij}(x,t)\frac{\partial}{\partial x_j}(\cdot)\right]_{i=1}^{n}.$$

With the above notations, we can state the main theorems in this section. We would like to skip the proof of those theorems, because the methods used here are very similar to those in the proof of convergence results for time-invariant case.

Theorem 7.6 *For any $T > 0$, let $u(x,t)$ be a solution of the robust DMZ equation (7.53) in $R^n \times [0,T]$. Let $r \gg 1$ and $u_r(x,t)$ is the solution to the robust DMZ equation on B_r. Assume the following conditions are satisfied, for all $(x,t) \in R^n \times [0,T]$:*

$$N(x,t) + \frac{3}{2}n\|GQG^\top\|_\infty + |f - D_w K| \leq C, \tag{7.57}$$

$$e^{-\sqrt{1+|x|^2}}[14n\|GQG^\top\|_\infty + 4|f - D_w K|] \leq \tilde{C}, \tag{7.58}$$

where C and $\tilde{C}$ are constants possibly depending on T. Let $v = u - u_r$; then $v \geq 0$ for all $(x,t) \in B_r \times [0,T]$ and

$$\int_{B_{\frac{r}{2}}} v(x,T) \leq \bar{C} e^{-\frac{9}{16}r} \int_{R^n} e^{\sqrt{1+|x|^2}} \sigma_0(x), \tag{7.59}$$

where $\bar{C}$ is some constant depending on T.

Theorem 7.7 *Let U be a bounded domain in R^n. Assume that*

$$|N(x,t)| \leq C, \quad |N(x,t) - N(x,t;\bar{t})| \leq \hat{C}|t-\bar{t}|^\alpha, \tag{7.60}$$

for all $(x,t) \in U \times [0,T]$, $\bar{t} \in [0,T]$ and for some $\alpha \in (0,1)$, where $N(x,t;\bar{t})$ denotes $N(x,t)$ with the observation $y_t = y_{\bar{t}}$.

Let u_U be the solution of the robust DMZ equation on $U \times [0,T]$ with zero Dirichlet boundary conditions. For any $0 \leq \tau \leq T$, let $\mathcal{P}_k^\tau = \{0 = \tau_0 < \tau_1 < \cdots < \tau_k = \tau\}$ be a partition of $[0,\tau]$. Let $u_{i,U}(x,t)$ be the approximate solution obtained by Yau-Yau algorithm restricted on $U \times [\tau_{i-1}, \tau_i]$. Then

$$u_U(x,\tau) = \lim_{k\to\infty} u_{k,U}(x,\tau), \tag{7.61}$$

in the L^1 sense in U and the following estimate holds:

$$\int_U |u_U(x,\tau) - u_{k,U}(x,\tau)| \leq \bar{C}\frac{1}{k^\alpha}, \tag{7.62}$$

where $\bar{C}$ is a constant depending on T, U, σ_0.

7.6 Numerical Methods for Solving Parabolic Differential Equations

After the discussion of the convergence results of the Yau-Yau algorithm, we would like to come back to the algorithm itself. In Sect. 7.2, we have proposed the procedure of this algorithm, that is,

1. Partition the interval $[0, \tau]$ by $0 = \tau_0 < \tau_1 < \cdots < \tau_k = \tau$.
2. At each time τ_i, use the value of $\widetilde{u}_{i-1}(\tau_{i-1}, x)$ to obtain the initial value (7.10) and solve the parabolic differential equation (7.9). We then get the value of $\widetilde{u}_i(\tau_i, x)$.
3. At time $\tau = \tau_k$, we get the value of $\widetilde{u}_k(\tau_k, x)$, and

$$u_k(\tau_k, x) = \exp\left(-\sum_{j=1}^{m} y_j(\tau_k) h_j(x)\right) \widetilde{u}_k(\tau_k, x),$$

serves to be an approximation to $u(\tau, x)$.

Therefore, for the practical use of this filtering algorithm, it remains to give an efficient numerical method for solving the parabolic partial differential equation (7.9), which is

$$\frac{\partial \widetilde{u}_i}{\partial t}(t, x) = \frac{1}{2} \Delta \widetilde{u}_i(t, x) - \sum_{i=1}^{n} f_i(x) \frac{\partial \widetilde{u}_i}{\partial x_i}(t, x)$$

$$- \left(\sum_{i=1}^{n} \frac{\partial f_i}{\partial x_i}(x) + \frac{1}{2} \sum_{i=1}^{m} h_i^2(x)\right) \widetilde{u}_i(t, x).$$

To this end, in this section, we will introduce the Galerkin spectral method to solve the parabolic partial differential equation in a suitable functional space and complete the Yau-Yau algorithm for general nonlinear filtering problems.

For the simplification of notations, we first restrict ourselves to the one-dimensional case and then give a brief discussion on the generalization of high-dimensional cases. Here, a general one-dimensional parabolic partial differential equation is of the following form:

$$\begin{cases} \frac{\partial}{\partial t} u(x, t) = \frac{\partial}{\partial x}\left(p(x, t) \frac{\partial}{\partial x} u(x, t)\right) + q(x, t) \frac{\partial}{\partial x} u(x, t) + w(x, t) u(x, t), \\ (x, t) \in R \times R_+ \\ u(x, 0) = u_0(x), \quad x \in R, \end{cases}$$

(7.63)

where p, q, w are functions that satisfy the following conditions:

1. $p, q, w, p_x, p_{xx}, q_x$ are locally Holder continuous in $[0, T] \times R$, where p_x, p_{xx} denotes the first and second partial derivatives of p with respect to x;
2. $p(t, x) \geq \lambda > 0, \forall (t, x) \in [0, T] \times R$;
3. $w(t, x) \leq 0, \forall (t, x) \in [0, T] \times R$;
4. $w - q_x + p_{xx} \leq 0, \forall (t, x) \in [0, T] \times R$.

Generally speaking, the idea of Galerkin spectral method is to get a best approximation of the solution in a certain sense from a finite-dimensional functional space. In this section, we first consider the finite-dimensional space spanned by the generalized Hermite functions.

Definition 7.1 (Luo and Yau [2]) The generalized Hermite functions in one dimension space are defined as

$$H_n^{\alpha,\beta}(x) = \frac{1}{\sqrt{2^n n!}} \mathcal{H}_n(\alpha(x - \beta)) e^{-\frac{1}{2}\alpha^2(x-\beta)^2}, \tag{7.64}$$

for $n \geq 0$, where $\alpha > 0, \beta \in R$ are some constants, namely, the scaling factor and the translating factor, respectively, and $\mathcal{H}_n(x)$ is the Hermite polynomials given by

$$\mathcal{H}_n(x) = (-1)^n e^{x^2} \frac{d^n}{dx^n} e^{-x^2}$$

and satisfies the following recursive formulae: $\mathcal{H}_0(x) \equiv 1$, $\mathcal{H}_1(x) = 2x$, and

$$\mathcal{H}_{n+1}(x) = 2x\mathcal{H}_n(x) - 2n\mathcal{H}_{n-1}(x), \quad n \geq 1. \tag{7.65}$$

From now on, when we fix $\alpha > 0$ and $\beta \in R$, and when there is no ambiguity in notations, we would like to drop the upper index α, β in $H_n^{\alpha,\beta}$ and simplify $H_n^{\alpha,\beta}(x)$ by $H_n(x)$.

The theory of functional analysis shows that the generalized Hermite functions $\{H_n\}_{n=0}^{\infty}$ form an orthogonal basis in the Hilbert space $L^2(R)$, with the inner product $\langle \cdot, \cdot \rangle$ defined by

$$\langle f, g \rangle = \int_R f(x)g(x)dx, \quad f, g \in L^2(R).$$

Then, each function $u \in L^2(R)$ can be written as

$$u(x) = \sum_{n=0}^{\infty} \widehat{u}_n H_n^{\alpha,\beta}(x),$$

where the series on the right-hand side converge in the L^2 sense. Because of the orthogonality of $\{H_n\}_{n=0}^{\infty}$, those $\widehat{u}_n$, which are called the Fourier-Hermite coefficients or generalized Fourier coefficients, can be calculated by

7.6 Numerical Methods for Solving Parabolic Differential Equations

$$\widehat{u}_n = \frac{\alpha}{\sqrt{\pi}} \int_R u(x) H_n^{\alpha,\beta}(x) dx. \tag{7.66}$$

The generalized Hermite functions have many simple but useful properties. We would like to demonstrate some of them as follows for later use. Hereafter in this chapter, we denote $\lambda_n = 2\alpha^2 n$, for $n \in \mathbb{N}$. Most of these properties can be derived through direct computations.

1. By the convention $H_n \equiv 0$, for $n < 0$. For $n \in \mathbb{Z}$ and $n \geq 0$, the three-term recurrence holds:

$$2\alpha(x - \beta) H_n(x) = \sqrt{2n} H_{n-1}(x) + \sqrt{2(n+1)} H_{n+1}(x) \quad \text{or}$$
$$2\alpha^2(x - \beta) H_n(x) = \sqrt{\lambda_n} H_{n-1}(x) + \sqrt{\lambda_{n+1}} H_{n+1}(x). \tag{7.67}$$

2. The derivative of $H_n(x)$ is a linear combination of $H_{n-1}(x)$ and $H_{n+1}(x)$:

$$\partial_x H_n(x) = \frac{1}{2}\sqrt{\lambda_n} H_{n-1}(x) - \frac{1}{2}\sqrt{\lambda_{n+1}} H_{n+1}. \tag{7.68}$$

3. The "quasi-orthogonality" of $\{\partial_x H_n(x)\}_{n=0}^{\infty}$

$$\int_R \partial_x H_n(x) \partial_x H_m(x) dx = \begin{cases} \dfrac{\sqrt{\pi}}{4\alpha}(\lambda_n + \lambda_{n+1}), & m = n \\[4pt] -\dfrac{\sqrt{\pi}}{4\alpha}\sqrt{\lambda_{m+1}\lambda_{m+2}}, & n - m = 2 \\[4pt] -\dfrac{\sqrt{\pi}}{4\alpha}\sqrt{\lambda_{n+1}\lambda_{n+2}}, & m - n = 2 \\[4pt] 0 & otherwise. \end{cases} \tag{7.69}$$

Let us denote the finite-dimensional linear space spanned by the first $N + 1$ generalized Hermite functions by

$$\mathcal{R}_N = \text{span}\{H_0^{\alpha,\beta}, H_1^{\alpha,\beta}, \cdots, H_N^{\alpha,\beta}\}.$$

For the Galerkin spectral method we consider here, we will use a function in $\mathcal{R}_N$ with suitable coefficients α and β to approximate the solution of a parabolic partial differential equation at each time $t \in [0, T]$.

In order to give this approximation a theoretical background, we need to introduce a new functional subspace of $L^2(R)$, denoted by $W_{\alpha,\beta}^r(R)$, or $W^r(R)$ for short, the members in which can be approximated by the elements in $\mathcal{R}_N$ for N large enough.

Definition 7.2 Fix $\alpha > 0$ and $\beta \in R$. For any integer $r \geq 0$, we define the function space $W^r(R)$ by

$$W^r(R) = \left\{ u \in L^2(R) : \sum_{k=0}^{\infty} \lambda_{k+1}^r \widehat{u}_k^2 < \infty \right\},$$

where $\lambda_k = 2\alpha^2 k$ and $\widehat{u}_k$ is the Hermite-Fourier coefficients defined in (7.66).

The norm on the space $W^r(R)$ is defined by

$$\|u\|_{r,\alpha,\beta}^2 = \sum_{k=0}^{\infty} \lambda_{k+1}^r \widehat{u}_k^2, \quad u \in W^r(R). \tag{7.70}$$

The following estimation for the derivatives of functions in $W^r(R)$ will be useful in the development of convergence analysis.

Lemma 7.3 *For any function $u \in W^{r_1+r_2}(R)$, with some integer $r_1, r_2 \geq 0$, we have*

$$\|x^{r_1} \partial_x^{r_2} u\|^2 \leq C\alpha^{-2r_1+1} \max\{(\alpha\beta)^{2r_1}, 1\} \|u\|_{r_1+r_2}^2, \tag{7.71}$$

where the operator $\partial_x^r \triangleq \frac{d^r}{dx^r}$, for any $r \in \mathbb{N}$.

Proof For any integers $r_1, r_2 \geq 0$, in the light of (7.67), (7.68), and (7.69), we have

$$\|x^{r_1}\partial_x^{r_2}\|^2 = \left\|\sum_{n=0}^{\infty} \widehat{u}_n x^{r_1} \partial_x^{r_2} H_n(x)\right\|^2 \leq C \left\|\alpha^{-2r_1} \sum_{n=0}^{\infty} \widehat{u}_n \sum_{k=-r_1-r_2}^{r_1+r_2} a_{n,k} H_{n+k}(x)\right\|^2, \tag{7.72}$$

where for each n fixed, $a_{n,k}$ is a product of $2(r_1+r_2)$ factors of $\alpha^2\beta$ or $\sqrt{\lambda_{n+j}}$, with $-r_2 - r_1 \leq j \leq r_2 + r_1$.

Let $n^* \geq 0$ be the biggest integer such that $\sqrt{\lambda_{n^*+1}} \leq \alpha^2\beta$, then, since $\lim_{n\to\infty} \frac{\lambda_{n+j}}{\lambda_{n+1}} = 1$ and $H_{n+j}(x) \equiv 0$ for $n+j < 0$. Hence,

$$\|x^{r_1}\partial_x^{r_2} u(x)\|^2 \leq C \left(\alpha^{-1}\beta^{2r_1} \sum_{n=0}^{n^*} \lambda_{n+1}^{r_1+r_2} \widehat{u}_n^2 + \alpha^{-2r_1-1} \sum_{n=n^*+1}^{\infty} \lambda_{n+1}^{r_1+r_2} \widehat{u}_n^2 \right)$$

$$\leq C\alpha^{-2r_1-1} \max\{(\alpha\beta)^{2r_1}, 1\} \|u\|_{r_1+r_2}^2, \tag{7.73}$$

for any integer $r_1, r_2 \geq 0$. □

For any $u \in L^2(R)$, we define the orthogonal projection of u on the finite-dimensional space $\mathcal{R}_N$ by

7.6 Numerical Methods for Solving Parabolic Differential Equations

$$P_N u(x) = \sum_{k=0}^{N} \widehat{u}_k H_k^{\alpha,\beta}(x).$$

The following theorem shows that any function $u \in W^r(R)$ as well as its weak derivatives can be approximated by an element in the finite-dimensional linear space $\mathcal{R}_N$.

Theorem 7.8 *For any $u \in W^r(R)$, for $N \gg 1$, and for any integer $0 \le \mu \le r$, we have*

$$|u - P_N u|_\mu \le C\alpha^{\mu-r-\frac{1}{2}} N^{\frac{\mu-r}{2}} \|u\|_r, \tag{7.74}$$

where $|u|_\mu = \|\frac{d^\mu}{dx^\mu} u\|$ is the L^2-norm of the μ-th weak derivative of u, and C is a generalized constant that does not depend on N.

Proof The theorem can be proved by induction. We first show that for $\mu = 0$, (7.74) holds.

For any integer $r \ge 0$,

$$\|u - P_N u\|^2 = \frac{\sqrt{\pi}}{\alpha} \sum_{n=N+1}^{\infty} \widehat{u}_n^2 = \frac{\sqrt{\pi}}{\alpha} \sum_{n=N+1}^{\infty} \lambda_{n+1}^{-r} \lambda_{n+1}^r \widehat{u}_n^2 \le C\alpha^{-2r-1} N^{-r} \|u\|_r^2. \tag{7.75}$$

Suppose that for $1 \le \mu \le r$, (7.74) holds for $\mu - 1$. Since

$$|u - P_N u|_\mu = |u_x - (P_N u)_x|_{\mu-1}, \tag{7.76}$$

where, as before, $u_x = \frac{du}{dx}$ denotes the derivative of u, then

$$|u - P_N u|_\mu \le |u_x - P_N u_x|_{\mu-1} + |P_N u_x - (P_N u)_x|_{\mu-1}. \tag{7.77}$$

By induction assumption, we have

$$|u_x - P_N u_x|_{\mu-1} \le C\alpha^{\mu-r-\frac{1}{2}} N^{\frac{\mu-r}{2}} \|u_x\|_{r-1} \le C\alpha^{\mu-r-\frac{1}{2}} N^{\frac{\mu-r}{2}} \|u\|_r. \tag{7.78}$$

The last inequality holds because

$$\|u_x\|_{r-1}^2 = \sum_{n=0}^{\infty} \lambda_{n+1}^{r-1} \widehat{(u_x)}_n^2,$$

and

$$\widehat{(u_x)}_n = \frac{\alpha}{\sqrt{\pi}} \int_R u_x H_n^{\alpha,\beta}(x) dx = -\frac{\alpha}{\sqrt{\pi}} \int_R u H_n'(x) dx = \frac{\sqrt{\lambda_{n+1}}}{2} \widehat{u}_{n+1} - \frac{\sqrt{\lambda_n}}{2} \widehat{u}_{n-1},$$

where the last equality follows from the properties of Hermite polynomials.

Next, we need to estimate $|P_N u_x - (P_N u)_x|_{\mu-1}$. In fact, applying the properties of Hermite polynomials again, we can obtain

$$P_N u_x - (P_N u)_x = P_N \sum_{n=0}^{\infty} \widehat{u}_n H_n'(x) - \sum_{n=0}^{N} \widehat{u}_n H_n'(x)$$

$$= \frac{1}{2}\sqrt{\lambda_{N+1}} \left[\widehat{u}_N H_{N+1}(x) + \widehat{u}_{N+1} H_N(x) \right].$$

Therefore,

$$|P_N u_x - (P_N u)_x|_{\mu-1}^2 \leq C \lambda_{N+1} \left(\widehat{u}_N^2 |H_{N+1}|_{\mu-1}^2 + \widehat{u}_{N+1}^2 |H_N|_{\mu-1}^2 \right). \tag{7.79}$$

In the meanwhile, by the virtue of (7.75) and (7.71), we have

$$\widehat{u}_N^2 \leq \sum_{n=N}^{\infty} \widehat{u}_n^2 \leq \frac{\alpha}{\sqrt{\pi}} \| u - P_{N-1} u \|^2 \leq C \alpha^{-2r} N^{-r} \| u \|_r^2, \tag{7.80}$$

$$|H_N|_{\mu-1}^2 = \| H_N^{(\mu-1)} \|^2 \leq C \alpha^{-1} \| H_N \|_{\mu-1}^2 = C \alpha^{-1} \lambda_N^{\mu-1} \leq C \alpha^{-1} \lambda_{N+1}^{\mu-1}. \tag{7.81}$$

Substitute (7.80) and (7.81) into (7.79), and we have

$$|P_N u_x - (P_N u)_x|_{\mu-1}^2 \leq C \alpha^{-2r-1} N^{-4} \lambda_{N+1}^{\mu} \| u \|_r^2 \leq C \alpha^{2\mu-2r-1} N^{\mu-r} \| u \|_r^2. \tag{7.82}$$

Combining (7.78) and (7.82), we have proven the inequality (7.74) for μ and therefore, (7.74) holds for all $1 \leq \mu \leq r$. □

Since all the functions in $W^r(R)$ can be approximated by elements in $\mathcal{R}_N$, we would like to consider the solution of a parabolic partial differential equation in the space $L^2(0, T; W^r(R))$, with suitable $r \geq 0$.

Definition 7.3 A function $u \in L^{\infty}(0, T; W^r(R)) \cap L^2(0, T; W^r(R))$ is called a weak solution of the parabolic partial differential equation (7.63) if $\forall \varphi \in C_0^{\infty}(R)$:

$$\begin{aligned} \langle u_t, \varphi \rangle &= -\langle p u_x, \varphi_x \rangle + \langle q u_x, \varphi \rangle + \langle w u, \varphi \rangle, \\ u(x, 0) &= u_0(x), \end{aligned} \tag{7.83}$$

where $\langle \cdot, \cdot \rangle$ denotes the inner product on $L^2(R)$.

If we restrict φ in $\mathcal{R}_N$, then we can find a function $u_N(t, x)$, with $u_N(t, \cdot) \in \mathcal{R}_N$ for all $t \in [0, T]$, such that the first equation in (7.83) is satisfied by all $\varphi \in \mathcal{R}_N$. The function u_N is the Galerkin approximation to the solution of the original Eq. (7.63).

7.6 Numerical Methods for Solving Parabolic Differential Equations

The next theorem shows that, under mild conditions, as $N \to \infty$, the Galerkin approximation will tend to the real solution of (7.63).

Theorem 7.9 *Assume that $u \in L^\infty(0, T; W^r(R)) \cap L^2(0, T; W^r(R))$ is a weak solution of (7.63) and $u_N(x, t)$ is the Galerkin approximation of u. Then, if we further assume that the coefficients p, q, w are all bounded, for r large enough, we have the following estimations:*

$$\|u - u_N\|^2(t) \le C_1 N^{1-r}, \tag{7.84}$$

where C_1 is a generalized constant depending on α, β, u, and the time period T.

Proof Let $U_N = P_N u$ be the truncated Hermitian series of u. Therefore, $u - U_N \perp \mathcal{R}_N$. Since u is a weak solution of Eq. (7.63), according to (7.83), for any $\varphi \in \mathcal{R}_N$,

$$0 = \langle (u - U_N)_t, \varphi \rangle = -\langle pu_x, \varphi_x \rangle + \langle qu_x, \varphi \rangle + \langle wu, \varphi \rangle - \langle (U_N)_t, \varphi \rangle, \tag{7.85}$$

i.e.,

$$\langle (U_N)_t, \varphi \rangle = -\langle pu_x, \varphi_x \rangle + \langle qu_x, \varphi \rangle + \langle wu, \varphi \rangle. \tag{7.86}$$

Since u_N is the Galerkin approximation to u, Eq. (7.83) also holds for u_N and $\varphi \in \mathcal{R}_N$.

Combining (7.86), we have

$$\langle (u_N - U_N)_t, \varphi \rangle = -\langle p(u_N - u)_x, \varphi_x \rangle + \langle q(u_N - u)_x, \varphi \rangle + \langle w(u_N - u), \varphi \rangle. \tag{7.87}$$

Denote $\rho_N = u_N - U_N$ and (7.87) becomes

$$\langle (\rho_N)_t, \varphi \rangle = -\langle p(\rho_N)_x, \varphi_x \rangle + \langle q(\rho_N)_x + w\rho_N, \varphi \rangle - \langle p(U_N - u)_x, \varphi_x \rangle$$
$$+ \langle q(U_N - u)_x + w(U_N - u), \varphi \rangle, \tag{7.88}$$

for all $\varphi \in \mathcal{R}_N$.

Now, let's take $\varphi = 2\rho_N$ in (7.88), and we obtain

$$\frac{d}{dt}\|\rho_N\|^2 = -2\langle p(\rho_N)_x, (\rho_N)_x \rangle + 2\langle q(\rho_N)_x + w\rho_N, \rho_N \rangle$$
$$- 2\langle p(U_N - u)_x, (\rho_N)_x \rangle \tag{7.89}$$
$$+ 2\langle q(U_N - u)_x + w(U_N - u), \rho_N \rangle.$$

Since $p(t, x) \ge \lambda > 0$ for all $(t, x) \in [0, T] \times R$,

$$-2\langle p(\rho_N)_x, (\rho_N)_x \rangle \le -2\lambda \|(\rho_N)_x\|^2.$$

Because p, q, w are bounded, assume that $\max\{p(t,x), q(t,x), w(t,x)\} = C$, and then the following estimations are followed directly from Cauchy-Schwartz inequality and Young's inequality:

$$2\langle q(\rho_N)_x, \rho_N \rangle \leq \lambda \|(\rho_N)_x\|^2 + \frac{C^2}{\lambda}\|\rho_N\|^2$$

$$2\langle w\rho_N, \rho_N \rangle \leq 2C\|\rho_N\|^2$$

$$-2\langle p(U_N - u)_x, (\rho_N)_x \rangle \leq \lambda \|(\rho_N)_x\|^2 + \frac{C^2}{\lambda}\|(U_N - u)_x\|^2$$

$$2\langle q(U_N - u)_x, \rho_N \rangle \leq \|\rho_N\|^2 + C^2\|(U_N - u)_x\|^2$$

$$2\langle w(U_N - u), \rho_N \rangle \leq \|\rho_N\|^2 + C^2\|(U_N - u)\|^2.$$

Therefore,

$$\frac{d}{dt}\|\rho_N\|^2 \leq C_1\|\rho_N\|^2 + C_2(\|(U_N - u)_x\|^2 + \|U_N - u\|^2). \tag{7.90}$$

From Lemma 7.3,

$$\|(U_N - u)_x\|^2 = |U_N - u|_1^2 \leq CN^{1-r}\|u\|_r^2$$

$$\|(U_N - u)\|^2 \leq CN^{-r}\|u\|_r^2.$$

Therefore,

$$\frac{d}{dt}\|\rho_N\|^2 \leq C_1\|\rho_N\|^2 + C_2\|u\|_r^2 N^{1-r}. \tag{7.91}$$

According to Gronwall's inequality, since $\rho_N(0) \equiv 0$, we have

$$\|\rho_N\|^2 \leq C_2 N^{1-r} \int_0^t e^{C_1(t-s)} \|u(s)\|_r^2 ds, \tag{7.92}$$

and then from Lemma 7.3,

$$\|u - u_N\|^2 \leq \|\rho_N\|^2 + \|u - U_N\|^2 \leq \widetilde{C}_1 N^{1-r}. \tag{7.93}$$

□

Theorem 7.9 together with Theorem 7.8 guarantees the convergence of the Galerkin spectral method under mild conditions. Thus, we can feel free to propose this method to solve the parabolic partial differential equations.

Since for each $t \in [0, T]$, the approximation $u_N(t, \cdot) \in \mathcal{R}_N$, we can assume that $u_N(t, x)$ is in the following form:

7.6 Numerical Methods for Solving Parabolic Differential Equations

$$u_N(t,x) = \sum_{k=0}^{N} \psi_k(t) H_k^{\alpha,\beta}(x), \tag{7.94}$$

where $\{\psi_k : 0 \leq k \leq N\}$ are coefficients to be determined.

Take Eq. (7.94) into (7.83) with $\varphi = H_k^{\alpha,\beta}$, for each $0 \leq k \leq N$. By the orthogonality of $\{H_k^{\alpha,\beta}\}_{k=0}^{N}$, we can get an ordinary differential system for $\{\psi_k : 0 \leq k \leq N\}$:

$$\psi_i'(t)\langle H_i, H_i\rangle = \sum_{k=0}^{N} \psi_k(t)\langle pH_k'' + qH_k' + wH_k, H_i\rangle, \quad i = 0, 1, \cdots, N. \tag{7.95}$$

The initial values of ψ_k are just the generalized Fourier coefficients of the function $u_0(x)$. After solving the ordinary differential equation system, we obtain the Galerkin approximation of the original parabolic partial differential equations.

Remark 7.1 For parabolic partial differential equations with space dimension higher than 1, we can also apply Galerkin spectral method with respect to an orthogonal basis of $L^2(R^d)$. The orthogonal basis of $L^2(R^d)$ can also be chosen as generalized Hermite functions [3].

In this case, the generalized Hermite functions are defined to be

$$\mathbf{H}_\mathbf{n}^{\alpha,\beta}(\mathbf{x}) = \prod_{j=1}^{d} H_{n_j}^{\alpha_j,\beta_j}(x_j), \quad \mathbf{x} = (x_1, \cdots, x_d) \in R^d, \tag{7.96}$$

where $\alpha = (\alpha_1, \cdots, \alpha_d)$, $\alpha_j > 0$, $\beta = (\beta_1, \cdots, \beta_d)$, $\mathbf{n} = (n_1, \cdots, n_d)$, $n_j \in \mathbb{N}$. Then, $\{\mathbf{H}_\mathbf{n}^{\alpha,\beta} : \mathbf{n} \in \mathbb{N}^d\}$ forms an orthogonal basis of $L^2(R^d)$, and under mild conditions, we can also prove that the Galerkin approximation with respect to this orthogonal system converges to the real solution of the original d-dimensional parabolic partial differential equation.

Remark 7.2 Apart from Hermite functions, other traditional orthogonal systems can also be used in Galerkin approximation. For instance, the Legendre functions defined by

$$L_0(x) \equiv 1; \; L_1(x) = x; \; L_{n+1}(x) = \frac{2n+1}{n+1} x L_n(x) - \frac{n}{n+1} L_{n-1}(x), \tag{7.97}$$

for $n = 1, 2, \cdots$ and $x \in [-1, 1]$ form an orthogonal system on $L^2([-1, 1])$ and can be generated to an orthogonal system on the square integrable spaces of every bounded intervals, $L^2([-M, M])$ through scaling and transforming.

The orthogonal system for bounded domain of dimension higher than 1 can also be constructed as we did in the case of Hermite functions. In fact,

$$\mathbf{L_k}(\mathbf{x}) = \prod_{j=1}^{d} L_{k_j}(x_j), \quad \mathbf{x} = (x_1, \cdots, x_d), \tag{7.98}$$

for $\mathbf{k} = (k_1, \cdots, k_d)$, $k_j \in \mathbb{N}$ form an orthogonal basis on $L^2([-1, 1]^d)$ and can be generated to arbitrary bounded rectangles in R^d through scaling and transforming.

As is shown in [1], under certain conditions, the convergence rate of Galerkin method using Legendre functions is twice faster than the method using Hermite functions.

Up to now, we have completely proposed the Yau-Yau algorithm for a general nonlinear filtering problem, and we would like to write down the entire procedure of this algorithm. Here, we use the same notations as we did in Sect. 7.2.

Algorithm 1 On-line algorithm

1: **Initialization:** Fix T, Δt, $y_0 = 0$ and σ_0. Let $k = \frac{T}{\Delta t}$ and $\{0 = \tau_0 < \tau_1 < \cdots < \tau_k = T\}$. Let $u_1(x, 0)$ be the normalization of σ_0, i.e.

$$u_1(x, 0) = \frac{\sigma_0(x)}{\int_{R^d} \sigma_0(x)dx}.$$

2: By the Off-Line Algorithm, obtain $u_1(x, \tau_1)$;
3: At time τ_1, the nuw observation y_{τ_1} comes and let

$$u_2(x, \tau_1) = \exp\left[h^\top(x, \tau_1)y_{\tau_1}\right]u_1(x, \tau_1).$$

4: **for** $i = 2$ to k **do**
5:    Obtain $u_{i-1}(x, \tau_{i-1})$ from the Off-Line Algorithm.
6:    Renew the initial value of the partial differential equation satisfied by $u_i(x, t)$,

$$u_i(x, \tau_{i-1}) = \exp\left[h^\top(x, \tau_{i-1})(y_{\tau_{i-1}} - y_{\tau_{i-2}})\right]u_{i-1}(x, \tau_{i-1}).$$

Algorithm 2 Off-line algorithm

1: **Initialization:** Given $u_1(x, \tau_0)$ in On-Line Algorithm.
2: **for** $i = 1$ to k **do**
3:    Solve the parabolic partial differential equation using Galerkin spectral method and get $u_i(x, t)$.
4:    Normalize $u_i(x, \tau_i) = \frac{u_i(x, \tau_i)}{\int_{R^d} u_i(x, \tau_i)dx}$.
5:    Obtain the unnormalized conditional probability density

$$\rho(x, \tau_i) = \exp\left[-h^\top(x, \tau_i)y_{\tau_{i-1}}\right]u_i(x, \tau_i)$$

   at each τ_i.

Finally, this chapter will end up with two explicit one-dimensional nonlinear filtering example, with the second one as an exercise. Readers are encouraged to conduct the whole process of Yau-Yau algorithm on this example, so that they will have a more intuitive understanding on this filtering algorithm for practical use.

7.7 Numerical Results

In this section, we will consider the following one-dimensional time-varying nonlinear filtering problem:

$$\begin{cases} dX_t = \frac{1}{2} dV_t \\ dY_t = X_t \left(1 + \frac{1}{4} \sin X_t \right) dt + dW_t, \end{cases} \quad (7.99)$$

where $\{W_t : t \geq 0\}$ and $\{V_t : t \geq 0\}$ are mutually independent one-dimensional standard Brownian motions.

Hereafter, we consider the dynamics (7.99) on the time interval $[0, 50]$, with initial value $X_0 = 0$ and time discretization step $\Delta t = 0.01 s$. The corresponding time step $K = \frac{50}{\Delta t} = 5000$. A typical trajectory of the state process X_t is shown in Fig. 7.1.

In order to conduct the Yau-Yau algorithm on this problem, we would like to use Hermite spectral method discussed in Sect. 7.6 with the number of basis $N = 15$. The performance of Yau-Yau algorithm is evaluated by the expected mean square error, which is defined by

$$EMSE = \frac{1}{K} E \sum_{k=1}^{K} (X_k - \hat{x}_k)^2, \quad (7.100)$$

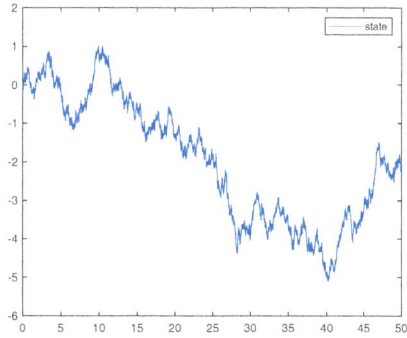

Fig. 7.1 A typical trajectory of the state process

where $\hat{x}_k$ is the estimation of the conditional expectation at time k. In practical implementation, the expected mean square error (7.100) is estimated by the average of $M = 100$ repeated experiments, which is defined by

$$MSE = \frac{1}{M}\frac{1}{K}\sum_{i=1}^{M}\sum_{k=1}^{K}(X_k^{(i)} - \hat{x}_k^{(i)})^2, \qquad (7.101)$$

where the superscript i in $X_k^{(i)}$ and $\hat{x}_k^{(i)}$ denotes the i-th experiment.

Based on the above settings, the estimated mean square error, MSE, of Yau-Yau algorithm is 1.4007, and a typical estimation result is shown in Fig. 7.2. In the meanwhile, the total online computation time is around 25 seconds. Because the total time interval we simulate is 50 seconds, we can use Yau-Yau algorithm to get real-time solutions of this nonlinear filtering problem in this setting.

Also, for a particular trajectory of the state process, we can calculate the average mean square error at each time $k \in \{1, \cdots, K\}$, which is defined by

$$MSE_k = \frac{1}{k}\sum_{j=1}^{k}(X_j - \hat{x}_j)^2. \qquad (7.102)$$

The evolution of the mean square error for the above typical trajectory of Yau-Yau algorithm is shown in Fig. 7.3. After several steps, the mean square error remains

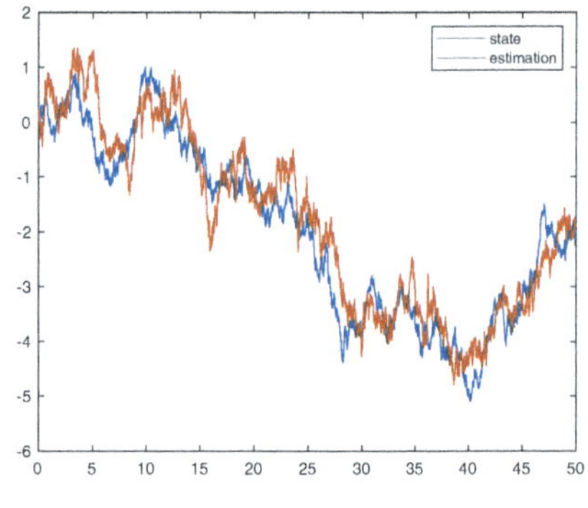

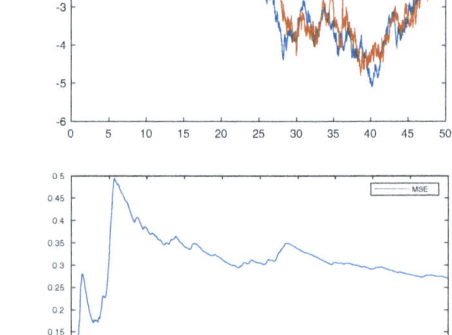

Fig. 7.2 A typical performance of Yau-Yau algorithm

Fig. 7.3 A typical trajectory of mean square errors

7.8 Exercises

around 0.3. This shows that Yau-Yau algorithm can help maintain a reliable mean square error and obtain accurate estimations to the state process.

7.8 Exercises

The exercises in this chapter are based on the following one-dimensional cubic sensor filtering problem:

$$\begin{cases} dX_t = dV_t, \\ dY_t = X_t^3 dt + dW_t, \end{cases} \quad (7.103)$$

with $\{W_t : t \geq 0\}$ and $\{V_t : t \geq 0\}$ mutually independent one-dimensional standard Brownian motions.

The purpose of the following exercises is to make readers more familiar with the entire procedure of Yau-Yau algorithm. Readers are also encouraged to simulate some trajectories of the system (7.103) by computer and test the performance of Yau-Yau algorithm.

1. Write down explicitly the DMZ equation satisfied by the unnormalized conditional probability density function $\sigma(t, x)$ of the filtering system (7.103).
2. Derive the robust DMZ equation satisfied by $u(t, x)$ given by

$$u(t, x) = \exp\left(-Y_t x_t^3\right) \sigma(t, x). \quad (7.104)$$

3. Let $0 = t_0 < t_1 < \cdots < t_K = T$ be a uniform partition of the time interval $[0, T]$. Derive the corresponding auxiliary equations in the Yau-Yau algorithm satisfied by $u_i(t, x)$ and $\tilde{u}_i(t, x)$, as Eqs. (7.7) and (7.8).
4. Use Hermite-Galerkin spectral method to solve the parabolic equation satisfied by $\tilde{u}_i(x, t)$ in Exercise 3. Let

$$\tilde{u}_i^{(N)}(t, x) = \sum_{j=1}^{N} a_{i,j}^{(N)}(t) H_j(x) \quad (7.105)$$

be the approximated solution to $\tilde{u}_i(x, t)$. Write down explicitly the equations satisfied by the generalized Fourier coefficients $\left(a_1^{(N)}, \cdots, a_N^{(N)}\right)$.
5. For practical implementations, if we are concerned with tracking the evolution of the position of the state process X_t, the conditional expectation $E[X_t|\mathcal{Y}_t]$ is a good estimator. Please provide an approximation of $E[X_t|\mathcal{Y}_t]$ based on the generalized Fourier coefficients $\left(a_1^{(N)}(t), \cdots, a_N^{(N)}(t)\right)$. (Hint: The expression may contain $\int_{\mathbb{R}} x H_j(x) dx, j = 1, \cdots, N$.)

6. Show that the properties of one-dimensional generalized Hermite functions (7.67), (7.68), and (7.69) hold. Hence, the integrals $\int_{\mathbb{R}} x H_j(x) dx$, $j = 1, \cdots, N$, can be calculated explicitly.
7. As for the convergence analysis, write down the explicit expression of the assumptions (7.13), (7.17), and (7.18). Discuss for what kind of observation trajectories, such assumptions will hold so that the convergence of Yau-Yau algorithm is guaranteed.

References

1. W. Dong, X. Luo and S. S.-T. Yau. Solving nonlinear filtering problems in real-time by Legendre Galerkin spectral method. *IEEE Transactions on Automatic Control*, 66(4):1559–1572, 2021.
2. X. Luo and S. S.-T. Yau. Hermite spectral method to 1-D forward Kolmogorov equation and its application to nonlinear filtering problems. *IEEE Transactions on Automatic Control*, 58(10):2495–2507, 2013.
3. X. Luo and S. S.-T. Yau. Hermite spectral method with hyperbolic cross approximations to high-dimensional parabolic PDEs. *SIAM Journal on Numerical Analysis*, 51(6):3186–3212, 2013.
4. S. T. Yau and S. S.-T. Yau. Real time solution of nonlinear filtering problem without memory I. *Mathematical Research Letters*, 7:671–693, 2000.
5. S. T. Yau and Stephen S.-T. Yau. Real time solution of the nonlinear filtering problem without memory II. *SIAM Journal on Control and Optimization*, 47(1):163–195, 2008.

Chapter 8
Direct Methods

In this chapter, we will focus on the explicit solution of DMZ equation for some specific nonlinear systems. First, we focus on continuous time-invariant finite-dimensional filtering systems. It shows that Yau filtering system can be solved explicitly with an arbitrary initial condition by solving a system of ordinary differential equations and a Kolmogorov equation. Second, we will extend the direct method to time-invariant Yau filtering system with nonlinear observations. Third, we mainly consider time-invariant Yau systems with a class of nonlinear observation and give explicit solution under the Yau-Yau algorithm framework. Finally, we extend the previous results to time-varying Yau systems.

8.1 Introduction

In this chapter, we consider continuous filtering systems and try to obtain the explicit solution of conditional density of state evolution $p(x|\mathcal{Y}_t)$, where $\mathcal{Y}_t := \sigma(y_\tau : 0 \leq \tau \leq t)$ is a sigma algebra. First, we recall time-invariant setting:

$$\begin{cases} dx(t) = f(x(t))dt + g(x(t))dv(t), & x(0) = x_0, \\ dy(t) = h(x(t))dt + dw(t), & y(0) = 0, \end{cases} \quad (8.1)$$

where x, v, y, and w are, respectively, $\mathbb{R}^n$, $\mathbb{R}^p$, $\mathbb{R}^m$, and $\mathbb{R}^m$-valued processes, and v and w have components that are independent, standard Brownian motion processes. We further assume that $n = p$, f and h are C^∞ smooth functions and that g is an orthogonal matrix.

By introducing a new unnormalized density $u(t, x) = \exp(-\sum_{i=1}^{m} h_i(x) y_i(t)) \sigma(t, x)$, DMZ equation will become following the so-called the robust DMZ equation:

$$\begin{cases} \dfrac{\partial u}{\partial t}(t,x) = \dfrac{1}{2}\sum_{i=1}^{n}\dfrac{\partial^2 u}{\partial x_i^2}(t,x) + \sum_{i=1}^{n}\left(-f_i(x) + \sum_{j=1}^{m} y_j(t)\dfrac{\partial h_j}{\partial x_i}(x)\right)\dfrac{\partial u}{\partial x_i} \\ \qquad - \left[\sum_{i=1}^{n}\dfrac{\partial f_i}{\partial x_i} + \dfrac{1}{2}\sum_{i=1}^{m} h_i^2 - \dfrac{1}{2}\sum_{i=1}^{m} y_i(t)\Delta h_i \right. \\ \qquad \left. + \sum_{i=1}^{m}\sum_{j=1}^{n} y_i(t)f_j\dfrac{\partial h_i}{\partial x_j} - \dfrac{1}{2}\sum_{i,j=1}^{m}\sum_{k=1}^{n} y_i y_j \dfrac{\partial h_i}{\partial x_k}\dfrac{\partial h_j}{\partial x_k}\right] u(t,x) \\ u(0,x) = \sigma_0(x). \end{cases}$$
(8.2)

We briefly recall the Yau-Yau algorithm [4] that solves the nonlinear filtering problem with arbitrary initial condition by reducing it to solve the Kolmogorov equation.

Suppose that $u(t,x)$ is the solution of the robust DMZ equation. First, we make partition for the time interval. Let $\mathcal{P}_k = \{0 = \tau_0 < \tau_1 < \cdots < \tau_k = \tau\}$ be a partition of $[0, \tau]$. Let $u_i(t,x)$ be a solution of the following partial differential equation for $\tau_{i-1} \leq t \leq \tau_i$, which satisfies the following equation:

$$\begin{cases} \dfrac{\partial u_i}{\partial t}(t,x) = \dfrac{1}{2}\Delta u_i(t,x) + \sum_{l=1}^{n}\left(-f_l(x) + \sum_{j=1}^{m} y_j(\tau_{i-1})\dfrac{\partial h_j}{\partial x_l}(x)\right)\dfrac{\partial u_i}{\partial x_l} \\ \qquad - \left[\sum_{l=1}^{n}\dfrac{\partial f_l}{\partial x_l} + \dfrac{1}{2}\sum_{l=1}^{m} h_l^2 - \dfrac{1}{2}\sum_{j=1}^{m} y_j(\tau_{i-1})\Delta h_j \right. \\ \qquad \left. + \sum_{j=1}^{m}\sum_{l=1}^{n} y_j(\tau_{i-1})f_l\dfrac{\partial h_j}{\partial x_l} - \dfrac{1}{2}\sum_{j,l=1}^{m}\sum_{p=1}^{n} y_j(\tau_{i-1})y_l(\tau_{i-1})\dfrac{\partial h_j}{\partial x_p}\dfrac{\partial h_l}{\partial x_p}\right] u_i(t,x) \\ u_i(\tau_{i-1}, x) = u_{i-1}(\tau_{i-1}, x). \end{cases}$$
(8.3)

In the work of Yau and Yau [4, 5], it has been proved that in both point-wise sense and L^2-sense, $u_k(t,x)$ will converge to the explicit solution $u(t,x)$, i.e., $u(\tau, x) = \lim_{|\mathcal{P}_k| \to 0} u_k(\tau, x)$.

Therefore, it remains to describe an algorithm to compute $u_k(\tau_k, x)$. In the work of Yau and Yau [4, 5], the novelty is by using an exponential-type transformation, solving robust DMZ equation can be reduced to solving a Kolmogorov equation at each time interval $[\tau_k, \tau_{k+1}]$, and observation term can be shifted to an update at each initial moment $t = \tau_k$. The details can be found in the following proposition.

Proposition 8.1 ([6]) *For each $\tau_{k-1} \leq t \leq \tau_k$, $1 \leq k \leq n$, $\tilde{u}_k(t,x)$ satisfies the following parabolic equation:*

8.1 Introduction

$$\frac{\partial \tilde{u}_k}{\partial t}(t, x) = \frac{1}{2}\Delta \tilde{u}_k(t, x) - \sum_{j=1}^{n} f_j(x)\frac{\partial \tilde{u}_k}{\partial x_j}(t, x)$$

$$- \left(\sum_{j=1}^{n} \frac{\partial f_j}{\partial x_j}(x) + \frac{1}{2}\sum_{j=1}^{m} h_j^2(x)\right)\tilde{u}_k(t, x), \quad (8.4)$$

for $\tau_{k-1} \leq t \leq \tau_k$, if and only if

$$\tilde{u}_k(t, x) = \exp\left(\sum_{i=1}^{m} y_i(\tau_{k-1})h_i(x)\right) u_k(t, x) \quad (8.5)$$

satisfies (8.3).

The initial condition for (8.4) on $\tau_{k-1} \leq t \leq \tau_k$

$$\tilde{u}_k(\tau_{k-1}, x) = \begin{cases} \sigma_0(x), k = 1, \\ \exp\left[\sum_{j=1}^{m}(y_j(\tau_{k-1}) - y_j(\tau_{k-2}))h_j(x)\right]\tilde{u}_{k-1}(\tau_{k-1}, x), k \geq 2. \end{cases} \quad (8.6)$$

Furthermore, the convergence theorem has also been proved.

Theorem 8.1 ([4]) *The unnormalized density σ can be computed via solution $\tilde{u}_i$ of Kolmogorov equation (8.4). More specifically,*

$$\sigma(\tau, x) = \lim_{|\mathcal{P}_k| \to 0} \tilde{u}_k(\tau_k, x). \quad (8.7)$$

In this chapter, we will focus on the time-invariant and time-variant Yau filtering system and propose to obtain the explicit solution by solving the Kolmogorov equation directly. Hence, this type of method is also called "direct method," which means explicit density evolution can be obtained. The arrangement of this chapter is listed as follows. In Sect. 8.2, for time-invariant finite-dimensional filtering system, the filtering problem is transformed to solving a Kolmogorov equation and a series of ODEs. In Sect. 8.3, by the Yau-Yau algorithm framework, we extend the work of Sect. 8.2 to solve the system with nonlinear observation and propose efficient so-called Gaussian approximation method. In Sect. 8.4, by the Yau-Yau algorithm framework and variable transformation, we transform the robust DMZ equation to a Schrödinger-type equation so-called time-varying Schrödinger equation. Then we can calculate the analytical fundamental solution that can be utilized to obtain explicit solution. In Sect. 8.5, the work of Sect. 8.4 will be extended to a special class of time-varying Yau filtering system, and the concrete procedure is similar.

8.2 Explicit Solution of DMZ Equation for Finite-Dimensional Filters

This section will focus on continuous time-invariant finite-dimensional filtering systems. It shows that the Yau filtering system can be solved explicitly with an arbitrary initial condition by solving a system of ordinary differential equations and a Kolmogorov equation. We shall show that there only require n sufficient statistics for solving the DMZ equation. This section is mainly referred to the work of Yau and Hu [7].

Definition 8.1 (Yau Filtering System) The filtering system (8.1) is called the Yau filtering system if the following condition holds:

$$(C_1') \quad \frac{\partial f_j}{\partial x_i} - \frac{\partial f_i}{\partial x_j} = \text{const}, \quad \forall 1 \leq i, j \leq n. \tag{8.8}$$

Remark 8.1 Yau's filtering systems include Kalman-Bucy filtering systems and Benés filtering systems as two special cases.

The following result describes the equivalent form of Yau filter.

Theorem 8.2 (C_1') holds if and only if

$$(f_1, \cdots, f_n) = (l_1, \cdots, l_n) + \left(\frac{\partial F}{\partial x_1}, \cdots, \frac{\partial F}{\partial x_n} \right), \tag{8.9}$$

where $l_1, \cdots, l_n$ are polynomials of degree 1 and F is a C^∞ function.

Proof **Sufficiency:** If we assume $(f_1, \cdots, f_n) = (l_1, \cdots, l_n) + \left(\frac{\partial F}{\partial x_1}, \cdots, \frac{\partial F}{\partial x_n} \right)$, then

$$\frac{\partial f_j}{\partial x_i} - \frac{\partial f_i}{\partial x_j} = \frac{\partial l_j}{\partial x_i} - \frac{\partial l_i}{\partial x_j} = \text{const}, \tag{8.10}$$

where const denotes a certain constant.

Necessity: Suppose that $\frac{\partial f_j}{\partial x_i} - \frac{\partial f_i}{\partial x_j} = c_{ij}$ and clearly there has property $c_{ij} = -c_{ji}$. In the following, we assume $l_i = \sum_{j=1}^{n} b_{ij} x_j$, where $b_{ij} = \frac{1}{2} c_{ji}$. Next we consider two exterior derivatives of first-order differential form $\sum_{j=1}^{n} f_j dx_j$ and $\sum_{j=1}^{n} l_j dx_j$, respectively:

$$d \left(\sum_{j=1}^{n} f_j dx_j \right) = \sum_{i<j} \left(\frac{\partial f_j}{\partial x_i} - \frac{\partial f_i}{\partial x_j} \right) dx_i \wedge dx_j$$
$$= \sum_{i<j} c_{ij} dx_i \wedge dx_j \tag{8.11}$$

8.2 Explicit Solution of DMZ Equation for Finite-Dimensional Filters

and

$$d\left(\sum_{j=1}^{n} l_j dx_j\right) = \sum_{i<j}(b_{ji} - b_{ij})dx_i \wedge dx_j$$
$$= \sum_{i<j} c_{ij} dx_i \wedge dx_j. \tag{8.12}$$

So $d(\sum_{j=1}^{n} l_j dx_j) = d(\sum_{j=1}^{n} f_j dx_j)$ holds, i.e., $d(\sum_{j=1}^{n} f_j dx_j - \sum_{j=1}^{n} l_j dx_j) = 0$. By Poincáre lemma, every closed form must be the exact form defined on the simply connected topological space. Then there exists function ψ such that

$$\sum_{j=1}^{n} f_j dx_j - \sum_{j=1}^{n} l_j dx_j = d\psi = \sum_{j=1}^{n} \frac{\partial \psi}{\partial x_j} dx_j. \tag{8.13}$$

It implies the desired result. □

From Theorem 8.2, we know that (C_1') is equivalent to the following condition:

$$(C_1) \quad f_i(x) = l_i(x) + \frac{\partial F}{\partial x_i}(x), \ 1 \leq i \leq n, \tag{8.14}$$

where $l_i(x) = \sum_{j=1}^{n} d_{ij} x_j + d_i$ for $1 \leq i \leq n$ and F is a C^∞ function.

Next we list the following condition by assuming the observation term is a linear function in a state:

$$(C_2) \quad h_i(x) = \sum_{j=1}^{n} c_{ij} x_j + c_i, \ 1 \leq i \leq m, \tag{8.15}$$

where c_{ij}, c_i are constants.

Moreover, we know that $\eta(x)$ is a quadratic polynomial in x for most interesting filtering systems [28, 29, 33]. Hence, we assume the following condition:

$$(C_3) \quad \eta(x) = \sum_{i,j=1}^{n} \eta_{ij} x_i x_j + \sum_{i=1}^{n} \eta_i x_i + \eta_0, \tag{8.16}$$

where $\eta_{ij}, \eta_i, \eta_0$ are constants.

In order to calculate the explicit solution of robust DMZ equation, first we need to simplify the robust DMZ equation (8.2).

Lemma 8.1 *Robust DMZ equation is equivalent to the following equation:*

$$\begin{cases} \dfrac{\partial u}{\partial t}(t,x) = \dfrac{1}{2}\Delta u(t,x) + \sum_{i=1}^{n}\theta_i(t,x)\dfrac{\partial u}{\partial x_i} + \theta(t,x)u(t,x) \\ u(0,x) = \sigma_0(x), \end{cases} \quad (8.17)$$

where

$$\theta_i(t,x) = -f_i(x) + \sum_{j=1}^{m} y_j(t)\dfrac{\partial h_j}{\partial x_i}(x),$$

$$\theta(t,x) = \dfrac{1}{2}\Big(\sum_{i=1}^{n}\theta_i^2(t,x) + \sum_{i=1}^{n}\dfrac{\partial \theta_i}{\partial x_i} - \eta(x)\Big). \quad (8.18)$$

Proof We only need to verify the right-hand side of (8.17) equals to the right-hand side of (8.2). By direct computations,

$$\theta_i^2(t,x) = \sum_{j,k=1}^{m} y_j y_k \dfrac{\partial h_j}{\partial x_i}\dfrac{\partial h_k}{\partial x_i} - 2f_i \sum_{j=1}^{m} y_j \dfrac{\partial h_j}{\partial x_i} + f_i^2$$

$$\dfrac{\partial \theta_i}{\partial x_i} = \sum_{j=1}^{m} y_j \dfrac{\partial^2 h_j}{\partial x_i^2} - \dfrac{\partial f_i}{\partial x_i}. \quad (8.19)$$

It implies

$$\theta(t,x) = \dfrac{1}{2}\sum_{i=1}^{n}\sum_{j,k=1}^{m} y_j y_k \dfrac{\partial h_j}{\partial x_i}\dfrac{\partial h_k}{\partial x_i} - \sum_{i=1}^{n} f_i \sum_{j=1}^{m} y_j \dfrac{\partial h_j}{\partial x_i}$$
$$+ \dfrac{1}{2}\sum_{i=1}^{n}\sum_{j=1}^{m} y_j \dfrac{\partial^2 h_j}{\partial x_i^2} - \sum_{i=1}^{n} \dfrac{\partial f_i}{\partial x_i} - \dfrac{1}{2}\sum_{i=1}^{m} h_i^2. \quad (8.20)$$

□

In the following result, we will introduce an exponential-type transformation so that the coefficients of gradient terms and function terms can become independent of time. Basic calculations of calculus imply the following result. Details can be found in Theorem 3.1 in [7].

Theorem 8.3 *Suppose $u(t,x)$ is a solution of (8.17) and $\tilde{u}(t,x) = e^{\Lambda(t,x)}u(t,x+b(t))$. Then $\tilde{u}(t,x)$ is the solution of the following Kolmogorov equation:*

$$\begin{cases} \dfrac{\partial \tilde{u}}{\partial t}(t,x) = \dfrac{1}{2}\Delta \tilde{u}(t,x) - \sum_{i=1}^{n} H_i(x)\dfrac{\partial \tilde{u}}{\partial x_i} - P(x)\tilde{u}(t,x), \\ \tilde{u}(0,x) = e^{\Lambda(0,x)}u(0,x+b(0)), \end{cases} \quad (8.21)$$

8.2 Explicit Solution of DMZ Equation for Finite-Dimensional Filters

if we can choose $H_i(x)$, $P(x)$ such that $\Lambda(t, x)$ satisfy the following equations:

$$b_i'(t) - \frac{\partial \Lambda(t,x)}{\partial x_i} + H_i(x) + \theta_i(t, x + b(t)) \equiv 0, \ 1 \leq i \leq n,$$

$$\frac{\partial \Lambda}{\partial t}(t, x) - \frac{1}{2}\sum_{i=1}^{n}(b_i'(t))^2 - \sum_{i=1}^{n}\theta_i(t, x + b(t))b_i'(t)$$

$$-\frac{1}{2}\eta(x + b(t)) + \frac{1}{2}\sum_{i=1}^{n}H_i^2(x) - \frac{1}{2}\sum_{i=1}^{n}\frac{\partial H_i}{\partial x_i} + P(x) \equiv 0.$$

(8.22)

Moreover, if u, $\frac{\partial u}{\partial x_1}, \cdots, \frac{\partial u}{\partial x_n}$ are linearly independent, then conditions (8.22) become necessary for system (8.21).

Proof Suppose $\tilde{u}(t, x) = e^{\Lambda(t,x)}u(t, x + b(t))$ and insert this equation to (8.21), we get an equation about u:

$$\frac{\partial \Lambda}{\partial t}u(t, x + b(t)) + \frac{\partial u}{\partial t}(t, x + b(t)) + \sum_{i=1}^{n} b_i' \frac{\partial u}{\partial x_i}(t, x + b(t))$$

$$-\frac{1}{2}\sum_{i=1}^{n}[\frac{\partial \Lambda}{\partial x_i}]^2 u(t, x + b(t)) - \frac{1}{2}\sum_{i=1}^{n}\frac{\partial^2 \Lambda}{\partial x_i^2}u(t, x + b(t))$$

$$-\sum_{i=1}^{n}\frac{\partial \Lambda}{\partial x_i}\frac{\partial u}{\partial x_i}(t, x + b(t)) - \frac{1}{2}\Delta u(t, x + b(t)) + \sum_{i=1}^{n}\frac{\partial \Lambda}{\partial x_i}H_i u(t, x + b(t))$$

$$+ \sum_{i=1}^{n}H_i \frac{\partial u}{\partial x_i}(t, x + b(t)) + P(x)u(t, x + b(t)) \equiv 0.$$

(8.23)

From (8.17), we know that

$$\frac{\partial u}{\partial t}(t, x + b(t)) = \frac{1}{2}\Delta u(t, x + b(t)) + \sum_{i=1}^{n}\theta_i(t, x + b(t))\frac{\partial u}{\partial x_i}(t, x + b(t))$$

$$+ \theta(t, x + b(t))u(t, x + b(t)).$$

(8.24)

By observing the coefficient $\frac{\partial u}{\partial x_i}(t, x + b(t))$ and $u(t, x + b(t))$ of (8.23), we have

$$\frac{\partial \Lambda}{\partial x_i} \equiv b_i'(t) + H_i + \theta_i(t, x + b(t)), \ 1 \leq i \leq n$$

(8.25)

and

$$\frac{\partial \Lambda}{\partial t} - \frac{1}{2}\sum_{i=1}^{n}\left(\frac{\partial \Lambda}{\partial x_i}\right)^2 - \frac{1}{2}\sum_{i=1}^{n}\frac{\partial^2 \Lambda}{\partial x_i^2} + \sum_{i=1}^{n}\frac{\partial \Lambda}{\partial x_i}H_i$$

$$+ \frac{1}{2}\sum_{i=1}^{n}\theta_i^2(t, x+b(t)) + \frac{1}{2}\sum_{i=1}^{n}\frac{\partial \theta_i}{\partial x_i}(t, x+b(t)) - \frac{1}{2}\eta(x+b(t)) + P(x) \equiv 0. \tag{8.26}$$

Equation (8.25) can be used to simplify Eq. (8.26) which together implies (8.22).

□

Now if we assume that condition (C_3) holds, then

$$\eta(x+b(t)) = \eta(x) + \sum_{i=1}^{n}B_i(t)x_i + B(t), \tag{8.27}$$

where $B_i(t) = \sum_{j=1}^{n}(\eta_{ij}+\eta_{ji})b_j(t)$ and $B(t) = \sum_{i,j=1}^{n}\eta_{ij}b_i(t)b_j(t) + \sum_{i=1}^{n}\eta_i b_i(t)$.

In order to verify the type of filtering systems that can be solved by formula (8.21), we prove that under certain mild condition, filters that can be solved by formula (8.21) must be a Yau type, which is stated in Theorem 3.2 from [7]. Furthermore, we can extend Theorem 3.2 from [7] to a more general result by dropping out the condition (C_2), which is stated in the following theorem:

Theorem 8.4 *Assume* $\eta(x) = \sum_{i,j=1}^{n}\eta_{ij}x_ix_j + \sum_{i=1}^{n}\eta_i x_i + \eta_0$ *is a quadratic polynomial in x. Suppose that $u(t,x)$ is a solution of (3.1) and $\tilde{u}(t,x) = e^{\Lambda(t,x)}u(t, x+b(t))$, where $b_i'(t), 1 \le i \le n$ are linearly independent and $\tilde{u}(t,x)$ is the solution of the following Kolmogorov equation:*

$$\begin{cases} \dfrac{\partial \tilde{u}}{\partial t}(t, x) = \dfrac{1}{2}\Delta \tilde{u}(t, x) - \sum_{i=1}^{n}H_i(x)\dfrac{\partial \tilde{u}}{\partial x_i}(t, x) - P(x)\tilde{u}(t, x), \\ \tilde{u}(0, x) = e^{\Lambda(0,x)}u(0, x+b(0)). \end{cases} \tag{8.28}$$

Furthermore, let $\Lambda(t,x) = c(t) + G(x) + \sum_{j=1}^{n}a_j(t)x_j - F(x+b(t))$ *and F is a C^∞ function. If we can choose C^∞ functions $H(x), G(x), P(x)$ such that*

$$\frac{1}{2}\sum_{i=1}^{n}H_i^2(x) - \frac{1}{2}\sum_{i=1}^{n}\frac{\partial H_i}{\partial x_i}(x) - \frac{1}{2}\eta(x) + P(x) \equiv 0, \tag{8.29}$$

then the filtering system must be a Yau-type system.

8.2 Explicit Solution of DMZ Equation for Finite-Dimensional Filters

Proof First, by definition of $\theta_i(t, x)$ and $\Lambda(t, x)$, we calculate

$$\theta_i(t, x+b(t)) = \sum_{j=1}^{m} y_j(t) \frac{\partial h_j}{\partial x_i}(x+b(t)) - f_i(x+b(t))$$

$$\frac{\partial \Lambda}{\partial x_i}(t, x) = \frac{\partial G}{\partial x_i}(x) + a_i(t) - \frac{\partial F}{\partial x_i}(x+b(t)), \quad 1 \le i \le n \quad (8.30)$$

$$\frac{\partial \Lambda}{\partial t}(t, x) = c'(t) + \sum_{j=1}^{n} a'_j(t) x_j - \sum_{j=1}^{n} \frac{\partial F}{\partial x_j}(x+b(t)) b'_j(t).$$

Then we can calculate (3.4) and (3.5) by using the above results:

$$b'_i(t) - a_i(t) + \sum_{j=1}^{m} y_j(t) \frac{\partial h_j}{\partial x_i}(x+b(t)) + H_i(x) \\ - \frac{\partial G}{\partial x_i}(x) + \frac{\partial F}{\partial x_i}(x+b(t)) - f_i(x+b(t)) \equiv 0, \quad 1 \le i \le n \quad (8.31)$$

$$c'(t) - \frac{1}{2} \sum_{i=1}^{n} (b'_i(t))^2 - \sum_{i=1}^{n} \sum_{j=1}^{m} y_j(t) \frac{\partial h_j}{\partial x_i}(x+b(t)) b'_i(t) \quad (8.32)$$

$$+ \sum_{j=1}^{n} a'_j(t) x_j + \left[\sum_{i=1}^{n} f_i(x+b(t)) b'_i(t) - \sum_{i=1}^{n} \frac{\partial F}{\partial x_i}(x+b(t)) b'_i(t) \right]$$

$$- \frac{1}{2} \eta(x+b(t)) + \frac{1}{2} \sum_{i=1}^{n} H_i^2(x) - \frac{1}{2} \sum_{i=1}^{n} \frac{\partial H_i}{\partial x_i}(x) + P(x) \equiv 0.$$

From (8.31), we have

$$f_i(x+b(t)) - \frac{\partial F}{\partial x_i}(x+b(t)) \equiv H_i(x) - \frac{\partial G}{\partial x_i}(x) + \sum_{j=1}^{m} y_j(t) \frac{\partial h_j}{\partial x_i}(x+b(t)) + b'_i(t) - a_i(t).$$

(8.33)

Putting (8.33) into (8.32), we get

$$c'(t) - \frac{1}{2}\sum_{i=1}^{n}(b_i'(t))^2 - \sum_{i=1}^{n}\sum_{j=1}^{m} y_j(t)\frac{\partial h_j}{\partial x_i}(x+b(t))b_i'(t)$$

$$+ \sum_{j=1}^{n} a_j'(t)x_j + \sum_{i=1}^{n}\left(H_i(x) - \frac{\partial G}{\partial x_i}(x)\right)$$

$$+ \sum_{j=1}^{m} y_j(t)\frac{\partial h_j}{\partial x_i}(x+b(t)) + b_i'(t) - a_i(t)\Big)b_i'(t)$$

$$- \frac{1}{2}\eta(x+b(t)) + \frac{1}{2}\sum_{i=1}^{n} H_i^2(x) - \frac{1}{2}\sum_{i=1}^{n}\frac{\partial H_i}{\partial x_i}(x) + P(x) \equiv 0.$$

(8.34)

By simplifying (8.34), we obtain

$$c'(t) - \frac{1}{2}\sum_{i=1}^{n}(b_i'(t))^2 + \sum_{i=1}^{n}(b_i'(t) - a_i(t))b_i'(t) + \sum_{j=1}^{n} a_j'(t)x_j$$

$$+ \sum_{i=1}^{n}\left(H_i(x) - \frac{\partial G}{\partial x_i}(x)\right)b_i'(t) - \frac{1}{2}\eta(x+b(t)) + \frac{1}{2}\sum_{i=1}^{n} H_i^2(x) \quad (8.35)$$

$$- \frac{1}{2}\sum_{i=1}^{n}\frac{\partial H_i}{\partial x_i}(x) + P(x) \equiv 0.$$

Then by using condition $\frac{1}{2}\sum_{i=1}^{n} H_i^2(x) - \frac{1}{2}\sum_{i=1}^{n}\frac{\partial H_i}{\partial x_i}(x) - \frac{1}{2}\eta(x) + P(x) \equiv 0$, we have

$$\sum_{j=1}^{n} a_j'(t)x_j + \sum_{i=1}^{n}\left(H_i(x) - \frac{\partial G}{\partial x_i}(x)\right)b_i'(t) + \frac{1}{2}\eta(x) - \frac{1}{2}\eta(x+b(t))$$

$$\equiv -c'(t) + \frac{1}{2}\sum_{i=1}^{n}(b_i'(t))^2 - \sum_{i=1}^{n}(b_i'(t) - a_i(t))b_i'(t).$$

(8.36)

Due to $\eta(x) = \sum_{i,j=1}^{n} \eta_{ij}x_i x_j + \sum_{i=1}^{n} \eta_i x_i + \eta_0$, then

$$\eta(x+b(t)) = \eta(x) + \sum_{i=1}^{n} B_i(t)x_i + B(t), \quad (8.37)$$

where $B_i(t) = \sum_{j=1}^{n}(\eta_{ij} + \eta_{ji})b_j(t)$ and $B(t) = \sum_{i,j=1}^{n} \eta_{ij}b_i(t)b_j(t) + \sum_{i=1}^{n} \eta_i b_i(t)$. Next we put (8.37) into (8.36) and get

8.2 Explicit Solution of DMZ Equation for Finite-Dimensional Filters 307

$$\sum_{j=1}^{n}\left(a'_j(t)-\frac{1}{2}B_j(t)\right)x_j+\sum_{i=1}^{n}\left(H_i(x)-\frac{\partial G}{\partial x_i}(x)\right)b'_i(t)$$

$$\equiv \frac{1}{2}B(t)-c'(t)+\frac{1}{2}\sum_{i=1}^{n}(b'_i(t))^2-\sum_{i=1}^{n}(b'_i(t)-a_i(t))b'_i(t). \tag{8.38}$$

Then we differentiate both sides by $\frac{\partial^2}{\partial x_k \partial x_l}$, $1 \le k, l \le n$ and get

$$\sum_{i=1}^{n}\frac{\partial^2}{\partial x_k \partial x_l}\left(H_i(x)-\frac{\partial G}{\partial x_i}(x)\right)b'_i(t) \equiv 0, \quad 1 \le k, l \le n. \tag{8.39}$$

Since $\{b'_i(t), 1 \le i \le n\}$ are linearly independent, we deduce

$$\frac{\partial^2}{\partial x_k \partial x_l}\left(H_i(x)-\frac{\partial G}{\partial x_i}(x)\right) \equiv 0, \quad 1 \le i, k, l \le n. \tag{8.40}$$

Then $H_i(x)-\frac{\partial G}{\partial x_i}(x)$, $1 \le i \le n$ are degree 1 polynomials and we can denote

$$H_i(x) = \frac{\partial G}{\partial x_i}(x)+P_1(x), \quad 1 \le i \le n. \tag{8.41}$$

Putting (8.41) into (8.31), we get

$$b'_i(t)-a_i(t)+\sum_{j=1}^{m}y_j(t)\frac{\partial h_j}{\partial x_i}(x+b(t))$$

$$+P_1(x)+\frac{\partial F}{\partial x_i}(x+b(t))-f_i(x+b(t)) \equiv 0, \quad 1 \le i \le n. \tag{8.42}$$

Then we differentiate both sides by $\frac{\partial}{\partial x_k}$, $1 \le k \le n$ and get

$$\sum_{j=1}^{m}y_j(t)\frac{\partial^2 h_j}{\partial x_k \partial x_i}(x+b(t))+const+\frac{\partial^2 F}{\partial x_k \partial x_i}(x+b(t)) \equiv \frac{\partial f_i(x+b(t))}{\partial x_k}, \quad 1 \le i, k \le n. \tag{8.43}$$

Exchange subscripts i and k, and we obtain

$$\sum_{j=1}^{m}y_j(t)\frac{\partial^2 h_j}{\partial x_i \partial x_k}(x+b(t))+const+\frac{\partial^2 F}{\partial x_i \partial x_k}(x+b(t)) \equiv \frac{\partial f_k(x+b(t))}{\partial x_i}, \quad 1 \le i, k \le n. \tag{8.44}$$

Considering h_j, F are C^∞ function, and letting (8.43) minus (8.44), we get

$$\frac{\partial f_i(x+b(t))}{\partial x_k} - \frac{\partial f_k(x+b(t))}{\partial x_i} \equiv const, \quad 1 \leq i, k \leq n, \quad (8.45)$$

i.e.,

$$\omega_{ik}(x+b(t)) \equiv const, \forall x, t, \quad 1 \leq i, k \leq n. \quad (8.46)$$

Thus, $\omega_{ik}(x) \equiv const$, $1 \leq i, k \leq n$ and Wong matrix has a constant structure. By Yau and Hu [7] Theorem 2.1, we deduce

$$f_i(x) = l_i(x) + \frac{\partial Q}{\partial x_i}(x), \quad 1 \leq i \leq n, \quad (8.47)$$

where $l_i(x)$ is a degree 1 polynomial and Q is a C^∞ function. By definition, the filtering system is a Yau type. □

Next we will give a concrete form of $\Lambda(t, x)$ and a constraint of $H_i(x)$, $G(x)$, $P(x)$ so that we can transform solving the original filtering problem to a procedure combining on-line and off-line steps.

Theorem 8.5 *Consider the filtering system (8.1) with condition* (C_1), (C_2), (C_3). *Then the solution of robust DMZ equation $u(t, x)$ is reduced to the solution $\tilde{u}(t, x)$ for the following Kolmogorov equation:*

$$\begin{cases} \frac{\partial \tilde{u}}{\partial t}(t, x) = \frac{1}{2}\Delta\tilde{u}(t, x) - \sum_{i=1}^{n} H_i(x)\frac{\partial \tilde{u}}{\partial x_i} - P(x)\tilde{u}(t, x) \\ \tilde{u}(0, x) = e^{G(x)-F(x)}\sigma_0, \end{cases} \quad (8.48)$$

where

$$\tilde{u}(t, x) = \exp\left[c(t) + G(x) + \sum_{i=1}^{n} a_i(t)x_i - F(x+b(t))\right] u(t, x+b(t)), \quad (8.49)$$

and $a_i(t), b(t), c(t)$ satisfy the following system of ODEs:

$$\begin{cases} b_i'(t) - a_i(t) - \sum_{j=1}^{n} d_{ij}b_j(t) + \sum_{j=1}^{m} c_{ji}y_j(t) = 0 \\ b_i(0) = 0 \end{cases} \quad (8.50)$$

$$\begin{cases} a_i'(t) - \frac{1}{2}\sum_{j=1}^{n}(\eta_{ij}+\eta_{ij})b_j(t) + \sum_{j=1}^{n} d_{ji}b_j'(t) = 0 \\ a_i(0) = 0 \end{cases} \quad (8.51)$$

$$\begin{cases} c'(t) - \frac{1}{2}\sum_{i=1}^{m}(b_i'(t))^2 + \sum_{i=1}^{n} a_i(t)b_i'(t) - \sum_{i=1}^{n} d_i b_i'(t) \\ + \frac{1}{2}\sum_{i,j=1}^{n} \eta_{ij}b_i(t)b_j(t) + \frac{1}{2}\sum_{i=1}^{n} \eta_i b_i = 0 \\ c(0) = 0 \end{cases} \quad (8.52)$$

8.2 Explicit Solution of DMZ Equation for Finite-Dimensional Filters

if we choose $H(x), G(x), P(x)$ satisfy

$$\frac{1}{2}\sum_{i=1}^{n} H_i^2(x) - \frac{1}{2}\sum_{i=1}^{n} \frac{\partial H_i}{\partial x_i} - \frac{1}{2}\eta(x) + P(x) \equiv 0, \tag{8.53}$$

where $H_i(x) - \frac{\partial G}{\partial x_i} = l_i(x)$.

Proof We choose $\Lambda(t, x) = c(t) + G(x) + \sum_{i=1}^{n} a_i(t)x_i - F(x + b(t))$. By using the condition $(C_1), (C_2), (C_3)$, we rewrite the first equation of (8.22) as

$$b_i'(t) - a_i(t) + l_i(x) + \sum_{j=1}^{m} c_{ji} y_j(t) - l_i(x + b(t)) \equiv 0, \ 1 \leq i \leq n, \tag{8.54}$$

which is exactly (8.50). By the similar way, we rewrite the second equation of (8.22) as

$$c'(t) - \frac{1}{2}\sum_{i=1}^{n}(b_i')^2 - \sum_{i=1}^{n}\sum_{j=1}^{m} c_{ji} y_j b_i'$$

$$- \frac{1}{2}B(t) + \sum_{j=1}^{n} a_j' x_j + \sum_{i=1}^{n} f_i(x + b(t))$$

$$- \frac{\partial F}{\partial x_i}(x + b(t))b_i' - \frac{1}{2}\sum_{i=1}^{n} B_i x_i \tag{8.55}$$

$$+ \frac{1}{2}\sum_{i=1}^{n} H_i^2 - \frac{1}{2}\sum_{i=1}^{n} \frac{\partial H_i}{\partial x_i}$$

$$- \frac{1}{2}\eta(x) + P(x) \equiv 0.$$

By inserting condition (8.53) into (8.55), we obtain

$$c'(t) - \frac{1}{2}\sum_{i=1}^{n}(b_i')^2 - \sum_{i=1}^{n}\sum_{j=1}^{m} c_{ji} y_j b_i' - \frac{1}{2}B(t)$$

$$+ \sum_{j=1}^{n} a_j' x_j + \sum_{i=1}^{n}\left(\sum_{j=1}^{n} d_{ij}(x_j + b_j) + d_i\right) b_i' \tag{8.56}$$

$$- \frac{1}{2}\sum_{i=1}^{n} B_i x_i \equiv 0.$$

By combining terms containing x_i, we obtain ODEs (8.51) and (8.52), the desired results. □

8.3 Direct Method for Yau Filtering System with Nonlinear Observations

In the previous section, we explore the direct method for time-invariant finite-dimensional Yau systems that contain linear observations. In this section, we will extend the direct method to time-invariant Yau filtering system with nonlinear observations. This type of method has two advantages: (i) the real-time computation of the solution of the DMZ equation is reduced to the computation of Kolmogorov equation. Based on Gaussian approximation of the initial condition, the Kolmogorov equation can be solved in terms of ordinary differential equations; (ii) for a given probability density function, we give a new and original approach to implement Gaussian decomposition technique which is very effective and simple especially in practice. The content of this section is mainly referred to Shi et al. [3].

We assume the filtering system (8.1) has the following three conditions:

$$(C_1) \quad f_i(x) = l_i(x) + \frac{\partial F}{\partial x_i}, \ 1 \le i \le n;$$

$$(C_2) \quad \sum_{i=1}^{m} h_i^2(x) = \sum_{i,j=1}^{n} q_{ij} x_i x_j + \sum_{i=1}^{n} q_i x_i + q_0; \tag{8.57}$$

$$(C_3) \quad \eta(x) = \sum_{i,j=1}^{n} \eta_{ij} x_i x_j + \sum_{i=1}^{n} \eta_i x_i + \eta_0,$$

where $l_i = \sum_{j=1}^{n} d_{ij} x_j + d_i$ and $d_{ij}, d_i, q_{ij} = q_{ji}, q_i, q_0, \eta_{ij}, \eta_i, \eta_0, 1 \le i, j \le n$ are constants. We remark that (C_2) means that observation terms are not necessary to be linear. However, (C_2) restricts that observation terms have linear growth.

Instead of starting at robust DMZ equation, we begin with the framework of Yau-Yau algorithm in which robust DMZ equation is reduced to a piece-wise Kolmogorov equation at each time interval, i.e., Proposition 8.1. Similarly to Lemma 8.1 in the previous section, we make some simplifications on notations.

Lemma 8.2 *For each k, $\tau_{k-1} \le t < \tau_k$, Eq. (8.4) is equivalent to the following equation:*

$$\frac{\partial \tilde{u}_k}{\partial t}(t, x) = \frac{1}{2} \Delta \tilde{u}_k(t, x) + \sum_{i=1}^{n} \theta_i(x) \cdot \frac{\partial \tilde{u}_k}{\partial x_i} + \theta(x) \tilde{u}_k(t, x), \tag{8.58}$$

where

$$\theta_i(x) = -f_i(x),$$

$$\theta(x) = \frac{1}{2} \left(\sum_{i=1}^{n} \theta_i^2(x) + \sum_{i=1}^{n} \frac{\partial \theta_i}{\partial x_i} - \eta(x) \right). \tag{8.59}$$

8.3 Direct Method for Yau Filtering System with Nonlinear Observations

Proof By direct calculations,

$$\theta(x) = \frac{1}{2}(\sum_{i=1}^{n}\theta_i^2(x) + \sum_{i=1}^{n}\frac{\partial \theta_i}{\partial x_i} - \eta(x))$$

$$= \sum_{1}^{2}\frac{\partial \theta_i}{\partial x_i} - \frac{1}{2}\eta + \sum_{i=1}^{n}\frac{\partial \theta_i}{\partial x_i} + \frac{1}{2}\frac{\partial f_i}{\partial x_i} \qquad (8.60)$$

$$= -(\sum_{i=1}^{n}\frac{\partial f_i}{\partial x_i} + \frac{1}{2}h_i^2).$$

It is easy to see that Eq. (8.58) is identical to Eq. (8.4). □

In order to solve the Kolmogorov equation (8.58) in terms of ordinary differential equations, we first introduce a new transformation in the following theorem.

Theorem 8.6 *For each k, $\tau_{k-1} \leq t < \tau_k$, suppose $\tilde{u}_k$ satisfies $\hat{u}_k(t,x) = e^{\Lambda(x)}\tilde{u}_k(t,x)$. Then $\hat{u}_k(t,x)$ is the solution of the following Kolmogorov equation:*

$$\begin{cases} \dfrac{\partial \hat{u}_k}{\partial t}(t,x) = \dfrac{1}{2}\Delta \hat{u}_k(t,x) - \sum_{i=1}^{n} H_i(x)\dfrac{\partial \hat{u}_k}{\partial x_i} - P(x)\hat{u}(t,x)_k \\ \hat{u}(\tau_{k-1},x) = e^{\Lambda(x)}\tilde{u}_k(\tau_{k-1},x), \end{cases} \qquad (8.61)$$

if we can choose H_i, $P(x)$, $\Lambda(x)$ satisfying the following condition:

$$\begin{aligned} -\frac{\partial \Lambda}{\partial x_i} + H_i(x) + \theta_i(x) &\equiv 0, \ 1 \leq i \leq n, \\ -\frac{1}{2}\eta(x) + \frac{1}{2}\sum_{i=1}^{n}H_i^2 - \frac{1}{2}\sum_{i=1}^{n}\frac{\partial H_i}{\partial x_i} + P(x) &\equiv 0. \end{aligned} \qquad (8.62)$$

Proof Direct computation yields that

$$\frac{\partial \hat{u}_k}{\partial t} = e^{\Lambda(x)}\frac{\partial \tilde{u}_k}{\partial t}$$

$$\frac{\partial \hat{u}_k}{\partial x_i} = e^{\Lambda(x)}(\frac{\partial \Lambda(x)}{\partial x_i}\tilde{u}_k + \frac{\partial \tilde{u}_k}{\partial x_i}) \qquad (8.63)$$

$$\frac{\partial^2 \hat{u}_k}{\partial x_i^2} = e^{\Lambda(x)}(\frac{\partial^2 \Lambda(x)}{\partial x_i^2} + (\frac{\partial \Lambda(x)}{\partial x_i})^2)\tilde{u}_k + 2\frac{\partial \Lambda(x)}{\partial x_i}\frac{\partial \tilde{u}_k}{\partial x_i} + \frac{\partial^2 \tilde{u}_k}{\partial x_i^2}.$$

If we substitute the above relations to Eq. (8.61) and compare the coefficients on both sides of $\tilde{u}_k$ and $\frac{\partial \tilde{u}_k}{\partial x_i}$, it can be obtained the following relation:

$$\theta_i(x) = -\frac{\partial \Lambda}{\partial x_i} + H_i(x)$$

$$\theta(x) = \frac{1}{2}\Delta\Lambda(x) + \frac{1}{2}|\nabla\Lambda(x)|^2 - H(x) \cdot \nabla\Lambda(x) - P(x).$$
(8.64)

Combining with the relation $\theta(x) = \frac{1}{2}(\theta_i^2(x) + \sum_{i=1}^{n} \frac{\partial \theta_i(x)}{\partial x_i} - \eta(x))$, the desired results can be obtained. □

Noting the special structure of the drift term, we can select a special Λ in the previous theorem, i.e., let $\Lambda(x) = G(x) - F(x)$. Then we obtain the following results.

Theorem 8.7 *For each k, $\tau_{k-1} \leq t < \tau_k$, suppose $\tilde{u}_k$ satisfies $\hat{u}_k(t, x) = e^{G(x)-F(x)}\tilde{u}_k(t, x)$. Then $\hat{u}_k(t, x)$ is the solution of following Kolmogorov equation:*

$$\begin{cases} \frac{\partial \hat{u}_k}{\partial t}(t, x) = \frac{1}{2}\Delta\hat{u}_k(t, x) - \sum_{i=1}^{n} H_i(x)\frac{\partial \hat{u}_k}{\partial x_i} - P(x)\hat{u}_k(t, x) \\ \hat{u}_k(\tau_{k-1}, x) = e^{G(x)-F(x)}\tilde{u}_k(\tau_{k-1}, x), \end{cases}$$
(8.65)

if we can choose $H_i(x)$, $P(x)$, $\Lambda(x)$ satisfying the following condition:

$$\begin{aligned} H_i - \frac{\partial G}{\partial x_i} &= l_i, \\ -\frac{1}{2}\eta(x) + \frac{1}{2}\sum_{i=1}^{n} H_i^2 - \frac{1}{2}\sum_{i=1}^{n}\frac{\partial H_i}{\partial x_i} + P(x) &\equiv 0. \end{aligned}$$
(8.66)

Proof Here we use $\Lambda(x) = G(x) - F(x)$ where $F(x)$ is the term of Theorem 8.2. Then

$$\frac{\partial \Lambda}{\partial x_i}(x) = \frac{\partial G}{\partial x_i}(x) - \frac{\partial F}{\partial x_i}(x).$$
(8.67)

Putting (8.67) into (8.62) and note that $\theta_i(x) = -f_i(x)$, we have $H_i - \frac{\partial G}{\partial x_i} = l_i$. □

In the following, we choose $H(x)$, $G(x)$, $P(x)$ which satisfy the conditions in Theorem 8.7 such that the coefficients of $\hat{u}_k(t, x)$, $\frac{\partial \hat{u}_k}{\partial x_i}$ are degree 1 polynomial and degree 2 polynomial, respectively. One of the convenient selections is

$$\begin{aligned} G(x) &\equiv 0 \\ H_i &= l_i, \ 1 \leq i \leq n \\ P(x) &= \frac{1}{2}\eta(x) - \frac{1}{2}\sum_{i=1}^{n} l_i^2 + \frac{1}{2}\sum_{i=1}^{n}\frac{\partial l_i}{\partial x_i}(x). \end{aligned}$$
(8.68)

8.3 Direct Method for Yau Filtering System with Nonlinear Observations

Observe if initial distribution of partial differential equation (8.65) is Gaussian; by using the method proposed by Yau and Lai [8], Kolmogorov equation can be solved by a system of ordinary differential equations.

Theorem 8.8 *Consider the following Kolmogorov equation with Gaussian initial condition:*

$$\begin{cases} \dfrac{\partial \hat{u}}{\partial t}(t,x) = \dfrac{1}{2}\Delta \hat{u}(t,x) - \sum_{i=1}^{n} l_i(x)\dfrac{\partial \hat{u}}{\partial x_i} + q(x)\hat{u}(t,x) \\ \hat{u}(t_0,x) = e^{x^T A(t_0)x + B^T(t_0)x + C(t_0)}, \end{cases} \quad (8.69)$$

where $l_i = \sum_{j=1}^{n} d_{ij}x_j + d_i$ *and* $q(x) = x^T Q x + p^T x + r$. $A(t_0)$ *is a symmetric matrix,* $B^T(t_0) = (B_1(t_0), \cdots, B_n(t_0))$, *and* $C(t_0)$ *is a scalar.*

Then the solution of Kolmogorov equation is of the following form:

$$\hat{u}(t,x) = e^{x^T A(t)x + B(t)^T x + C(t)}, \quad (8.70)$$

where $A(t), B(t), C(t)$ *satisfy the following ordinary differential equations:*

$$\begin{aligned} \dfrac{dA}{dt} &= 2A^2 - [AD + D^T A] + Q, \\ \dfrac{dB}{dt} &= 2B^T A - B^T D - 2d^T A + p^T, \\ \dfrac{dC}{dt} &= tr(A) + \dfrac{1}{2}B^T B - d^T B + r, \end{aligned} \quad (8.71)$$

where $D = (d_{ij}) \in \mathbb{R}^{n \times n}$ *and* $d^T = (d_1, \cdots, d_n) \in \mathbb{R}^n$.

Proof Suppose $\hat{u}(t,x) = e^{x^T A(t)x + B(t)^T x + C(t)}$ and by basic calculus,

$$\dfrac{\partial \hat{u}}{\partial t} = \left(x^T \dfrac{dA}{dt}x + \dfrac{dB^T}{dt}x + \dfrac{dC}{dt} \right)\hat{u} \quad (8.72)$$

and

$$\dfrac{1}{2}\Delta \hat{u}(t,x) - \sum_{i=1}^{n} l_i(x)\dfrac{\partial \hat{u}}{\partial x_i} + q(x)\hat{u}(t,x)$$

$$= [x^T(\dfrac{1}{2}A^T A + \dfrac{1}{2}AA^T + A^2)x + B^T(A + A^T)x + tr(A) + \dfrac{1}{2}B^T B]\hat{u} \quad (8.73)$$

$$- [x^T(A^T + A)Dx + (B^T D + d^T A + d^T A^T)x + d^T B]\hat{u}.$$

$$+ (x^T Q x + p^T x + r)\hat{u}.$$

Comparing the above two equations, we obtain a system of ordinary differential equations satisfied by A, B, C. □

Next we introduce idea of Gaussian approximation proposed in [3]. Based on Yau and Lai's idea in dealing with Kolmogorov equation, the initial state has to be Gaussian. However, in solving Eq. (8.65), we cannot guarantee the initial state in each interval $[\tau_{k-1}, \tau_k]$ is Gaussian. In order to solve this problem, Shi et al. proposed to approximate initial state $\hat{u}(\tau_{k-1}, x)$ in each interval $[\tau_{k-1}, \tau_k]$ with sum of Gaussian distribution functions.

In the following algorithm, we introduce the details of Gaussian approximation method.

Algorithm 3 Gaussian approximation

Step 1. Let $f(x) = \phi(x)$ and the threshold $E = \alpha \max \phi(x)$, where α is a given small number.
Step 2. Fitting the peaks of $f(x)$ which are larger than E with Gaussian distributions. Specifically, for a peak $P_i(x_i, y_i)$ of $f(x)$ with $y_i \geq E$, we use the function $g_i(x) = y_i \exp(-\frac{(x-x_i)^2}{2\sigma_i^2})$ to fit $P(x, y)$ with points in a neighborhood of $P_i(x_i, y_i)$ where no other peaks exists, and the best fitting parameter σ_i is obtained by fitting. Suppose the sum of Gaussian distributions $g_i(x)$ in this step is $g(x)$.
Step 3. Let $f_1(x) = f(x) - g(x)$. If $f_1(x)$ has no peaks whose values are larger than E, then go to step 4. Otherwise, let $f(x) = f_1(x)$ and go to step 2.
Step 4. Let $f_2(x) = -f_1(x)$. If $f_2(x)$ has no peaks that are larger than E, then done. Otherwise, let $f(x) = f_2(x)$ and go to step 2.

In [3], some numerical examples are shown and we can find the direct method performs better than EKF. Therefore, Shi et al. extended the direct method proposed by Yau and Hu [7] to system with specific nonlinear observation terms successfully.

8.4 Nonlinear Filtering and Time-Varying Schrödinger Equation I

In this section, we mainly consider time-invariant Yau systems with a class of nonlinear observation and give explicit solution under the Yau-Yau algorithm framework. Details can be found in [9]. We consider the filtering system (8.1) with drift term:

$$f(x) = Lx + l + \nabla \phi, \qquad (8.74)$$

where $L = (l_{ij}), 1 \leq i, j \leq n, l^T = (l_1, \cdots, l_n)$ and ϕ is a C^∞ function on $\mathbb{R}^n$. Recall that L can be uniquely decomposed as $L = L_1 + L_2$, where $L_1^T = L_1$ and $L_2^T = -L_2$. Observe that $L_1 x = \nabla \phi_1 = \nabla(\frac{1}{2} x^T L_1 x)$. It follows that $f(x) = L_2 x + l + \nabla \tilde{\phi}$, where $\tilde{\phi} = \phi + \phi_1$. Hence, without loss of generality, we assume that in (8.74) $L^T = -L$. Let

8.4 Nonlinear Filtering and Time-Varying Schrödinger Equation I

$$q(x) := \Delta\phi(x) + |\nabla\phi|^2 + 2(Lx+l)\cdot\nabla\phi + \sum_{i=1}^{m} h_i^2(x) + 2\mathrm{tr}L. \tag{8.75}$$

In order to solve the filtering problem with nonlinear observation, it suffices to solve the following Kolmogorov equation in real time, for $\tau_{i-1} \leq t \leq \tau_i$:

$$\begin{cases} \dfrac{\partial \tilde{u}}{\partial t}(t,x) = \dfrac{1}{2}\Delta\tilde{u}(t,x) - \sum_{j=1}^{n} f_j(x)\dfrac{\partial \tilde{u}}{\partial x_j}(t,x) \\ \qquad - \left(\sum_{j=1}^{n}\dfrac{\partial f_j}{\partial x_j}(x) + \dfrac{1}{2}\sum_{j=1}^{m}h_j^2(x)\right)\tilde{u}(t,x) \\ \tilde{u}(\tau_{i-1}, x) = \sigma_i. \end{cases} \tag{8.76}$$

In order to simplify the coefficients of $\frac{\partial \tilde{u}}{\partial x_j}(t,x)$, let $\tilde{u}(t,x) = e^{\phi(x)}\tilde{v}(t,x)$. Then we calculate

$$\begin{aligned}\dfrac{\partial \tilde{u}}{\partial t} &= e^{\phi(x)}\dfrac{\partial \tilde{v}}{\partial t} \\ \nabla\tilde{u} &= (\nabla\phi)e^{\phi(x)}\tilde{v} + e^{\phi(x)}\nabla\tilde{v} \\ \Delta\tilde{u} &= (\Delta\phi)e^{\phi(x)}\tilde{v} + |\nabla\phi|^2 e^{\phi(x)}\tilde{v} \\ &\quad + 2e^{\phi(x)}(\nabla\phi\cdot\nabla\tilde{v}) + e^{\phi(x)}\Delta\tilde{v}. \end{aligned} \tag{8.77}$$

Put Eqs. (8.77) to (8.76), and we get the following equation about $\tilde{v}$, for $\tau_{i-1} \leq t \leq \tau_i$:

$$\begin{cases} \dfrac{\partial \tilde{v}}{\partial t}(t,x) = \dfrac{1}{2}\Delta\tilde{v}(t,x) - (Lx+l)\cdot\nabla\tilde{v}(t,x) \\ \qquad - \left(\dfrac{1}{2}\Delta\phi(x) + \dfrac{1}{2}|\nabla\phi|^2 + (Lx+l)\cdot\nabla\phi + \dfrac{1}{2}\sum_{i=1}^{m}h_i^2(x) + \mathrm{tr}L\right)\tilde{v}(t,x) \\ \tilde{v}(\tau_{i-1}, x) = \sigma_i e^{-\phi(x)}. \end{cases}$$
$$\tag{8.78}$$

In the following, in order to eliminate the gradient term of $\tilde{v}$, i.e., $\nabla\tilde{v}(t,x)$, we need to make a translation in terms of variable x:

$$\tilde{v}(t,x) = v(t, B(t)x + b(t)) = v(t, \tilde{x}), \tag{8.79}$$

where $\tilde{x} = B(t)x + b(t)$ and $B(t) = (b_{ij}(t))$, $1 \leq i,j \leq n$ and $b^T(t) = (b_1(t), \cdots, b_n(t))$ such that

$$\frac{dB(t)}{dt} = -B(t)L \quad \text{and} \quad \frac{db(t)}{dt} = -B(t)l. \tag{8.80}$$

Then

$$B(t) = e^{-Lt} \quad \text{and} \quad b(t) = -\int_0^t e^{-Ls} l\, ds \tag{8.81}$$

and $B(t)$ is an orthogonal matrix since $BB^T = e^{-Lt}e^{-L^T t} = e^{-Lt}e^{Lt} = I$. Then we calculate

$$\frac{\partial \tilde{v}}{\partial x_i} = \sum_{j=1}^n \frac{\partial v}{\partial \tilde{x}_j}(t, B(t)x + b(t))b_{ji}(t)$$

$$\Delta_x \tilde{v}(t, x) = \Delta_{\tilde{x}} v(t, B(t)x + b(t))$$

$$\frac{\partial \tilde{v}}{\partial t}(t, x) = \frac{\partial v}{\partial t}(t, B(t)x + b(t)) \tag{8.82}$$

$$+ \sum_{i,j=1}^n \frac{\partial v}{\partial x_i}(t, B(t)x + b(t)) \frac{db_{ij}}{dt} x_j$$

$$+ \sum_{i=1}^n \frac{\partial v}{\partial x_i}(t, B(t)x + b(t)) \frac{db_i}{dt}.$$

Putting Eq. (8.82) in (8.78) and using (8.80), we get the following equation about v, for $\tau_{i-1} \leq t \leq \tau_i$:

$$\begin{cases} \dfrac{\partial v}{\partial t}(t, B(t)x + b(t)) = \dfrac{1}{2}\Delta_{\tilde{x}} v(t, B(t)x + b(t)) \\ \qquad - \left(\dfrac{1}{2}\Delta\phi(x) + \dfrac{1}{2}|\nabla\phi|^2 + (Lx + l) \cdot \nabla\phi \right. \\ \qquad \left. + \dfrac{1}{2}\sum_{i=1}^m h_i^2(x) + \mathrm{tr}L \right) v(t, B(t)x + b(t)) \\ v(\tau_{i-1}, x) = \sigma_i e^{-\phi(x)}. \end{cases} \tag{8.83}$$

We summarize the previous results in the following theorem:

Theorem 8.9 *In order to solve the nonlinear filtering problem with nonlinear observations, it suffices to solve the Schrödinger equation (8.83), which is equivalent to the following equation:*

8.4 Nonlinear Filtering and Time-Varying Schrödinger Equation I

$$\begin{cases} \dfrac{\partial v}{\partial t}(t,\tilde{x}) = \dfrac{1}{2}\Delta_{\tilde{x}} v(t,\tilde{x}) - \dfrac{1}{2}q(B^{-1}(t)\tilde{x} - B^{-1}(t)b(t))v(t,\tilde{x}) \\ v(\tau_{i-1},\tilde{x}) = \sigma_i(\tilde{x})e^{-\phi(\tilde{x})}. \end{cases} \quad (8.84)$$

In the following, we need to calculate the explicit solution of Schrödinger equation (8.84). First, we assume $q(x)$ in (8.75) is a quadratic polynomial, i.e.,

$$q(x) = x^T Q x + P^T x + r, \quad (8.85)$$

where $Q = Q^T = (q_{ij})$, $1 \le i, j \le n$, $P^T = (p_1, \cdots, p_n)$ and r is a scalar. Observe here that $h_i(x)$ may not be linear (i.e., degree 1 polynomial). Since $q(x)$ is quadratic, h_i, $1 \le i \le m$ are of linear growth.

Next we introduce the fundamental solution of parabolic PDE.

Definition 8.2 $K(t, x, y)$ is said to be the fundamental solution of the parabolic equation:

$$\begin{cases} \dfrac{\partial u}{\partial t}(t, x) = L_x u(t, x), \ 0 \le t < \infty, \ x \in \mathbb{R}^n, \\ u(0, x) = \phi(x), \end{cases} \quad (8.86)$$

if $\dfrac{\partial K}{\partial t}(t, x, y) = L_x K(t, x, y)$ and $\lim_{t \to 0} \int_{\mathbb{R}^n} K(t, x, y)\phi(y)dy = \phi(x)$.

In view of the above definition, solution of (8.86) can be written as

$$u(t, x) = \int_{\mathbb{R}^n} K(t, x, y)\phi(y)dy. \quad (8.87)$$

We are now going to solve (8.84). For simplicity, we proceed with (8.84) for $0 \le t \le \tau$.

Theorem 8.10 *Let $K(t, \tilde{x}, \tilde{y})$ be the fundamental solution of*

$$\begin{cases} \dfrac{\partial v}{\partial t}(t,\tilde{x}) = \dfrac{1}{2}\Delta v(t,\tilde{x}) - \dfrac{1}{2}q(B^{-1}(t)\tilde{x} - B^{-1}(t)b(t))v(t,\tilde{x}) \\ v(0,\tilde{x}) = \sigma_1(\tilde{x})e^{-\phi(\tilde{x})}, \end{cases} \quad (8.88)$$

where

$$q(B^{-1}(t)\tilde{x} - B^{-1}(t)b(t)) = \tilde{x}^T B(t) Q B(t)^T \tilde{x}$$
$$- [2b(t)^T B(t)^T Q B(t)^T - P^T B(t)^T] x \tilde{x}$$
$$+ b(t)^T B(t) Q B(t)^T b(t) - P^T B(t)^T b(t) + r. \quad (8.89)$$

Assume that the fundamental solution $K(t, \tilde{x}, \tilde{y})$ *is written as*

$$K(t, \tilde{x}, \tilde{y}) = (2\pi t)^{-n/2} \exp\left\{\tilde{x}^T \tilde{A}(t)\tilde{x} + \tilde{x}^T \tilde{B}(t)\tilde{y}\right.$$

$$\left. + \tilde{y}^T \tilde{C}(t)\tilde{y} + \tilde{D}(t)^T \tilde{x} + \tilde{E}(t)^T \tilde{y} + s(t)\right\},$$
(8.90)

where $\tilde{A}(t) = \tilde{A}^T(t) = (\tilde{a}_{ij}(t))$, $\tilde{b}(t) = (\tilde{b}_{ij}(t))$, $\tilde{C}(t) = \tilde{C}^T(t) = (\tilde{c}_{ij}(t))$, $1 \leq i, j \leq n$, $\tilde{D}^T(t) = (d_1(t), \cdots, d_n(t))$, $\tilde{E}^T(t) = (e_1(t), \cdots, e_n(t))$.

Then $\tilde{A}(t), \tilde{B}(t), \tilde{C}(t), \tilde{D}(t), \tilde{E}(t)$ *should satisfy the following ODEs:*

$$\frac{d\tilde{A}(t)}{dt} = 2\tilde{A}(t)^2 - \frac{1}{2}B(t)QB(t)^T \quad (8.91)$$

$$\frac{d\tilde{B}(t)}{dt} = 2\tilde{A}(t)\tilde{b}(t) \quad (8.92)$$

$$\frac{d\tilde{C}(t)}{dt} = \frac{1}{2}\tilde{B}(t)^T \tilde{B}(t) \quad (8.93)$$

$$\frac{d\tilde{D}(t)}{dt} = 2\tilde{A}(t)\tilde{D}(t) + \tilde{B}(t)Q\tilde{B}(t)^T - \frac{1}{2}\tilde{B}(t)P \quad (8.94)$$

$$\frac{d\tilde{E}(t)}{dt} = \tilde{B}(t)^T \tilde{D}(t) \quad (8.95)$$

$$\frac{ds(t)}{dt} = \frac{1}{2}\tilde{D}(t)^T \tilde{D}(t) + tr\tilde{A}(t)$$
$$- \frac{1}{2}[b^T(t)B(t)QB^T(t)b(t) - P^T B^T(t)b(t) + r] + \frac{n}{2t}. \quad (8.96)$$

Proof

$$\frac{\partial K}{\partial t}(t, \tilde{x}, \tilde{y}) = \left[\tilde{x}^T \frac{d\tilde{A}}{dt}\tilde{x} + \tilde{x}^T \frac{d\tilde{B}}{dt}\tilde{y}\right.$$

$$\left. + \tilde{y}^T \frac{d\tilde{C}}{dt}\tilde{y} + \frac{d\tilde{D}^T}{dt}\tilde{x} + \frac{d\tilde{E}^T}{dt}\tilde{y} - \frac{ds}{dt} - \frac{n}{2t}\right] K(t, \tilde{x}, \tilde{y})$$
(8.97)

$$\nabla_{\tilde{x}} K(t, \tilde{x}, \tilde{y}) = [(\tilde{A} + \tilde{A}^T)\tilde{x} + \tilde{B}\tilde{y} + \tilde{D}]K(t, \tilde{x}, \tilde{y}) \quad (8.98)$$

8.4 Nonlinear Filtering and Time-Varying Schrödinger Equation I

$$\frac{1}{2}\Delta_{\tilde{x}} K(t,\tilde{x},\tilde{y}) = [\frac{1}{2}\tilde{x}^T(\tilde{A}+\tilde{A}^T)^2\tilde{x} + \tilde{x}^T(\tilde{A}+\tilde{A}^T)\tilde{B}\tilde{y}$$
$$+ \frac{1}{2}\tilde{y}\tilde{B}^T\tilde{B}\tilde{y} + \tilde{x}^T(\tilde{A}+\tilde{A}^T)\tilde{D} + \tilde{y}^T\tilde{R}^T\tilde{D} + \frac{1}{2}\tilde{D}^T\tilde{D} \quad (8.99)$$
$$+ tr(\tilde{A})]K(t,\tilde{x},\tilde{y}).$$

For $K(t,\tilde{x},\tilde{y})$ to satisfy (8.88), it is easy to see that we need (8.91)–(8.96) by putting (8.97) and (8.99) in (8.88). □

Proposition 8.2 *Suppose that*

$$\tilde{A}(t) = \sum_{n=-1}^{\infty} \tilde{A}_n t^n, \ \tilde{B}(t) = \sum_{n=-1}^{\infty} \tilde{B}_n t^n, \ \tilde{C}(t) = \sum_{n=-1}^{\infty} \tilde{C}_n t^n$$

$$\tilde{D}(t) = \sum_{n=-1}^{\infty} \tilde{D}_n t^n, \ \tilde{E}(t) = \sum_{n=-1}^{\infty} \tilde{E}_n t^n, \ s(t) = \sum_{n=-1}^{\infty} s_n t^n \quad (8.100)$$

$$b(t) = \sum_{n=0}^{\infty} b^n t^n, \ B(t) = \sum_{n=0}^{\infty} B_n t^n.$$

Then the following holds:

(1) Equation (8.91) is equivalent to

$$-\tilde{A}_{-1} = 2\tilde{A}_{-1}^2$$
$$0 = 2(\tilde{A}_{-1}\tilde{A}_0 + \tilde{A}_0\tilde{A}_{-1})$$
$$n\tilde{A}_n = 2(\tilde{A}_{-1}\tilde{A}_n + \tilde{A}_0\tilde{A}_{n-1} + \cdots + \tilde{A}_n\tilde{A}_{-1}) \quad (8.101)$$
$$- \frac{1}{2}(\tilde{B}_0 Q \tilde{B}_{n-1}^T + \cdots + \tilde{B}_{n-1} Q B_0^T).$$

(2) Equation (8.92) is equivalent to

$$-\tilde{B}_{-1} = 2\tilde{A}_{-1}\tilde{B}_{-1}$$
$$0 = 2(\tilde{A}_{-1}\tilde{B}_0 + \tilde{A}_0\tilde{B}_{-1}) \quad (8.102)$$
$$n\tilde{B}_n = 2(\tilde{A}_{-1}\tilde{B}_n + \cdots + \tilde{A}_n\tilde{B}_{-1}).$$

(3) Equation (8.93) is equivalent to

$$-\tilde{C}_{-1} = \frac{1}{2}\tilde{B}_{-1}^T\tilde{B}_{-1}$$
$$0 = (\frac{1}{2}\tilde{B}_{-1}^T\tilde{B}_0 + \tilde{B}_0^T\tilde{B}_{-1}) \quad (8.103)$$
$$n\tilde{C}_n = \frac{1}{2}(\tilde{B}_{-1}^T\tilde{B}_n + \cdots + \tilde{B}_n^T\tilde{B}_{-1}).$$

(4) Equation (8.94) is equivalent to

$$-\tilde{D}_{-1} = 2\tilde{A}_{-1}\tilde{D}_{-1}$$
$$0 = 2(\tilde{A}_{-1}\tilde{D}_0 + \tilde{A}_0\tilde{D}_{-1})$$
$$n\tilde{D}_n = 2(\tilde{A}_{-1}\tilde{D}_n + \cdots + \tilde{A}_n\tilde{D}_{-1})$$
$$+ (B_0 Q B_{n-1}^T + B_1 Q B_{n-2}^T + \cdots + B_{n-1} Q B_0^T)b^0 \quad (8.104)$$
$$\cdots$$
$$+ B_0 Q B_0^T b^{n-1}$$
$$- \frac{1}{2} B_{n-1} P.$$

(5) Equation (8.95) is equivalent to

$$-\tilde{E}_{-1} = 2\tilde{B}_{-1}^T\tilde{D}_{-1}$$
$$0 = 2(\tilde{B}_{-1}^T\tilde{D}_0 + \tilde{B}_0^T\tilde{D}_{-1}) \quad (8.105)$$
$$n\tilde{E}_n = \tilde{B}_{-1}^T\tilde{D}_n + \cdots + \tilde{B}_n^T\tilde{D}_{-1}.$$

(6) Equation (8.96) is equivalent to

$$s_{-1} = -\frac{1}{2}\tilde{D}_{-1}^T\tilde{D}_{-1}$$
$$0 = \frac{1}{2}(\tilde{D}_{-1}^T\tilde{D}_0 + \tilde{D}_0\tilde{D}_{-1}) + tr\tilde{A}_{-1} + \frac{n}{2}$$
$$s_1 = \frac{1}{2}(\tilde{D}_{-1}^T\tilde{D}_1 + \tilde{D}_0^T\tilde{D}_0 + \tilde{D}_1^T\tilde{D}_{-1}) + tr\tilde{A}_0 \quad (8.106)$$
$$- \frac{1}{2}(b^{0T} B_0 Q B_0 b^0 - P^T B_0^T b^0 + r)$$
$$\cdots.$$

Proof Direct computation can derive the results and details can be found in [9]. □

8.4 Nonlinear Filtering and Time-Varying Schrödinger Equation I

In the following, we give construction of fundamental solution by power series method and prove the existence.

Theorem 8.11 *The fundamental solution* $K(t, \tilde{x}, \tilde{y})$ *of*

$$\begin{cases} \frac{\partial v}{\partial t}(t, \tilde{x}) = \frac{1}{2}\Delta v(t, \tilde{x}) - \frac{1}{2}q(B^{-1}(t)\tilde{x} - B^{-1}(t)b(t))v(t, \tilde{x}) \\ v(0, \tilde{x}) = \sigma_1(\tilde{x})e^{-\phi(\tilde{x})}, \end{cases} \quad (8.107)$$

where

$$q(B^{-1}(t)\tilde{x} - B^{-1}(t)b(t)) = \tilde{x}^T B(t) Q B(t)^T \tilde{x}$$
$$- [2b(t)^T B(t)^T Q B(t)^T - P^T B(t)^T] x \tilde{x}$$
$$+ b(t)^T B(t) Q B(t)^T b(t) - P^T B(t)^T b(t) + r$$
$$(8.108)$$

exists and is of the following form:

$$K(t, \tilde{x}, \tilde{y}) = (2\pi t)^{-n/2} \exp\left\{-\frac{|\tilde{x} - \tilde{y}|^2}{2t} + \tilde{x}^T \tilde{A}(t)\tilde{x} + \tilde{x}^T \tilde{B}(t)\tilde{y} \right.$$
$$\left. + \tilde{y}^T \tilde{C}(t)\tilde{y} + \tilde{D}(t)^T \tilde{x} + \tilde{E}(t)^T \tilde{y} + s(t)\right\}, \quad (8.109)$$

where $\tilde{A}(t) = \sum_{n=1}^{\infty} \tilde{A}_n t^n$, $\tilde{B}(t) = \sum_{n=1}^{\infty} \tilde{B}_n t^n$, $\tilde{C}(t) = \sum_{n=1}^{\infty} \tilde{C}_n t^n$, $\tilde{D}(t) = \sum_{n=1}^{\infty} \tilde{D}_n t^n$, $\tilde{E}(t) = \sum_{n=1}^{\infty} \tilde{E}_n t^n$, $s(t) = \sum_{n=1}^{\infty} s_n t^n$, $b(t) = \sum_{n=0}^{\infty} b^n t^n$, $B(t) = \sum_{n=0}^{\infty} B_n t^n$.

Moreover, $\tilde{A}_n, \tilde{B}_n, \tilde{C}_n, \tilde{D}_n, \tilde{E}_n$ *and* s_n *can be computed by the following formulas:*

$$\tilde{A}_1 = -\frac{1}{6} B_0 Q_0^T$$
$$\tilde{A}_2 = -\frac{1}{8}(B_0 Q B_1^T + B_1 Q B_0^T)$$
$$\tilde{A}_n = \frac{2}{n+2}(\tilde{A}_1 \tilde{A}_{n-2} + \cdots + \tilde{A}_{n-2}\tilde{A}_1) - \frac{1}{2(n+2)} \quad (8.110)$$
$$\times (B_0 Q B_{n-1}^T + \cdots + B_{n-1} Q B_0^T)$$

$$\tilde{B}_1 = \tilde{A}_1$$
$$\tilde{B}_2 = \frac{2}{3}\tilde{A}_2 \quad (8.111)$$
$$\tilde{B}_n = \frac{2}{n+1}(\tilde{A}_1 \tilde{B}_{n-2} + \cdots + \tilde{A}_{n-2}\tilde{B}_1 + \tilde{A}_n)$$

$$\tilde{C}_1 = \frac{1}{2}(\tilde{B}_1 + \tilde{B}_1^T)$$
$$\tilde{C}_2 = \frac{1}{4}(\tilde{B}_2 + \tilde{B}_2^T) \qquad (8.112)$$
$$\tilde{C}_n = \frac{1}{2n}(\tilde{B}_n + \tilde{B}_1^T \tilde{B}_{n-2} + \cdots + \tilde{B}_{n-2}^T \tilde{B}_1 + \tilde{B}_n^T)$$

$$\tilde{D}_1 = \frac{1}{2} B_0 Q B_0^T b^0 - \frac{1}{4} B_0 P$$
$$\tilde{D}_2 = \frac{1}{3}(B_0 Q B_1^T + B_1 Q B_0^T) b^0 + \frac{1}{3} B_0 Q B_0^T b^1 - \frac{1}{6} B_1 P$$
$$\tilde{D}_n = \frac{2}{n+1}(\tilde{A}_1 \tilde{D}_{n-2} + \cdots + \tilde{A}_{n-2} \tilde{D}_1)$$
$$\quad + \frac{1}{n+1}(B_0 Q B_{n-1}^T + \cdots + B_{n-1} Q B_0^T) b^0 \qquad (8.113)$$
$$\quad + \cdots$$
$$\quad + \frac{1}{n+1} B_0 Q B_0^T b^{n-1}$$
$$\quad - \frac{1}{2(n+1)} B_{n-1} P$$

$$\tilde{E}_1 = \tilde{D}_1$$
$$\tilde{E}_2 = \frac{1}{2} \tilde{D}_2 \qquad (8.114)$$
$$\tilde{E}_n = \frac{1}{n}(\tilde{D}_n + \tilde{B}_1^T \tilde{D}_{n-2} + \cdots + \tilde{B}_{n-2}^T \tilde{D}_1)$$

$$s_1 = -\frac{1}{2}(b^{0T} B_0 Q B_0^T b^0 - P^T B_0^T b^0 + r)$$
$$s_2 = \frac{1}{2} tr \tilde{A}_1 - \frac{1}{4}[b^{0T} B_0 Q (B_0^T b^1 + B_1^T b^0) + (b^{0T} B_1 + b^{1T} B_0) Q B_0^T b^0]$$
$$\quad + \frac{1}{4} P^T (B_0^T b^1 + B_1^T b^0)$$
$$s_n = \frac{1}{2n}(\tilde{D}_1^T \tilde{D}_{n-2} + \cdots + \tilde{D}_{n-2}^T \tilde{D}_1) + \frac{1}{n} tr \tilde{A}_{n-1}$$
$$\quad + \cdots$$
$$\quad + \frac{1}{2n} P^T (B_0^T b^{n-2} + B_1^T b^{n-1} + \cdots + B_{n-1}^T b^0).$$
$$\qquad (8.115)$$

8.4 Nonlinear Filtering and Time-Varying Schrödinger Equation I

Proof Observe that if we let

$$\tilde{A}_{-1} = \tilde{C}_{-1} = -\frac{1}{2}I, \ \tilde{B}_{-1} = I$$
$$\tilde{A}_0 = \tilde{B}_0 = \tilde{C}_0 = \tilde{D}_{-1} = \tilde{D}_0 = \tilde{E}_{-1} = \tilde{E}_0 = 0 \tag{8.116}$$
$$s_0 = s_{-1} = 0.$$

and $\tilde{A}_n, \tilde{B}_n, \tilde{C}_n, \tilde{D}_n, \tilde{E}_n, s_n, n \geq 1$ as in (8.110), then relations in Proposition 8.2 are satisfied. Next we show that $K(t, \tilde{x}, \tilde{y})$ in Eq. (8.109) is a fundamental solution. We only need to verify

$$\lim_{t \to 0} \int_{\mathbb{R}^n} K(t, \tilde{x}, \tilde{y}) v(0, \tilde{y}) d\tilde{y} = v(0, \tilde{x}). \tag{8.117}$$

By replacing $\tilde{y}$ by $\tilde{x} - \tilde{y}$, we see that

$$\lim_{t \to 0} \int_{\mathbb{R}^n} K(t, \tilde{x}, \tilde{y}) v(0, \tilde{y}) d\tilde{y}$$
$$= \lim_{t \to 0} \int_{\mathbb{R}^n} K(t, \tilde{x}, \tilde{x} - \tilde{y}) v(0, \tilde{x} - \tilde{y}) d\tilde{y}$$
$$= \lim_{t \to 0} \int_{\mathbb{R}^n} (2\pi t)^{-n/2} \exp\{-\frac{|\tilde{y}|^2}{2t} + \tilde{x}^T \tilde{A}(t)\tilde{x} + \tilde{x}^T \tilde{B}(t)(\tilde{x} - \tilde{y})$$
$$+ (\tilde{x} - \tilde{y})^T \tilde{C}(t)(\tilde{x} - \tilde{y}) + \tilde{D}(t)^T \tilde{x} + \tilde{E}(t)^T(\tilde{x} - \tilde{y}) + s(t)\} v(0, \tilde{x} - \tilde{y}) d\tilde{y}. \tag{8.118}$$

Let $\tilde{y} = \sqrt{2t} z$ where $z = (z_1, \cdots, z_n)$. Then

$$\lim_{t \to 0} \int_{\mathbb{R}^n} K(t, \tilde{x}, \tilde{y}) v(0, \tilde{y}) d\tilde{y}$$
$$= \lim_{t \to 0} \int_{\mathbb{R}^n} (\pi)^{-n/2} \exp\{-|z|^2 + \tilde{x}^T \tilde{A}(t)\tilde{x} + \tilde{x}^T \tilde{B}(t)(\tilde{x} - \sqrt{2t}z)$$
$$+ (\tilde{x} - \sqrt{2t}z)^T \tilde{C}(t)(\tilde{x} - \sqrt{2t}z) + \tilde{D}(t)^T \tilde{x} \tag{8.119}$$
$$+ \tilde{E}(t)^T(\tilde{x} - \sqrt{2t}z) + s(t)\} v(0, \tilde{x} - \sqrt{2t}z) d\tilde{z}.$$
$$= \int_{\mathbb{R}^n} \pi^{-n/2} e^{-|z|^2} v(0, \tilde{x}) dz$$
$$= v(0, \tilde{x}).$$

$\square$

Finally, by summarizing previous results, we obtain the explicit solution of Schrödinger equation (8.83).

Theorem 8.12 *The solution of*

$$\begin{cases} \dfrac{\partial v}{\partial t}(t,\tilde{x}) = \dfrac{1}{2}\Delta v(t,\tilde{x}) - \dfrac{1}{2}q(B^{-1}(t)\tilde{x} - B^{-1}(t)b(t))v(t,\tilde{x}), \tau_{i-1} \le t \le \tau_i, \\ v(\tau_{i-1},\tilde{x}) = \sigma_i(\tilde{x})e^{-\phi(\tilde{x})}, \end{cases}$$
(8.120)

where

$$q(B^{-1}(t)\tilde{x} - B^{-1}(t)b(t)) = \tilde{x}^T B(t) Q B(t)^T \tilde{x} - [2b(t)^T B(t)^T Q B(t)^T - P^T B(t)^T]\tilde{x}$$
$$+ b(t)^T B(t) Q B(t)^T b(t) - P^T B(t)^T b(t) + r$$
(8.121)

is given by

$$v(t,\tilde{x}) = \int_{\mathbb{R}^n} K(t,\tilde{x},\tilde{y}) v(\tau_{i-1},\tilde{y}) d\tilde{y}.$$
(8.122)

Here

$$K(t,\tilde{x},\tilde{y}) = (2\pi(t-\tau_{i-1}))^{-n/2} \exp\left\{ -\dfrac{|\tilde{x}-\tilde{y}|^2}{2(t-\tau_{i-1})} + \tilde{x}^T \tilde{A}(t)\tilde{x} + \tilde{x}^T \tilde{B}(t)\tilde{y} \right.$$
$$\left. + \tilde{y}^T \tilde{C}(t)\tilde{y} + \tilde{D}(t)^T \tilde{x} + \tilde{E}(t)^T \tilde{y} + s(t) \right\},$$
(8.123)

where $\tilde{A}(t-\tau_{i-1}) = \sum_{n=1}^{\infty} \tilde{A}_n(t-\tau_{i-1})^n$, $\tilde{B}(t-\tau_{i-1}) = \sum_{n=1}^{\infty} \tilde{B}_n(t-\tau_{i-1})^n$, $\tilde{C}(t-\tau_{i-1}) = \sum_{n=1}^{\infty} \tilde{C}_n(t-\tau_{i-1})^n$, $\tilde{D}(t-\tau_{i-1}) = \sum_{n=1}^{\infty} \tilde{D}_n(t-\tau_{i-1})^n$, $\tilde{E}(t-\tau_{i-1}) = \sum_{n=1}^{\infty} \tilde{E}_n(t-\tau_{i-1})^n$, $s(t-\tau_{i-1}) = \sum_{n=1}^{\infty} s_n(t-\tau_{i-1})^n$, $b(t-\tau_{i-1}) = \sum_{n=0}^{\infty} b^n(t-\tau_{i-1})^n$, $B(t-\tau_{i-1}) = \sum_{n=0}^{\infty} B_n(t-\tau_{i-1})^n$ *can be computed via (8.110).*

In brief, we show that in order to solve the nonlinear filtering problem for the time-invariant Yau filtering system with arbitrary initial condition, it suffices to solve a time-varying Schrödinger equation with arbitrary initial condition. We actually solve the time-varying Schrödinger equation with arbitrary initial condition by constructing the fundamental solution explicitly in case the potential is quadratic in state variables (which include the case that the observation $h_i(x)$, $1 \le i \le m$, are nonlinear but with linear growth). The fundamental solution is constructed via a system of nonlinear ODEs. This system of nonlinear ODEs is solved explicitly by power series method.

8.5 Nonlinear Filtering and Time-Varying Schrödinger Equation II

In this section, we extend the result of Sect. 8.4 to time-varying Yau systems. This section mainly refers to the work of Chen et al. [1]. Similar to the previous section, robust DMZ equation is changed to a Kolmogorov equation by exponential transformations in each time interval, and then under some assumptions, the Kolmogorov can be transformed into time-varying Schrödinger equation which can be solved explicitly.

The continuous time-varying filtering problem considered in this section can be stated as follows:

$$\begin{cases} dx_t = f(x_t, t)dt + g(t)dv_t, & x(0) = x_0, \\ dy_t = h(x_t, t)dt + dw_t, & y(0) = 0, \end{cases} \quad (8.124)$$

where $x_t, f \in R^n$, $g \in \mathbb{R}^{n \times r}$, $v_t \in \mathbb{R}^r$ is a Brownian motion process with $E[dv_t dv_t^T] = \tilde{Q}(t)dt$ and $\tilde{Q}(t) > 0$, $y_t, h \in \mathbb{R}^m$ and $w_t \in \mathbb{R}^m$ is a Brownian motion process with $E[dv_t dv_t^T] = S(t)dt$ and $S(t) > 0$. Here we refer x_t as the state of the system at time t, $f(x_t, t)$ as the drift term, $\tilde{Q}(t)$, $S(t)$ as the variance of the noises, and y_t as the observation at time t.

First, we give some assumptions in terms of system (8.124). We assume that $G(t) := g(t)\tilde{Q}(t)g(t)^T$ is C^∞ smooth, and $f(x, t)$, $h(x, t)$ are C^∞ smooth in both state and time. For the sake of clarity, we state some notations first: $*_{ij}$ denotes the ij-entry of the matrix $*$, $*_i$ denotes the i-th element of the vector $*$, and $*^T$ denotes the transposition of $*$.

Next we derive the robust DMZ equation for time-varying system, and we recall the DMZ equation can be written as follows:

$$\begin{cases} d\sigma(t, x) = L\sigma(t, x)dt + \sigma(t, x)h^T(x, t)S^{-1}(t)dy_t, \\ \sigma(0, x) = \sigma_0(x), \end{cases} \quad (8.125)$$

where $\sigma_0(x)$ is the probability density of the initial state x_0 and

$$L(*) := \frac{1}{2} \sum_{i,j=1}^n \frac{\partial^2}{\partial x_i \partial x_j}[G_{ij}(t)*] - \sum_{i=1}^n \frac{\partial(f_i *)}{\partial x_i}. \quad (8.126)$$

For each arrived observation, we make an invertible exponential transformation:

$$u(t, x) = \exp[-h^T(x, t)S^{-1}y_t]\sigma(t, x). \quad (8.127)$$

Then we transform DMZ equation to robust DMZ equation, which is a deterministic PDE with stochastic coefficients:

$$\begin{cases} \dfrac{\partial u}{\partial t}(t,x) = \dfrac{1}{2}\sum_{i=1}^{n} G_{ij}(t) \dfrac{\partial^2 u}{\partial x_i \partial x_j}(t,x) - \sum_{i=1}^{n} f_i(x)\dfrac{\partial u}{\partial x_i} + \dfrac{\partial}{\partial t}(h^T S^{-1})^T y_t \cdot \nabla u(t,x) \\ \qquad + \left\{ \dfrac{1}{2} \sum_{i,j=1}^{n} G_{ij}(t) \left[\dfrac{\partial^2 \tilde{K}}{\partial x_i \partial x_j} \right] - \sum_{i=1}^{n} f_i \dfrac{\partial \tilde{K}}{\partial x_i}(t,x) - \sum_{i=1}^{n} \dfrac{\partial f_j}{\partial x_i}(t,x) \right. \\ \qquad \left. - \dfrac{1}{2}(h^T S^{-1} h) \right\} u(t,x) \\ u(0,x) = \sigma_0(x), \end{cases} \qquad (8.128)$$

where

$$\tilde{K}(x,t) = h^T(x,t) S^{-1}(t) y_t. \qquad (8.129)$$

Similarly to the previous section, we can apply Yau-Yau algorithm. We assume the observations arrive at discrete instants and we denote the observation time sequence as $\mathcal{P}_k = \{0 = \tau_0 < \tau_1 < \cdots < \tau_k = \tau\}$, $|\mathcal{P}_k| := \sup_{1 \le i \le k}(\tau_i - \tau_{i-1})$. In each time interval $[\tau_{i-1}, \tau_i]$, $1 \le i \le k$, we assume observation signal y_t is taken $y_{\tau_{i-1}}$ and then solve the robust DMZ equation. Yau-Yau algorithm can guarantee convergence of the solution in L^2 and point-wise sense when $|\mathcal{P}_k| \to 0$,

In the second and third authors proposed an on- and off-line algorithm to solve the NLF problems in real time which has been verified numerically as an effective tool in very low dimension. The key idea of Luo et al. is that the heavy computation of solving PDE can be moved to off-line by the following proposition.

Proposition 8.3 *For each $\tau_{k-1} \le t \le \tau_k$, $k = 1, 2, \cdots$, $u_k(t,x)$ satisfies (8.128) if and only if*

$$\tilde{u}_k(t,x) = \exp[h^T(x,t) S^{-1}(t) y_{\tau_{k-1}}] u_k(t,x), \qquad (8.130)$$

satisfies the Kolmogorov equation:

$$\dfrac{\partial \tilde{u}_k}{\partial t}(t,x) = \left(L - \dfrac{1}{2} h^T S^{-1} h \right) \tilde{u}_k(t,x), \qquad (8.131)$$

i.e.,

8.5 Nonlinear Filtering and Time-Varying Schrödinger Equation II

$$\begin{cases} \dfrac{\partial \tilde{u}_k}{\partial t}(t,x) = \dfrac{1}{2}\sum_{i,j=1}^{n} G_{ij}(t)\dfrac{\partial^2 \tilde{u}_k}{\partial x_i \partial x_j} - \sum_{j=1}^{n} f_j(x)\dfrac{\partial \tilde{u}_k}{\partial x_j}(t,x) \\ \qquad - \left(\displaystyle\sum_{j=1}^{n} \dfrac{\partial f_j}{\partial x_j}(x) + \dfrac{1}{2}h^T S^{-1} h \right) \tilde{u}_k(t,x) \\ \tilde{u}_1(0,x) = \sigma_0(x), \\ \tilde{u}_k(\tau_{k-1},x) = \exp\left[h^T(x,\tau_{k-1}) S^{-1}(\tau_{k-1})(y_{\tau_{k-1}} - y_{\tau_{k-2}}) \right] \tilde{u}_{k-1}(\tau_{k-1},x), k \geq 2. \end{cases}$$
(8.132)

In this section, we aim to extend the results to the more general time-varying Yau systems:

$$f(x,t) = L(t)x + l(t) + G(t)\nabla_x \phi(t,x), \tag{8.133}$$

where $L(t) = (l_{ij}(t))$, $1 \leq i, j \leq n$, $l^T = (l_1(t), \cdots, l_n(t))$ and $\phi(t,x)$ is a smooth function on $\mathbb{R}^n$.

In the following proposition, we make an exponential transformation from $\tilde{u}_k$ to $\tilde{v}_k$ so that the coefficients of gradient of $\tilde{v}_k$ become linear.

Proposition 8.4 ([1]) *Suppose $\tilde{u}_k(t,x)$ is the solution to (8.132) in the interval $\tau_{k-1} \leq t \leq \tau_k$, $k = 1, 2, \cdots$ and $f(x,t)$ is of the form (8.133). Let*

$$\tilde{u}_k(t,x) = e^{\phi(t,x)} \tilde{v}_k(t,x); \tag{8.134}$$

then we have the following equation for $\tilde{v}_k(t,x)$:

$$\begin{cases} \dfrac{\partial \tilde{v}_k}{\partial t}(t,x) = \dfrac{1}{2}\sum_{i,j=1}^{n} G_{ij}(t)\dfrac{\partial^2 \tilde{v}_k}{\partial x_i \partial x_j} - (Lx+l)^T \nabla \tilde{v}_k \\ \qquad - \dfrac{1}{2}q(t,x)\tilde{v}_k(t,x) \\ \tilde{v}_1(0,x) = \sigma_0(x)e^{-\phi(0,x)}, \\ \tilde{v}_k(\tau_{k-1},x) = \exp\left[h^T(x,\tau_{k-1}) S^{-1}(\tau_{k-1})(y_{\tau_{k-1}} - y_{\tau_{k-2}}) \right] \tilde{v}_{k-1}(\tau_{k-1},x), k \geq 2, \end{cases}$$
(8.135)

where

$$q(t, x) = \sum_{i,j=1}^{n} G_{ij}(t) \frac{\partial^2 \phi}{\partial x_i \partial x_j}(t, x) + \nabla_x \phi^T(t, x) G(t) \nabla_x \phi(t, x)$$
$$+ 2(Lx + l)^T \nabla_x \phi(t, x) \quad (8.136)$$
$$+ \sum_{p,l=1}^{n} S_{pl}^{-1}(t) h_p(x, t) h_l(x, t) + 2tr(L) + 2\frac{\partial \phi}{\partial t}.$$

In [9], the third author and his co-worker changed the Kolmogorov forward equation of $\tilde{v}_k$ into Schrödinger equation. However, the transformation is much more difficult here since the coefficients G_{ij} in front of the second derivative are time-varying rather than the identity matrix I. Some assumptions on the system are stated below.

Assumption 1 $G(t)$ is a positive definite matrix.

Since $G(t)$ is positive definite, then we can find an invertible matrix $F(t)$ such that

$$G(t) = F(t)F(t)^T. \quad (8.137)$$

Assumption 2 $L(t)$ can be expressed as follows:

$$L(t) = G(t)\Omega(t) + \frac{dF(t)}{dt} F^{-1}(t), \quad (8.138)$$

where $\Omega(t) \in R^{n \times n}$ is an arbitrary symmetric matrix.

Remark 8.2 If the state of system is scalar or the state is a vector and $G(t), L(t)$ are diagonal, it is obvious that Assumption 2 is naturally satisfied.

Under Assumption 1 and 2, we introduce a transformation to eliminate the gradient term $\nabla \tilde{v}_k$ in (8.135), so that the Schrödinger equation can be naturally connected to the NLF problems later. Details can be found in [1].

Theorem 8.13 *Under Assumptions 1 and 2, suppose $\tilde{v}_k(t, x)$ is a solution of (8.135) and let*

$$\tilde{v}_k(t, x) = e^{x^T D(t) x} v_k(t, z), \quad (8.139)$$

where

8.5 Nonlinear Filtering and Time-Varying Schrödinger Equation II

$$z = B(t)x + b(t)$$
$$B(t) = F^{-1}(t)$$
$$b(t) = \int_0^t B(s)l(s)ds \qquad (8.140)$$
$$D(t) = \frac{1}{2}\Omega(t).$$

Then $v_k(t,z)$ is the solution of the following Schrödinger equation:

$$\begin{cases} \dfrac{\partial v_k}{\partial t}(t,z) = \dfrac{1}{2}\Delta v_k(t,z) \\ \qquad - \dfrac{1}{2}\tilde{q}(t, F(t)z - F(t)b(t))v_k(t,z) \\ v_1(0,z) = \sigma_0(F(0)z)\exp[-\phi(0, F(0)z) - (F(0)z)^T D(0)(F(0)z)], \\ v_k(\tau_{k-1}, z) = \exp[h^T(F(\tau_{k-1})z - F(\tau_{k-1})b(\tau_{k-1}), \tau_{k-1}) \\ \qquad S^{-1}(\tau_{k-1})(y_{\tau_{k-1}} - y_{\tau_{k-2}})]v_{k-1}(\tau_{k-1}, z), k \geq 2, \end{cases} \qquad (8.141)$$

where

$$\begin{aligned}\tilde{q}(t,x) = {} & q(t,x) + 2x^T \frac{dD(t)}{dt}x - \sum_{i,j=1}^n G_{ij}(t)(D_{ij} + D_{ji}) \\ & - x^T(D(t) + D^T(t))G(t)(D(t) + D^T(t))x \\ & + 2(L(t)x + l)^T(D(t) + D^T(t))x.\end{aligned} \qquad (8.142)$$

Proof Direct computation and the details can be found in [1]. □

Next we discuss to solve the Schrödinger equation (8.141) explicitly. First, we make an assumption that $\tilde{q}$ is a quadratic polynomial in variable x.

Assumption 3 $\tilde{q}(t,x)$ defined in (8.142) is quadratic with respect to x.

Actually, Assumption 3 includes Kalman-Bucy filter and Benes filtering. Notice that observation term $h_i(x,t)$ can be nonlinear which extends the Kalman-Bucy filtering system. $\tilde{q}$ is quadratic in x under Assumption 3. Thus, we can assume that

$$\tilde{q}(t,x) = x^T Q(t)x + p^T(t)x + r(t). \qquad (8.143)$$

In the following theorem, in order to solve the Schrödinger equation, we need to find the fundamental solution of (8.141). Similar to heat kernel, we assume fundamental solution has the exponential quadratic form and find the relations between specific coefficients.

Theorem 8.14 Let $K(t, x, y)$ be the fundamental solution of

$$\frac{\partial v_k}{\partial t}(t, x) = \frac{1}{2}\Delta v_k(t, x) - \frac{1}{2}\tilde{q}(t, F(t)x - F(t)b(t))v_k(t, x), \tag{8.144}$$

where

$$\begin{aligned}\tilde{q}(t, F(t)x - F(t)b(t)) =& x^T F^T(t)Q(t)F(t)x \\ &- [2b^T(t)F^T(t)Q(t)F(t) - p^T(t)F(t)]x \\ &+ b^T(t)F^T(t)Q(t)F(t)b(t) - p^T(t)F(t)b(t) + r(t).\end{aligned} \tag{8.145}$$

Assume the fundamental solution $K(t, x, y)$ can be written as

$$K(t, x, y) = (2\pi t)^{-n/2}\exp\{x^T\tilde{A}(t)x + x^T\tilde{B}(t)y + y^T\tilde{C}(t)y + \tilde{D}^T(t)x + \tilde{E}^T(t)y + s(t)\}, \tag{8.146}$$

where $\tilde{A}(t), \tilde{C}(t)$ are $n \times n$ symmetric matrices, $\tilde{B}(t)$ is a $n \times n$ matrix, and $\tilde{D}(t)$ and $\tilde{E}(t)$ are column n-vector. Then coefficient $\tilde{A}(t)$–$\tilde{E}(t)$ satisfy the following ODEs:

$$\frac{d\tilde{A}}{dt}(t) = 2\tilde{A}(t)^2 - \frac{1}{2}F^T(t)Q(t)F(t)$$

$$\frac{d\tilde{B}}{dt}(t) = 2\tilde{A}(t)\tilde{B}(t)$$

$$\frac{d\tilde{C}}{dt}(t) = \frac{1}{2}\tilde{B}^T\tilde{B}$$

$$\frac{d\tilde{E}}{dt}(t) = 2\tilde{A}\tilde{D} + F^T(t)Q(t)F(t)b(t) - \frac{1}{2}F^T(t)p(t)$$

$$\frac{d\tilde{E}}{dt}(t) = \tilde{B}^T\tilde{D}$$

$$\frac{ds}{dt}(t) = \frac{1}{2}\tilde{D}^T\tilde{D} + tr(\tilde{A})$$

$$- \frac{1}{2}[b^T(t)F^T(t)Q(t)F(t)b(t) - p^T(t)F(t)b(t) + r(t)] + \frac{n}{2t}. \tag{8.147}$$

Proof The proof is similar to Theorem 8.10. □

In the following theorem, we use power series method to solve the ODEs in Theorem 8.14 and obtain the explicit solution of Schrödinger equation on v_k. The method is similar to Sect. 8.4 and details can be referred to the work of Chen et al. [1].

8.5 Nonlinear Filtering and Time-Varying Schrödinger Equation II

Theorem 8.15 *Under Assumptions 1–3, the solution $v_k(t, z)$ in $\tau_{k-1} \leq t \leq \tau_k$ of (8.141) is given by*

$$v_k(t, x) = \int_{\mathbb{R}^n} K(t, x, y) v_k(\tau_{k-1}, y) dy, \qquad (8.148)$$

where

$$K(t, x, y) = (2\pi(t - \tau_{k-1}))^{-n/2} \exp\{-\frac{|x-y|^2}{2(t - \tau_{k-1})} + x^T \tilde{A}(t - \tau_{k-1})x$$
$$+ x^T \tilde{B}(t - \tau_{k-1})y + y^T \tilde{C}(t - \tau_{k-1})y + \tilde{D}^T(t - \tau_{k-1})x$$
$$+ \tilde{E}^T(t - \tau_{k-1})y + s(t - \tau_{k-1})\},$$
$$(8.149)$$

where $\tilde{A}(t - \tau_{k-1}) = \sum_{n=1}^{\infty} \tilde{A}_n(t - \tau_{k-1})^n$, $\tilde{B}(t - \tau_{k-1}) = \sum_{n=1}^{\infty} \tilde{B}_n(t - \tau_{k-1})^n$, $\tilde{C}(t - \tau_{k-1}) = \sum_{n=1}^{\infty} \tilde{C}_n(t - \tau_{k-1})^n$, $\tilde{D}(t - \tau_{k-1}) = \sum_{n=1}^{\infty} \tilde{D}_n(t - \tau_{k-1})^n$, $\tilde{E}(t - \tau_{k-1}) = \sum_{n=1}^{\infty} \tilde{E}_n(t - \tau_{k-1})^n$, $s(t - \tau_{k-1}) = \sum_{n=1}^{\infty} s_n(t - \tau_{k-1})^n$, $b(t - \tau_{k-1}) = \sum_{n=0}^{\infty} b_n(t - \tau_{k-1})^n$, $F(t - \tau_{k-1}) = \sum_{n=0}^{\infty} F_n(t - \tau_{k-1})^n$, $Q(t - \tau_{k-1}) = \sum_{n=0}^{\infty} Q_n(t - \tau_{k-1})^n$, $p(t - \tau_{k-1}) = \sum_{n=0}^{\infty} p_n(t - \tau_{k-1})^n$, $r(t - \tau_{k-1}) = \sum_{n=0}^{\infty} r_n(t - \tau_{k-1})^n$, *where*

$$\tilde{A}_{n+1} = \frac{2}{n+3} \sum_{i=0}^{n} \tilde{A}_i \tilde{A}_{n-i} - \frac{1}{2(n+3)} \sum_{j=0}^{n} \sum_{i=0}^{j} F_i^T Q_{j-i} F_{n-j}$$

$$\tilde{B}_{n+1} = \frac{2}{n+2} \sum_{i=0}^{n+1} \tilde{A}_i \tilde{B}_{n-i}$$

$$\tilde{C}_{n+1} = \frac{1}{2(n+1)} \sum_{i=-1}^{n+1} \tilde{B}_i^T \tilde{B}_{n-i}$$

$$\tilde{D}_{n+1} = \frac{2}{n+2} \sum_{i=0}^{n+1} \tilde{A}_i \tilde{D}_{n-i}$$
$$- \frac{1}{n+2} \sum_{i=0}^{n} F_i^T p_{n-i} - \frac{1}{2(n+2)} \sum_{j=0}^{n} \sum_{i=0}^{j} \sum_{l=0}^{i} F_l^T Q_{i-l} F_{j-l} b_{n-j}$$

$$\tilde{E}_{n+1} = \frac{2}{n+1} \sum_{i=-1}^{n+1} \tilde{B}_i \tilde{D}_{n-i}$$

$$s_{n+1} = \frac{1}{2(n+1)} \sum_{i=-1}^{n+1} \tilde{D}_i^T \tilde{D}_{n-i} + \frac{1}{n+1} tr(\tilde{A}_n)$$

$$- \frac{1}{2(n+1)} \left[\sum_{i=0}^{n} \sum_{j=0}^{i} \sum_{m=0}^{j} \sum_{l=0}^{m} b_l^T F_{m-l}^T Q_{j-m} F_{i-j} b_{n-i} \right.$$

$$\left. - \sum_{j=0}^{n} \sum_{i=0}^{j} p_i^T F_{j-i} b_{n-j} + r_n \right]. \tag{8.150}$$

So far, we extend explicit solution of a class of time-varying system under the Yau-Yau algorithm framework.

8.6 Nonlinear Filtering and Time-Varying Schrödinger Equation III

In this section, we extend the result of Sect. 8.5 to a broader context. This section mainly refers to the work of Chen et al. [2].

The filtering problem addressed here is the same as in the previous section. With the same approach, it simplifies to deal with Eq. (8.128). Unlike the previous section, Assumption 2 is no longer required.

Theorem 8.16 *Under Assumption 1, suppose $\tilde{v}_k(t, x)$ is a solution of (8.128) and let*

$$\tilde{v}_k(t, x) = v_k(t, z), \tag{8.151}$$

where

$$z = B(t)x$$
$$B(t) = F^{-1}(t). \tag{8.152}$$

Then $v_k(t, z)$ is the solution of the following equation:

8.6 Nonlinear Filtering and Time-Varying Schrödinger Equation III

$$\begin{cases} \dfrac{\partial v_k}{\partial t}(t,z) = \dfrac{1}{2}\Delta v_k(t,z) - \dfrac{1}{2}q(t,F(t)z)\,v_k(t,z) \\ \qquad - \left[\left(\dfrac{dB}{dt}B^{-1} + BLB^{-1}\right)z + Bl\right]^T \nabla v_k(t,z) \\ v_1(0,z) = \sigma_0\,(F(0)z)\exp(-\phi\,(0,F(0)z)) \\ v_k(\tau_{k-1},z) = \exp\left[h^T\,(F(\tau_{k-1})z,\tau_{k-1})\,S^{-1}(\tau_{k-1})\right. \\ \qquad \left.\cdot\left(y_{\tau_{k-1}} - y_{\tau_{k-2}}\right)\right] v_{k-1}(\tau_{k-1},z) \\ k = 2, 3, \ldots, N. \end{cases} \quad (8.153)$$

Proof Direct computation and the details can be found in [2]. □

Let

$$\tilde{q}(t,z) = q\,(t,F(t)z) \quad (8.154)$$

and modify Assumption 3 to the following form:

Assumption 3' $\tilde{q}(t,z)$ defined in (8.154) is quadratic with respect to z.

Similarly, we can assume that

$$-\dfrac{1}{2}\tilde{q}(t,z) = z^T Q(t)z + p^T z + r(t). \quad (8.155)$$

Theorem 8.17 *Under Assumption 1 and 3', consider the following KFE with Gaussian initial distribution:*

$$\begin{cases} \dfrac{\partial v_k}{\partial t}(t,z) = \dfrac{1}{2}\Delta v_k(t,z) - \dfrac{1}{2}\tilde{q}(t,z)v_k(t,z) \\ \qquad - \left[\left(\dfrac{dB}{dt}B^{-1} + BLB^{-1}\right)z + Bl\right]^T \nabla v_k(t,z) \\ v_k(\tau_{k-1},z) = \exp\left\{z^T A(\tau_{k-1})z + b^T(\tau_{k-1})z + c(\tau_{k-1})\right\}, \end{cases} \quad (8.156)$$

where $A(\tau_{k-1})$ is an $n \times n$ symmetric matrix, $b(\tau_{k-1})$ is an $n \times 1$ vector, $x^T = (x_1, x_2, \ldots, x_n)$ is a row vector, and $c(\tau_{k-1})$ is a scalar. Then the solution of (8.156) is of the following form:

$$v_k(t,z) = \exp\left\{z^T A(t)z + b^T(t)z + c(t)\right\} \quad (8.157)$$

where $A(t)$ is an $n \times n$ symmetric matrix-valued function of t, $b(t)$ is an $n \times 1$ vector-valued function of t, and $c(t)$ is a scalar-valued function of t satisfying the following system of nonlinear ODEs:

$$\frac{dA(t)}{dt} = 2A^2(t) - 2A(t)D(t) + Q(t)$$

$$\frac{db^T(t)}{dt} = 2b^T(t)A(t) - b^T(t)D(t) - 2d^T(t)A(t) + p^T(t) \quad (8.158)$$

$$\frac{dc(t)}{dt} = trA(t) + \frac{1}{2}b^T b(t) - d^T(t)b(t) + r(t),$$

with

$$D(t) = \frac{dB}{dt}B^{-1} + BLB^{-1} \quad (8.159)$$

$$d(t) = B(t)l(t).$$

Proof The proof is similar to Theorem 8.8. □

Then we can give an outline of direct method based on the framework presented in [2].

Algorithm 4 Direct method

1: **for** $k \in \{1, \ldots, N\}$ **do**
2:     Calculate $v_k(z, \tau_{k-1})$ by initial condition of (8.153)
3:     Using Algorithm 3 to get the Gaussian approximation

$$v_k(z, \tau_{k-1}) \approx \sum_i^{N_k} \alpha_{k,i} \exp\left(z^T A_{k,i} z + b_{k,i}^T z + c_{k,i}\right).$$

4:     **for** $i \in \{1, \ldots, N_k\}$ **do**
5:         Solve (8.156) with the initial condition

$$v_{k,i}(z, \tau_{k-1}) = \exp\left(z^T A_{k,i} z + b_{k,i}^T z + c_{k,i}\right)$$

    by Theorem 8.17.
6:     Calculate the approximate solution v_k to (8.153) by $v_k \approx \sum_i^{N_k} \alpha_{k,i} v_{k,i}$.
7:     Calculate $\tilde{v}_k(x, \tau_k)$ by (8.151).
8:     Calculate $\tilde{u}_k(x, \tau_k)$ by (8.134).
9:     Calculate $u_k(x, \tau_k)$ by (8.130).
10:    Calculate $\sigma(x, \tau_k)$ by (8.127).
11:    Calculate the conditional expectation of the state x_{τ_k}.

Here we use two numerical examples to demonstrate the efficiency of this algorithm. The filtering system here is as follows:

8.6 Nonlinear Filtering and Time-Varying Schrödinger Equation III

$$\begin{cases} dx_t = \left(c\left(x_t + 1\right) + \phi'\left(x_t\right)\right) dt + dv_t \\ dy_t = \begin{pmatrix} x_t \sin x_t \\ x_t \cos x_t \end{pmatrix} dt + dw_t \\ \sigma_0(x) = \dfrac{\exp\left(-x \sin x - 0.5x \cos x - x^2 + 3x + 2\right)}{\int_{\mathbb{R}} \exp\left(-x \sin x - 0.5x \cos x - x^2 + 3x + 2\right) dx}. \end{cases} \quad (8.160)$$

Here, $t \in [0, T]$, where $T > 0$ is a fixed termination, with a sampling interval $\Delta t = 0.01$, and v_t and w_t are independent Brownian motions, with $E(v_t v_t) = 1$ and $E\left(w_t w_t^T\right) = \begin{pmatrix} (1 + \sin(0.2t))^2 & 0 \\ 0 & (1 + \sin(0.2t))^2 \end{pmatrix}$.

We introduce the mean of the squared estimation error (MSE) to demonstrate the average performance of direct method. The MSE for m repeated realizations at instant $\tau_k := k\Delta t$ is defined as follows:

$$\text{MSE}(\tau_k) := \frac{1}{mk} \sum_{i=1}^{m} \sum_{j=0}^{k} \left(x_{\tau_j}^i - \hat{x}_{\tau_j}^i\right)^2, \quad (8.161)$$

where $x_{\tau_j}^i$ is the real state at instant τ_j in the ith realization and $\hat{x}_{\tau_j}^i$ is the estimation of $x_{\tau_j}^i$ via direct method.

Example 8.1 In this example, we set $T = 2$, $c = 0.1$, and $\phi = 0$. The average CPU running time is 3.3273s (Fig. 8.1).

Example 8.2 In this example, we set $T = 5$, $c = 0.3$, and

$$\phi(x) = \int_{-\infty}^{x} \left[\frac{e^{-(0.3z - 0.15)^2}}{\int_{-\infty}^{z} e^{-(0.3y - 0.15)^2} dy} - 0.45 \right] dz. \quad (8.162)$$

The average CPU running time is 5.1736s (Fig. 8.2).

It can be obviously seen that direct method demonstrates notable efficacy in addressing nonlinear filtering problems. More precisely, its performance is characterized by a low time consumption and a consistently low MSE, which

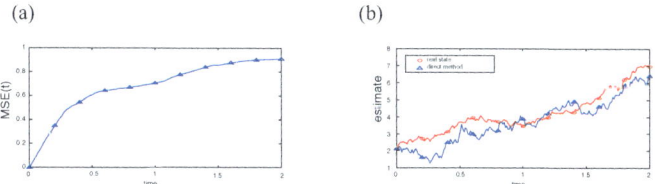

Fig. 8.1 Simulation results of Example 8.1. (**a**) MSE based on 50 simulations. (**b**) A typical simulation of Example 8.1

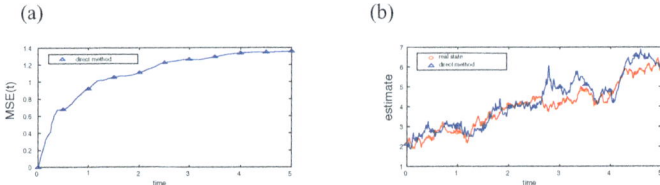

Fig. 8.2 Simulation results of Example 8.2. (**a**) MSE based on 50 simulations. (**b**) A typical simulation of Example 8.2

remains relatively stable as time increases. Results of a typical simulation reveal that the estimate trajectory provided by direct method closely tracks the actual state trajectory, despite the state trajectory exhibiting considerable oscillations over time. Therefore, direct method is a good choice in applications.

8.7 Exercises

1. Prove the following vector field identity, where f, g are smoothing functions and F is a vector field in $\mathbb{R}^n$.

   (1) $\nabla \cdot (fF) = \nabla f \cdot F + f \nabla \cdot F$
   (2) $\nabla \cdot (\nabla(fg)) = g\nabla \cdot \nabla f + 2\nabla f \cdot \nabla g + f \nabla \cdot \nabla g$

2. Verify the following choice of $H(x), G(x), P(x)$ satisfying the condition of Theorem 8.5:

$$\begin{cases} G(x) \equiv 0 \\ P(x) = \frac{1}{2}\eta(x) - \frac{1}{2}\sum_{i=1}^n l_i^2 + \frac{1}{2}\sum_{i=1}^n \frac{\partial l_i}{\partial x_i} \\ H_i = l_i, 1 \le i \le n. \end{cases} \quad (8.163)$$

3. Choose a C^∞ function $G(x)$ such that $\frac{\partial G}{\partial x_i} = -l_i$, if $d_{ij} = d_{ji}$. Let $P(x) = \frac{1}{2}\eta$ and $H_i \equiv 0, 1 \le i \le n$. Verify the choice of $H(x), G(x), P(x)$ satisfying the condition of Theorem 8.5.

4. Verify the following choice of $H(x), G(x), P(x)$ satisfying the condition of Theorem 8.5:

$$\begin{cases} G(x) = F(x) \\ P(x) = \frac{1}{2}\eta - \frac{1}{2}\sum_{i=1}^n f_i^2 + \frac{1}{2}\sum_{i=1}^n \frac{\partial f_i}{\partial x_i} \\ H_i = f_i. \end{cases} \quad (8.164)$$

5. Prove that Ω is a constant matrix if and only if

$$(f_1, f_2, \cdots, f_n) = (l_1, l_2, \cdots, l_n) + \left(\frac{\partial F}{\partial x_1}, \cdots, \frac{\partial F}{\partial x_n}\right), \tag{8.165}$$

where l_i, $1 \leq i \leq n$ are polynomials of degree 1 and F is a smooth function.

6. Achieve a numerical example of decomposing a non-Gaussian distribution to sum of Gaussian distribution based on Algorithm 3.
7. Construct an example of filtering system that satisfies conditions $(C_1), (C_2), (C_3)$.
8. Consider function $q(x) := \Delta \phi(x) + |\nabla \phi|^2 + 2(Lx+l) \cdot \nabla \phi + \sum_{i=1}^{m} h_i^2(x) + 2\mathrm{tr}L$ of Eq. (8.75) and time-invariant Yau filtering setting $f(x) = Lx + l + \nabla \phi$ and $h(x) = Hx$. Prove that assumption that $q(x)$ is a quadratic polynomial in x is equivalent to that $\eta := |f|^2 + \nabla \cdot f + |h|^2$ is quadratic in x.
9. Consider the following Kolmogorov forward equation:

$$\frac{\partial u}{\partial t}(t,x) = \frac{1}{2} \sum_{i,j=1}^{n} G_{ij} \frac{\partial^2 u}{\partial x_i \partial x_j} - f \cdot \nabla u - (\nabla \cdot f + \frac{1}{2}|h|^2)u, \tag{8.166}$$

with $f = L(t)x + l(t) + \nabla_x \tilde{\phi}(t,x)$. If we take an invertible transformation $u = e^{\phi} v$ with $\nabla \phi = G^{-1} \nabla \tilde{\phi}$, prove that the equation satisfied by v becomes as below:

$$\frac{\partial v}{\partial t} = \frac{1}{2} \sum_{i,j=1}^{n} G_{ij} \frac{\partial^2 v}{\partial x_i \partial x_j} - (Lx+l) \cdot \nabla v - \frac{1}{2} q(t,x)v, \tag{8.167}$$

with

$$q(t,x) = -\sum_{i,j=1}^{n} G_{ij} \frac{\partial^2 \phi}{\partial x_i \partial x_j} + \nabla \phi^\top G \nabla \phi + 2(Lx+l) \cdot \nabla \phi$$
$$+ 2\Delta \tilde{\phi} + 2\frac{\partial \phi}{\partial t} + |h|^2 + 2\mathrm{tr}(L). \tag{8.168}$$

References

1. X. Chen, X. Luo and S. S.-T. Yau. Direct method for time-varying nonlinear filtering problems. *IEEE Transactions on Aerospace and Electronic Systems*, 53(2):630–639, 2017.
2. X. Chen, J. Shi and S. S.-T. Yau. Real-time Solution of Time-varying Yau Filtering Problems via Direct Method and Gaussian Approximation. *IEEE Transactions on Automatic Control*, 64(4):1648–1654, 2019.
3. J. Shi, Z. Yang and S. S.-T. Yau. Direct method for Yau filtering system with nonlinear observations. *International Journal of Control*, 91(3):678–687, 2018.
4. S. S.-T. Yau and S.-T. Yau. Solution of filtering problem with nonlinear observations. *SIAM Journal on Control and Optimization*, 44(3):1019–1039, 2005.

5. S.-T. Yau and S. S.-T. Yau. Real time solution of nonlinear filtering problem without memory I. *Mathematical Research Letters*, 7(6):671–693, 2000.
6. S. S.-T. Yau, C. Yan and S.-T. Yau. Linear filtering with nonlinear observations *43rd IEEE Conference on Decision and Control (CDC)*, 2:2112–2117, 2004.
7. S. S.-T. Yau and G.-Q. Hu. Finite-dimensional filters with nonlinear drift X: Explicit solution of DMZ equation. *IEEE Transactions on Automatic Control*, 46(1): 142–148, 2001.
8. S. S.-T. Yau and Y. T. Lai. Explicit solution of DMZ equation in nonlinear filtering via solution of ODEs. *IEEE Transactions on Automatic Control*, 48(3):505–508, 2003.
9. S.-T. Yau and S. S.-T. Yau. Nonlinear filtering and time varying Schrodinger equation. *IEEE Transactions on Aerospace and Electronic Systems*, 40(1):284–292, 2004.

Chapter 9
Classical Filtering Methods

In this chapter, we shall introduce several important filtering algorithms. In Sect. 9.2, we shall introduce the filtering algorithms based on the Bayesian framework, in which the system equation and observation equation can be consider as discrete functions. In Sect. 9.3, we shall introduce filtering algorithms based on DMZ equation, in which the system equation and observation equation can be considered as continuous functions. In Sect. 9.4, we shall introduce another filtering topic called robust filtering. In both sections, we shall start with the linear filtering problems and extend them for general nonlinear system.

9.1 Introduction

In 1960, Rudolf E. Kalman proposed a linear quadratic estimation technique which is well-known as the Kalman filter (KF) [6]. KF is also one of the very few algorithms that completely solve the state estimation for a kind of system in an optimal way.

With the development, the industrial pays great attention to the problems of state estimation, and many nonlinear filter algorithms motivated by KF have been designed. However, there is no nonlinear filtering algorithm that can "completely" (as KF) solve the nonlinear filtering problem.

Even if many central problems of nonlinear filtering are still open, many important filtering algorithms have played an extremely important role in various applications. Nonlinear filtering transformed the world of signal processing. By using filtering, many researchers developed highly sophisticated navigation systems [18], camera tracking [20], fault diagnosis, chemical processes [22], vision-based systems, target tracking [15, 21], biomedical systems [23, 24], robotics [25], predictive economics, and stock forecasting systems [16, 17].

In nonlinear filtering, there are two major problems of research interest:

- Numerical methods of solving the nonlinear filtering problem.
- Nonlinear filtering in case of uncertainty in model and data association.

With the prosperity and development of the deep learning field, more and more data estimation methods are proposed, which constantly give birth to more filtering algorithms based on deep learning. In this chapter, we shall focus on the first problem which is numerical methods for solving the nonlinear filtering problem.

For different assumptions of the system, we can roughly divide the filtering problem into two categories: continuous filtering system and discrete filtering system. For discrete filtering problems, the transfer and update of the probability density function depend on the Bayesian formula, so we also regard the discrete filtering system as the Bayesian framework of filtering. Similarly, for a continuous filtering system, the posterior distributions are determined by the DMZ equation or Kushner equation.

We shall summarize the nonlinear filtering algorithms of the Bayesian framework into the following two categories:

1. The structure of system functions is used for extending KF to nonlinear filtering systems.
2. The structure of posterior distribution is used for extending KF to a nonlinear filtering system.

As for 1, there is a well-known algorithm called the extended Kalman filter (EKF) [26]. This filter linearizes a nonlinear system using approximation techniques. After EKF was proposed, there were many works focused on approximating the system functions with some kind of functions such as polynomials.

As for 2, there are three ways of using the structure of posterior distribution.

- The first one is to use the statistical characteristics of distribution, and unscented Kalman filter (UKF) [27] is a standard example of it. Since the nonlinear system model is approximated using Jacobian matrices in the EKF, the calculation may be costly, and it may not be easy to obtain accurate results for the highly nonlinear systems due to linearization. The UKF uses a deterministic sampling approach to capture the statistical points which makes the UKF to be considered more robust and more accurate.
- The second one is to project the posterior distribution function into a finite-dimensional space, which is started from the projection filter (PrF) [31]. Soon, the idea of PrF was combined with the information geometry which gave birth to many related filtering algorithms.
- Finally, the third one is to use the sampling particles for the reconstruction of the posterior distribution instead of solving the equations (ODE, PDE, SDE). The idea, of sequence Monte Carlo methods, gave the birth of the particle filter (PF) [29]. PF uses a set of weighted samples (called particles) to approximate the Bayesian prior and posterior.

It needs to be pointed out that the classifications mentioned above are not independent of each other. Many good algorithms are designed by combining the advantages of many ideas, such as the iteration EKF, mix-Gaussian method, projection PF, and so on.

Besides the Bayesian framework, there are many interests in extending the successful filtering algorithms to the DMZ framework. From the perspective of the algorithm, the Yau-Yau filtering algorithm provides a way to systematically solve nonlinear filtering problems by using numerical schemes of the parabolic partial differential equation. At the same time, Yau-Yau algorithms can be combined with a variety of numerical methods, such as projecting the system to some function spaces (the idea of EKF) or projecting the density function to some spaces (PrF).

In addition to the Yau-Yau algorithm, traditional PF including other simulation-based approaches does not have the (innovation error-based) feedback structure used in the KF. An important breakthrough of PF is the feedback particle filter (FPF) [11] which uses a feedback structure based on the mean-field game theory. The FPF equalizes the weights of particles, thus avoiding the disadvantages of traditional particle methods related to weights.

Besides, another interesting filtering method called robust filtering is considered here. The motivation is that these filters are derived by MMSE criterion making them very sensitive to heavy-tailed observation noises, which are frequently encountered in practical applications. How to improve state estimate robustness against non-Gaussian heavy-tailed observation noises has been the focus of robust filtering.

The organization of the remainder of this chapter is as follows. Section 9.2.1 presents the mathematical problem of nonlinear filtering considered in the Bayesian framework and the discrete KF. Section 9.2.2 presents the EKF in the Bayesian framework which linearizes a nonlinear system using approximation techniques. Section 9.2.3 presents the UKF as an extension of EKF. Section 9.2.4 presents the PF and EnKF [28] as sampling-based filtering algorithms. Section 9.3.1 presents the mathematical problem of nonlinear filtering considered in the DMZ framework. We introduce the continuous KF which is well-known as Kalman-Bucy filter (KBF) [30] in this section. Section 9.3.2 presents the EKF in the DMZ framework. Section 9.3.3 presents a control view of the PF framework and the FPF algorithm including several different simulation methods. Section 9.4 presents the iterative outlier-robust EKF [36] framework with three robust cost functions.

9.2 Filtering Algorithm Based on Bayesian Framework

At the beginning of this section, we shall introduce the discrete filtering system which is modeled as the following stochastic differential equation (SDE):

$$\begin{cases} X_{k+1} = f_k(X_k) + \Sigma_k(X_k)V_k, \\ Y_k = h_k(X_k) + W_k, \end{cases} \quad (9.1)$$

where X_k, V_k, Y_k, W_k are, respectively, R^n-, R^n-, R^m-, and R^m-valued processes and V_k and W_k are independent Gaussians with proper dimensions and covariance matrices I_n, S_k, respectively. We further assume that f_k, Σ_k, and h_k are C^∞ smooth functions for any $k \geq 1$. And X_0 follows some given distributions.

Remark 9.1 In general, the noise in the state equation is correlated with the noise in the observation equation. And according to the actual applications, there are mainly two cases:

- $E[V_k^\top W_k] = C_k$,
- $E[V_k^\top W_{k+1}] = C_k$.

But in this paper, we mainly focus on the cases where V and W are independent. The derivation of the noise-related case is similar to the noise-independent case, but there are differences in details. We provide two references [12, 13] for readers.

In the rest of this section, we use $p(X_k)$ as the density function of X_k, $p(X_k|Y_k)$ as the conditional density of X_k given Y_k and $p(X_k|X_{1:k-1}) := p(X_k|X_1, \cdots, X_{k-1})$. Under this framework, we have the following assumptions for the system:

- The state process $\{X_k\}_{k=1}^\infty$ is a Markov process, that is

$$p(X_k|X_{1:k-1}) = p(X_k|X_{k-1}),$$

for any $k \geq 1$.
- The state X_k is independent of $\{Y_1, \cdots, Y_{k-1}\}$ given X_{k-1}, which means $p(X_k|X_{k-1}, Y_{1:k-1}) = p(X_k|X_{k-1})$.

At the time step k, the conditional density function of the state X_k given observations $\{Y_i\}_{i=1}^k$ is $p(X_k|Y_{1:k})$. The conditional density function at k and $k+1$ is connected by the Bayesian formula as follows:

$$p(X_{k+1}|Y_{1:k+1}) = \frac{p(Y_{k+1}|X_{k+1}, Y_{1:k})p(X_{k+1}|Y_{1:k})}{p(Y_{k+1}|Y_{1:k})}. \tag{9.2}$$

Next, we shall introduce the general process of the Bayesian filtering framework.

Bayesian filtering framework is given as follows:
- Prediction step is to get $p(X_{k+1}|Y_{1:k})$. The prior density function $p(X_{k+1}|Y_{1:k})$ satisfies the following Chapman-Kolmogorov equation:

$$p(X_{k+1}|Y_{1:k}) = \int p(X_{k+1}|X_k)p(X_k|Y_{1:k})dX_k. \tag{9.3}$$

- Update step is to get $p(X_{k+1}|Y_{1:k})$ by using $p(Y_{k+1}|Y_{1:k})$.
  By using the Markov property, we can have

$$p(Y_{k+1}|X_{k+1}, Y_{1:k}) = p(Y_{k+1}|X_{k+1}). \tag{9.4}$$

Then $p(X_{k+1}|Y_{1:k+1})$ can be calculated by using (9.2). In update step, we still need to normalize $p(Y_{k+1}|X_{k+1}, Y_{1:k})p(X_{k+1}|Y_{1:k})$ in order to calculate $p(X_{k+1}|Y_{1:k+1})$.

9.2.1 Linear System and KF

As described in Chap. 5, the density evolution of the continuous filtering system is characterized by a family of SPDE, which is difficult to solve in real applications. In the continuous nonlinear filtering problem, the change of the system at two consecutive adjacent times is small.

9.2.2 Discrete KF

A stochastic time-variant linear system is described by the difference equation and the observation model

$$\begin{cases} X_k = A_{k-1}X_{k-1} + B_{k-1} + \Sigma_k W_{k-1}, \\ Y_k = H_k X_k + V_k, \end{cases} \quad (9.5)$$

which can be considered as a special case of (9.1) by assuming $f_k(x) = A_k \cdot x + B_k$ with $A_k \in R^{n \times n}$, $B_k \in R^n$, $\Sigma_k \in R^{n \times n}$, and $h_k(x) = H_k \cdot x$ with $H_k \in R^{m \times n}$ for any $k \geq 1$. Furthermore, V_k and W_k are independent Gaussians with corresponding dimensions and covariance matrices $I_n \in R^{n \times n}$, $S_k \in R^{m \times m}$, respectively, for any $k \geq 1$. And the initial state $X_0 \in R^n$ is a Gaussian with known mean $\mu_0 = E[X_0]$ and covariance $P_0 = E[(X_0 - \mu_0)(X_0 - \mu_0)^\top]$.

It is well-known that, for system (9.5), the prior and posterior densities of state are Gaussians, i.e.,

$$p(X_{k-1}|Y_{1:k-1}) = \mathcal{N}(\mu_{k-1|k-1}, P_{k-1|k-1}),$$
$$p(X_k|Y_{1:k-1}) = \mathcal{N}(\mu_{k|k-1}, P_{k|k-1}). \quad (9.6)$$

Next, we shall introduce the result first proposed by Kalman.

Theorem 9.1 ([6]) *For the discrete filtering system* (9.5), *if we assume the initial distribution* $X_0 \sim \mathcal{N}(\mu_0, P_0)$, *then the optimal (minimum variance unbiased) estimate and its covariance at step k are $\mu_{k|k}$ and $P_{k|k}$, respectively, which satisfy the following equations:*

Step of prediction:

$$\mu_{k|k-1} = A_{k-1}\mu_{k-1|k-1} + B_{k-1},$$
$$P_{k|k-1} = A_{k-1}P_{k-1|k-1}A_{k-1}^\top + Q_{k-1}, \tag{9.7}$$

Step of update:

$$\mu_{k|k} = \mu_{k|k-1} + K_k(Y_k - H_k\mu_{k|k-1}),$$
$$P_{k|k} = P_{k|k-1} - K_k H_k P_{k|k-1}, \tag{9.8}$$

where $K_k = P_{k|k-1}H_k^\top(H_k P_{k|k-1}H_k^\top + S_k)^{-1}$ is called as Kalman gain.

Proof The optimal estimate is the conditional mean and is computed in two steps: the forecast step using the model difference equations and the data assimilation step. Next, we shall prove it by using the Bayesian filtering framework.

Step of prediction

$$\mu_{0|0} = \mu_0 = E[X_0]$$
$$P_{0|0} = E[(X_0 - \mu_0)(X_0 - \mu_0)^\top] \tag{9.9}$$

Assume now that we have an optimal estimate $\mu_{k-1|k-1} = E[X_{k-1}|Y_{1:k-1}]$ with $P_{k-1|k-1}$ covariance at time $k-1$. The predictable part of $\mu_{k|k-1}$ is given by

$$\mu_{k|k-1} = E[X_k|Y_{1:k-1}]$$
$$= E[A_{k-1}X_{k-1} + B_{k-1} + W_{k-1}|Y_{1:k-1}]$$
$$= A_{k-1}\mu_{k-1|k-1} + B_{k-1}. \tag{9.10}$$

The forecast error is given as follows:

$$e_{k|k-1} = X_k - \mu_{k|k-1}$$
$$= A_{k-1}(X_{k-1} - \mu_{k-1|k-1}) + W_{k-1}$$
$$= A_{k-1}e_{k-1|k-1} + W_{k-1}. \tag{9.11}$$

The forecast error covariance is given by

$$P_{k|k-1} = E[e_{k|k-1}e_{k|k-1}^\top]$$
$$= E[(A_{k-1}e_{k-1} + W_{k-1})(A_{k-1}e_{k-1} + W_{k-1})^\top]$$
$$= A_{k-1}E[e_{k-1}e_{k-1}^\top]A_{k-1}^\top + Q_{k-1}$$
$$= A_{k-1}P_{k-1}A_{k-1}^\top + Q_{k-1}. \tag{9.12}$$

Step of update

At the time k we have two pieces of information: the forecast value $\mu_{k|k-1}$ with the covariance $P_{k|k-1}$ and the measurement Y_k with the covariance S_k. Then, by using the (9.2), there is

$$p(X_k|Y_{1:k}) = \frac{p(Y_k|X_k, Y_{1:k-1})p(X_k|Y_{1:k-1})}{p(Y_k|Y_{1:k-1})}. \tag{9.13}$$

On the right-hand side of (9.13), the denominator part is independent with X_k, which is only a normalization coefficient. The molecular part is the product of two Gaussian densities, so the exponential part of the density function is still a quadratic polynomial, where the variance and expectation are determined by the coefficient of the quadratic term and the coefficient of the linear term.

Let us focus on the coefficient of quadratic term in the molecular part of (9.13) and the basic definition of Gaussian densities, which yields

$$-\frac{1}{2}H_k^\top (S_k)^{-1} H_k - \frac{1}{2}P_{k|k-1}^{-1} =: -\frac{1}{2}P_{k|k}^{-1} \tag{9.14}$$

Since

$$(H_k^\top S_k^{-1} H_k + P_{k|k-1}^{-1}) = P_{k|k-1}^{-1}(I_n + P_{k|k-1} H_k^\top S_k^{-1} H_k)$$
$$= ((I_n + P_{k|k-1} H_k^\top S_k^{-1} H_k)^{-1} P)^{-1}, \tag{9.15}$$

then we consider the following matrix equation:

$$(I_n + P_{k|k-1} H_k^\top S_k^{-1} H_k) \cdot (I - P_{k|k-1} H_k^\top (H_k P_{k|k-1} H_k^\top + S_k)^{-1} H_k)$$
$$= I_n + P_{k|k-1} H_k^\top S_k^{-1} H_k - P_{k|k-1} H_k^\top (H_k P_{k|k-1} H_k^\top + S_k)^{-1} H_k$$
$$- P_{k|k-1} H_k^\top S_k^{-1} H_k P_{k|k-1} H_k^\top (H_k P_{k|k-1} H_k^\top + S_k)^{-1} H_k \tag{9.16}$$
$$= I_n,$$

which means that

$$P_{k|k} = ((I_n + P_{k|k-1} H_k^\top S_k^{-1} H_k)^{-1} P)^{-1} = (I_n - K_k H_k^\top) P_{k|k-1}, \tag{9.17}$$

with $K_k = P_{k|k-1} H_k^\top (H_k P_{k|k-1} H_k^\top + S_k)^{-1}$.

Then, similarly, consider the linear term which is $2P_{k|k-1}^{-1}\mu_{k|k-1} + 2H_k^\top S_k^{-1} Y_k$ and using the (9.17), we can have

$$\mu_{k|k} := P_{k|k}(P_{k|k-1}^{-1}\mu_{k|k-1} + H_k^\top S_k^{-1} Y_k)$$
$$= \mu_{k|k-1} + K_k(Y_k - H_k \mu_{k|k-1}). \tag{9.18}$$

So, we finish the proof. □

Next, we will reduce the KF from the perspective of linear control. From the previous proof, it is not difficult to see that the optimal estimation at time k is determined by the optimal estimation at the previous time and the observation k. So we shall consider the linear control system as follows:

$$\bar{X}_k = A_{k-1}\bar{X}_{k-1} + B_{k-1} + K_k Y_k + U_k(\bar{X}_{k-1}), \quad (9.19)$$

where Y_k is given in (9.5).

We can understand Eq. (9.19) as the original system equation and add the extra control functions K_k, U_k and which are assumed to be linear functions of X_{k-1} and Y_k.

In control theory, we have two important properties, controllability and optimality.

- **Controllability** is about whether the desired state can be achieved.
- **Optimality** is to find the optimal solution to all controllable targets, which is often in the sense of a loss function such as mean square error.

It is trivial to see that $\bar{X}_{k|k}$ in system (9.19) can become any Gaussians at a given time k and it cannot become any non-Gaussians at any time k. And for controllability, we shall further assume

$$E[\bar{X}_{k|k}|\bar{X}_{k-1|k-1}] = A_{k-1}\bar{X}_{k-1|k-1}. \quad (9.20)$$

So, $U(\cdot)$ will be

$$U(\bar{X}_{k-1|k-1}) = -K_k H_k A_{k-1}\bar{X}_{k-1|k-1}. \quad (9.21)$$

Furthermore, by the means of optimality, that is to choose the suitable control functions K_k for the following equation to minimize the covariance of such estimator:

$$\min_{K_k, 1\leq k\leq n} \sum_{k=1}^{n} J_k(\bar{X}_k),$$
$$\bar{X}_k = A_{k-1}\bar{X}_{k-1|k-1} + B_{k-1} + K_k(Y_k - H_k A_{k-1}\bar{X}_{k-1|k-1}), \quad (9.22)$$

The cost functional to be minimized is given by

$$J_k(\bar{X}_{k|k}) := E[(\bar{X}_{k|k} - E[\bar{X}_{k|k}])^\top (\bar{X}_{k|k} - E[\bar{X}_{k|k}])] \in R. \quad (9.23)$$

We can denote that $\bar{P}_{k|k}$ as the covariance matrix of $\bar{X}_{k|k}$. By using the $\mathrm{tr}(ab) = \mathrm{tr}(ba) \forall a, b^\top \in R^{n\times m}$, we shall have

$$J_k = \mathrm{tr}(\bar{P}_{k|k}).$$

9.2 Filtering Algorithm Based on Bayesian Framework

Furthermore, let $\bar{X}_{k|k-1} = A_{k-1}\bar{X}_{k-1|k-1} + B_{k-1}$ and $\bar{P}_{k|k-1} := E[(\bar{X}_{k|k-1} - E[\bar{X}_{k|k-1}])(\bar{X}_{k|k-1} - E[\bar{X}_{k|k-1}])^\top]$, Eq. (9.22) can be simplified into the following:

$$\min_{K_k, 1 \le k \le n} \sum_{k=1}^n J_k(\bar{X}_k), \quad (9.24)$$

$$\bar{X}_k = \bar{X}_{k|k-1} + K_k(Y_k - H_k\bar{X}_{k|k-1}).$$

We assume that

$$K_k = P_{k|k-1} H_k^\top (H_k P_{k|k-1} H_k^\top + S_k)^{-1} + \hat{K}_k, \quad (9.25)$$

and we need to choose a $\hat{K}_k$ to minimize Eq. (9.22)

$$\begin{aligned} J_k &= tr(\bar{P}_{k|k}) = tr(\bar{P}_{k|k-1} - \bar{P}_{k|k-1} H_k^\top (H_k \bar{P}_{k|k-1} H_k^\top + S_k)^{-1} H_k \bar{P}_{k|k-1}) \\ &\quad + tr(\hat{K}_k (H_k \bar{P}_{k|k-1} H_k^\top + S_k)\hat{K}_k^\top). \end{aligned} \quad (9.26)$$

Now, $H_k \bar{P}_{k|k-1} H_k^\top + S_k$ is positive-defined matrix. So,

$$tr(\hat{K}_k(H_k \bar{P}_{k|k-1} H_k^\top + S_k)\hat{K}_k^\top) \ge 0, \quad (9.27)$$

for all $\hat{K}_k \in R^{n \times m}$. Therefore, the minimizing $\hat{K}_k = 0_{n \times m}$,

So we can summarize the above analysis into the following theorem.

Theorem 9.2 *For the discrete filtering system (9.5), if we assume the initial distribution $X_0 \sim \mathcal{N}(\mu_0, P_0)$. Consider the following control problem:*

$$\min_{K_k, U_k, 1 \le k \le n} \sum_{k=1}^n E[(\bar{X}_k - X_k)^\top (\bar{X}_k - X_k)]$$

Subject to as follows: $\bar{X}_k = A_{k-1}\bar{X}_{k-1} + B_{k-1} + K_k(Y_k - H_k A_{k-1}\bar{X}_{k-1})$, $1 \le k \le n$. (9.28)

And the optimal solution pair K_k of Eq. (9.28) is the Kalman gain.

9.2.3 From KF to EKF

It is almost impossible to encounter completely linear systems in practical problems. We need a set of algorithms that can be used for nonlinear systems. From the perspective of analysis, the system function and the observation function can be approximated as some linear functions. Then, an extended Kalman filter [26] motivated by the theory of KF is proposed, and such an idea is called local linearization.

For a general nonlinear filtering system, (9.1) and any k, the following Taylor-like expansion is expected:

$$f_k(x) = \nabla f_k(X_k)(x - X_k) + f_k(X_k) + O(\|x - X_k\|_2^2), \tag{9.29}$$

$$h_k(x) = \nabla h_k(X_k)(x - X_k) + h_k(X_k) + O(\|x - X_k\|_2^2), \tag{9.30}$$

where the $\nabla f_k(\cdot)$ and $\nabla h_k(\cdot)$ are the Jaccobi matrices of f and h, the $O(\|\cdot\|_2^m)$ represent the residual term of at least order m in the Taylor expansion in Taylor expansion, and $\|\cdot\|_2$ is the 2−norm of vectors in Euclidean space.

Theorem 9.3 *For a nonlinear and discrete filtering system (9.1), if we assume the initial distribution $X_0 \sim \mathcal{N}(\mu_0, P_0)$, then there are sequential suboptimal estimation called extended Kalman filter (EKF). We assume they are $\mu_{k|k}$ and $P_{k|k}$ which satisfy the following equations:*

The prediction step

$$\begin{aligned} \mu_{k|k-1} &= f(\mu_{k-1|k-1}), \\ P_{k|k-1} &= \nabla f_{k-1}(\mu_{k-1|k-1}) P_{k-1|k-1} \nabla f_{k-1}^\top(\mu_{k-1|k-1}) + Q_{k-1}. \end{aligned} \tag{9.31}$$

The update step

$$\begin{aligned} \mu_{k|k} &= \mu_{k|k-1} + K_k(Y_k - h_k(\mu_{k|k-1})), \\ P_{k|k} &= P_{k|k-1} - K_k \nabla h_k(\mu_{k|k-1}) P_{k|k-1}, \end{aligned} \tag{9.32}$$

where $K_k = P_{k|k-1} \nabla h_k^\top(\mu_{k|k-1})(\nabla h_k(\mu_{k|k-1}) P_{k|k-1} \frac{\partial h_k^\top}{\partial x}(\mu_{k|k-1}) + S_k)^{-1}$.

Proof It follows from the Theorem 9.1 with the Eqs. (9.29) and (9.30). □

9.2.4 UKF

Notice that it is easier to approximate a probability distribution than it is to approximate an arbitrary nonlinear transformation. Based on intuition, the unscented transformation (UT) is proposed, which aims to calculate the statistics of a random variable that undergoes a nonlinear transformation. It is a deterministic sampling technique, which uses a minimal set of sample points, called sigma points, which are chosen such that the weighted sample mean and covariance of the sigma points are close enough to the real mean and covariance. It has been proven that UT gives the third-order accuracy for Gaussian inputs and at least the second accuracy for non-Gaussian inputs. More details about UT can be found in [14].

9.2 Filtering Algorithm Based on Bayesian Framework

Step of prediction

A set of sigma points is derived from the estimated state:

$$\mathcal{X}^0_{k-1|k-1} = \mu_{k-1|k-1},$$
$$\mathcal{X}^i_{k-1|k-1} = \mu_{k-1|k-1} + \left(\sqrt{(n+\lambda)P_{k-1|k-1}}\right)_i, i = 1 \ldots n, \quad (9.33)$$
$$\mathcal{X}^i_{k-1|k-1} = \mu_{k-1|k-1} - \left(\sqrt{(n+\lambda)P_{k-1|k-1}}\right)_i, i = n+1 \ldots 2n,$$

where $\left(\sqrt{(n+\lambda)P_{k-1|k-1}}\right)_i$ is the i-th column of the matrix square root of $(n+\lambda)P_{k-1|k-1}$.

Then the predicted mean and covariance are computed by propagating the sigma points through the transition function f.

$$\mathcal{X}^i_{k|k-1} = f\left(\mathcal{X}^i_{k-1|k-1}\right), i = 0 \ldots 2n,$$

$$\mu_{k|k-1} = \sum_{i=0}^{2n} \mathcal{W}^i_m \mathcal{X}^i_{k|k-1},$$

$$P_{k-1|k-1} = \sum_{i=0}^{2n} \mathcal{W}^i_c \left[\mathcal{X}^i_{k|k-1} - \mu_{k|k-1}\right] \times \left[\mathcal{X}^i_{k|k-1} - \mu_{k|k-1}\right]^T + Q_{k-1},$$
$$(9.34)$$

where the associated weights for the state and covariance estimation are given by

$$\mathcal{W}^0_m = \frac{\lambda}{(n+\lambda)},$$
$$\mathcal{W}^0_c = \lambda/(n+\lambda) + \left(1 - \alpha^2 + \beta\right), \quad (9.35)$$
$$\mathcal{W}^i_m = \frac{1}{2(n+\lambda)}, \quad i = 1 \ldots 2n,$$
$$\mathcal{W}^i_c = \frac{1}{2(n+\lambda)}, \quad i = 1 \ldots 2n.$$

Parameter λ is a scaling factor, which is defined as

$$\lambda = \alpha^2(n+\kappa) - n, \quad (9.36)$$

where α and κ control the spread of the sigma points and β is related to the distribution of x_k. If noises are Gaussian, $\beta = 2$ is optimal.

Step of update. Similar to the prediction step, a set of $2n + 1$ sigma points are derived to compute the filtered mean and covariance.

$$\begin{aligned}
\mathcal{X}^0_{k|k-1} &= \mu_{k|k-1}, \\
\mathcal{X}^i_{k|k-1} &= \mu_{k|k-1} + \left(\sqrt{(n+\lambda)P_{k|k-1}}\right)_i, i = 1\ldots n, \\
\mathcal{X}^i_{k|k-1} &= \mu_{k|k-1} - \left(\sqrt{(n+\lambda)P_{k|k-1}}\right)_i, i = n+1\ldots 2n.
\end{aligned} \quad (9.37)$$

After that the sigma points are projected through the observation function h as follows:

$$Y^i_k = h\left(\mathcal{X}^i_{k|k-1}\right), i = 0\ldots 2n. \quad (9.38)$$

Then the weighted sigma points are recombined to produce the predicted measurement mean and covariance as follows:

$$\begin{aligned}
\hat{y}_k &= \sum_{i=0}^{2n} \mathcal{W}^i_m Y^i_k, \\
P_{zz} &= \sum_{i=0}^{2n} \mathcal{W}^i_c \left[Y^i_k - \hat{y}_k\right]\left[Y^i_k - \hat{y}_k\right]^T + R_k.
\end{aligned} \quad (9.39)$$

The state-measurement cross-covariance matrix is given as

$$P_{xz} = \sum_{i=0}^{2n} \mathcal{W}^i_c \left[\mathcal{X}^i_{k|k-1} - \mu_{k|k-1}\right]\left[\mathcal{X}^i_{k|k-1} - \mu_{k|k-1}\right]^T, \quad (9.40)$$

which is used to compute the UKF gain by

$$K_k = P_{xz} P_{zz}^{-1}. \quad (9.41)$$

The updated state is the predicted state plus the innovation weighted by the UKF gain:

$$\mu_{k|k} = \mu_{k|k-1} + K_k \left(y_k - \hat{y}_k\right). \quad (9.42)$$

Moreover, the updated covariance is the predicted covariance minus the predicted measurement covariance weighted by the UKF gain:

$$P_{k|k} = P_{k|k-1} - K_k P_{zz} K_k^T. \quad (9.43)$$

9.2.5 Discrete Particle Methods for Filtering

Next, in this subsection, we shall introduce the filtering algorithm based on the simulation approach, which is well-known as particle filter (PF). A discrete PF is a filtering system based on a Bayesian framework.

For filtering problems, it is often necessary to solve some characteristics of density functions, i.e.,

$$E[\phi(X_k)|Y_{1:k}] = \int \phi(X_k) p(X_k|Y_{1:k}) dX_k, \qquad (9.44)$$

where the $\phi(X_k)$ is a test function.

9.2.5.1 Monte Carlo Method

For the above-expected integral operation, we use the Monte Carlo method to give a numerical approximation. We use the empirical distribution of N_p particles to approximate the posterior distribution

$$p(X_k|Y_{1:k}) \approx \hat{p}(X_k|Y_{1:k}) := \frac{1}{N_p} \sum_{i=1}^{N_p} \int \delta(X_k - X_k^i), \qquad (9.45)$$

where the particles $\{X_k^i\}_{i=1}^{N_p}$ are independently sampling from the $p(X_k|Y_{1:k})$. As the $N_p \to \infty$, the $\hat{p}(X_k|Y_{1:k}) \to p(X_k|Y_{1:k})$, so that we can get a good approximation by using the particles $\{X_k^i\}_{i=1}^{N_p}$:

$$E[\phi(X_k)|Y_{1:k}] \approx \int \phi(X_k) \hat{p}(X_k|Y_{1:k}) dX_k = \frac{1}{N_p} \sum_{i=1}^{N_p} \phi(X_k^i). \qquad (9.46)$$

9.2.5.2 Sequential Importance Sampling

So now, we need to propose an effective sampling method to sample the $p(X_k|Y_{1:k})$. But the posterior distribution $pp(X_k|Y_{1:k})$ is very complex, and it is hard to sample from it. Usually, we will consider a simpler function $q(X_k|Y_{1:k})$, and the density function is the so-called proposal distribution or the importance distribution.

$$\begin{aligned} E[\phi(X_k)|Y_{1:k}] &= \int \phi(X_k) \frac{p(X_k|Y_{1:k})}{q(X_k|Y_{1:k})} q(X_k|Y_{1:k}) dX_k \\ &= \int \phi(X_k) \frac{\omega_k(X_k)}{p(Y_{1:k})} q(X_k|Y_{1:k}) dX_k \end{aligned} \qquad (9.47)$$

$$= \frac{1}{p(Y_{1:k})} \int \phi(X_k)\omega_k(X_k)q(X_k|Y_{1:k})dX_k,$$

where

$$\omega_k(X_k) = \frac{p(Y_{1:k}|X_k)p(X_k)}{q(X_k|Y_{1:k})} \propto \frac{p(X_k|Y_{1:k})}{q(X_k|Y_{1:k})}, \quad (9.48)$$

which is known as the unnormalized importance weight. So we can reform Eq. (9.47) as follows:

$$E[\phi(X_k)|Y_{1:k}] = \frac{\int \phi(X_k)\omega_k(X_k)q(X_k|Y_{1:k})dX_k}{\int p(Y_{1:k}|X_k)p(X_k)dX_k}$$

$$= \frac{\int \phi(X_k)\omega_k(X_k)q(X_k|Y_{1:k})dX_k}{\int \omega_k(X_k)q(X_k|Y_{1:k})dX_k}$$

$$= \frac{E_{q(X_k|Y_{1:k})}[\omega_k(X_k)\phi(X_k)]}{E_{q(X_k|Y_{1:k})}[\omega_k(X_k)]}. \quad (9.49)$$

Equation (9.49) means that we can use the particles which are sampled from another distribution to approximate the expectation of original distributions. This approach is well-known as importance sampling.

$$E[\phi(X_k)|Y_{1:k}] \approx \frac{\frac{1}{N_p}\sum_{i=1}^{N_p}\omega_k(X_k^i)\phi(X_k^i)}{\frac{1}{N_p}\sum_{i=1}^{N_p}\omega_k(X_k^i)} = \sum_{i=1}^{N_p}\tilde{\omega}_k(X_k^i)\phi(X_k^i), \quad (9.50)$$

where $\tilde{\omega}_k(X_k^i) = \frac{\omega_k(X_k^i)}{\sum_{i=1}^{N_p}\omega_k(X_k^i)}$ holds.

So when we use PF, what we need is a stream of evolving particles and normalized weights for each step $\{X_k^i, \tilde{\omega}_k^i\}_{i=1}^{N_p}$.

Similar to the UKF and EKF, we shall summarize the PF algorithm in the Bayesian framework as follows. In the following, we can assume sampling from the proposed distributions $q(X_k|Y_{1:k})$ instead of $p(X_k|Y_{1:k})$.

Step of prediction Prediction involves using a system function to push forward the particles, i.e.,

$$X_{k+1}^i \sim \tilde{q}(X_{k+1}^i|X_k^i, Y_{1:k+1}) := \frac{q(X_{k+1}^i|Y_{1:k+1})}{q(X_k^i|Y_{1:k})}, \quad (9.51)$$

where the weights of particles are unchanged in this step.

Step of update The update step involves the application of Bayes' formula to update the weights. Given a new observation, the unnormalized importance weight

9.2 Filtering Algorithm Based on Bayesian Framework

$\{\omega_k^i\}_{i=1}^{N_p}$ evolve as

$$\begin{aligned}
\omega_{k+1}^i &= \frac{p(X_K^i|Y_{1:k})p(X_{k+1}^i|X_k^i)}{q(X_k^i|Y_{1:k})\tilde{q}(X_{k+1}^i|X_k^i,Y_{1:k+1})} \\
&\propto \omega_k^i \frac{p(X_K^i|Y_{1:k})p(X_{k+1}^i|X_k^i)}{p(X_k^i|Y_{1:k})\tilde{q}(X_{k+1}^i|X_k^i,Y_{1:k+1})} \\
&= \omega_k^i \frac{p(Y_{k+1}|X_k^i)p(X_{k+1}^i|X_k^i)}{p(X_k^i|Y_{1:k})\tilde{q}(Y_{k+1}|Y_{1:k})} \qquad (9.52) \\
&\propto \omega_k^i \frac{p(Y_{k+1}|X_k^i)p(X_{k+1}^i|X_k^i)}{\tilde{q}(Y_{k+1}|Y_{1:k})}.
\end{aligned}$$

The algorithm using formula (9.51) and (9.52) to iterate the weighted particles is sequential importance sampling.

9.2.5.3 Advantage and Disadvantage for PF

As we know, the PF is a design based on the Monte Carlo method which means that there is a standard method for error analysis, please refer to [1–4].

The particles approximates the true posterior p_0 in R^n by the empirical measure $p_0^{(N)}$ defined as $\sum_{i=1}^N \delta(x - X_i)$ where $\{X_i\}_{i=1}^N$ are the particles sampling from p_0, for any test function $\phi : R^n \to R$, then

$$E_{p_0}(\phi) = \int_{R^n} \phi(x)p_0(x)dx, \quad E_{p_0}^{(N)}(\phi) = \frac{1}{N}\sum_{i=1}^N \phi(X_i). \qquad (9.53)$$

The variance of this estimator is defined as $Var_{p_0}(\phi) = (E_{p_0}(\phi^2)) - E_{p_0}(\phi)^2$, furthermore we shall have the following error estimation:

$$E[(E_{p_0}^{(N)}(\phi) - E_{p_0}(\phi))^2] \le C\frac{Var_{p_0}(\phi)}{N}, \qquad (9.54)$$

where C is some constant.

We summarize the advantages of PF as follows:

- By Eq. (9.54), we can obtain an explicit PF convergence order on N.
- Any improvements in sampling methods can give birth to new PFs.
- The distribution is completely constructed by PF.

However, if we want to extend the result into some global convergence of PF, we must know the difference between weighted particles and equally weighted particles. In the next, we shall introduce the effective number of samples which is important to understand how to characterize weight degradation.

Definition 9.1 ([3]) An indicator of the degree of depletion is the the effective number of samples, N_{eff} defined in terms of the coefficient of variation Var as

$$N_{eff} = \frac{N}{1 + Var(\{\tilde{\omega}_k^i\}_{i=1}^N)}, \qquad (9.55)$$

where the N is the particle number in the simulation and the $\{\tilde{\omega}_k^i\}_{i=1}^N$ are the weights.

Remark 9.2 A logical computable approximation of N_{eff} is provided by

$$\hat{N}_{eff} = \frac{1}{\sum_{i=1}^N (\omega_k^i)^2}$$

This approximation shares the property $1 \leq \hat{N}_{eff} \leq N$ with the definition. The upper bound $N_{eff} = N$ is attained when all particles have the same weight and the lower bound $N_{eff} = 1$ when all the probability mass is devoted to a single particle. The resampling condition in the PF can now be defined as $N_{eff} < N_{th}$. The threshold can for instance be chosen as $\hat{N}_{th} = \frac{2N}{3}$.

It has been shown that the variance of the importance weights can only increase over time, and thus, it is impossible to avoid the degeneracy phenomenon.

So, it is important to avoid the degeneracy phenomenon, and a common solution is resampling.

Step of resampling
This step involves sampling X_k^i with replacement from the set of particle positions according to the probability vector of normalized weights $\{\tilde{w}_k^i\}_{i=1}^{N_p}$. After resampling, the particles are set to be equally weighted for the next iteration.

- The standard version of the PF algorithm is termed sampling importance resampling (SIR) [5], or bootstrap PF, and is obtained by resampling each time.
- The alternative is to use importance sampling, in which case resampling is performed only when needed. This is called sampling importance sampling (SIS). Usually, resampling is done when the effective number of samples, as will be defined in the next section, becomes too small.

As an alternative, the resampling step can be replaced with a sampling step from a distribution that is fitted to the particles after both the time and measurement update. The Gaussian PF (GPF) fits a Gaussian distribution to the particle after which a new set of particles is generated from this distribution. The Gaussian sum PF (GSPF) uses a Gaussian sum instead of a distribution.

However, in the traditional PF framework, the PF suffers from weight degeneracy and the number of particles increases exponentially as the dimension increases. The weight degeneracy problem can only be alleviated by resampling steps under ten dimensions, while rapid increases in computational capacity have provided an impetus to the development of particle methods.

9.2.5.4 An Ensemble Method for EKF

As a successful algorithm, EKF is widely used in many different applications. In contrast to the standard EKF, which works with the entire distribution of the state explicitly, the ensemble Kalman filter (EnKF) stores, propagates, and updates an ensemble of vectors that approximates the state distribution. The EnKF has been highly successful in many extremely high-dimensional, nonlinear, and non-Gaussian data assimilation applications [19], which can be considered a good extension of EKF.

The basic idea of EnKF is very similar to PF. However, the EnKF uses the particle to obtain the conditional means instead of constructing the distribution like PF. So it's more efficient and requires fewer particles.

We shall use the same linear approximation for the system that appeared in Theorem 9.3 as the starting point of EnKF.

We assume the particles $\left\{X_{k|k}^i\right\}_{i=1}^N$ is the simulation of $p(X_k|Y_{1:k})$ and $\bar{P}_{k|k}$ is the conditional variances of system (9.1) and we assume $\left\{X_{0|0}^i\right\} \sim X_0$ as the initial condition,

Step of prediction

$$X_{k|k-1}^i = f_{k-1}(X_{k-1|k-1}^i) + \Sigma_{k-1} W_{k-1}^i,$$
$$\bar{P}_{k|k-1} = \nabla f_{k-1}(\bar{\mu}_{k-1|k-1}) \bar{P}_{k-1|k-1} \nabla f_{k-1}^\top(\bar{\mu}_{k-1|k-1}) + Q_{k-1},$$
(9.56)

where the $\bar{\mu}_{k-1|k-1} = \frac{1}{N} \sum_{i=1}^N X_{k-1|k-1}^i$.

Step of update

$$X_{k|k}^i = X_{k|k-1}^i + \bar{K}_k(Y_k - h_k(X_{k|k-1}^i) + V_k^i),$$
$$\bar{P}_{k|k} = \bar{P}_{k|k-1} - \bar{K}_k \nabla h_k(\bar{\mu}_{k|k-1}) \bar{P}_{k|k-1},$$
(9.57)

where $\bar{K}_k = \bar{P}_{k|k-1} \nabla h_k^\top(\bar{\mu}_{k|k-1})(\nabla h_k(\bar{\mu}_{k|k-1}) \bar{P}_{k|k-1} \frac{\partial h_k^\top}{\partial x}(\bar{\mu}_{k|k-1}) + S_k)^{-1}$, the $\bar{\mu}_{k|k-1} = \frac{1}{N} \sum_{i=1}^N X_{k|k-1}^i$, the $\{W_k^i\}_{i=1}^N$ and $\{V_k^i\}_{i=1}^N$ are the sampling particle from W_k, V_k for any $k \geq 1$, respectively.

Finally, we use the $\bar{\mu}_{k|k} = \frac{1}{N} \sum_{i=1}^N X_{k|k}^i$ as the approximation of conditional means.

9.3 Filtering Algorithm of DMZ Framework

9.3.1 DMZ Equation After Applying the Hopf-Cole Transformation

Let us focus on the following $(n+m)$-dimensional system of non-linear filtering system:

$$\begin{cases} dX_t = f(t, X_t)dt + \sigma_V(t, X_t)dV_t, \\ dY_t = h(t, X_t)dt + dW_t. \end{cases} \quad (9.58)$$

where $X_t \in R^n$ is the state at time t, $Y_t \in R^m$ is the observation vector, and $\{W_t\}$, $\{V_t\}$ are two mutually independent Wiener processes taking values in R^d and R^m. The mappings $f(\cdot) : R^+ \times R^n \to R^n$, $h(\cdot) : R^+ \times R^n \to R^m$ and $\sigma_V(\cdot) : R^+ \times R^n \to R^{n \times n}$ are C^2 functions, and $\sigma_V(\cdot) : R^n \to R^{n \times n}$. The covariance matrix, $S(t)$, of the observation noise $\{W_t\}$ is assumed to be positive definite. The function f, h are column vectors whose j-th coordinate are denoted as f_i, h_j, i.e., $f = (f_1, f_2, \cdots, f_n)^\top$, $h = (h_1, h_2, \cdots, h_m)^\top$. The objective of the filtering problem is to estimate the posterior distribution of X_t given the history $\mathcal{Y}_t := \sigma(Y_s : 0 \leq s \leq t)$.

The unnormalized conditional density function $\sigma(t, x)$ of (9.59) on the observation history $\mathcal{Y}_t$ satisfies as follows:

$$\begin{cases} d\sigma(t, x) = L_0 \sigma(t, x)dt + h^\top(t, x)S^{-1}(t)\sigma(t, x) \circ dY_t, \\ \sigma(0, x) = \sigma_0(x), \end{cases} \quad (9.59)$$

where

$$L_0(*) := \frac{1}{2} \sum_{i,j=1}^n \frac{\partial^2}{\partial x_i \partial x_j} \left(Q_{i,j}(t, x) \cdot * \right) - \sum_{i=1}^n \frac{\partial}{\partial x_i} (f_i(t, x) \cdot *) \\ + \frac{1}{2} h(t, x)^\top S^{-1} h(t, x) \cdot (*), \quad (9.60)$$

and $Q(t, x) := \sigma_V(t, x)\sigma_V^\top(t, x)$.

The posterior density $p(t, x|\mathcal{Y}_t)$ can be easily obtained by

$$p(t, x|\mathcal{Y}_t) := \frac{\sigma(t, x)}{\int_{R^n} \sigma(t, x)dx}. \quad (9.61)$$

In the following of this section, we shall denote that $\pi_t(\cdot) := \int_{R^n}(\cdot)p(t, x)dx$, for any $t \geq 0$.

9.3 Filtering Algorithm of DMZ Framework

Now, we need to transform for the density $\sigma(t,x)$ into the $u(t,x)$ by using the Hopf-Cole transformation as follows:

$$u(t,x) := -\log \sigma(t,x), \tag{9.62}$$

Theorem 9.4 *Consider the filtering system of (9.58), and we donate the $u(t,x)$ to be the solution of DMZ equation after applying the Hopf-Cole transformation. Then, $u(t,x)$ is the solution of the following SPDE:*

$$\begin{aligned}
du = & \frac{1}{2} \sum_{i,j=1}^{n} \left(Q_{i,j} \frac{\partial^2 u}{\partial x_i \partial x_j} + \left(\frac{\partial Q_{i,j}}{\partial x_i} \frac{\partial u}{\partial x_j} + \frac{\partial Q_{i,j}}{\partial x_j} \frac{\partial u}{\partial x_i} \right) - Q_{i,j} \frac{\partial u}{\partial x_i} \frac{\partial u}{\partial x_j} \right) dt \\
& - \sum_{i=1}^{n} f_i(t,x) \frac{\partial u}{\partial x_i} dt - \frac{1}{2} \left(\sum_{i,j=1}^{n} \frac{\partial^2 Q_{i,j}}{\partial x_i \partial x_j} + h^\top S^{-1} h \right) dt \\
& + \sum_{i=1}^{n} \frac{\partial f_i}{\partial x_i} dt - h^\top(t,x) S^{-1} \circ dy_t, \\
u(0,x) = & -\log \sigma(0,x).
\end{aligned} \tag{9.63}$$

Proof Submit the $\sigma(t,x) = e^{-u(t,x)}$ in Eq. (9.59); then the rest of the proof is trivial. □

9.3.2 Linear Filtering System and KBF

Similar to the previous section. Let us consider the simplest model of the continuous filtering system in which the state equation and the observation equation are both linear.

$$dX_t = A_t X_t dt + B_t dt + \sigma_V(t) dV_t, \tag{9.64}$$

$$dY_t = H_t X_t dt + D_t dt + dW_t, \tag{9.65}$$

where the $f(t,x) := A_t x + B_t$, $h(t,x) := H_t x + D_t$, $\sigma_V(t,x) := \sigma_V(t)$ holds in Eq. (9.58). We assume $X_0 \sim N(\mu_0, P_0)$.

Corollary 9.1 *Consider the linear filtering system in (9.64), the $u(t,x)$ in Theorem 9.4 can be simplified as follows:*

$$du = \frac{1}{2} \sum_{i,j=1}^{n} Q_{i,j} \left(\frac{\partial^2 u}{\partial x_i \partial x_j} - Q_{i,j} \frac{\partial u}{\partial x_i} \frac{\partial u}{\partial x_j} \right) dt$$

$$- (A_t x + B_t)^\top (t) \nabla u \, dt + tr(A) dt$$

$$- \frac{1}{2} (H_t x + D_t - \hat{h}_t)^\top S^{-1} (H_t x + D_t - \hat{h}_t) dt \qquad (9.66)$$

$$- x^\top H^\top (t) S^{-1}(t) \circ dY_t.$$

$$u(0, x) := \frac{1}{2} (x - \mu_0)^\top \Sigma_0^{-1} ((x - \mu_0)),$$

where $\hat{h} = \int_{R^n} h(t, x) p(t, x | \mathcal{Y}_t) dx$.

Furthermore, Eq. (9.66) has a closed subspace which is the space of a polynomial with a degree no more than 2. Since the solution of DMZ equation is sort of unnormalized density, we shall only consider coefficients of the quadratic term and the linear term. And from (9.66), it is easy to see that the quadratic term is independent of the observation process Y_t.

Let $u(t, x) = \frac{1}{2} x^\top (t) C(t) x + b(t)^\top x + e(t)$ hold. Now, we submit this representation into (9.66), then there are two evolution equations as follows:

$$\frac{dC(t)}{dt} = -C(t) Q(t) C(t) - A^\top (t) C(t) - C(t) A_t + H^\top (t) S^{-1}(t) H_t, \qquad (9.67)$$

and

$$db(t) = -C(t) Q(t) b(t) dt - A^\top b(t) dt - B^\top (t) b(t) - H^\top (t) S^{-1}(t) \circ dY_t. \qquad (9.68)$$

If we apply that $\mu(t) = -C(t) b(t)$, $P(t) = C^{-1}(t)$, and $\hat{h}_t = H_t \mu_t + D_t$, then (9.68) and (9.67) can be transformed into follows:

$$\begin{cases} d\mu(t) = A_t \mu(t) dt + B_t dt + K(t) \circ (dY_t - H_t \mu_t - D_t) \\ \frac{dP(t)}{dt} = A_t P(t) + P(t) A_t^\top + Q(t) - H^\top (t) S^{-1}(t) H_t, \end{cases} \qquad (9.69)$$

where $K(t) := P(t) H^\top (t) S^{-1}(t)$ is the Kalman gain.

9.3.3 From KBF to Continuous EKF

In this subsection, we shall consider the general continuous filtering system (9.58). Similarly, with the EKF for discrete systems, we shall consider approximating the system function and observation function with linear functions. Thus, the first question is how to choose the initial point to do Taylor's expansion. In the discrete

9.3 Filtering Algorithm of DMZ Framework

EKF algorithm, we consider using Taylor's expansion at the point of the prediction means which is governed by the state equation without noise term.

So, we shall consider the solution of the zero-noise limit case as the starting point:

$$\frac{d\bar{x}_t}{dt} = f(\bar{x}_t), \quad \bar{x}_0 = x_0. \tag{9.70}$$

Intuitively, if we need to approximate some nonlinear systems with the statistics of linear systems, these nonlinear systems should be close enough to the linear systems; otherwise, it is difficult to get a good result. Therefore, the following Taylor-like expansion is expected, and it should be close to the original nonlinear system:

$$\begin{cases} dX_t \approx (f'(\bar{x}_t)(X_t - \bar{x}_t) + f(\bar{x}_t))dt + \sigma_V(\bar{x}_t)dV_t \\ dY_t \approx (h'(\bar{x}_t)(X_t - \bar{x}_t) + h(\bar{x}_t))dt + dW_t, \end{cases} \tag{9.71}$$

where the $\bar{x}_0 = X_0$. From the above analysis, we can directly notice that if the noise term in the system equation is relatively large, then the estimation is not accurate. Furthermore, there is an improved method called bilinear extended Kalman filter (BEKF) which considers the Taylor expansion to up to the first-order term for the noise terms, and it is much better than EKF.

Proposition 9.1 *Let $\mu_t = \{\mu_t, t \geq 0\}$ be the conditional mean of the signal of (9.71). We denote conditional covariance matrix, P_t, is the $n \times n$-dimensional process with components whose $i, j-$ entry is defined below:*

$$P_t^{i,j} = E[X_t^i X_t^j | Y_t] - E[X_t^i | Y_t] E[X_t^j | Y_t] \quad i, j = 1, \ldots, d, \ t \geq 0. \tag{9.72}$$

Then μ_t satisfies the stochastic differential equation

$$d\mu_t = (f'(\bar{x}_t)(\mu_t - \bar{x}_t) + f(\bar{x}_t))dt + R_t H_t^\top (dY_t - (H_t \mu_t + h_t)dt)$$
$$+ R_t h'^\top(\bar{x}_t)[dY_t - (h'(\mu_t)(\mu_t - \bar{x}_t) + h(\bar{x}_t))dt]. \tag{9.73}$$

and P satisfies the deterministic matrix Riccati equation

$$\frac{dP_t}{dt} = f'(\bar{x}_t) P_t + P_t f'(\bar{x}_t)^\top + \sigma(\bar{x}_t) \sigma^\top(\bar{x}_t) - P_t h'(\bar{x}_t)^\top h'(\bar{x}_t) R_t. \tag{9.74}$$

with $\mu_0 = x_0$ and $R_0 = P_0$. Hence, we can estimate the position of the signal by using μ_t as computed above. We can use the same procedure, but instead of x_t, we can use any Y_t-adapted estimator process μ_t.

Proof It is the direct consequence of KBF. □

From the process of deriving the EKF algorithm, it is obvious that the effectiveness of the EKF algorithm depends on the following factors [8, 12]:

- The f function and h function can be well approximated by linear functions.
- The initial position of the signal is well approximated.
- System noise is small enough.

9.3.4 Control-Oriented Particle Filtering for Multidimensional System

In this section, we shall introduce a continuous particle filter called feedback particle filter (FPF) which is motivated by the mean-fields control theory [10]. Firstly, consider the previous filtering problem (9.58), and we assume that the system function and observation function are all time-invariant.

Consider the equation of state

$$dX_t = f(X_t)dt + dV_t, X_0 \text{ is some distribution}, \tag{9.75}$$

where we assume $\sigma_V(\cdot) := I_n$ in (9.58) and all the function are time-invariant. There is a quite easy way to sample from the X_t at time t which is

$$dX_t^i = f(X_t^i)dt + dV_t^i, X_0^i \sim X_0, 1 \leq i \leq N, \tag{9.76}$$

where $\{X_t^i\}_{i=1}^N$ are the particles at time t and $\{V_t^i\}_{i=1}^N$ are sampling from standard Brownian motion.

As with traditional PF, we are not sampling from the posterior distribution but updating the weight of the particles by observations. However, such methods cause weight and particle degeneracy problems. Inspired by the idea of control, the main idea of FPF is adding extra feedback control terms for the evolution of the state equation which makes the posterior distribution at any time t exactly the posterior distribution of the filtering problem. The dynamics of the i-th particle have the following gain feedback form:

$$dX_t^i = f(X_t^i)dt + dV_t^i + u(t, X_t^i)dt + K(t, X_t^i) \circ dY_t, \tag{9.77}$$

where $dY_t = h(x)dt + dW_t$ and W_t is the standard Brownian motion.

The different control inputs $u(t, x)$ and the $K(t, x)$ will determine the different process, so there is a natural question that need to be answered:

Is there a control input (u, K) to make the posterior densities such that the posterior density function of Eq. (9.77) is the solution of the Kushner equation?

To answer the question, there are three steps:

9.3 Filtering Algorithm of DMZ Framework

1. We denote the posterior distribution of (9.77) as $p^*(t, x)$. And we analyze the evolution equation of posterior $p^*(t, x)$.
2. Compare the evolution equation of posterior $p^*(t, x)$ with the Kushner equation of filtering. Then, we construct control inputs $u(t, x)$ and the $K(t, x)$ making the two equations be same.
3. Validate the control inputs $u(t, x)$ and the $K(t, x)$.

First, we shall introduce admissible input as the start.

Definition 9.2 (Admissible Input) The control input $u(t, x)$ and the $K(t, x)$ are admissible if the random variables $u(t, X_t)$ and $K(t, X_t)$ in (9.77) are $\mathcal{Y}_t = \sigma(Y_s : s \leq t)$ measurable for each t. And for each t,

$$E[|u|] := E[|u(t, X_t)|] < \infty,$$

and

$$E[|K|^2] := E\left[\sum_{1 \leq l \leq n, 1 \leq j \leq m} |K_{lj}(t, X_t)|^2\right] < \infty.$$

Next, for a control dynamical system, the posterior distributions $p^*(t, x)$ are given as follows.

Proposition 9.2 Consider the process $\{X_t^i\}_{i=1}^N$ that evolves according to the PF model (9.77). The conditional distribution of X_t^i given the filtration $\mathcal{Y}_t$, $p^*(t, x)$, satisfies the forward equation

$$dp^* = \mathcal{L}p^* dt - \sum_{i=1}^{m} \nabla \cdot (p^*[K]_i) \circ dY_t^i - \nabla \cdot (p^* u) dt. \tag{9.78}$$

where $\mathcal{L}(*) = -\nabla(f \cdot *) + \frac{1}{2} \sum_{i=1}^{n} \frac{\partial^2}{\partial x_i^2}(*)$ and $[K]_i$ is i-the column of K.

We denote as $p(x, t)$ the conditional distribution of X_t given $\mathcal{Y}_t = \sigma(Y_s : s \leq t)$. The evolution of $p(x, t)$ is described by the Kushner equation

$$dp = L_0 p \, dt + p(h - \hat{h}) \circ (dY_t - \hat{h}_t dt), \tag{9.79}$$

where $\hat{h} = \int_{R^n} h(x) p(t, x) dx$, and

$$L_0(*) := \frac{1}{2} \sum_{i=1}^{n} \frac{\partial^2}{\partial x_i^2}(\cdot *) - \sum_{i=1}^{n} \frac{\partial}{\partial x_i}(f_i(x) \cdot *)$$
$$+ \frac{1}{2}(h(x) - \hat{h}_t)^\top (h(x) - \hat{h}_t) \cdot (*), \tag{9.80}$$

Next, we need to select the control input so that the two SPDEs (9.79) and (9.78) are the same.

Theorem 9.5 ([11]) *Consider the two evolution equations for p and p^*, defined according to the solution of the forward equation and the K-S equation, respectively. We select u, K as follows:*

$$\nabla \cdot (p[K]_i) = -p(h^i - \hat{h}^i), \forall 1 \leq i \leq m, \quad (9.81)$$

where $[K]_i$ is i-the column of K. And, we get

$$u = -\frac{1}{2}K(h - \hat{h}) + \frac{1}{2}(h - \hat{h})^T(h - \hat{h}). \quad (9.82)$$

Then, provided $p(\cdot, 0) = p^*(\cdot, 0)$, we have for all $t \geq 0$,

$$p(t, \cdot) = p^*(t, \cdot).$$

Proof It is only necessary to show that with the choice of $\{u, K\}$ given by, we have $dp(t, x) = dp^*(x, t)$.

Using the (9.81) in the forward equation (9.77), then we can have (9.79). □

However, it is easy to see that there are infinitely many K satisfies the condition (9.81). So, we need to choose the K minimized some cost function, and it is natural to consider the cost function as $E[|K|^2]$ which is considered as an Euler-Lagrange boundary value problem (E-L BVP).

The gain function K is obtained as a solution to E-L BVP: For $j = 1, 2, \cdots, m$, the function ϕ_j is a solution to the second-order differential equation

$$\nabla \cdot (p(t, x)\nabla\phi_j(x, t)) = -(h_j(x) - \hat{h}_j)p(t, x),$$
$$\int \phi_j(x, t)p(t, x)dx = 0, \quad (9.83)$$

where p denotes the conditional distribution of X_t^i given $\mathcal{Y}_t$. The gain function is given by

$$[K]_{i,j} = \frac{\partial \phi_j}{\partial x_i}. \quad (9.84)$$

Note that the gain function needs to be obtained for each value of time t. So, we can summarize the FPF algorithm as follows:

$$dX_t^i = f(X_t^i)dt + dV_t^i + K(t, X_t^i) \circ dI_t^i \quad (9.85)$$

9.4 Robust Filtering

Table 9.1 The development of FPF

Methods for (9.83)	Name	Reference
Constant approximation	constant FPF	[11]
Galerkin method	Galerkin FPF	[9, 11]
Kernel method	Diffusion FPF	[9]
Gaussian approximation	Gaussian FPF	[11]
Deep learning method	Deep FPF	[32]

The innovation process that appears in the nonlinear filter

$$dI_t^i = dY_t - \frac{1}{2}(h(X_t^i) + \hat{h})dt, \tag{9.86}$$

with $\hat{h} := E[h(X_t^i)|\mathcal{F}_t]$. In a numerical implementation, we approximate $\hat{h} \approx \hat{h}^N := \frac{1}{N}\sum_{i=1}^N h(X_t^i)$.

Theorem 9.6 (FPF for Linear System) *Consider the linear filtering system with the initial distribution to be Gaussian, which we assume $f(x) := Ax$, $h(x) := Hx$ (9.77), then (9.83) has an explicit solution such that $K(t,x) = P_t H^\top$ where P_t is the conditional covariance matrix of state and defined in KBF (9.69). So, (9.77) can be rewritten as follows:*

$$dX_t^i = AX_t^i dt + dV_t^i + P_t H^\top \circ \left(dY_t - \frac{HX_t^i + H\mu_t}{2}\right), \tag{9.87}$$

where μ_t is the conditional mean in KBF and $X_0^i \sim$ A Gaussian distribution.

Generally, $\phi(x)$ in (9.83) is hard to calculate explicitly for a general nonlinear filtering system. The different ways to approximate (9.83) represent the different FPF algorithms. We shall summarize the different FPFs in Table 9.1.

9.4 Robust Filtering

We consider the nonlinear autonomous system with state $x_k \in \mathbb{R}^n$ and observation $y_k \in \mathbb{R}^m$. It is given by the following state and observation equations:

$$x_k = f(x_{k-1}) + w_k \quad \text{(state equation)}, \tag{9.88a}$$

$$y_k = h(x_k) + v_k \quad \text{(observation equation)}, \tag{9.88b}$$

where $f : \mathbb{R}^n \to \mathbb{R}^n$ and $h : \mathbb{R}^n \to \mathbb{R}^m$ are nonlinear functions called state function and observation function, respectively. State noise w_k and observation noise v_k are uncorrelated multivariate Gaussian with zero means and nominal covariance matrices $Q_k \in \mathbb{R}^{n \times n}$ and $R_k \in \mathbb{R}^{m \times m}$, respectively. In what follows, we assume that the real distribution of observation noise is unknown to us. And the real distribution of observation noise is $(1 - \epsilon) \, \mathcal{N}(0, R_k) + \epsilon \, \mathcal{S}(0, S_k)$ rather than $\mathcal{N}(0, R_k)$, where $0 < \epsilon \ll 1$ is the unknown probability, and $\mathcal{S}(0, S_k)$ is an arbitrary unknown distribution with large covariance S_k. Here R_k is known to us, so it is called the nominal covariance matrix. Let $y_{1:k}$ denote the σ-algebra generated by noisy observations $\{y_1, \ldots, y_k\}$ induced by unknown large outliers. The outlier-robust filtering problem refers to solving the following conditional estimation problem:

$$\hat{\varphi}_k = \arg\min_{\varphi_k} \mathbb{E}\left[\|\varphi_k - \varphi(x_k)\|^2 \mid y_{1:k}\right], \tag{9.89}$$

where $\varphi(x)$ is a function of estimation interest (e.g., $\varphi(x) = x$ or $\varphi(x) = xx^\top$).

9.4.1 Nonlinear Regression Form and Robust Optimization Framework

To facilitate the subsequent discussion, $\hat{x}_{k|k-1}$ and $\hat{x}_{k|k}$ will be denoted as desired estimation means of prediction at time steps $k-1$ and k, respectively. $P_{k|k-1}$ and $P_{k|k}$ will be corresponding prediction covariances, respectively.

9.4.1.1 Nonlinear Regression Form

Here we shall first illustrate how to view the update step of extended Kalman filtering as a nonlinear regression problem. Assume that we have obtained $\hat{x}_{k|k-1}$ and observation y_k. Then let us first consider the augmented model, which is given by

$$\begin{bmatrix} \hat{x}_{k|k-1} \\ y_k \end{bmatrix} = \begin{bmatrix} x_k \\ h(x_k) \end{bmatrix} + v_k, \tag{9.90}$$

where v_k is given by

$$v_k = \begin{bmatrix} -(x_k - \hat{x}_{k|k-1}) \\ V_k \end{bmatrix}. \tag{9.91}$$

Then it is easy to see that

$$\mathbb{E}\left[V_k V_k^\top\right] = \begin{bmatrix} P_{k|k-1} & 0 \\ 0 & R_k \end{bmatrix} = B_k B_k^\top, \tag{9.92}$$

with

9.4 Robust Filtering

$$B_k = \begin{bmatrix} B_{k|k-1}^p & 0 \\ 0 & B_k^r \end{bmatrix}, \tag{9.93}$$

where $B_{k|k-1}^p$ and B_k^r are Cholesky decompositions of $P_{k|k-1}$ and R_k, respectively. Left multiplying both sides of (9.90) by B_k^{-1}, we obtain

$$d_k = m_k(x_k) + e_k, \tag{9.94}$$

where $d_k = \begin{bmatrix} \left(B_{k|k-1}^p\right)^{-1} \hat{x}_{k|k-1} \\ \left(B_k^r\right)^{-1} y_k \end{bmatrix}$, $m_k(x_k) = \begin{bmatrix} \left(B_{k|k-1}^p\right)^{-1} x_k \\ \left(B_k^r\right)^{-1} h(x_k) \end{bmatrix}$. Note that $e_k = B_k^{-1} v_k$, which implies $\mathbb{E}\left[e_k e_k^\top\right] = I_{n+m}$. Hence the residual error e_k is white noise, which makes Eq.(9.94) become a nonlinear regression function.

9.4.1.2 Robust Optimization Framework

With the help of regression function (9.94), we can formulate a optimization-based filtering update step. It is given by

$$\hat{x}_{k|k} = \arg\min_{x_k} \mathcal{L}(x_k), \tag{9.95}$$

where the cost function $\mathcal{L}(\cdot)$ is regression-induced in nature. It is given by

$$\mathcal{L}(x_k) = \sum_{i=1}^{n+m} \rho\left(e_{k,i}\right), \tag{9.96}$$

with $e_{k,i}$ is i-th component of the residual vector e_k. ρ is a robust cost function that is used to cut off the outliers. Note that for $\mathcal{L}(x_k)$, we have

$$\nabla_{x_k} \mathcal{L}(x_k) = \sum_{i=1}^{m+n} \frac{\partial \rho\left(e_{k,i}\right)}{\partial e_{k,i}} \frac{\partial e_{k,i}}{\partial x_k}. \tag{9.97}$$

Let us consider the following diagonal matrices:

$$\begin{aligned} \Psi_x(e_k) &= \mathrm{diag}\left[\psi(e_{k,1}), \psi(e_{k,2}), \ldots, \psi(e_{k,n})\right] \\ \Psi_y(e_k) &= \mathrm{diag}\left[\psi(e_{k,n+1}), \psi(e_{k,2}), \ldots, \psi(e_{k,m+n})\right], \end{aligned} \tag{9.98}$$

and

$$\Psi(e_k) = \begin{bmatrix} \Psi_x(e_k) & 0 \\ 0 & \Psi_y(e_k) \end{bmatrix} \tag{9.99}$$

with weight function $\psi(e_{k,i}) = \frac{\partial \rho(e_{k,i})}{\partial e_{k,i}} / e_{k,i}$. Notice that d_k does not depend on x_k. We can denote

$$m(x_k) = \nabla_{x_k} e_k = \frac{\partial (d_k - m_k(x_k))}{\partial x_k} = -\frac{\partial m_k(x_k)}{\partial x_k} \quad (9.100)$$

$$\nabla_{e_k} \mathcal{L}(x_k) = \Psi(e_k)(d_k - m_k(x_k)).$$

Then the gradient (9.97) can be rewritten by

$$\nabla_{x_k} \mathcal{L}(x_k) = \left(\frac{\partial (d_k - m_k(x_k))}{\partial x_k}\right)^\top \Psi(e_k)(d_k - m_k(x_k)) \quad (9.101)$$

$$= m(x_k)^\top \nabla_{e_k} \mathcal{L}(x_k).$$

The solution of $\nabla_{x_k} \mathcal{L}(x_k) = 0$ can be derived using iteratively reweighted least squares (IRLS) [34], which is of the form

$$\begin{aligned} x_k^{(j+1)} &= x_k^{(j)} - \left(m(x_k^{(j)})^\top \Psi(e_k^{(j)}) m(x_k^{(j)})\right)^{-1} \\ &\quad \times m(x_k^{(j)})^\top \Psi(e_k^{(j)}) \left(d_k - m_k(x_k^{(j)})\right), \end{aligned} \quad (9.102)$$

where the superscript (j) refers to the iteration index.

Remark 9.3 Eq.(9.102) is similar to the derivation process of standard IEKF in [33], where the IEKF is equivalent to Gauss-Newton method from an optimization perspective.

9.4.2 Iterative Outlier-Robust Extended Kalman Filtering

In this section, the novel OR-IEKF framework is presented. Its prediction step is the same with the common EKF, i.e.,

$$\begin{aligned} \hat{x}_{k|k-1} &= f\left(\hat{x}_{k-1|k-1}\right) \\ P_{k|k-1} &= f_k P_{k-1|k-1} f_k^\top + Q_k, \end{aligned} \quad (9.103)$$

where $f_k = \frac{\partial f}{\partial x_k}\Big|_{x_k = \hat{x}_{k-1|k-1}}$. The update step is given in Eq.(9.102). With initial point $x_k^{(0)} = \hat{x}_{k|k-1}$, we can further simplify Eq.(9.102) by using the matrix inversion lemma [35]. The result is

9.4 Robust Filtering

$$x_k^{(j+1)} = \hat{x}_{k|k-1} + K_k^{(j)} \Big[y_k - h\left(x_k^{(j)}\right) \\ - h\left(x_k^{(j)}\right)\left(\hat{x}_{k|k-1} - x_k^{(j)}\right) \Big], \qquad (9.104)$$

where

$$K_k^{(j)} = P_{k|k-1}^{(j)} h\left(x_k^{(j)}\right)^\top \left(h\left(x_k^{(j)}\right) P_{k|k-1}^{(j)} h\left(x_k^{(j)}\right)^\top + R_k^{(j)} \right), \qquad (9.105)$$

with $P_{k|k-1}^{(j)} = B_{k|k-1}^p \left(\Psi_x(e_k^{(j)})\right)^{-1} B_{k|k-1}^p$, $R_k^{(j)} = B_k^r \left(\Psi_y(e_k^{(j)})\right)^{-1} B_k^r$ and $h\left(x_k^{(j)}\right) = \frac{\partial h}{\partial x_k}\Big|_{x_k = x_k^{(j)}}$. For numerical stability, when $k \geq 1$, we introduce the step parameter $0 < \alpha \leq 1$ for modification. This means that we first use Eq.(9.104) to compute $x_k^{(1)}$ without using step parameter α. For $k \geq 1$, we shall use step parameter α with initial guess $x_k^{(1)}$, which means that Eq.(9.104) can be rewritten as

$$x_k^{(j+1)} = \hat{x}_{k|k-1} + \alpha P_k^{(j)}, \qquad (9.106)$$

where the direction $P_k^{(j)}$ is given by

$$P_k^{(j)} = K_k^{(j)} \Big[y_k - h\left(x_k^{(j)}\right) - h\left(x_k^{(j)}\right)\left(\hat{x}_{k|k-1} - x_k^{(j)}\right) \Big]. \qquad (9.107)$$

After iterations, we shall obtain converged solutions x_k and e_k. The recursion of filtering covariance can be computed by

$$P_{k|k} = (I_n - K_k h(x_k)) B_{k|k-1}^p \left(\Psi_x(e_k)\right)^{-1} B_{k|k-1}^p. \qquad (9.108)$$

Here we summarise the steps of OR-IEKF in Algorithm 5.

Algorithm 5 OR-IEKF

1: **Input:** $f(\cdot), h(\cdot), Q_k, R_k, \rho(\cdot), \epsilon, \alpha$
2: **Output:** $\hat{x}_{k|k}$ for $k = 1, 2, \ldots, N$
3: **Intitialization.** Start with initial filtering mean $\hat{x}_{0|0}$ and filtering covariance $P_{0|0}$.
4: **for** $k = 1, 2, \ldots, N$ **do**
5:     Compute prior mean $\hat{x}_{k|k-1}$ and prior covariance $P_{k|k-1}$ via Eq.(9.103).
6:     Let $x_k^{(0)} = \hat{x}_{k|k-1}$ and compute $x_k^{(1)}$ via Eq.(9.104).
7:     **while** $\frac{\|x_k^{(j+1)} - x_k^{(j)}\|}{\|x_k^{(j)}\|} > \epsilon$ **do**
8:         Compute iterative solution $x_k^{(j+1)}$ via Eq.(9.106) with corresponding update of $K_k^{(j+1)}$ in Eq.(9.105).
9:     Update filtering covariance matrix $P_{k|k}$ using Eq.(9.108).

Table 9.2 Three robust cost function examples

Name	ρ_τ	w_τ
Huber	$\begin{cases} \frac{1}{2}x^2 & \text{if } |x| < \tau, \\ \tau|x| - \frac{1}{2}\tau^2 & \text{if } |x| \geq \tau. \end{cases}$	$\begin{cases} 1 & \text{if } |x| < \tau, \\ \tau \times \frac{\text{sgn}(x)}{x} & \text{if } |x| \geq \tau. \end{cases}$
t-likelihood	$\frac{\tau^2}{2} \log(1 + \frac{x^2}{\tau^2})$	$\frac{1}{1+(\frac{x}{\tau})^2}$
Correntropy-induced cost	$\frac{\tau^2}{2}\left(1 - \exp\left(-\left(\frac{x}{\tau}\right)^2\right)\right)$	$\exp\left(-\left(\frac{x}{\tau}\right)^2\right)$

Remark 9.4 Note that, if $\Psi(e_k) = I_{n+m}$, the above iterations reduces to the IEKF solution. In addition, when the observation is linear, the solution reduces to the standard linear KF update.

9.4.2.1 Robust Cost Functions

Note that in Eq.(9.99), the item $\psi(e_{k,i}) = \frac{\partial \rho(e_{k,i})}{\partial e_{k,i}} / e_{k,i}$ plays an important role in cutting of large outliers. We shall call it the weight function. The scalar function $\rho(x)$ is called robust cost function. Let us denote ρ, ψ by ρ_τ, w_τ, where $\tau \in \mathbb{R}$ is a tuning factor. Here we consider three examples, which are given in Table 9.2. They correspond to the three new iterative filtering algorithms, namely, Huber-IEKF, t-likelihood-IEKF, and Correntropy-IEKF.

9.5 Numerical Results

In this section, we will present the numerical results of the KF, EKF, UKF, and PF. These filters are widely used in state estimation and target tracking applications and are capable of effectively handling nonlinear systems.

9.5.1 Linear Filtering Problem

Consider a linear system describing the position and velocity of an object in one-dimensional motion

$$\begin{aligned} x_{k+1} &= F x_k + w_k \\ y_k &= H x_k + v_k, \end{aligned} \tag{9.109}$$

9.5 Numerical Results

where $x_k = [p_k, v_k]^T$ represents the state vector comprising position p_k and velocity v_k at time step k. The system matrices are defined as

$$F = \begin{bmatrix} 1, & \Delta t \\ 0, & 1 \end{bmatrix}, \quad H = \begin{bmatrix} 1, & 0 \end{bmatrix} \quad (9.110)$$

Here, F is the state transition matrix, and H is the measurement matrix. We set the time step $\Delta t = 0.1$ seconds. The process noise w_k and measurement noise v_k are assumed to be zero-mean Gaussian white noise processes with covariances Q and R, respectively:

$$w_k \sim \mathcal{N}(0, Q), \quad v_k \sim \mathcal{N}(0, R) \quad (9.111)$$

To comprehensively analyze the KF's performance under various conditions, we consider three scenarios that explore the impact of different noise parameters and initial state distributions:

- Scenario 1: Varying Process Noise.

  In this scenario, we investigate the effect of different levels of process noise on the filter's performance. We fix the measurement noise covariance at $R = 0.1$ and vary the process noise covariance Q as follows:

$$Q_1 = \begin{bmatrix} 0.01 & 0 \\ 0 & 0.01 \end{bmatrix}, \quad Q_2 = \begin{bmatrix} 0.1 & 0 \\ 0 & 0.1 \end{bmatrix}, \quad Q_3 = \begin{bmatrix} 1 & 0 \\ 0 & 1 \end{bmatrix} \quad (9.112)$$

- Scenario 2: Varying Measurement Noise

  This scenario examines how different levels of measurement noise affect the filter's estimation accuracy. We fix the process noise covariance at $Q = 0.1 I_2$ and vary the measurement noise covariance R:

$$R_1 = 0.01, \quad R_2 = 0.1, \quad R_3 = 1 \quad (9.113)$$

These values correspond to low, medium, and high measurement noise levels, respectively.

- Scenario 3: Varying Initial State Distribution

  In this final scenario, we explore the impact of different initial state uncertainties on the filter's convergence. We fix the process and measurement noise covariances at $Q = 0.1 I_2$ and $R = 0.1$, respectively, while varying the initial state covariance P_0:

$$P_0^{(1)} = \begin{bmatrix} 0.1 & 0 \\ 0 & 0.1 \end{bmatrix}, \quad P_0^{(2)} = \begin{bmatrix} 1 & 0 \\ 0 & 1 \end{bmatrix}, \quad P_0^{(3)} = \begin{bmatrix} 10 & 0 \\ 0 & 10 \end{bmatrix} \quad (9.114)$$

We conduct $M = 100$ Monte Carlo simulations for each scenario and evaluate the performance using two metrics for multidimensional position estimates: the root

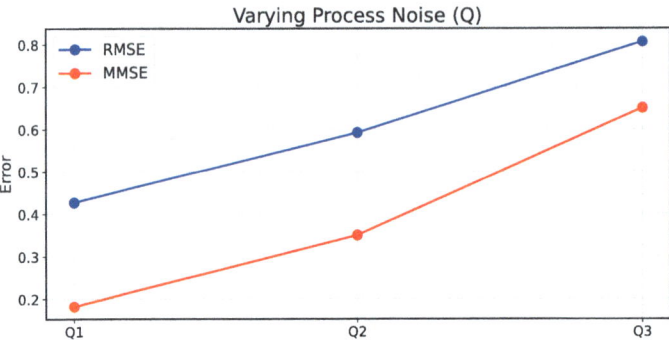

Fig. 9.1 Results for Scenario 1: Varying Process Noise. The plots show the RMSE and MMSE for different process noise covariance matrices Q_1, Q_2, and Q_3

mean square error (RMSE) and the mean square error (MSE). These metrics are defined as follows:

$$\text{RMSE} = \frac{1}{M} \sum_{m=1}^{M} \sqrt{\frac{1}{N} \sum_{k=1}^{N} \|\mathbf{p}_k^m - \hat{\mathbf{p}}_k^m\|^2}$$

$$\text{MSE (for step} k) = \frac{1}{M} \sum_{m=1}^{M} \|\mathbf{p}_k^m - \hat{\mathbf{p}}_k^m\|^2, \tag{9.115}$$

where $\mathbf{p}_k^m$ represents the true position vector, $\hat{\mathbf{p}}_k^m$ is the estimated position vector, M is the number of Monte Carlo simulations, N is the number of time steps in each simulation, and $\|\cdot\|$ denotes the Euclidean norm. And the MMSE denote the mean of MSE for all steps.

The results of the numerical experiments are presented in the following figures. Each figure corresponds to one of the scenarios described in the previous section.

Figure 9.1 illustrates the outcomes of Scenario 1, which investigates the effect of varying the process noise covariance matrix Q. As anticipated, higher process noise corresponds to increased RMSE and MMSE values, indicating larger estimation errors. Figures 9.2, 9.3, and 9.4 depict the KF's performance under different Q values. While the MSE converges to a stable value over time for each noise level, increasing Q leads to a more unstable convergence process and a significant rise in MMSE. These results highlight the KF's ability to adapt to varying levels of process uncertainty, albeit with degraded performance as uncertainty increases.

Figure 9.5 presents the results of Scenario 2, which examines the impact of varying the measurement noise covariance matrix R. The findings reveal that increased measurement noise leads to higher RMSE and MMSE values, indicating greater estimation error. As illustrated in Figs. 9.6, 9.7, and 9.8, the MSE converges to a stable value over time for each noise level. However, higher observation noise results in significantly increased system MSE and more pronounced fluctuations

9.5 Numerical Results

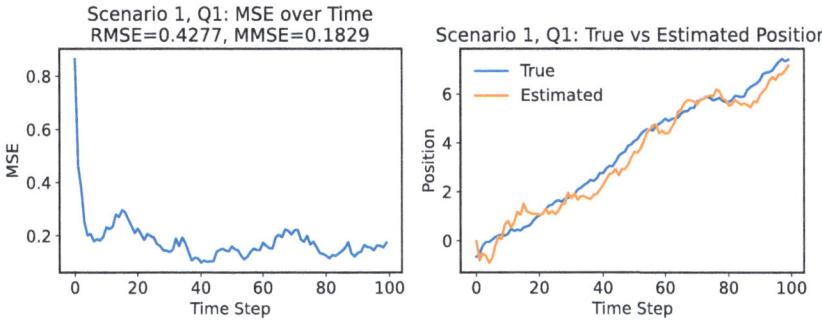

Fig. 9.2 MSE and tracking result with Q_1

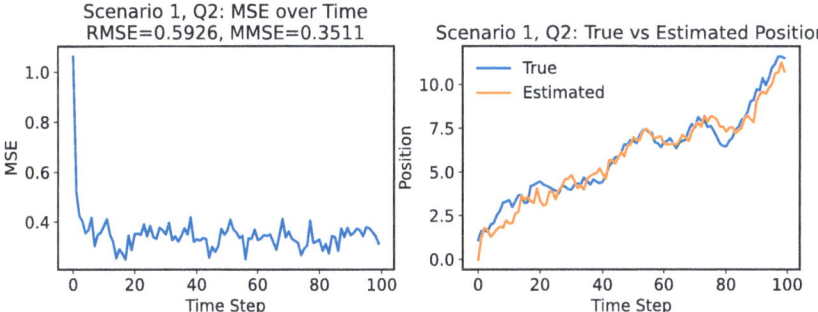

Fig. 9.3 MSE and tracking result with Q_2

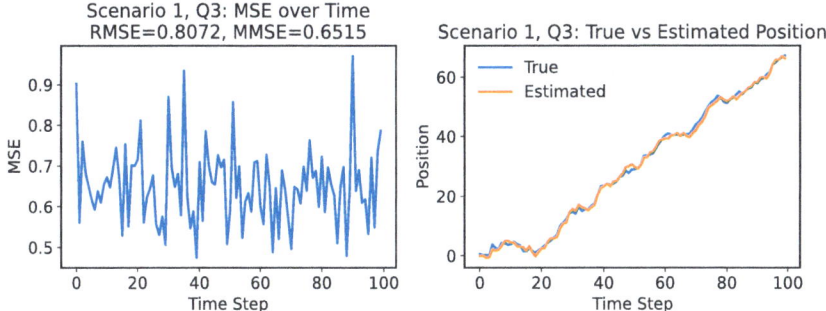

Fig. 9.4 MSE and tracking result with Q_3

in the KF's tracking results, demonstrating the filter's sensitivity to measurement quality.

Figure 9.9 illustrates the results of Scenario 3, which examines the effect of varying the initial estimation error covariance matrix P_0 on KF performance. The findings demonstrate that higher P_0 values lead to increased initial RMSE and MMSE, reflecting greater initial uncertainty. However, as evident in Figs. 9.10, 9.11,

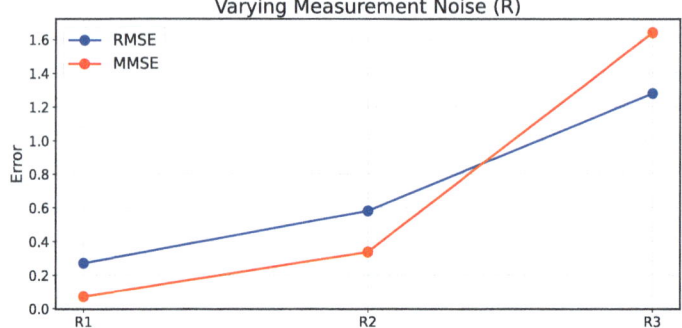

Fig. 9.5 Results for Scenario 2: Varying Measurement Noise. The plots show the RMSE and MMSE different measurement noise covariance matrices R_1, R_2, and R_3

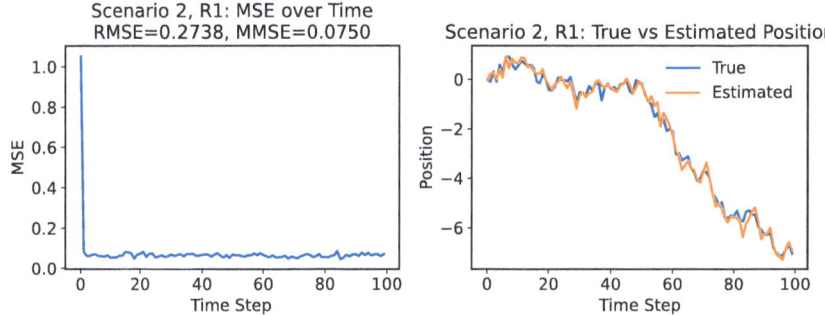

Fig. 9.6 MSE and tracking result with R_1

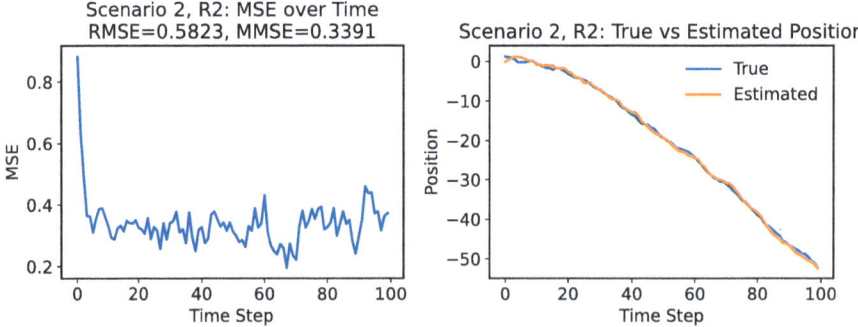

Fig. 9.7 MSE and tracking result with R_2

and 9.12, the KF consistently converges to the true state over time, regardless of the initial P_0. This convergence is reflected in the MSE trajectories, which show initial disparities that diminish as the estimation process progresses. These results underscore the KF's robustness and its ability to overcome poor initial estimates, a

9.5 Numerical Results

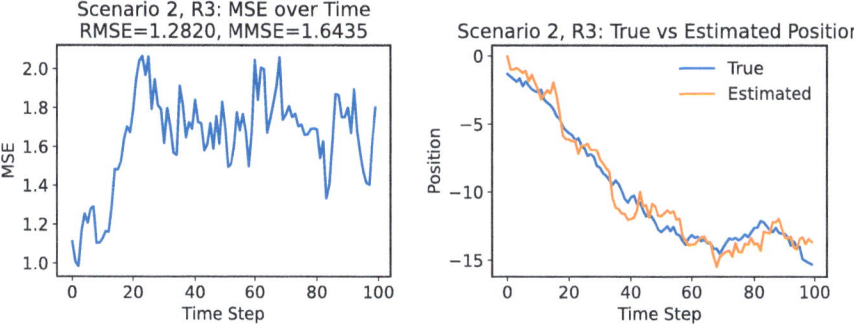

Fig. 9.8 MSE and tracking result with R_3

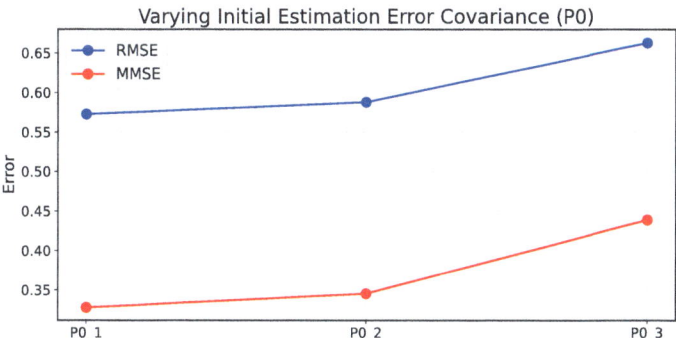

Fig. 9.9 Results for Scenario 3: Varying Initial Estimation Error Covariance. The plots show the RMSE and MMSE over time for different initial estimation error covariance matrices $P_0^{(1)}$, $P_0^{(2)}$, and $P_0^{(3)}$

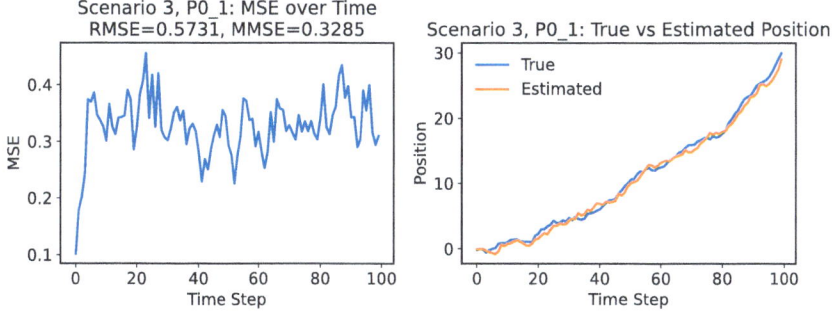

Fig. 9.10 MSE and tracking result with $P_0^{(2)}$

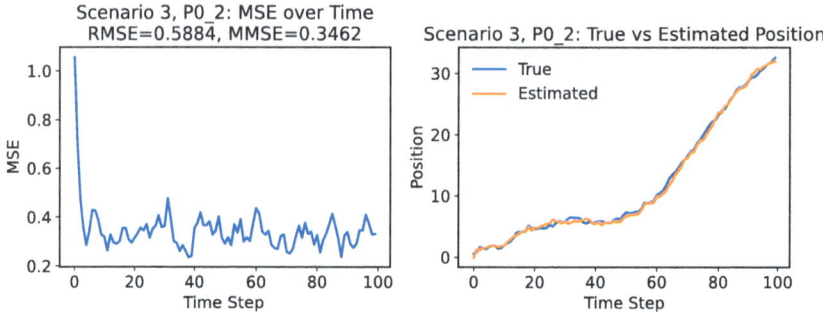

Fig. 9.11 MSE and tracking result with $P_0^{(2)}$

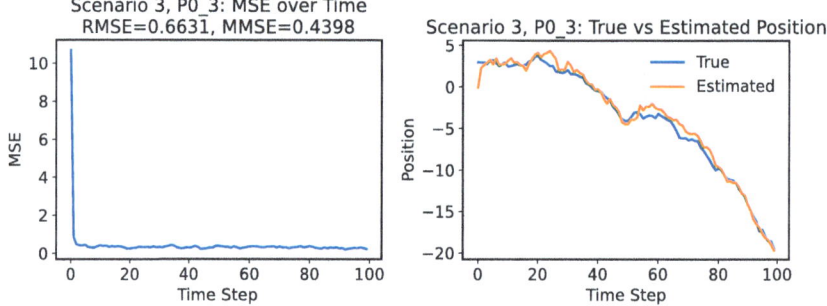

Fig. 9.12 MSE and tracking result with $P_0^{(3)}$

valuable characteristic in practical applications where precise initialization may be challenging.

In all scenarios, the KF demonstrated its ability to estimate the true state accurately, with the estimation error decreasing as more measurements were incorporated. However, the performance was influenced by the noise levels and the initial estimation error covariance, as expected from the theoretical analysis.

9.5.2 Nonlinear Filtering Problem

This section presents a detailed numerical experiment designed to evaluate the performance of various filtering techniques applied to a nonlinear system model with measurement noise.

$$\begin{bmatrix} x(k+1) \\ v_x(k+1) \\ y(k+1) \\ v_y(k+1) \end{bmatrix} = F \begin{bmatrix} x(k+1) \\ v_x(k+1) \\ y(k+1) \\ v_y(k+1) \end{bmatrix} + \begin{bmatrix} \alpha \sin(v_x(k)) \\ 0 \\ \alpha \cos(v_y(k)) \\ 0 \end{bmatrix} + w_k \quad (9.116)$$

9.5 Numerical Results

where the variables $x(k)$ and $y(k)$ represent spatial coordinates at step k; variables $x(k)$ and $y(k)$ represent velocity component of x, y at step k, respectively; and the parameter α scales the nonlinear term and $\mathbf{w}_k \sim \mathcal{N}(\mathbf{0}, Q)$ is the process noise. The state transition matrix $F \in \mathbb{R}^{4 \times 4}$ is given by

$$F = \begin{bmatrix} 1 & \Delta t & 0 & 0 \\ 0 & 1 & 0 & 0 \\ 0 & 0 & 1 & \Delta t \\ 0 & 0 & 0 & 1 \end{bmatrix}, \tag{9.117}$$

where Δt is the time step. The process noise covariance matrix $Q \in R^{4 \times 4}$ is defined as $Q = 0.02 \cdot I_4$, where I_4 is the 4×4 identity matrix. The measurement model is given by

$$\begin{bmatrix} z_1(k) \\ z_2(k) \end{bmatrix} = \begin{bmatrix} \sqrt{x(k)^2 + y(k)^2} \\ \arctan\left(\frac{y(k)}{x(k)}\right) \end{bmatrix} + v_k \tag{9.118}$$

where $v_k \sim \mathcal{N}(\mathbf{0}, R)$ is a two-dimensional Gaussian distribution.

In this experiment, we evaluate the performance of the EKF, UKF, and PF under various conditions. The parameters of interest are:

- Nonlinearity parameter α: We test different values of α to assess its impact on filter performance, with α values chosen from the set $\{0.01, 0.1, 1\}$. This allows us to observe how varying levels of nonlinearity affect the accuracy of each filter.
- Number of particles for PF: For the PF, we vary the number of particles to examine how this impacts the filter's accuracy, $N \in \{100, 200, 500, 1000\}$. This is crucial as the PF's performance is highly dependent on the number of particles used in the approximation process.
- Time steps: The simulation is consistently run for 100 time steps across all scenarios with $\Delta t = 0.1$, enabling us to observe the behavior of each filter over a fixed period and how their performance evolves throughout the simulation.

The UKF is configured using its unscented transform parameters, which include the scaling parameters $a = 0.002$, $\beta = 2$ and $\kappa = 0$. These parameters control the spread of the sigma points around the mean and influence the accuracy of the filter in capturing nonlinearities. For this experiment, the standard values for these parameters are used, ensuring a fair comparison across all scenarios.

The simulation is consistently run for 100 time steps across all scenarios, enabling us to observe the behavior of each filter over a fixed period and how their performance evolves throughout the simulation. The performance of each filter is assessed using metrics such as MSE and RMSE defined in (9.115). These metrics provide a quantitative measure of the filters' accuracy and their ability to track the true state over time (Fig. 9.13).

First, we present the two-dimensional tracking plots of the different algorithms compared to the true trajectory. In our case, the first and third dimensions represent

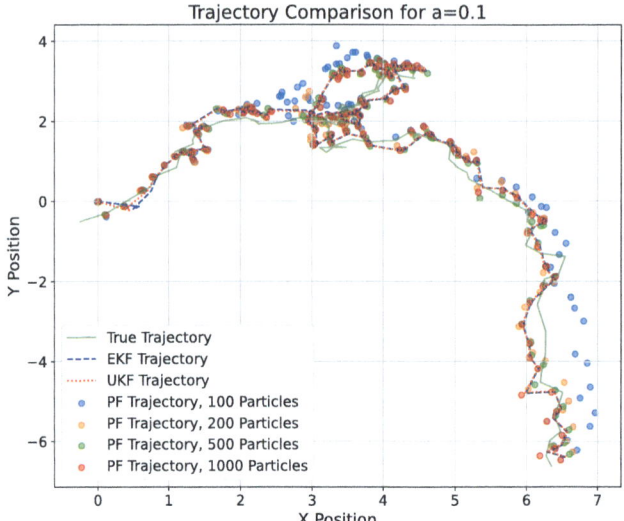

Fig. 9.13 Trajectory comparison with nonlinearity paramete $\alpha = 0.1$

Table 9.3 Execution times for different filters and nonlinearities

Filter	Execution time (seconds)		
	Nonlinearity 0.01	Nonlinearity 0.1	Nonlinearity 1
EKF	0.0077	0.0076	0.0071
UKF	0.0460	0.0434	0.0383
PF (100 particles)	0.0957	0.0879	0.0822
PF (200 particles)	0.1386	0.1552	0.1402
PF (500 particles)	0.2963	0.2830	0.2814
PF (1000 particles)	0.5300	0.5181	0.5259

position variables, while the second and fourth dimensions represent velocity variables. Therefore, we naturally chose a two-dimensional position plane to visualize the specific experimental results as shown below: Then, we focused particularly on the execution time of these filters to evaluate their computational efficiency in practical applications in Table 9.3.

The results in Table 9.3 show several trends in filtering algorithm performance. Notably, EKF and UKF execution times decrease slightly with increased system nonlinearity, likely due to experimental conditions. In contrast, the PF's execution time remains stable across nonlinearity levels, indicating robustness to complexity changes. EKF consistently has the fastest execution times, followed by the UKF, which is approximately five times slower due to its advanced sampling strategy. As expected, the PF is the slowest, with execution time increasing as the number of particles grows.

9.5 Numerical Results

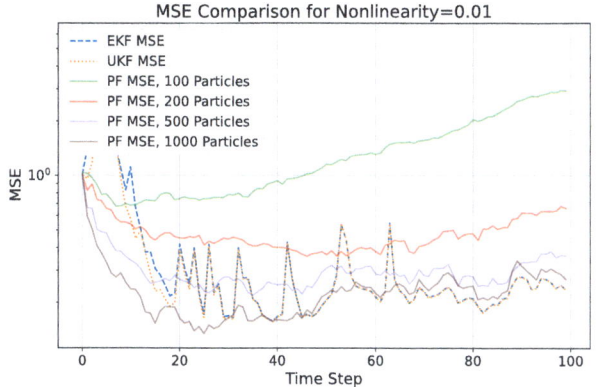

Fig. 9.14 MSE comparison for $\alpha = 0.01$

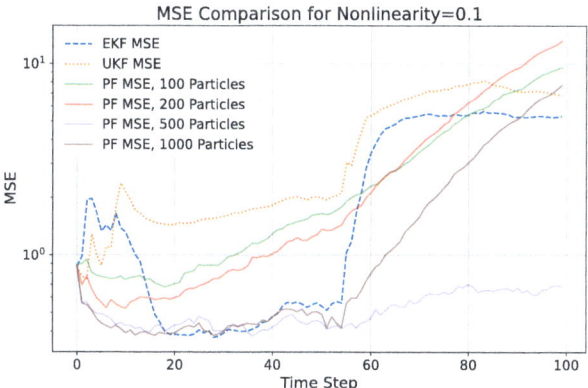

Fig. 9.15 MSE comparison for $\alpha = 0.1$

Figures 9.14, 9.15, and 9.16 present the MSE comparison for EKF, UKF, and PF with different numbers of particles under varying nonlinearity parameters $\alpha \in \{0.01, 0.1, 1\}$. These comparisons illustrate how each filter performs as the level of system nonlinearity and the number of particles (for PF) are varied $N \in \{100, 200, 500, 1000\}$.

From Figs. 9.14, 9.15, and 9.14, we can observe the following: In scenarios with low nonlinearity, the performance of the EKF and the UKF is comparable, and both algorithms perform well. However, as time progresses, there are significant fluctuations in the MSE for both algorithms, indicating an inherent lack of robustness. Under conditions of moderate to high nonlinearity, the performance of the EKF deteriorates noticeably. In contrast, the UKF demonstrates superior performance over the EKF in highly nonlinear situations due to its more effective handling of nonlinearities via the unscented transform. For the PF, with a lower number of particles (e.g., 100 or 200 as in this example), the filtering performance is suboptimal

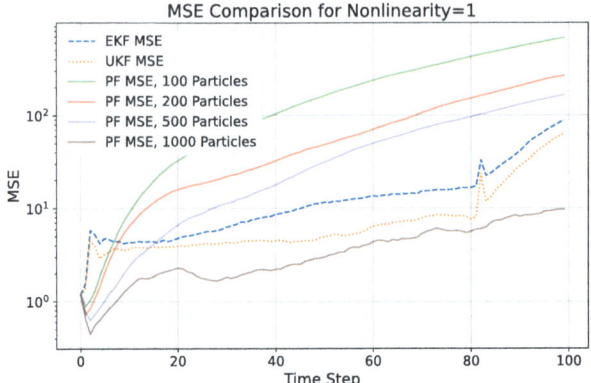

Fig. 9.16 MSE Comparison for $\alpha = 1$

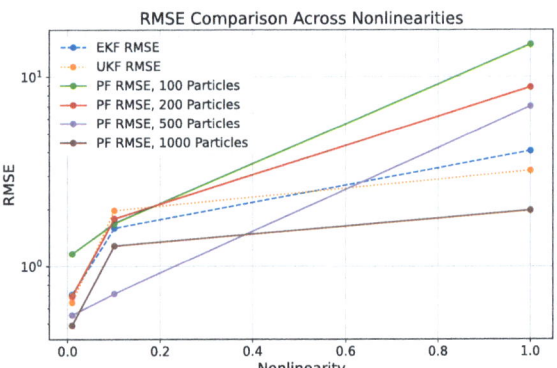

Fig. 9.17 RMSE comparison across nonlinearities

regardless of the level of nonlinearity. However, with an increased number of particles (e.g., 1000), the PF exhibits robust predictive capabilities across various nonlinear scenarios, with stable MSE values showing no significant fluctuations.

To better visualize the numerical results of different algorithms under varying levels of nonlinearity, we present the nonlinearity as the horizontal axis and the RMSE performance of the algorithms as the vertical axis. Each algorithm was independently run 100 times, and the average RMSE results are plotted in Fig. 9.17.

Key observations from Fig. 9.17 include the following: as nonlinearity increases, the performance of all filters generally declines, as evidenced by rising RMSE values. The EKF experiences the significant performance degradation, underscoring its limitations in handling highly nonlinear systems. In contrast, the UKF consistently outperforms the EKF in high levels of nonlinearity, demonstrating greater robustness to nonlinear dynamics. The PF, particularly with a higher number of particles, exhibits the best performance in highly nonlinear scenarios, although this comes with increased computational demands. However, with a lower number of particles (e.g., 100 or 200), the PF's performance is the worst among all the algorithms, regardless of the nonlinearity level.

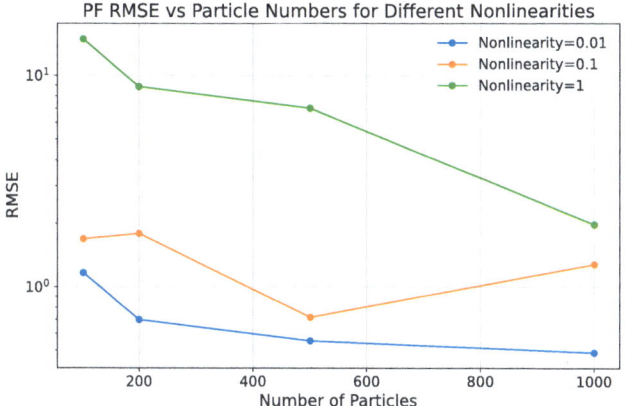

Fig. 9.18 PF RMSE vs. number of particles for different nonlinearities

Figure 9.18 shows how the RMSE of the PF changes with the number of particles for different nonlinearity parameters.

From Fig. 9.18, we can conclude that regardless of the level of nonlinearity, the RMSE of the PF decreases rapidly as the number of particles increases, which aligns with our expectations of the PF's convergence behavior. In both low and high nonlinearity scenarios, the improvement in performance becomes more pronounced as the number of particles increases, indicating that the filter more accurately captures the posterior distribution with a larger particle set. This result confirms that, as expected, increasing the particle count enhances the filter's ability to approximate the true state, leading to a faster reduction in RMSE across different levels of system nonlinearity.

9.6 Exercises

1. Consider matrices where $A \in R^{n \times n}$, $C \in R^{n \times n}$ are nonsingular, $B \in R^{n \times m}$, and $D \in R^{m \times n}$. The following identity holds:

$$(A + BCD)^{-1} = A^{-1} - A^{-1}B(C^{-1} + DA^{-1}B)^{-1}DA^{-1}$$

2. Given an example of a linear filtering system and derive the equations for the prediction and update steps of the KF.
3. Compute the Kalman Gain for a system with the following parameters:

$$A = \begin{pmatrix} 1 & 1 \\ 0 & 1 \end{pmatrix}, \quad H = \begin{pmatrix} 1 & 0 \end{pmatrix}, \quad Q = \begin{pmatrix} 0.001 & 0 \\ 0 & 0.001 \end{pmatrix}, \quad R = 0.01.$$

4. Given a sequence of measurements $z = [1.0, 1.3, 1.6, 1.9]$, apply the KF to estimate the state of a system described by the above matrices, assuming an initial state of $\mathbf{x}_0 = \begin{pmatrix} 0 \\ 0 \end{pmatrix}$ and $P_0 = \begin{pmatrix} 1 & 0 \\ 0 & 1 \end{pmatrix}$.
5. Explain the resampling process in PF and its necessity.
6. For a simple one-dimensional random walk model $x_{t+1} = x_t + w_t$, where $w_t \sim \mathcal{N}(0, 1)$, use a PF to estimate the state given measurements $z_t = x_t + v_t$ where $v_t \sim \mathcal{N}(0, 1)$. Calculate this scenario for five time steps.
7. Consider a simple nonlinear oscillator whose state equation is given by

$$\dot{x} = \begin{pmatrix} x_2 \\ -x_1 - 0.1x_1^3 + 0.5\cos(1.2t) \end{pmatrix},$$

and the measurement is directly the position:

$$z = x_1 + v,$$

where v is the measurement noise with a normal distribution having zero mean and variance of 0.1^2.

a. Derive the discrete-time state equations assuming a sampling time $\Delta t = 0.1$ seconds.
b. Calculate the Jacobians of the state transition and measurement models required for EKF.

8. Consider a nonlinear system described by the state equation:

$$x_{t+1} = \sin(x_t) + 0.1\omega_t,$$

where $\omega_t \sim \mathcal{N}(0, 1)$ is the process noise. The measurements are given by

$$z_t = x_t^2 + v_t,$$

where $v_t \sim \mathcal{N}(0, 1)$ is the measurement noise.

a. Set up the UKF for this system. Describe how you would choose the sigma points.
b. Implement the UKF to estimate x_t over 10 time steps given noisy measurements. Assume an initial estimate of $x_0 = 1$ with an initial covariance $P_0 = 1$.

9. Consider a continuous-time linear Gaussian system modeled by the following state and measurement differential equations:

$$dx_t = Ax_t\, dt + dw_t,$$
$$dy_t = Hx_t\, dt + dv_t,$$

where dw_t and dV_t are Brownian motions with intensities Q and R, respectively.

a. Discuss how the FPF can be applied to this continuous system. Describe the form of the feedback control law in this context.
b. Simulate this system over a given period and implement the FPF. Assess its performance and compare it to a continuous-time Kalman filter.

10. Assume a scenario where two sensors continuously measure the position of a moving object along a straight line. The state and measurement differential equations are

$$dx_t = v_t \, dt,$$
$$dz_{t,1} = x_t \, dt + dw_{t,1},$$
$$dz_{t,2} = x_t \, dt + dw_{t,2},$$

where v_t is the process noise modeled as Brownian motions with intensity Q and $dw_{t,1}$ and $dw_{t,2}$ are independent Wiener process increments with intensities R_1 and R_2.

a. Formulate the FPF for this continuous multisensor setup. Determine the continuous-time feedback gains for each sensor.
b. Discuss how the feedback gains change with varying noise intensities R_1 and R_2.
c. Implement the FPF for this scenario and analyze the benefits of using two sensors on the estimation accuracy compared to a single sensor setup.

References

1. S. M. Arulampalam, S., Maskell, N. J. Gordon, and T. Clapp. A tutorial on particle filters for on-line nonlinear/non-Gaussian Bayesian tracking. *IEEE Transactions on Signal Processing*, 50(2):174–188, 2002.
2. N. Chopin. Central limit theorem for sequential Monte Carlo and its application to Bayesian inference. *The Annals of Statistics*, 32, 2385-2411, 2004.
3. A. Doucet, N. de Freitas, and N. J. Gordon. An introduction to sequential Monte Carlo methods. *Sequential Monte Carlo Methods in Practice*,3–14, 2001.
4. R. Doucet, O. Capp, and E. Moulines. Comparison of resampling schemes for particle filtering. *Proceedings of the 4th International Symposium on Image and Signal Processing and Analysis*, 2005.
5. N. J. Gordon, D. J. Salmond and A. F. M. Smith. Novel approach to nonlinear/non-Gaussian Bayesian state estimation. *IEE Proceedings F (Radar and Signal Processing)*, 140:107–113, 1993.
6. R. E. Kalman. A new approach to linear filtering and prediction problems. *American Society of Mechanical Engineers Transactions*, 82:35-45, 1960.
7. G. Kitagawa. Monte Carlo filter and smoother for non-Gaussian non-linear state space models. *Journal of Computational and Graphical Statistics*, 5:1-25, 1996.

8. M. I. Ribeiro. Kalman and extended Kalman filters: concept, derivation, and properties. *Institute for Systems and Robotics*, 43(46):3736-3741, 2004.
9. A. Taghvaei, P. G. Mehta, S. P. Meyn. Diffusion map-based algorithm for gain function approximation in the feedback particle filter. *SIAM/ASA Journal on Uncertainty Quantification*, 8(3):1090-1117, 2020.
10. T. Yang, P. G. Mehta, S. P. Meyn. Feedback particle filter. *IEEE Transactions on Automatic Control*, 58(10):2465-2480, 2013.
11. T. Yang, R. S. Laugesen, P. G. Mehta, S. P. Meyn. Multivariable feedback particle filter. *Automatica*, 71:10-23, 2016.
12. A. Bain and D. Crisan. *Fundamentals of stochastic filtering*. Springer-Verlag, New York, 2009.
13. D. Crisan. Robust filtering: correlated noise and multidimensional observation. *The Annals of Applied Probability*, 23(5):2139–2160, 2013.
14. S. J. Julier and J. K. Uhlmann. New extension of the Kalman filter to nonlinear systems. *Signal Processing, Sensor Fusion, and Target Recognition VI*, 3068:182-193, 1997.
15. P. R. Gunjal, B. R. Gunjal, H. A. Shinde, S. M. Vanam and S. S. Aher. Moving object tracking using Kalman filter. *International Conference on Advances in Communication and Computing Technology*, 2018.
16. Y. Xu and G. Zhang. Application of Kalman filter in the prediction of stock price. *In International Symposium on Knowledge Acquisition and Modeling (KAM)*, 197–198, 2015.
17. H. Haleh, A. Moghaddam, and S. Ebrahimjam. A new approach to forecasting stock price with EKF data fusion. *International Journal of Trade*, 2(2):109, 2011.
18. B. Azimi-Sadjadi and P. S. Krishnaprasad. Approximate nonlinear filtering and its application in navigation. *Automatica*, 41:945–956, 2005.
19. Y. Bar-Shalom and T. E. Fortmann. *Tracking and data association*. Academic Press, 1988.
20. D. A. Forsyth and J. Ponce. Computer vision: a modern approach. *Prentice Hall Professional Technical Reference*, 2002.
21. I. Hwang, H. Balakrishnan, K. Roy, and C. Tomlin. Multiple-target tracking and identity management in clutter for air traffic control. *Proceedings of American Control Conference*, 4:3422–3428, 2004.
22. K. Doya, S. Ishii, A. Pouget, and R. P. N. Rao. *Bayesian brain*. MIT Press, 2007.
23. R. P. N. Rao, B. A. Olshausen, and M. S. Lewicki. *Probabilistic models of the brain*. MIT Press, 2002.
24. Y. Sato, T. Toyoizumi, and K. Aihara. Bayesian inference explains the perception of unity and ventriloquism aftereffect: identification of common sources of audiovisual stimuli. *Neural Computation*, 19:3335–3355, 2007.
25. S. Thrun. Probabilistic robotics. *Communications of the ACM*, 45:52–57, 2002.
26. L. Ljung. Asymptotic behavior of the extended Kalman filter as a parameter estimator for linear systems. *IEEE Transactions on Automatic Control*, 24(1):36–50, 1979.
27. E. A. Wan and R. van der Merwe. The unscented Kalman filter for nonlinear estimation. *Proceedings of the IEEE 2000 Adaptive Systems for Signal Processing, Communications, and Control Symposium*, 153–158, 2000.
28. G. Evensen. Sequential data assimilation with nonlinear quasi-geostrophic model using Monte Carlo methods to forecast error statistics. *Journal of Geophysical Research*, 99(C5):143–162, 1994.
29. A. Budhiraja, L. Chen, and C. Lee. A survey of numerical methods for nonlinear filtering problems. *Physica D: Nonlinear Phenomena*, 230(1-2):27–36, 2007.
30. R. E. Kalman and R. S. Bucy. New results in linear filtering and prediction theory. *American Society of Mechanical Engineers Transactions*, 83:95–107, 1961.
31. D. Brigo, B. Hanzon, and F. LeGland. A differential geometric approach to nonlinear filtering: the projection filter. *IEEE Transactions on Automatic Control*, 43(2):247–252, 1998.
32. O., S. Yagiz, A. Taghvaei, and P. G. Mehta. Deep FPF: Gain function approximation in the high-dimensional setting. *59th IEEE Conference on Decision and Control (CDC)*, 2020.
33. B. M. Bell and F. W. Cathey. The iterated Kalman filter update as a Gauss-Newton method. *IEEE Transactions on Automatic Control*, 38(2):294-297, 1993.

34. R. Wolke and H. Schwetlick. Iteratively reweighted least squares: algorithms, convergence analysis, and numerical comparisons. *SIAM Journal on Scientific and Statistical Computing*, 9(5):907-921, 1988.
35. W. W. Hager. Updating the inverse of a matrix. *SIAM Review*, 31(2):221-239, 1989.
36. Y. Tao and S. S.-T. Yau. Outlier-robust iterative extended Kalman filtering. *IEEE Signal Processing Letters*, 30:743-747, 2023.

Chapter 10
Estimation Algorithms Based on Deep Learning

In this chapter, we shall first review the estimation problem [10] from the perspective of the machine learning [39]. Then we shall revisit the classical neural networks models including feedforward neural networks (FNNs) [16] and recurrent neural networks (RNNs) [9] in details. Specifically, we're going to introduce the specific mathematical form of their network architecture. Then we shall discuss their approximation ability, the universal approximation theorem, and give an elegant proof. And we also introduce the mathematics on how to train the neural networks, i.e., the backpropagation algorithm [32] and currently popular optimization algorithms for deep learning optimization problems. Finally, we shall introduce how to use deep learning method to solve state estimation problems, i.e., filtering problems.

10.1 Overview

In the previous chapters, we introduced the nonlinear filtering algorithms and related theories in detail. With the advent of data-driven approaches, more and more hybrid approaches are emerging across industries, such as computer vision, natural language processing, and computational biology. This makes deep learning and deep neural network models [22] receive widespread attention. Now we turn to this popular topic, and it can be viewed as the modification of the data assimilation [21] and the control theory [38] in the era of big data.

10.2 Estimation Problems

In this section, we shall focus on the state-space model (SSM) and introduce the basic estimation problems for SSM. They are state estimation, parameter estimation, and dual estimation. At the same time, we also explain some of their relationships with machine learning problems. The following introduction to these three issues refers to [36].

10.2.1 State Estimation

The state estimation is about recovering state signal given noisy observation signal, which can be formulated as the estimation of the state of a discrete-time nonlinear dynamic system

$$\begin{aligned} X_{k+1} &= f(X_k, u_k) + W_k, \\ Y_k &= h(X_k) + V_k, \end{aligned} \quad (10.1)$$

where X_k represents the state signal of the system which cannot be observed directly, u_k is a known exogenous input signal, and Y_k is the noisy measurement signal. The process noise W_k drives the dynamic system, and the observation noise is given by V_k. It is necessary to note that the system dynamic model f and h are assumed known.

10.2.2 Parameter Estimation

Parameter estimation is a fundamental issue in the machine learning, also known as system identification [25], which involves determining a nonlinear mapping

$$Y_k = \mathbf{G}(X_k, \mathbf{w}), \quad (10.2)$$

where X_k is the input, Y_k is the output, and the nonlinear map $\mathbf{G}(\cdot)$ is parameterized by the vector $\mathbf{w}$. The classic one is viewing training neural networks as a parameter estimation problem, where $\mathbf{w}$ represents the weights of the neural networks. There are some series of works on this topic [13, 24, 34]. Such a methodology has numerous applications in regression, classification, and dynamic modeling.

Learning corresponds to estimating the parameters w. Typically, a training set is provided with sample pairs consisting of known input and desired outputs, $\{X_k, Y_k\}$. The error is defined as $\mathbf{e}_k = Y_k - \mathbf{G}(X_k, \mathbf{w})$, and the goal of learning involves solving for the parameters $\mathbf{w}$ in order to minimize the expectation of some given function of the error.

While a number of optimization approaches exist, the EKF may be used to estimate the parameters by writing a new state-space representation

10.3 Feedforward Neural Networks

$$\begin{aligned}\mathbf{w}_{k+1} &= \mathbf{w}_k + \mathbf{r}_k, \\ Y_k &= \mathbf{G}(X_k, \mathbf{w}_k) + \mathbf{e}_k,\end{aligned} \quad (10.3)$$

where the parameters $\mathbf{w}_k$ correspond to a stationary process with identity state transition matrix, driven by process noise $\mathbf{r}_k$. The output Y_k corresponds to a nonlinear observation on W_k. The EKF can then be applied directly as an efficient "second-order" technique for learning the parameters.

10.2.3 Dual Estimation

A special case of machine learning arises when the input X_k is unobserved and requires coupling both state estimation and parameter estimation. For these dual estimation problems, we again consider

$$\begin{aligned}X_{k+1} &= f(X_k, u_k, \mathbf{w}) + W_k, \\ Y_k &= h(X_k, \mathbf{w}) + V_k,\end{aligned} \quad (10.4)$$

where both the system states X_k and the set of model parameters $\mathbf{w}$ for the dynamic system must be simultaneously estimated from only the observed noisy signal Y_k.

Example applications include adaptive nonlinear control, noise reduction (e.g., speech or image enhancement), determining the underlying price of financial time-series, etc. The classic method is usually to first use the EM algorithm [27] to identify the system equation and then use the filtering algorithms to estimate the state of the system. The modern method is using variational learning to learn the system parameters and the posterior [20] simultaneously.

10.3 Feedforward Neural Networks

In this section, we shall introduce the simplest neural networks. It is known as the FNNs and also referred to the multilayer perceptrons (MLPs). The mathematical form of the network architecture will be given first. Then we shall state the universal approximation theorem for FNNs, which reflects its computing power. The detailed proof of the related theorems will also be given.

10.3.1 Mathematical Forms for FNNs

The schematic of the FNN is as follows:

Fig. 10.1 A two-layer FNN example

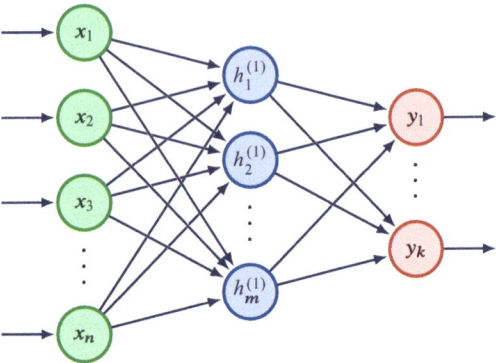

Usually, this sort of networks are described in terms of layers which are chained together to create the output function, where a layer is a collection of neurons that can be thought of as a unit of computation. We denote σ as the sigmoid function which is defined by

$$\sigma(x) = \frac{1}{1+e^{-x}}. \tag{10.5}$$

In the simplest case, there is a single input layer and a single output layer, just as presented in Fig 10.1. In this case, j-th neuron in hidden layer is connected to the input vector $x = (x_1, \ldots, x_d) \in R^d$ via a biased weighted sum and an activation function σ_j:

$$h_j^{(1)} = \sigma_j \left(b_j + \sum_{i=1}^{d} w_{i,j} x_i \right), \tag{10.6}$$

where $w_{i,j} \in R$ and $b_j \in R$. The activation function σ_j is the sigmoid function defined above.

It is also possible to incorporate additional hidden layers between the input and output layers. For example, the output of the MLP with one hidden layer will be

$$y_k = \sigma_k^{(2)} \left[b_k^{(2)} + \sum_{j=1}^{d_2} \omega_{j,k}^{(2)} \cdot \sigma_j^{(1)} \left(b_j^{(1)} + \sum_{i=1}^{d_1} \omega_{i,j}^{(1)} x_j \right) \right], \tag{10.7}$$

where $\sigma_k^{(2)}, \sigma_j^{(1)}$ are nonlinear activation functions for each layer and the bracketed superscripts refer to the corresponding layer. Such a neural network is usually called a two-layer FNN.

For the convenience of stating our universal approximation theorem, we can formulate the abovementioned FNN as the several classes of functions below.

10.3 Feedforward Neural Networks

Definition 10.1 For any $r \in \mathbb{N} \equiv \{1, 2, \ldots\}$, $\mathbf{A}^r$ is the set of all affine functions from R^r to R, that is

$$\mathbf{A}^r := \left\{ A(x) = w^\mathsf{T} x + b : w, x \in R^{r \times 1}, b \in R \right\}. \tag{10.8}$$

In the FNN, x, w, and b represent the input, weight, and bias of the network, respectively. $A(x)$ is the linear operator in FNNs.

Definition 10.2 A function $\kappa : R \to [0, 1]$ is a squashing function if it is nondecreasing, $\lim_{\lambda \to +\infty} \kappa(\lambda) = 1$, and $\lim_{\lambda \to -\infty} \kappa(\lambda) = 0$.

Here, κ represents a activation function.

Definition 10.3 ([11]) $\Sigma^r(\kappa)$ be the class of functions

$$\left\{ \bar{\zeta} : R^r \to R : \bar{\zeta}(x) = \sum_{j=1}^{q} \beta_j \kappa \left(A_j(x) \right), x \in R^r, \beta_j \in R, A_j \in \mathbf{A}^r, q = 1, 2, \ldots \right\}. \tag{10.9}$$

Apparently, $\bar{\zeta}$ represents the standard two-layer FNN with r input neurons, q hidden neurons, and one output neuron.

10.3.2 Universal Approximation Theorem for FNNs

We begin with some frequently used notations and definitions in sequel. Let $I_n = [0, 1]^n$ is the n-dimensional unit cube. $M(I_n)$ be the space of finite, signed regular Borel measures on I_n. Let C^r be the set of continuous functions from R^r to R with the supremum norm $\| \cdot \|$, and the supremum norm of $f : A \to B$ is defined by

$$\|f\| = \sup\{|f(x)| : x \in A\}. \tag{10.10}$$

Definition 10.4 A subset S of a metric space (X, ρ) is ρ-dense in a subset T if for every $\epsilon > 0$ and for every $t \in T$, there is an $s \in S$ such that $\rho(s, t) < \epsilon$.

Definition 10.5 A subset S of C^r is said to be uniformly dense on compacta in C^r if for every compact subset $K \subset R^r$, S is ρ-dense in C^r, where for $f, g \in C^r$ and $\rho(f, g) = \sup_{x \in K} |f(x) - g(x)|$.

Definition 10.6 (Discriminatory) We say that a function σ is discriminatory if given a measure $\mu \in M(I_n)$ such that

$$\int_{I_n} \sigma \left(w^\mathsf{T} x + b \right) d\mu(x) = 0, \quad \forall w \in R^n, b \in R, \tag{10.11}$$

implies that $\mu = 0$.

Here we give the Kolmogorov-Arnold representation theorem (or superposition theorem) [19], which states that every multivariate continuous function can be represented as a superposition of continuous functions of one variable. The Kolmogorov-Arnold representation theorem is given as follows.

Theorem 10.1 (Kolmogorov-Arnold Representation Theorem) *For any integer $n \geq 2$ there are continuous real functions $\psi^{p,q}(x)$ on the closed unit interval I_1 such that each continuous real function $f(x_1, \ldots, x_n)$ on the n-dimensional unit cube I_n can be written as*

$$f(x_1, \ldots, x_n) = \sum_{q=1}^{q=2n+1} \chi_q \left[\sum_{p=1}^{n} \psi^{pq}(x_p) \right],$$

where $\chi_q(y)$ are continuous real functions.

Remark 10.1 Theorem 10.1 implies, among other things, that if we could chose the nonlinearity of each unit we can represent any continuous function exactly with a FNN with 1 hidden layer. Therefore, the following universal approximation theorem for FNNs we will state can be seen the special case of Theorem 10.1.

Now we shall present some well-known results of FNNs presented in [16]. It is well-known that this class of FNN functions is capable to approximate any continuous function over a compact set to any desired degree of accuracy. For the following content about theorem proof, we need readers to be familiar with the basic content of functional analysis. And details of these concepts and theorems can be found in [7].

Lemma 10.1 ([8]) *Any bounded, measurable squashing function σ is discriminatory. In particular, any continuous squashing function is discriminatory.*

Proof For any x, y, θ, φ we have

$$\sigma_\lambda(x) = \sigma\left(\lambda(y^T x + \theta)\right) + \varphi = \begin{cases} \to 1, & \text{for } y^T x + \theta > 0 \text{ as } \lambda \to \infty \\ \to 0, & \text{for } y^T x + \theta < 0 \text{ as } \lambda \to \infty \\ \sigma(\varphi), & \text{for } y^T x + \theta = 0 \text{ for all } \lambda. \end{cases} \quad (10.12)$$

This can be seen as applying the properties of the squashing function to its input for varying values of λ. In other words, as $\lambda \to \infty$ for $y^T x + \theta > 0$, we are in essence calculating $\sigma(t)$ for $t \to \infty$. Similarly, as $\lambda \to \infty$ for $y^T x + \theta < 0$ we get $\sigma(t)$ for $t \to -\infty$. The third case is obvious. Thus the functions parameterized by λ, $\sigma_\lambda(x)$ converge in the sense of pointwise and is bounded by

10.3 Feedforward Neural Networks

$$\gamma(x) = \begin{cases} 1, & \text{for } y^T x + \theta > 0 \\ 0, & \text{for } y^T x + \theta < 0 \\ \sigma(\varphi), & \text{for } y^T x + \theta = 0, \end{cases} \quad (10.13)$$

as $\lambda \to \infty$. This follows directly from the above.
Let $\Pi_{y,\theta} = \{x \mid y^T x + \theta = 0\}$ be an affine hyperplane and

$$H_{y,\theta} = \{x \mid y^T x + \theta > 0\}$$

be the open half-space defined. Note that $|\sigma_\lambda(x)| \leq \max(1, \sigma(\varphi))$ for all x. We shall first fix y. Then for all φ, θ, y, we can apply the dominated convergence theorem to get

$$\lim_{\lambda \to \infty} \int_{I_n} \sigma_\lambda(x) d\mu(x) = \int_{I_n} \lim_{\lambda \to \infty} \sigma_\lambda(x) d\mu(x)$$
$$= \int_{I_n} \gamma(x) d\mu(x) \quad (10.14)$$
$$= \sigma(\varphi) \mu\left(\Pi_{y,\theta}\right) + \mu\left(H_{y,\theta}\right).$$

For a bounded, measurable function h, define a linear functional F

$$F(h) = \int_{I_n} h(y^T x) d\mu(x). \quad (10.15)$$

Note that F is a bounded linear functional on $L^\infty(R)$ since μ is a finite signed measure. This is because when we integrate with respect to a finite measure, we can't get an infinite result. Let h be the indicator function for the interval $[\theta, \infty)$ so that

$$F(h) = \int_{I_n} h(y^T x) d\mu(x) = \mu\left(\Pi_{y,-\theta}\right) + \mu\left(H_{y,-\theta}\right) = 0. \quad (10.16)$$

To see why this is true, recall that the indicator function $\chi_A : X \to \{0, 1\}$ for a set $A \subseteq X$ is defined as follows

$$\chi_A(x) = \begin{cases} 1, & \text{if } x \in A \\ 0, & \text{otherwise.} \end{cases} \quad (10.17)$$

Thus, if $y^T x \in [\theta, \infty)$, then $y^T x - \theta \geq 0$. For the integral, this decomposes into the measure of two disjoint sets, the hyperplane $\Pi_{y,-\theta} = \{x \mid y^T x - \theta = 0\}$ and the half-space $H_{y,-\theta} = \{x \mid y^T x - \theta > 0\}$. Similarly, $F(h) = 0$ if h is the indicator function for the open interval (θ, ∞). By linearity, F is 0 for the indicator function on any interval and hence for any simple function. Simple functions are dense in $L^\infty(R)$, so $F = 0$. □

Theorem 10.2 (Hahn-Banach) *Let V be a normed vector space and $R \subset V$ a subspace of V. Let $L \in R^*$. Then there exists $\hat{L} \in V^*$ that extends L to V and satisfies $\|\hat{L}\|_{V^*} = \|L\|_{R^*}$.*

Corollary 10.1 *Let V be a normed vector space, $R \subset V$ a subspace of V. Let $x_0 \in V$ such that $d(x_0, R) = \gamma > 0$. Then there exists $L \in V^*$ such that*

(1) $\|L\|_{V^*} = 1$.
(2) $L(x_0) = \gamma$.
(3) $L(R) = 0$.

Theorem 10.3 (Riesz Representation Theorem) *Let L be a bounded linear functional on $C(I_n)$. Then there exists a unique $\mu \in M(I_n)$ such that*

$$L(h) = \int_{I_n} h(x) \mathrm{d}\mu(x), \quad \forall h \in C(I_n). \tag{10.18}$$

We now state the universal approximation theorem of FNN and give an elegant proof. This proof is based on [8].

Theorem 10.4 (Universal Approximation of FNNs [16]) *For every squashing function κ, every $r \in \mathbb{N}$, $\Sigma^r(\kappa)$ is uniformly dense on compacta in C^r.*

Proof For simplicity, we consider $C^r = C(I_r)$ and $\kappa = \sigma$. However, it is easy to extend to the general case. Let R be the closure of the $\Sigma^r(\kappa)$, our goal is to show $R = C(I_r)$.

Note that $\Sigma^r(\kappa)$ is a linear subspace of $C(I_r)$. By contradiction suppose $R \subsetneq C(I_r)$, that is $\exists f \in C(I_r)$ such that $d(f, R) > 0$. By the Corollary 10.1, $\exists L$ bounded linear functional on $C(I_r)$ such that $L \neq 0$, but $L(I_r) = L(R) = 0$.

By Theorem 10.3, there exists unique $\mu \in M(I_r)$ such that

$$L(h) = \int_{I_n} h(x) \mathrm{d}\mu(x), \quad \forall h \in C(I_r). \tag{10.19}$$

Since $L(R) = 0$ and $\kappa(w^\top x + b) \in R, \forall w, b$. Then

$$0 = L(\kappa(w^\top x + b)) = \int_{I_n} \kappa(w^\top x + b) \mathrm{d}\mu(x) \quad , \forall w, b. \tag{10.20}$$

Notice that κ is discriminatory; (10.20) implies $\mu = 0$, which in turn implies $L = 0$; and this is a contradiction. □

This theorem tells us that standard FNNs with only a single hidden layer can approximate any continuous function uniformly on any compact set.

Naturally, Theorem 10.4 can be extended to the approximation of vector-valued functions. Let $C^{r,N}$ be the set of continuous functions from R^r to R^N and $\Sigma^{r,N}(\kappa)$ be the class of functions

$$\left\{\begin{array}{l}\bar{\zeta} = \left(\bar{\zeta}_1, \ldots, \bar{\zeta}_N\right)^{\mathrm{T}} : R^r \to R^N : \bar{\zeta}_l(x) = \sum_{j=1}^{q} \beta_{l,j} \kappa \left(A_j(x)\right), \\ x \in R^r, \beta_{l,j} \in R, A_j \in \mathbf{A}^r, 1 \leq l \leq N, q = 1, 2, \ldots.\end{array}\right\} \quad (10.21)$$

Then we have the following corollary.

Corollary 10.2 *Theorem 10.4 holds for the approximation of functions in $C^{r,N}$ by the extended function class $\Sigma^{r,N}(\kappa)$. Thereby the metric $\rho_S^N(f, g)$ is given by*

$$\rho_S^N(f, g) := \sup_{x \in S} \sum_{l=1}^{N} |f_l(x) - g_l(x)|. \quad (10.22)$$

10.4 Optimization and Backpropagation

In this section, we shall review the popular optimization methods for training neural networks and explain backpropagation mechanism based on FNNs.

10.4.1 Optimization Algorithms for Neural Networks

Before proceeding with the backpropagation of FNN, we shall briefly introduce the optimization algorithms used in deep learning. For more details, we recommend [2]. The most commonly used approach for estimating the parameters of a neural network is based on gradient descent (GD) which is a simple methodology for optimizing a function.

For a given function $f : R^d \to R$, we wish to determine the value of x that achieves the minimum value of f. GD method begins with an initial value x_0 and computes the gradient of f at this point. Then we iterate x_n according to

$$x_{n+1} = x_n - \eta \nabla_x f(x_n), \quad (10.23)$$

where η is the step size known as the learning rate. The algorithm converges to a critical point when the gradient is equal to zero, though it should be noted that this is not necessarily a global minimum.

In the context of neural networks, we would compute the derivatives of the loss functional with respect to the parameter set θ and follow the procedure (10.23).

10.4.1.1 Stochastic Gradient Descent (SGD)

The main difficulty with the use of GD to train neural networks is the computational cost associated with the procedure when training sets are large. Therefore, an extension of the GD known as stochastic gradient descent (SGD) is proposed. For the loss function $L(\theta; x, y)$, it can be written as

$$\nabla L(\theta; x, y) = \frac{1}{m} \sum_{i=1}^{m} \nabla_\theta L_i(\theta; x_i, y_i), \qquad (10.24)$$

where m is the size of the training set and L_i is the loss function for each example. The approach in SGD is to view the gradient as an expectation and approximate it with a random subset of the training set called a mini-batch. That is, for a fixed mini-batch of size $m' << m$ the gradient is estimated as

$$\nabla L(\theta; x, y) \approx \frac{1}{m'} \sum_{i=1}^{m'} \nabla_\theta L_i(\theta; x_i, y_i). \qquad (10.25)$$

10.4.1.2 Adaptive Moment Estimation (Adam)

SGD is the basic optimizer used in optimizing loss function for deep neural network; it is too slow for a deep neural network with a large number of data. So the following optimizer called Adaptive Moment Estimation (Adam) [17] is introduced.

Adam is another method that computes adaptive learning rates for each parameter. In addition to storing an exponentially decaying average of past squared gradients v_t, Adam also keeps an exponentially decaying average of past gradients m_t, similar to momentum. Whereas momentum can be seen as a ball running down a slope, Adam behaves like a heavy ball with friction, which thus prefers flat minima in the error surface. We compute the decaying averages of past and past squared gradients m_t and v_t, respectively, as follows:

$$\begin{aligned} m_t &= \beta_1 m_{t-1} + (1 - \beta_1) g_t, \\ v_t &= \beta_2 v_{t-1} + (1 - \beta_2) g_t^2, \end{aligned} \qquad (10.26)$$

where g_t is the gradient at time t and m_t and v_t are estimates of the first moment (the mean) and the second moment (the uncentered variance) of the gradients, respectively, hence the name of the method. As m_t and v_t are initialized as zeros vectors, the authors of Adam observe that they are biased toward zero, especially during the initial time steps, and especially when the decay rates are small (i.e., β_1 and β_2 are close to 1). They counteract these biases by computing bias-corrected first and second moment estimates:

$$\begin{aligned} \hat{m}_t &= \frac{m_t}{1 - \beta_1^t}, \\ \hat{v}_t &= \frac{v_t}{1 - \beta_2^t}. \end{aligned} \qquad (10.27)$$

They then use these to update the parameters as the following:

10.4 Optimization and Backpropagation

$$\theta_{t+1} = \theta_t - \frac{\eta}{\sqrt{\hat{v}_t} + \epsilon} \hat{m}_t. \quad (10.28)$$

The authors propose default values of 0.9 for β_1, 0.999 for β_2, and 10^{-8} for ϵ. They show empirically that Adam works well in practice and compares favorably to other adaptive learning-method algorithms.

10.4.2 Backpropagation

The SGD optimization approach and Adam optimization approach described in the previous section requires repeated computations of the gradients of a highly nonlinear function, e.g., FNN neural network [11]. Backpropagation [23, 32] provides a computationally efficient means by which this can be achieved. It is based on recursively applying the chain rule and on defining computational graphs to understand which computations can be run in parallel.

Let's illustrate the computational graphs with a simple example. We consider $x = f(w), y = f(x), z = f(y)$. Then we could compute $\frac{\partial z}{\partial w}$ as follows:

$$\begin{aligned}
\frac{\partial z}{\partial w} &= \frac{\partial z}{\partial y} \frac{\partial y}{\partial x} \frac{\partial x}{\partial w} \\
&= f'(y) f'(x) f'(w) \\
&= f'(f(f(w))) f'(f(w)) f'(w).
\end{aligned} \quad (10.29)$$

(10.29) can be formulated in Fig 10.2. It is called the computational graph for $z = f(f(f(w)))$, which contains the forward pass and backward pass.

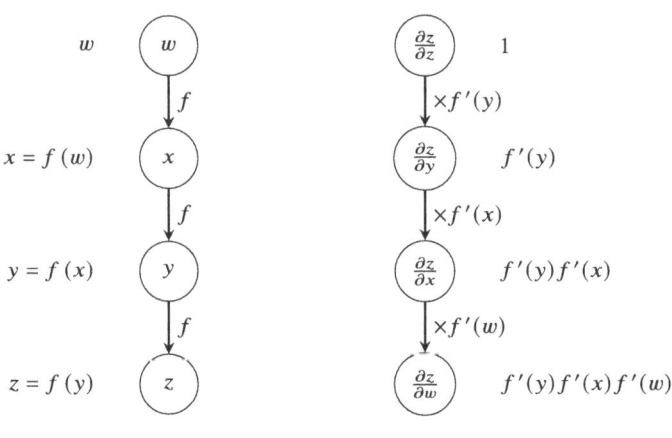

Fig. 10.2 Visualization of backpropagation algorithm via computational graphs. The left panel shows the composition of functions connecting input to output; the right panel shows the use of the chain rule to compute the derivative

Figure 10.2 is simple, but it illustrates that the composition of simple functions like FNN can be viewed as operations between nodes in the graph. And the derivative of output y with respect to x can be computed analytically by repeating applications of the chain rule like the right panel in Fig 10.2. Then the backpropagation algorithm could compute the derivative of the output y with respect to x, which is represented symbolically along this computation graph (backward). It is very important in the implementation of the deep learning frameworks with automatic differentiation, such as Pytorch [29].

After introducing the computational graphs, we shall take a multilayer FNN as an example to briefly introduce the principle of backpropagation. The multilayer FNN (ignoring bias b) is defined by

$$\mathbf{x}^{(1)} = \sigma(\mathbf{W}^{(1)}\mathbf{x}^{(0)}),$$
$$\mathbf{x}^{(2)} = \sigma(\mathbf{W}^{(2)}\mathbf{x}^{(1)}),$$
$$\vdots$$
$$\mathbf{x}^{(l)} = \sigma(\mathbf{W}^{(l)}\mathbf{x}^{(l-1)}).$$
(10.30)

The output $\mathbf{x}^{(l)}$ of the neural network is the prediction made by the neural network. Here H denotes the loss function defined by

$$z = H(\mathbf{y}, \mathbf{x}^{(l)}),$$
(10.31)

where $\mathbf{y}$ is the true label. To use GD updating the parameters $\mathbf{W}^{(1)}, \ldots, \mathbf{W}^{(l)}$, we need to calculate the gradient of the loss z with respect to each variable:

$$\frac{\partial z}{\partial \mathbf{W}^l}, \ldots, \frac{\partial z}{\partial \mathbf{W}^1}.$$
(10.32)

The core of backpropagation is the chain rule of derivation. The chain rule can be used to do backpropagation to get the gradient of the loss with respect to the parameters of the neural network. Specifically, compute the gradient $\frac{\partial z}{\partial \mathbf{x}^{(l)}}$ first. Then do the loop, starting from $i = l, \ldots, 1$:

(1) Compute the gradient of the loss z with respect to the parameter $\mathbf{W}^{(l)}$ that can be obtained according to the chain rule:

$$\frac{\partial z}{\partial \mathbf{W}^{(i)}} = \frac{\partial \mathbf{x}^{(i)}}{\partial \mathbf{W}^{(i)}} \frac{\partial z}{\partial \mathbf{x}^{(i)}}.$$
(10.33)

This gradient is used to update the parameter $\mathbf{W}^{(i)}$.

(2) Compute the gradient of the loss z with respect to the parameter $\mathbf{x}^{(i-1)}$ that can be obtained according to the chain rule again:

$$\frac{\partial z}{\partial \mathbf{x}^{(i-1)}} = \frac{\partial \mathbf{x}^{(i)}}{\partial \mathbf{x}^{(i-1)}} \frac{\partial z}{\partial \mathbf{x}^{(i)}}. \tag{10.34}$$

This gradient is propagated to the next layer (i.e., layer $i-1$), and the loop continues.

10.5 Recurrent Neural Networks

Recurrent neural networks (RNNs) are able to learn features and long-term dependencies from sequential data. Elman [9] popularized simple RNNs (Elman network). RNNs have various applications in many fields, such as language modeling [28], speech recognition [12], image processing [3], and machine translation [1]. A typical RNN architecture has many advantages, such as possibility of processing input of any length, model size not increasing with size of input, and taking into account historical information and weights shared across time. However, it has some drawbacks too, computation being slow, difficulty of accessing information from a long time ago and not considering any future input for the current state. To solve these problems, the new extensions of RNN has been proposed, such as LSTM [15] and GRU [6]. In the following, we only focus on the basic RNNs.

10.5.1 Mathematical Forms for Networks

RNNs are a class of neural networks that allow previous outputs to be used as inputs while having hidden states. They are typically as follows:

$$\begin{aligned} h_t &= f\left(U x_t + W h_{t-1} + b_1\right), \\ y_t &= g\left(V h_t + b_2\right), \end{aligned} \tag{10.35}$$

where $h_t \in R^h$ is the hidden variable at step t with input $x_t \in R^d$ and last hidden variable $h_{t-1} \in R^h$ and f is typically tanh. $y_t \in R^r$ is the output at step t with the readout map g, which is typically identity map. $U \in R^{h \times d}$, $W \in R^{h \times h}$, $V \in R^{r \times h}$, $b_1 \in R^h$, $b_2 \in R^r$ are parameters to be adjusted. A schematic diagram of the structure of a simple RNN forward propagation is shown in Fig. 10.3. It takes the inputs $x_1, \ldots, x_6$ and outputs $y_1, \ldots, y_6$ with hidden states $h_1, \ldots, h_6$.

Fig. 10.3 A two-layer RNN example

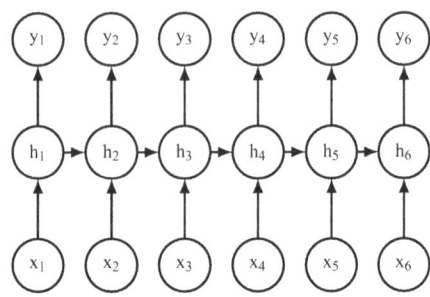

10.5.2 Universal Approximation Theorem for RNNs

For the convenience of stating our universal approximation theorem for RNNs, we shall introduce the open dynamical system and formulate the above mentioned RNNs as the class of functions below.

Definition 10.7 An open dynamical system in discrete time can be represented by the following equations:

$$\begin{cases} s_{k+1} = \eta(s_k, \alpha_{k+1}), & \text{(state transition)} \\ \beta_k = \xi(s_k), & \text{(output equation)} \end{cases} \quad (10.36)$$

where α_k is the stochastic external input, s_k is the state, and β_k is the observable output for $\forall k \geq 1$.

Definition 10.8 For any Borel-measurable function $\sigma(\cdot) : R^h \to R^h$ and $r, d, T \in \mathbb{N}$ be $\text{RNN}^{r,d}(\sigma)$ the class of functions

$$\begin{aligned} h_{t+1} &= \sigma(Uh_t + Wx_t + b), \\ y_t &= Vh_t, \end{aligned} \quad (10.37)$$

where $x_t \in R^r$, $h_t \in R^h$, and $y_t \in R^d$, with $t = 1, \ldots, T$. And the matrices $U \in R^{h \times h}$, $W \in R^{h \times l}$, and $V \in R^{d \times h}$ and the bias $b \in R^h$.

Here, we give the universal approximation theorem for RNN. The statement and the proof are based on [33]. This theorem tells us that RNNs in state space model form (10.37) are universal approximators and able to approximate every open dynamical system (10.36) with an arbitrary accuracy.

Theorem 10.5 (Universal Approximation Theorem for RNN [33]) *Let $g(\cdot) : R^h \times R^r \to R^h$ be measurable and $h(\cdot) : R^h \to R^d$ be continuous, the external inputs $x_t \in R^r$, the inner states $s_t \in R^h$, and the outputs $y_t \in R^d (t = 1, \ldots, T)$. Then, any open dynamical system of the form*

10.5 Recurrent Neural Networks

$$s_t = g(s_{t-1}, x_t),$$
$$y_t = h(s_t),$$
(10.38)

which can be approximated by an element of the function class $RNN^{r,d}(\sigma)$ with an arbitrary accuracy.

Proof The main idea of this proof is that we can use two different FNNs with single hidden layer to approximate the state equation and output equation of an open dynamical system with an arbitrary accuracy in terms of Theorem 10.4. Thereby we can approximate an open dynamical system with an arbitrary accuracy.

First, the state transition function of the nonlinear dynamical system $s_t = g(s_{t-1}, x_t)$ can be approximated by a two-layer FNN $s'_t = Cf(As_{t-1} + Bx_t + b)$ with an arbitrary accuracy, where A, B, C is the weight matrix and b is the bias vector. This two-layer FNN can be decomposed into

$$\begin{aligned} s'_t &= \sigma(As_{t-1} + Bx_t + b) \\ &= \sigma\left(ACs'_{t-1} + Bx_t + b\right), \\ s_t &= Cs'_t. \end{aligned}$$
(10.39)

Similarly, the output function of the nonlinear dynamical system $y_t = h(s_t) = h(g(s_{t-1}, x_t))$ can also be approximated by a two-layer FNN

$$y_t = D\sigma\left(A's_{t-1} + B'x_t + b'\right)$$

with an arbitrary accuracy, where A', B' and D is the weight matrix, b' is the bias vector. The second two-layer FNN is given by

$$\begin{aligned} y'_t &= \sigma\left(A's_{t-1} + B'x_t + b'\right) \\ &= \sigma\left(A'Cs'_{t-1} + B'x_t + b'\right), \\ y_t &= Dy'_t. \end{aligned}$$
(10.40)

Notice that

$$\begin{bmatrix} s'_t \\ y'_t \end{bmatrix} = \sigma\left(\begin{bmatrix} AC & 0 \\ A'C & 0 \end{bmatrix} \begin{bmatrix} s'_{t-1} \\ y'_{t-1} \end{bmatrix} + \begin{bmatrix} B \\ B' \end{bmatrix} x_t + \begin{bmatrix} b \\ b' \end{bmatrix}\right),$$
(10.41)

and

$$y_t = \begin{bmatrix} 0 & D \end{bmatrix} \begin{bmatrix} s'_t \\ y'_t \end{bmatrix}.$$
(10.42)

Let $h_t = [s_t', y_t']$. Then the nonlinear dynamical system can be approximated by the following RNN

$$h_t = \sigma\left(\mathbf{U}h_{t-1} + \mathbf{W}x_t + \mathbf{b}\right),$$
$$y_t = \mathbf{V}h_t, \tag{10.43}$$

where $\mathbf{U} = \begin{bmatrix} AC & 0 \\ A'C & 0 \end{bmatrix}$, $\mathbf{W} = \begin{bmatrix} B \\ B' \end{bmatrix}$, $\mathbf{b} = \begin{bmatrix} b \\ b' \end{bmatrix}$, $\mathbf{V} = \begin{bmatrix} 0 & D \end{bmatrix}$. □

10.6 The Application of Deep Learning in Nonlinear Filtering

Finally, as the application of the deep learning, we shall introduce how to use deep learning techniques mentioned above to solve nonlinear filtering problems. The filtering algorithm based on neural networks is originated from Lo [26] called neural filtering, which uses several correlated observations with corresponding states to train the recurrent MLP to an optimal filter. Recently, nonlinear filtering algorithms based on recurrent neural network (RNN) have been developed and implemented in both discrete-time and continuous-time settings, with a more comprehensive theoretical background.

For discrete-time filtering problems, Chen et al. [4] formulate the finite dimensional filter as the dynamical system with stochastic inputs and show that the open dynamical systems with stochastic inputs can be approximated by a class of RNNs in state space model form with an arbitrary accuracy. Moreover, they construct an RNN-based filter and prove that it can well-approximate finite dimensional filters which include Kalman filter (KF) and Beneš filter as special cases.

For continuous-time filtering problems, Chen et al. [5] develop a uniform formulation of Yau-Yau algorithm and introduce RNN for implementation. Leveraging the powerful representation capability of neural network, the proposed RNN Yau-Yau algorithm outperforms classical methods especially for nonlinear filtering systems in high-dimensional space. In fact, with a pre-trained RNN, this deep learning-based filtering algorithm can deal with nonlinear filtering problems with state space dimension up to 100 in real time. Moreover, it is rigorously proved that the proposed RNN Yau-Yau algorithm has the capability of overcoming the curse of dimensionality, in the sense that the magnitude of the neural network employed in the algorithm only needs to increase polynomially (rather than exponentially) with respect to the dimension of the system.

There are also other types of deep learning-based filtering algorithms which have been proposed recently. For example, instead of RNNs, the neural projection filter proposed by Tao et al. [35] engages neural stochastic differential equations in the algorithm, and this data-driven method, in comparison with other model-based methods, is especially useful in application scenarios where an accurate filtering

10.6 The Application of Deep Learning in Nonlinear Filtering

model may not exist; the deep filters proposed in [30, 31, 37] train fully connected neural networks for state estimation.

Nevertheless, we would like to focus on the RNN-based filtering algorithms here, because they preserve the recursive structure of the filtering problem and are convenient for practical implementation and theoretical analysis.

Here in this section, we shall introduce the universal approximation theorem for RNN with stochastic inputs and illustrate the algorithm details of the RNN-based algorithms for both discrete-time and continuous-time filtering problems. Convergence analysis of the algorithms will also be presented here, and especially, we will theoretically interpret why RNN Yau-Yau algorithm can overcome the curse of dimensionality.

10.6.1 Preliminaries

Before proceeding, we shall introduce the preliminaries about sufficient statistics and finite dimensional filter (FDF). And we also introduce the uniform integrability which is the key part in the proofs of the main results.

10.6.1.1 FDF

Recall the nonlinear filtering problem, its purpose is to compute the target posterior $p(X_k \mid Y_k)$. In addition to probability density functions, there is another concept in statistics to describe distributions called sufficient statistics. For example, let us denote $\mathbf{S}_{k|k}$ as the sufficient statistics of $p(X_k \mid Y_k)$. Obviously, $p(X_k \mid Y_k)$ can be completely determined by $\mathbf{S}_{k|k}$. The definition of the sufficient statistics is given as follows.

Definition 10.9 (Sufficient Statistic) If the conditional distribution $p(X_k \mid Y_k)$ can be completely determined by a vector-valued function $\mathbf{S}_{k|k} \in R^{n_s}$ of the observation sequence Y_k, where $n_s \in \mathbb{N}$, then we say $\mathbf{S}_{k|k}$ is a sufficient statistic for $p(X_k \mid Y_k)$.

Hence, there exists a function $\gamma : R^{n_s} \to R^n$, such that

$$\mathbb{E}[X_k \mid Y_k] = \gamma\left(\mathbf{S}_{k|k}\right), \tag{10.44}$$

since the optimal estimate $\mathbb{E}[X_k \mid Y_k]$ is determined by $p(X_k \mid Y_k)$, which is completely determined by the sufficient statistic $\mathbf{S}_{k|k}$.

Similarly, we use vector $\mathbf{S}_{k|k}$ to denote the finite dimensional sufficient statistics of the posterior distribution $p(X_k \mid Y_k)$. The evolution function of the statistics is denoted as Φ, and the map from $\mathbf{S}_{k|k}$ to conditional mean $\mathbb{E}[X_k \mid Y_k]$ is denoted as Γ, that is

$$\mathbf{S}_{k|k} = \Phi\left(\mathbf{S}_{k-1|k-1}, Y_k\right)$$
$$\mathbb{E}[X_k \mid Y_k] = \Gamma\left(\mathbf{S}_{k|k}\right). \tag{10.45}$$

As we know, in most cases, it is not easy to write down the explicit forms of the map functions Φ and Γ. However, by taking advantage of neural networks, we can approximate these functions just using the input and output data, which motivates us to use neural networks to solve the FDF problems.

10.6.1.2 Uniform Integrability

Before we start the analysis of RNN, we need to introduce an important concept, i.e., uniform integrability. Here, we define the truncation operator $\mathcal{T}_K$ with level $K > 0$ as

$$\mathcal{T}_K(x_i) = \begin{cases} x_i, & \text{if } |x_i| \leq K \\ K \cdot \text{sign}(x_i), & \text{otherwise} \end{cases} \tag{10.46}$$

and

$$\mathcal{T}_K(x) := (\mathcal{T}_K(x_1), \ldots, \mathcal{T}_K(x_n))^T \tag{10.47}$$

for $x = (x_1, \ldots, x_n)^T \in R^n$. It can be easily checked that $\mathcal{T}_K x = x$ when $|x| \leq K$, and $|\mathcal{T}_K x| \leq |x|$ for all $x \in R^n$. In addition

$$\left\|\mathcal{T}_K X - \mathcal{T}_K \tilde{X}\right\|_1 \leq \|X - \tilde{X}\|_1 \quad \forall X, \tilde{X} \in L^1(\Omega; R^n), \tag{10.48}$$

Definition 10.10 A collection of random variables $\{X_i \in R, i \in I\}$ in $L^1(\Omega; R)$ is said to be uniformly integrable if

$$\lim_{M \to +\infty} \left(\sup_{i \in I} \mathbb{E}\left[|X_i| \mathbb{I}_{|X_i| > M}\right]\right) = 0. \tag{10.49}$$

Similarly, this definition can be extended to random vectors.

Definition 10.11 A collection of random vectors $\{X_i \in R^n, i \in I\}$ in $L^1(\Omega; R^n)$ is said to be uniformly integrable if

$$\lim_{M \to +\infty} \left(\sup_{i \in I} \mathbb{E}\left[|X_i| \mathbb{I}_{|X_i| > M}\right]\right) = 0. \tag{10.50}$$

A common way to check the uniform integrability is listed in the following lemma.

Lemma 10.2 *Let $\{X_i \in R^n, i \in I\}$ be a collection of random vectors. If*

10.6 The Application of Deep Learning in Nonlinear Filtering

$$\sup_{i \in I} \mathbb{E}\left[|X_i|^p\right] < \infty, \text{ for some } p > 1, \tag{10.51}$$

then $\{X_i i \in I\}$ is uniformly integrable.

Following Lemma 10.2, we can obtain the following two useful results which will be used in sequel.

Lemma 10.3 *Assume a collection of random vectors $\{X_i \in R^n, i \in I\}$ is uniformly integrable. Then for any $\varepsilon > 0$, there exists a positive $K > 0$, such that*

$$\sup_{i \in I} \|X_i - \mathcal{T}_K X_i\|_1 < \varepsilon, \tag{10.52}$$

where the truncation operator $\mathcal{T}_K$ is defined in (2).

Proof Since $\{X_i : i \in I\}$ is uniformly integrable, that is,

$$\lim_{M \to +\infty} \left(\sup_{i \in I} \mathbb{E}\left[|X_i| \mathbb{I}_{|X_i| > M}\right]\right) = 0 \tag{10.53}$$

there exists $K > 0$, such that

$$\sup_{i \in I} \mathbb{E}\left[|X_i| \mathbb{I}_{|X_i| > K}\right] < \frac{\varepsilon}{2}. \tag{10.54}$$

Then we have

$$\sup_{i \in I} \|X_i - \mathcal{T}_K X_i\|_1 \leq \sup_{i \in I} \mathbb{E}\left[|X_i - \mathcal{T}_K X_i| \mathbb{I}_{|X_i| \leq K}\right] + \sup_{i \in I} \mathbb{E}\left[|X_i - \mathcal{T}_K X_i| \mathbb{I}_{|X_i| > K}\right]$$

$$= 0 + \sup_{i \in I} \mathbb{E}\left[|X_i - \mathcal{T}_K X_i| \mathbb{I}_{|X_i| > K}\right]$$

$$\leq \sup_{i \in I} \left(\mathbb{E}\left[|X_i| \mathbb{I}_{|X_i| > K}\right] + \mathbb{E}\left[|\mathcal{T}_K X_i| \mathbb{I}_{|X_i| > K}\right]\right)$$

$$\leq 2\mathbb{E}\left[|X_i| \mathbb{I}_{|X_i| > K}\right]$$

$$< \varepsilon. \tag{10.55}$$

□

Remark 10.2 According to Lemma 10.3, it is known that we can find a sufficiently large cube, such that most of the densities of the uniformly integrable random vectors fall in this bounded set. In other words, if $\{X_i \in R^n, i \in I\}$ is uniformly integrable, then we can choose a sufficient large $K > 0$, such that uniformly over $X_i \in \{X_i \in R^n, i \in I\}$, the random vector $\mathcal{T}_K X_i$ is a good approximation of X_i in terms of the L_1-norm. Crucially, every $\mathcal{T}_K X_i$ is a bounded random vector, which is the desired property allowing us to approximate functions in RNN with infinite time steps.

Combing Lemmas 10.2 and 10.3, we can easily obtain the following lemma.

Lemma 10.4 *Assume that a collection of random vectors $X_i \in R^n, i \in I$, satisfy $\sup_{i \in I} \|X_i\|_2 < \infty$. Then for any $\varepsilon > 0$, there exists a positive $K > 0$, such that*

$$\sup_{i \in I} \|X_i - \mathcal{T}_K X_i\|_1 < \varepsilon, \tag{10.56}$$

where the truncation operator $\mathcal{T}_K$ is defined in (10.47).

Proof It is apparent that

$$\sup_{i \in I} \mathbb{E}\left[|X_i|^2\right] < \infty. \tag{10.57}$$

Then according to Lemma 10.2, we know that $\{X_i, i \in I\}$ uniformly integrable. Using Lemma 10.3, we obtain the desire result. □

10.6.2 Universal Approximation Theorem for RNN with Stochastic Inputs

In the previous section, we have seen that while FNNs can be used to approximate continuous functions in compact set, RNN can be mapped to an open dynamical system with sequential external inputs.

Now we aim to approximate the open dynamical system (10.36) with stochastic inputs by a class of RNNs. More explicitly, we investigate $\text{RNN}^{r_1, r_2, r_3}(\kappa)$, which is defined as follows.

Definition 10.12 For any squashing function κ, and $r_1, r_2, r_3 \in \mathbb{N}$, $\text{RNN}^{r_1, r_2, r_3}(\kappa)$ is a class of functions with the following state space model form:

$$\begin{cases} \tilde{s}_{k+1} = \tilde{\eta}(\tilde{s}_k, \alpha_{k+1}) \\ \tilde{\beta}_k = \tilde{\xi}(\tilde{s}_k), \end{cases} \tag{10.58}$$

where $\alpha_k \in R^{r_1}$ is the input, $\tilde{s}_k \in R^{r_2}$ is the hidden state, $\tilde{\beta}_k \in R^{r_3}$ is the output, and

$$\tilde{\eta}(\tilde{s}, \alpha) = \bar{\eta}(\mathcal{T}_{K^s}\tilde{s}, \mathcal{T}_{K^a}\alpha) \tag{10.59}$$

$$\tilde{\xi}(\tilde{s}) = \bar{\xi}(\mathcal{T}_{K^s}\tilde{s}), \tag{10.60}$$

in which $\bar{\eta} \in \Sigma^{r_1+r_2, r_2}(\kappa), \bar{\xi} \in \Sigma^{r_2, r_3}(\kappa)$, K^s and K^α are two positive numbers which are the parameters of RNN, and $\mathcal{T}$ is the truncation operator defined in (10.47).

Theorem 10.6 (Universal Approximation Theorem for RNN with Stochastic Inputs) *Let $\eta(\cdot) : R^{r_2} \times R^{r_1} \to R^{r_2}$ and $\xi(\cdot) : R^{r_2} \to R^{r_3}$ be continuous,*

10.6 The Application of Deep Learning in Nonlinear Filtering

the external stochastic inputs $\alpha_k \in R^{r_1}$, the inner state $s_k \in R^{r_2}$, and the output $\beta_k \in R^{r_3}$, $k = 1, 2, \ldots$ For any open dynamical system of the form

$$\begin{cases} s_{k+1} = \eta(s_k, \alpha_{k+1}) \\ \beta_k = \xi(s_k), \end{cases} \quad (10.61)$$

if the following conditions hold:

(i) $\{\alpha_k, k \geq 1\}$ and $\{s_k, k \geq 1\}$ are uniformly integrable;
(ii) for $\forall s, \bar{s} \in L^1(\Omega; R^{r_2})$ and $\forall \alpha, \bar{\alpha} \in L^1(\Omega; R^{r_1})$, $\|\eta(s, \alpha) - \eta(\bar{s}, \bar{\alpha})\|_1 \leq C_{\eta 1}\|s - \bar{s}\|_1 + C_{\eta 2}\|\alpha - \bar{\alpha}\|_1$, and the Lipschitz constant $C_{\eta 1}$ satisfies $|C_{\eta 1}| < 1$;
(iii) for $\forall \epsilon > 0$, there exists $\delta > 0$, such that for any $s, \bar{s} \in L^1(\Omega; R^{r_2})$ satisfying $\|s - \bar{s}\|_1 < \delta$, we have $\|\xi(s) - \xi(\bar{s})\|_1 < \epsilon$,

then (10.61) can be approximated by the functions in $\text{RNN}^{r_1, r_2, r_3}(\kappa)$ with an arbitrary accuracy, i.e., for $\forall \varepsilon > 0$, there exist functions $\tilde{\eta}$ and $\tilde{\xi}$ of forms (10.59) and (10.60), which determine the RNN system (10.58) with the same input $\{\alpha_k, k \geq 1\}$ of (10.61), such that

$$\begin{aligned} \lim_{k \to \infty} \|s_k - \tilde{s}_k\|_1 &< \varepsilon, \\ \lim_{k \to \infty} \|\beta_k - \tilde{\beta}_k\|_1 &< \varepsilon, \end{aligned} \quad (10.62)$$

where $\tilde{s}_k$ and $\tilde{\beta}_k$ are the state and the output of the RNN system (24), respectively.

Proof The theorem is proven in three steps. We first construct appropriate approximated RNN functions using the universal approximation of FNNs. Then we try to obtain the iterative inequalities for errors. Finally, we compute the upper bounds of the accumulated errors.

Step 1: In this step, we will construct functions in $\text{RNN}^{r_1, r_2, r_3}(\kappa)$ to approximate system (10.61).

Since $\{\alpha_k, k \geq 1\}$ and $\{s_k, k \geq 1\}$ are uniformly integrable, for $\forall \varepsilon_1 > 0$, we can find $K_1 > 0$ and $K_2 > 0$, such that

$$\begin{aligned} \sup_{k \geq 1} \|s_k - \mathcal{T}_{K_1} s_k\|_1 &< \varepsilon_1, \\ \sup_{k \geq 1} \|\alpha_k - \mathcal{T}_{K_2} \alpha_k\|_1 &< \varepsilon_1, \end{aligned} \quad (10.63)$$

according to Lemma 10.3. Let

$$\begin{aligned} B_1 &:= \left\{ x = (x_1, \ldots, x_{r_2})^T \in R^{r_2} : |x_i| \leq K_1, 1 \leq i \leq r_2 \right\}, \\ B_2 &:= \left\{ x = (x_1, \ldots, x_{r_1})^T \in R^{r_1} : |x_i| \leq K_2, 1 \leq i \leq r_1 \right\}. \end{aligned} \quad (10.64)$$

Observing B_1 and B_2 are compact sets, and by Corollary 10.2, we know that for $\forall \varepsilon_2 > 0$, there exist functions $\bar{\eta} \in \Sigma^{r_1+r_2,r_2}$ and $\bar{\xi} \in \Sigma^{r_2,r_3}$ represented by FNNs, such that

$$\sup_{s \in B_1} |\xi(s) - \bar{\xi}(s)| < \varepsilon_2$$
$$\sup_{s \in B_1, \alpha \in B_2} |\eta(s,\alpha) - \bar{\eta}(s,\alpha)| < \varepsilon_2. \tag{10.65}$$

Set

$$\tilde{\xi}(s) := \bar{\xi}\left(\mathcal{T}_{K_1} s\right)$$
$$\tilde{\eta}(s,\alpha) := \bar{\eta}\left(\mathcal{T}_{K_1} s, \mathcal{T}_{K_2} \alpha\right). \tag{10.66}$$

Step 2: Define $e_k := \|s_k - \tilde{s}_k\|_1$, where $\tilde{s}_k$ is the state of system (10.58) with $\tilde{\eta}$ and $\tilde{\xi}$ defined in (10.66). Now we derive the evolution equation of the error e_k. Comparing (10.58) and (10.61), we have

$$\begin{aligned}
e_{k+1} &= \|s_{k+1} - \tilde{s}_{k+1}\|_1 \\
&= \|\eta(s_k, \alpha_{k+1}) - \tilde{\eta}(\tilde{s}_k, \alpha_{k+1})\|_1 \\
&= \|\eta(s_k, \alpha_{k+1}) - \bar{\eta}(\mathcal{T}_{K_1}\tilde{s}_k, \mathcal{T}_{K_2}\alpha_{k+1})\|_1 \\
&\leq \|\eta(s_k, \alpha_{k+1}) - \eta(\mathcal{T}_{K_1} s_k, \mathcal{T}_{K_2}\alpha_{k+1})\|_1 \\
&\quad + \|\eta(\mathcal{T}_{K_1} s_k, \mathcal{T}_{K_2}\alpha_{k+1}) - \eta(\mathcal{T}_{K_1}\tilde{s}_k, \mathcal{T}_{K_2}\alpha_{k+1})\|_1 \\
&\quad + \|\eta(\mathcal{T}_{K_1}\tilde{s}_k, \mathcal{T}_{K_2}\alpha_{k+1}) - \bar{\eta}(\mathcal{T}_{K_1}\tilde{s}_k, \mathcal{T}_{K_2}\alpha_{k+1})\|_1 \\
&\triangleq \Pi_1 + \Pi_2 + \Pi_3.
\end{aligned} \tag{10.67}$$

Now we analyze these three terms separately. As for Π_1, we have

$$\begin{aligned}
\Pi_1 &= \|\eta(s_k, \alpha_{k+1}) - \eta(\mathcal{T}_{K_1} s_k, \mathcal{T}_{K_2}\alpha_{k+1})\|_1 \\
&\leq C_{\eta 1}\|s_k - \mathcal{T}_{K_1} s_k\|_1 + C_{\eta 2}\|\alpha_{k+1} - \mathcal{T}_{K_2}\alpha_{k+1}\|_1 \\
&< (C_{\eta 1} + C_{\eta 2})\varepsilon_1,
\end{aligned} \tag{10.68}$$

where the first inequality is due to the second condition and the second inequality comes from (10.63). In terms of Π_2, using the Lipschitz property of η and (10.48), we have

$$\begin{aligned}
\Pi_2 &= \|\eta(\mathcal{T}_{K_1} s_k, \mathcal{T}_{K_2}\alpha_{k+1}) - \eta(\mathcal{T}_{K_1}\tilde{s}_k, \mathcal{T}_{K_2}\alpha_{k+1})\|_1 \\
&\leq C_{\eta 1}\|\mathcal{T}_{K_1} s_k - \mathcal{T}_{K_1}\tilde{s}_k\|_1 \\
&\leq C_{\eta 1} e_k.
\end{aligned} \tag{10.69}$$

10.6 The Application of Deep Learning in Nonlinear Filtering

As for Π_3, according to the second inequality in (10.65), we know that

$$\Pi_3 = \left\| \eta \left(\mathcal{T}_{K_1} \tilde{s}_k, \mathcal{T}_{K_2} \alpha_{k+1} \right) - \bar{\eta} \left(\mathcal{T}_{K_1} \tilde{s}_k, \mathcal{T}_{K_2} \alpha_{k+1} \right) \right\|_1 < \varepsilon_2, \tag{10.70}$$

since $\mathcal{T}_{K_1} \tilde{s}_k \in B_1$ and $\mathcal{T}_{K_2} \alpha_{k+1} \in B_2$. Substituting (10.68), (10.69), and (10.70) into (10.67), we can obtain

$$e_{k+1} < C_{\eta 1} e_k + \left(C_{\eta 1} + C_{\eta 2} \right) \varepsilon_1 + \varepsilon_2. \tag{10.71}$$

Step 3: Now we analyze the accumulated errors. Using (10.71) repeatedly, it follows that

$$\begin{aligned}
e_{k+1} &< C_{\eta 1} e_k + \left(C_{\eta 1} + C_{\eta 2} \right) \varepsilon_1 + \varepsilon_2 \\
&< C_{\eta 1}^2 e_{k-1} + \left(C_{\eta 1} + 1 \right) \left(\left(C_{\eta 1} + C_{\eta 2} \right) \varepsilon_1 + \varepsilon_2 \right) \\
&\vdots \\
&< C_{\eta 1}^k e_1 + \left(\left(C_{\eta 1} + C_{\eta 2} \right) \varepsilon_1 + \varepsilon_2 \right) \sum_{i=0}^{k-1} C_{\eta 1}^i \\
&= C_{\eta 1}^k e_1 + \frac{C_{\eta 1}^k - 1}{C_{\eta 1} - 1} \left(\left(C_{\eta 1} + C_{\eta 2} \right) \varepsilon_1 + \varepsilon_2 \right).
\end{aligned} \tag{10.72}$$

Thus, we have

$$\varlimsup_{k \to \infty} e_k \leq \frac{1}{1 - C_{\eta 1}} \left(\left(C_{\eta 1} + C_{\eta 2} \right) \varepsilon_1 + \varepsilon_2 \right), \tag{10.73}$$

once the condition $\left| C_{\eta 1} \right| < 1$ holds. Based on the third condition, we know that for $\forall \varepsilon > 0$, there exists $\delta > 0$, such that for any $s, \bar{s} \in L^2 \left(\Omega; R^{r_2} \right)$ satisfying $\| s - \bar{s} \|_1 < \delta$, we have $\| \xi(s) - \xi(\bar{s}) \|_1 < \varepsilon / 6$. Apparently, we can choose small enough ε_1 and ε_2, so that

$$\begin{cases} \varlimsup_{k \to \infty} e_k \leq \frac{1}{1 - C_{\eta 1}} \left(\left(C_{\eta 1} + C_{\eta 2} \right) \varepsilon_1 + \varepsilon_2 \right) < \min \left\{ \varepsilon, \frac{\delta}{2} \right\} \\ \| s_k - \mathcal{T}_{K_1} s_k \|_1 < \varepsilon_1 < \delta \\ \sup_{s \in B_1} | \xi(s) - \bar{\xi}(s) | < \varepsilon_2 < \varepsilon / 6, \end{cases} \tag{10.74}$$

based on (10.63) and the first inequality in (10.65). It follows that there exists $N_0 > 0$, such that

$$e_k = \| s_k - \tilde{s}_k \|_1 < \delta \quad \forall k \geq N_0. \tag{10.75}$$

Therefore, for any $k \geq N_0$, we have

$$\left\|\beta_k - \tilde{\beta}_k\right\|_1 = \left\|\xi(s_k) - \tilde{\xi}(\tilde{s}_k)\right\|_1$$

$$= \left\|\xi(s_k) - \bar{\xi}(\mathcal{T}_{K_1}\tilde{s}_k)\right\|_1$$

$$\leq \left\|\xi(s_k) - \xi(\mathcal{T}_{K_1}s_k)\right\|_1 + \left\|\xi(\mathcal{T}_{K_1}s_k) - \xi(\mathcal{T}_{K_1}\tilde{s}_k)\right\|_1 \qquad (10.76)$$

$$+ \left\|\xi(\mathcal{T}_{K_1}\tilde{s}_k) - \bar{\xi}(\mathcal{T}_{K_1}\tilde{s}_k)\right\|_1$$

$$< \varepsilon/6 + \varepsilon/6 + \varepsilon/6 = \varepsilon/2.$$

Since $\left\|s_k - \mathcal{T}_{K_1}s_k\right\|_1 < \delta$ and $\left\|\mathcal{T}_{K_1}s_k - \mathcal{T}_{K_1}\tilde{s}_k\right\|_1 < \delta$. Then

$$\overline{\lim_{k\to\infty}} \left\|\beta_k - \tilde{\beta}_k\right\|_1 < \varepsilon. \qquad (10.77)$$

It is obvious that we obtain the desired results from the first inequality of (10.74) and (10.77). □

Remark 10.3 As for the three conditions in Theorem 10.6, we have the following discussions.

(i) In terms of the first condition, if $\sup_{k\geq 1} \mathbb{E}[|s_k|^{p_1}] < \infty$ and $\sup_{k\geq 1} \mathbb{E}[|a_k|^{p_2}] < \infty$ for some $p_1, p_2 > 1$. Then by Lemma 10.2, we know that this condition is satisfied. We put this condition since we need to find a big enough high-dimensional cube, which can capture most of the densities of all the input and state random vectors. Then we can approximate the functions on the bounded domain using the approximation ability of FNNs and neglect the unbounded parts. This is why we can approximate functions on the whole space.

(ii) The second condition implies that the system (10.61) is stable [14], which is natural and useful in practice. This condition is used to ensure that the accumulated error will not blow up.

(iii) The third condition means that ξ is continuous in the given metric space. So that we can estimate the approximation error of the outputs from the approximation error of the hidden state.

Now we give an example which satisfies the three conditions in Theorem 10.6.

Example 10.1 Consider the following linear scalar system:

$$\begin{cases} s_{k+1} = c_0 s_k + c_1 \alpha_{k+1}, \\ \beta_k = c_2 s_k, \end{cases} \qquad (10.78)$$

where $|c_0| < 1, c_1 \neq 0$, and $\{\alpha_k, k \geq 0\}$ is a white Gaussian random sequence which is independent of s_1. By iterations, we can easily get

$$s_k = c_0^{k-1} s_1 + c_1 \sum_{i=0}^{k-2} c_0^i \alpha_{k-i} \quad \forall k \geq 2, \qquad (10.79)$$

then we have

$$\mathbb{E}\left[|\alpha_k|^2\right] = 1$$

$$\mathbb{E}\left[|s_k|^2\right] = c_0^{2(k-1)}\mathbb{E}\left[|s_1|^2\right] + c_1^2 \sum_{i=0}^{k-2} c_0^{2i} \qquad (10.80)$$

$$= c_0^{2(k-1)}\mathbb{E}\left[|s_1|^2\right] + c_1^2 \frac{1 - c_0^{2(k-1)}}{1 - c_0^2}.$$

Apparently, $\sup_{k\geq 1}\mathbb{E}\left[|s_k|^2\right] < \infty$ and $\sup_{k\geq 1}\mathbb{E}\left[|\alpha_k|^2\right] < \infty$. It can be easily checked that the three conditions in Theorem 10.6 are satisfied.

10.6.3 RNN-Based Filtering for Discrete-Time Systems

10.6.3.1 The Main Idea

Observing that, in FDFs, we have the following evolution functions of the sufficient statistics and the estimation:

$$\begin{cases} S_{k|k} = \Phi\left(S_{k-1|k-1}, y_k\right), \\ \mathbb{E}\left[x_k \mid Y_k\right] = \Gamma\left(S_{k|k}\right). \end{cases} \qquad (10.81)$$

It is obvious that (10.81) is an open dynamical system with the stochastic inputs $\{y_k, k \geq 0\}$ and the stochastic outputs $\{\mathbb{E}[x_k \mid Y_k]\}$ which are the desired optimal estimates of the states.

Naturally, using the universal approximation of RNN with stochastic inputs as shown in Theorem 10.6, we can approximate the open dynamical system (10.81) by RNN functions as detailed in last subsection. Following Theorem 10.6, it is known that we can approximate Φ and Γ by functions $\tilde{\Phi}$ and $\tilde{\Gamma}$ represented by FNNs, respectively, that is

$$\begin{aligned} \bar{\Phi}(s, y) &= \bar{\Phi}\left(\mathcal{T}_{K_1} s, \mathcal{T}_{K_2} y\right), \\ \bar{\Gamma}(s) &= \bar{\Gamma}\left(\mathcal{T}_{K_1} s\right), \end{aligned} \qquad (10.82)$$

where $\bar{\Phi} \in \Sigma^{n_s+m, n_s}(\kappa)$, $\bar{\Gamma} \in \Sigma^{n_s, n}(\kappa)$, K_1 and K_2 are two positive numbers which are the parameters of RNN, and $\mathcal{T}$ is the truncation operator defined in (10.47). Then we can obtain an RNN system which is as follows:

$$\begin{cases} \tilde{S}_{k|k} = \tilde{\Phi}\left(\tilde{S}_{k-1|k-1}, y_k\right), \\ \hat{x}_{k|k} = \tilde{\Gamma}\left(\tilde{S}_{k|k}\right), \end{cases} \qquad (10.83)$$

where $\tilde{S}_{k|k}$ and $\hat{x}_{k|k}$ are defined as the state and the output of the RNN system (10.83), respectively. We need to remark here that $\hat{x}_{k|k}$ is a function of Y_k.

Using the data $\{y_k, \mathbb{E}[x_k \mid Y_k]\}_{k \geq 0}$, we can train the RNN system (10.83) such that $\mathbb{E}[x_k \mid Y_k]$ can be well-approximated by the output $\hat{x}_{k|k}$, which can be regarded as the estimate of the state x_k based on observation history Y_k. We call this filtering method as RNN-based filter (RNNF).

Theorem 10.7 *Consider a discrete filtering system* (10.1) *with optimal FDF. Let $S_{k|k}, k \geq 0$ be the theoretical statistics evolving according to* (10.81) *and $\tilde{S}_{k|k}, k \geq 0$ be the statistics generated by our RNNF which evolve according to* (10.83). *We need the following assumptions:*

(i) *The sufficient statistics $\{S_{k|k}\}_{k \geq 0}$ and the observations $\{y_k\}_{k \geq 0}$ are uniformly integrable.*
(ii) *Function Φ is Lipschitz, i.e., for any $S, \bar{S} \in R^{n_S}$ and $y, \bar{y} \in R^m$, such that*

$$\|\Phi(S, y) - \Phi(\bar{S}, \bar{y})\|_1 \leq C_{\Phi 1} \|S - \bar{S}\|_1 + C_{\Phi 2} \|y - \bar{y}\|_1, \tag{10.84}$$

where n_S is the dimension of $S_{k|k}$, $C_{\Phi 1}$ and $C_{\Phi 2}$ are Lipschitz constants, and $C_{\Phi 1}$ satisfies $|C_{\Phi 1}| < 1$.
(iii) *For $\forall \epsilon > 0$, there exists $\delta > 0$, such that for any $s, \bar{s} \in L^1(\Omega; R^{n_s})$ satisfying $\|s - \bar{s}\|_1 < \delta$, we have $\|\Gamma(s) - \Gamma(\bar{s})\|_1 < \epsilon$. then for any $\varepsilon > 0$, there exists an RNNF* (10.83), *i.e., there exist $\tilde{\Phi}$ and $\tilde{\Gamma}$ of the forms* (10.82), *respectively, such that*

$$\varlimsup_{k \to \infty} \left\| S_{k|k} - \tilde{S}_{k|k} \right\|_1 < \varepsilon, \tag{10.85}$$

and

$$\varlimsup_{k \to \infty} \left\| \hat{x}_{k|k} - \mathbb{E}[x_k \mid Y_k] \right\|_1 < \varepsilon. \tag{10.86}$$

Proof The proof is similar to that of Theorem 10.6, we leave it as an exercise for the readers. □

10.6.3.2 Algorithm Implementation

The RNN-based filter (10.82), which is also denoted as RNNF$(y; \theta)$, consists of two parts

$$\begin{aligned} \tilde{S}_{k|k} &= \tilde{\Phi}\left(\tilde{S}_{k-1|k-1}, y_k; \theta_1\right), \\ \hat{x}_{k|k} &= \tilde{\Gamma}\left(\tilde{S}_{k|k}; \theta_2\right), \end{aligned} \tag{10.87}$$

10.6 The Application of Deep Learning in Nonlinear Filtering

where $\theta^T = [\theta_1^T, \theta_2^T]$ is all the trainable parameters in RNNF, $\tilde{\Phi}$ is represented by a single-layer feedforward network with l neurons, l is a hyperparameter to be determined, and $\tilde{\Gamma}$ is a linear function with input dimension l and output dimension n equal to the dimension of state x_k. Naturally, we aim to minimize

$$L_0(\theta) := \frac{1}{K_1+1} \mathbb{E}\left[\sum_{k=0}^{K_1} |\hat{x}_{k|k} - \mathbb{E}[x_k \mid Y_k]|^2\right], \tag{10.88}$$

where $K_1 \in \mathbb{N}$ is the total time step in training. Observing that

$$\mathbb{E}\left[|x_k - \hat{x}_{k|k}|^2\right] = \mathbb{E}\left[|x_k - \mathbb{E}[x_k \mid Y_k] + \mathbb{E}[x_k \mid Y_k] - \hat{x}_{k|k}|^2\right]$$

$$= \mathbb{E}\left[|x_k - \mathbb{E}[x_k \mid Y_k]|^2\right] + \mathbb{E}\left[|\mathbb{E}[x_k \mid Y_k] - \hat{x}_{k|k}|^2\right]$$

$$+ 2\mathbb{E}\left[(x_k - \mathbb{E}[x_k \mid Y_k])^T (\mathbb{E}[x_k \mid Y_k] - \hat{x}_{k|k})\right]$$

$$= \mathbb{E}\left[|x_k - \mathbb{E}[x_k \mid Y_k]|^2\right] + \mathbb{E}\left[|\mathbb{E}[x_k \mid Y_k] - \hat{x}_{k|k}|^2\right]$$

$$+ 2\mathbb{E}\left[\mathbb{E}\left((x_k - \mathbb{E}[x_k \mid Y_k])^T (\mathbb{E}[x_k \mid Y_k] - \hat{x}_{k|k}) \mid Y_k\right)\right]$$

$$= \mathbb{E}\left[|x_k - \mathbb{E}[x_k \mid Y_k]|^2\right] + \mathbb{E}\left[|\mathbb{E}[x_k \mid Y_k] - \hat{x}_{k|k}|^2\right]$$

$$+ 2\mathbb{E}\left[\mathbb{E}(x_k - \mathbb{E}[x_k \mid Y_k] \mid Y_k)^T (\mathbb{E}[x_k \mid Y_k] - \hat{x}_{k|k})\right]$$

$$= \mathbb{E}\left[|x_k - \mathbb{E}[x_k \mid Y_k]|^2\right] + \mathbb{E}\left[|\mathbb{E}[x_k \mid Y_k] - \hat{x}_{k|k}|^2\right], \tag{10.89}$$

where the third equality comes from the tower property of conditional expectation and the fourth equality is due to the fact that $\hat{x}_{k|k}$ is $\sigma(Y_k)$-measurable; it follows that

$$\arg\min_{\theta} L_0(\theta) = \arg\min_{\theta} L(\theta), \tag{10.90}$$

where

$$L(\theta) := \frac{1}{K_1+1} \mathbb{E}\left[\sum_{k=0}^{K_1} |\hat{x}_{k|k} - x_k|^2\right]. \tag{10.91}$$

Therefore, instead of data $\{y_k, \mathbb{E}[x_k \mid Y_k]\}_{k \geq 0}$ where $\mathbb{E}[x_k \mid Y_k]$ cannot be obtained in most cases, we only need data $\{y_k, x_k\}_{k \geq 0}$ which can be easily generated from the system (10.1). We need to remark that this step is crucial since it allows us to get accessible data.

In real computations, the expectation in $L(\theta)$ is approximated by the average of the results obtained from a large number of trials. Hence, we define the loss function as follows:

$$L^{(N)}(\theta) := \frac{1}{N} \frac{1}{K_1+1} \sum_{n=1}^{N} \left(\sum_{k=0}^{K_1} |x_k(\omega_n) - \hat{x}_{k|k}(\omega_n)|^2 \right), \quad (10.92)$$

where $\hat{x}_{k|k}(\omega_n) = \text{RNNF}(y_k(\omega_n); \theta)$ is the output of RNNF with input $y_k(\omega_n)$, and N and K_1 are the numbers of Monte Carlo paths and total time steps in training, respectively.

The detailed procedures of RNNF are listed as follows:

Algorithm 6 RNNF training algorithm

Require:
1: Train data: $\left\{ \{(y_k(\omega_n), x_k(\omega_n))\}_{k=0}^{K_1} \right\}_{n=1}^{N}$;
2: Batch size: M;
3: Total epochs: I;
4: Learning rate: λ;

Ensure:
5: RNNF output: $\left\{ \{\text{RNNF}(y_k(\omega_n)); \theta\}_{k=0}^{K_1} \right\}_{n=1}^{N}$;
6: **for** $i = 1, \ldots I$ **do**
7:     Sample batch $\left\{ \{(y_k(\omega_n), x_k(\omega_n))\}_{k=0}^{K_1} \right\}_{n=1}^{N}$ from Train data;
8:     Compute loss $L(\theta)$ via (10.92);
9:     Update θ via $\theta \leftarrow \theta - \lambda \nabla_\theta L(\theta)$.

10.6.4 RNN-Based Yau-Yau Algorithm for Continuous-Time Systems

Now, let us turn to the filtering problems in continuous-time setting. Just as in Chap. 7, a continuous-time filtering system can be described by the following couple of stochastic differential equations:

$$\begin{cases} dX_t = f(X_t)dt + GdV_t, & X_0 \sim \sigma_0, \\ dY_t = h(X_t)dt + dW_t, & Y_0 = 0, \end{cases} \quad (10.93)$$

where X, V, Y, W are R^n-, R^n-, R^m-, R^m-valued stochastic processes, respectively; V, W are independent standard Brownian motions and are independent with the initial value X_0. Hereafter, we further assume that the coefficients f and h are C^2 functions.

Utilizing an orthonormal basis in some Hilbert spaces of functions such as $L^2(R^n)$, we can convert the Yau-Yau algorithm discussed in Chap. 7 into the propagation and evolution of parameters, and we would like to call it **the uniform framework of Yau-Yau algorithm**. In this subsection, we will show that the

10.6 The Application of Deep Learning in Nonlinear Filtering

propagation of parameters can be approximated with arbitrary accuracy by a recurrent neural network (RNN) and provide a practical implementation of Yau-Yau algorithm for continuous-time systems.

Moreover, we can theoretically prove that the number of the parameters required in the uniform framework of Yau-Yau algorithm only grow polynomially (rather than exponentially) with respect to the dimension of the filtering system, which shows that this framework of Yau-Yau algorithm has the capability of overcoming the curse of dimensionality.

10.6.4.1 The Uniform Framework of Yau-Yau Algorithm

Just as in Chap. 7, in this uniform framework of Yau-Yau algorithm, we also need first consider the uniform partition $\mathcal{P}: 0 = \tau_0 < \tau_1 < \cdots < \tau_N = T$ of the time interval $[0, T]$, with time-discretization step $\delta = \frac{T}{N}$.

At each time interval $[\tau_{i-1}, \tau_i]$, $1 \leq i \leq N$ the Yau-Yau algorithm solves the observation independent parabolic partial differential equation (7.9) in Chap. 7, which is

$$\frac{\partial \widetilde{u}_i}{\partial t}(t, x) = \frac{1}{2} \Delta \widetilde{u}_i(t, x) - \sum_{i=1}^{n} f_i(x) \frac{\partial \widetilde{u}_i}{\partial x_i}(t, x)$$
$$- \left(\sum_{i=1}^{n} \frac{\partial f_i}{\partial x_i}(x) + \frac{1}{2} \sum_{i=1}^{m} h_i^2(x) \right) \widetilde{u}_i(t, x), \tag{10.94}$$

with initial value given by

$$\widetilde{u}_i(\tau_{i-1}, x) = \exp\left(\sum_{j=1}^{m} (y_{\tau_{i-1}}^j - y_{\tau_{i-2}}^j) h_j(x) \right) \widetilde{u}_{i-1}(\tau_{i-1}, x), \ i \geq 2, \tag{10.95}$$

and

$$\widetilde{u}_1(0, x) = \sigma_0(x) \exp\left(\sum_{j=1}^{m} y_j(\tau_0) h_j(x) \right). \tag{10.96}$$

Here we also use lower-case $\{y_t : 0 \leq t \leq T\}$ to represent a particular observation trajectory.

The main idea of this uniform framework is to convert the evolution of functions $\widetilde{u}_i(\tau_{i-1}, x)$ for $1 \leq i \leq N$ (which, according to the exponential transformation processes in Chap. 7, can be used to approximate the unnormalized conditional probability density function $\sigma(\tau_{i-1}, x)$), to the propagation of finitely many parameters which can be computed recursively. Because in general, $\{\widetilde{u}_i(\tau_{i-1}, x) : 1 \leq i \leq N\}$ evolves in an infinite dimensional functional space which cannot be characterized by finitely many parameters, such conversion is not equivalent and some approximation techniques are required to fulfill this idea.

The approximation techniques we choose here is based on a set of orthonormal basis functions $\{\phi_l\}_{l=1}^{\infty}$ of some Hilbert spaces of functions, such as $L^2(R^n)$, which contains every $\tilde{u}_{i,t}(x) \triangleq \tilde{u}_i(t,x)$, $1 \leq i \leq N$, $t \in [\tau_{i-1}, \tau_i]$. The functions $\tilde{u}_i(t,x)$ can thus be written in the form of variable separation given by

$$\tilde{u}_i(t,x) = \sum_{l=1}^{\infty} a_{i,l}(t)\phi_l(x), \ t \in [\tau_{i-1}, \tau_i] \tag{10.97}$$

If the orthonormal basis functions possess some kind of *good* properties, we believe (and rigorously prove later) that it is reasonable to find an element in S_M given by

$$\tilde{u}_{M,i}(t,x) = \sum_{l=1}^{M} \lambda_{i,l}(t)\phi_l(x), \tag{10.98}$$

which can approximate $\tilde{u}_i(t,x)$ well. Examples of this kind of orthonormal basis functions include classical ones such as Hermite functions and Legendre functions discussed in Chap. 7.

For the convenience of notations, let us denote $\lambda_{i,l} = \lambda_{i,l}(\tau_{i-1})$, which is the value of the parameter at the left endpoint of $[\tau_{i-1}, \tau_i]$. In this uniform framework, we also require that there exists a recursive formula for the M-dimensional parameter vector $\lambda_i = [\lambda_{i,1}, \cdots, \lambda_{i,M}]^\top$, with respect to i, i.e., there exists a continuous function $\eta : R^M \times R^m \to R^M$, such that

$$\lambda_{i+1} = \eta(\lambda_i, y_{\tau_i} - y_{\tau_{i-1}}), \ 0 \leq i \leq N-1. \tag{10.99}$$

Therefore, we cannot simply take the seemingly straightforward choice of

$$\tilde{u}_{M,i}(\tau_{i-1}, x),$$

with each

$$\lambda_{i,l} = a_{i,l}(\tau_{i-1}), \ 1 \leq l \leq M.$$

In fact, for general systems, the value of $a_{i,l}(t)$ ($1 \leq l \leq M$) depends on the tail terms $\{a_{i,l}(t) : l \geq M\}$ and there does not exist a recursive formula for the finite dimensional parameter $\{(a_{i,l}(\tau_{i-1})) : 1 \leq l \leq M\}$.

Nevertheless, with the linearity of (10.94) and the projection of functions on the finite dimensional space S_M, we can find proper parameters satisfying a recursive formula of the form (10.99), such that $\tilde{u}_{M,i}(\tau_{i-1}, x)$ approximates $\tilde{u}_i(\tau_{i-1}, x)$ well with sufficiently large M.

10.6 The Application of Deep Learning in Nonlinear Filtering

Let us denote the semigroup generated by the parabolic partial differential equation (10.94) by $\{U_t : t \geq 0\}$, and we can write $\tilde{u}_i(\tau_i, x) = U_\delta \tilde{u}_i(\tau_{i-1}, x)$. With the linearity of (10.94), at least formally, we have

$$\tilde{u}_i(\tau_i, x) = U_\delta \sum_{l=1}^{\infty} a_{i,l}(\tau_{i-1}) \phi_l(x) = \sum_{l=1}^{\infty} a_{i,l}(\tau_{i-1}) U_\delta \phi_l(x). \qquad (10.100)$$

If (10.100) converges in some sense, we can truncate the right-hand side at the M-th term and obtain

$$\tilde{u}_i(\tau_i, x) \approx \sum_{l=1}^{M} a_{i,l}(\tau_{i-1}) U_\delta \phi_l(x) \qquad (10.101)$$

Projecting each $U_\delta \phi_l$ onto the finite dimensional subspace S_M, we have

$$U_\delta \phi_l(x) \approx \sum_{j=1}^{M} d_{l,j} \phi_j(x) \qquad (10.102)$$

with

$$d_{l,j} = \langle U_\delta \phi_l, \phi_j \rangle, \ 1 \leq l, j \leq M. \qquad (10.103)$$

Therefore,

$$\tilde{u}_i(\tau_i, x) \approx \sum_{l=1}^{M} \sum_{j=1}^{M} a_{i,l}(\tau_{i-1}) d_{l,j} \phi_j(x). \qquad (10.104)$$

At time $t = \tau_i$, according to the exponential transform (10.95), after projection on S_M, we have

$$\tilde{u}_{i+1}(\tau_i, x) = \exp\left(\sum_{k=1}^{m}(y^k_{\tau_{i-1}} - y^k_{\tau_{i-2}}) h_k(x)\right) \tilde{u}_i(\tau_i, x)$$

$$\approx \sum_{l=1}^{M} \sum_{j=1}^{M} a_{i,l}(\tau_{i-1}) d_{l,j} \exp\left(\sum_{k=1}^{m}(y^k_{\tau_{i-1}} - y^k_{\tau_{i-2}}) h_k(x)\right) \phi_j(x)$$

$$\approx \sum_{l=1}^{M} \sum_{j=1}^{M} \sum_{j_1=1}^{M} a_{i,l}(\tau_{i-1}) d_{l,j} r_{j,j_1}(y_{\tau_i} - y_{\tau_{i-1}}) \phi_{j_1}(x),$$

$$\qquad (10.105)$$

with

$$r_{j,j_1}(y_{\tau_i} - y_{\tau_{i-1}}) = \left\langle \exp\left(\sum_{k=1}^{m}(y_{\tau_{i-1}}^k - y_{\tau_{i-2}}^k)h_k(x)\right)\phi_j, \phi_{j_1} \right\rangle, \ 1 \le j, j_1 \le M. \tag{10.106}$$

Combining (10.98) with (10.105), we obtain a candidate parameter λ_i, which evolves according to

$$\lambda_{i+1,j_1} = \sum_{l=1}^{M}\sum_{j=1}^{M}\lambda_{i,l}d_{l,j}r_{j,j_1}(y_{\tau_i} - y_{\tau_{i-1}}), \ 0 \le i \le N-1, \ 1 \le j_1 \le M. \tag{10.107}$$

Finally, since $\tilde{u}_{M,i+1}(\tau_i, x) = \sum_{l=1}^{M}\lambda_{i+1,l}\phi_l(x)$ is an approximator of $\tilde{u}_{i+1}(\tau_i, x)$, it is also an approximator of the unnormalized conditional probability density function $\sigma(\tau_i, x)$. In the filtering problem, if we are concerned with the conditional expectation $E[\varphi(X_t)|\mathcal{Y}_t]$, we can compute the normalized integral

$$E[\varphi(X_t)|\mathcal{Y}_t] = \frac{\int \varphi(x)\sigma(t,x)dx}{\int \sigma(t,x)dx}. \tag{10.108}$$

This normalized integral can also be parametrized in this uniform framework of Yau-Yau algorithm. In fact, the conditional expectation at each $t = \tau_i$ can be approximated as follows:

$$E[\varphi(X_{\tau_i})|\mathcal{Y}_{\tau_i}] \approx \frac{\int \varphi(x)\tilde{u}_{M,i+1}(\tau_i, x)dx}{\int \tilde{u}_{M,i+1}(\tau_i, x)dx} = \frac{\sum_{j=1}^{M}\beta_{\varphi,j}\lambda_{i+1,j}}{\sum_{j=1}^{M}\beta_{1,j}\lambda_{i+1,j}}, \tag{10.109}$$

with

$$\beta_{\varphi,j} = \int \varphi(x)\phi_j(x)dx, \ \beta_{1,j} = \int \varphi(x)\phi_j(x)dx, \ 1 \le j \le M. \tag{10.110}$$

We are ready to theoretically prove that $\tilde{u}_{M,i}(\tau_{i-1}, x)$ defined in (10.98) with candidate parameter λ_i satisfying (10.107) approximate $\tilde{u}_i(\tau_{i-1}, x)$ well. Before that, let us summarize this uniform framework of Yau-Yau algorithm.

10.6.4.2 Convergence Analysis and the Capability of Overcoming the Curse of Dimensionality

For the convergence analysis of this uniform framework of Yau-Yau algorithm, we can first state that if the orthonormal basis $\{\phi_l(x)\}_{l=1}^{\infty}$ is chosen to be classical ones such as Hermite functions or Legendre functions, the convergence results have already been obtained through the analysis in Chap. 7. However, from the proofs in Chap. 7, one may notice that in order to obtain a similar estimation error,

10.6 The Application of Deep Learning in Nonlinear Filtering

Algorithm 7 The uniform framework of Yau-Yau algorithm

1: Off-line computation
2: Compute the semigroup U_δ on the finite dimensional subspace S_M spanned by a given orthonormal function set $\{\phi_l(x)\}_{l=1}^M$, i.e., Compute $\{d_{l,j}\}_{1 \leq l,j \leq M}$.
3: Compute $\beta_{\varphi,j}$ and $\beta_{1,j}$ for $1 \leq j \leq M$.
4: Initialization
5: Compute $\lambda_0 = (\lambda_{0,1}, \cdots, \lambda_{0,M})^T$ by projecting $\widetilde{u}_1(x, \tau_0) = \sigma_0(x)$ on S_M.
6: On-line computation
7: **for** $i = 1$ to N **do**
8:     Updating λ_i according to

$$\lambda_{i+1,j_1} = \sum_{l=1}^{M}\sum_{j=1}^{M} \lambda_{i,l} d_{l,j} r_{j,j_1}(y_{\tau_i} - y_{\tau_{i-1}}), \; 1 \leq i \leq N, \; 1 \leq j_1 \leq M.$$

9:     Compute the approximated conditional expectation

$$\frac{\sum_{j=1}^{M} \beta_{\varphi,j} \lambda_{i+1,j}}{\sum_{j=1}^{M} \beta_{1,j} \lambda_{i+1,j}}.$$

---

theoretically, the number of the basis functions required will increase exponentially with respect to the dimension of the system. This phenomenon is always referred to as the *curse of dimensionality* and will cause inefficiency in high-dimensional problems.

The phenomenon, curse of dimensionality, occurs because the representation capability of classical orthonormal bases is not sufficient for efficiently approximating high-dimensional functions. Therefore, we would like to conduct convergence analysis here with orthonormal basis which is elaborately chosen and related to the specific models, so that the representation capability for the conditional probability density functions is strengthened and we can prove that the Yau-Yau algorithm under this uniform framework can overcome the curse of dimensionality in the sense that the number of basis functions we need only increase polynomially (rather than exponentially) with respect to the system dimension.

The main convergence result of this uniform framework of Yau-Yau algorithm is stated as follows. Remember that in Chap. 7, we have proved that most of the densities of the conditional distribution will lie inside a big ball $B_r = \{x \in R^n : |x| \leq r\}$ with radius $r \gg 1$. Therefore, this convergence result will focus on B_r, rather than the whole space R^n.

Theorem 10.8 *Consider the square-integrable, $L^2(B_r)$-valued random variables $\{\widetilde{u}_{i+1}(x, \tau_i)\}_{i=0}^N$ defined by (10.94) and (10.95) in the Yau-Yau algorithm. There exists a set of M normalized square-integrable functions, $\{\phi_j\}_{j=1}^M \subset L^2(B_r)$, which are orthogonal to each other, such that for each $i = 0, \cdots, N$, we can find a function $\widetilde{v}_i(x)$ in the M-dimensional vector space S_M spanned by $\{\phi_j\}_{j=1}^M$, which satisfies*

$$\tilde{E}\int_{B_r}|\tilde{v}_i(x)-\tilde{u}_{i+1}(\tau_i,x)|dx \leq C_T\sqrt{\delta}, \ \forall\, i=0,1,\cdots,N, \tag{10.111}$$

where C_T is a constant which depends on T, R, and the coefficients in the filtering system, but does not depend directly on the dimension of the filtering system, n, m, or the time discretization step δ. Here, the notation $\tilde{E}$ means that the expectation is taken with respect to the reference probability measure $\tilde{P}$.

Next, if we represent $\tilde{v}_i$ by

$$\tilde{v}_i(x) = \sum_{j=1}^{M} \lambda_{i+1,j}\phi_j(x), \tag{10.112}$$

then the evolution of $\lambda_{i+1} = (\lambda_{i+1,1},\cdots,\lambda_{i+1,M})^\top \in \mathbb{R}^M$ satisfies an open dynamical system

$$\lambda_{i+1} = \eta(\lambda_i, y_{\tau_i} - y_{\tau_{i-1}}), \quad i=1,\cdots,N, \tag{10.113}$$

with a given initial value λ_1, where $\eta : \mathbb{R}^M \times \mathbb{R}^m \to \mathbb{R}^M$ is a continuous function with respect to $\lambda_i \in \mathbb{R}^M$ and $y_{\tau_i} - y_{\tau_{i-1}} \in \mathbb{R}^m$ and is time-invariant.

Moreover, the number M of functions in the set $\{\phi_j\}_{j=1}^M$ can be chosen to grow at most polynomially with respect to the dimension m, which shows the capability of this framework of the Yau-Yau algorithm to overcome the curse of dimensionality.

Here we would like to sketch the main idea in the proof of Theorem 10.8, and readers can refer to [5] for a detailed proof.

Proof (A Sketch of the Proof of Theorem 10.8) The main idea of the proof consists of the following two parts.

Firstly, we would like to introduce an auxiliary partial differential equation with initial conditions slightly different from (10.94) and prove that the difference between the solution of the original PDE and the auxiliary one is small. The auxiliary equation is defined as follows:

$$\begin{cases} \dfrac{\partial v_i}{\partial t}(t,x) = \left(\mathcal{L} - \dfrac{1}{2}h^\top h\right)v_i(t,x), \ (t,x)\in[\tau_{i-1},\tau_i]\times B_r, \\ v_i(t,x) = 0, \quad (t,x)\in[\tau_{i-1},\tau_i]\times\partial B_r, \\ v_i(\tau_{i-1},x) = b_{i-1}(x)v_{i-1}(\tau_{i-1},x), \ x\in B_r. \end{cases} \tag{10.114}$$

with $b_i(x)$ the truncated Taylor series of $\exp\left(\sum_{k=1}^m (y_{\tau_i}^k - y_{\tau_{i-1}}^k)h_k(x)\right)$ at order two given by

10.6 The Application of Deep Learning in Nonlinear Filtering

$$b_i(x) = 1 + \sum_{k=1}^{m} h_k(x)(y_{\tau_i}^k - y_{\tau_{i-1}}^k)$$
$$+ \frac{1}{2} \sum_{j=1}^{m} \sum_{k=1}^{m} h_j(x) h_k(x)(y_{\tau_i}^k - y_{\tau_{i-1}}^k)(y_{\tau_i}^j - y_{\tau_{i-1}}^j). \quad (10.115)$$

Use the fact that $\{y_t : 0 \leq t \leq T\}$ is a standard Brownian motion under the reference probability $\tilde{P}$, as well as the property of parabolic partial differential equation; we can obtain the following recursive estimation of the difference between $\tilde{u}_{i+1}(\tau_i, x)$ and $v_{i+1}(\tau_i, x)$:

$$\tilde{E} \int_{B_r} |\tilde{u}_{i+1}(\tau_i, x) - v_{i+1}(\tau_i, x)| dx$$
$$\leq M_1 \delta^{\frac{3}{2}} + (1 + M_2 \delta) \tilde{E} \int_{B_r} |\tilde{u}_i(\tau_{i-1}, x) - v_i(\tau_{i-1}, x)| dx, \quad (10.116)$$

for all $1 \leq i \leq N$, where M_1, M_2 are constants independent of δ. Hence,

$$\tilde{E} \int_{B_r} |\tilde{u}_{i+1}(\tau_i, x) - v_{i+1}(\tau_i, x)| dx$$
$$\leq (1 + M_2 \delta)^i \int_{B_r} |\tilde{u}_1(\tau_0, x) - v_1(\tau_0, x)| dx + M_1 \delta^{\frac{3}{2}} \sum_{j=0}^{i-1} (1 + M_2 \delta)^j$$
$$= \frac{M_1}{M_2} \sqrt{\delta} \left((1 + M_2 \delta)^i - 1\right) \leq \frac{M_1}{M_2} \sqrt{\delta} \left(e^{M_2 T} - 1\right). \quad (10.117)$$

Thus, the solution of the auxiliary equation $v_{i+1}(x, \tau_i)$ provides a good estimation to $\tilde{u}_{i+1}(\tau_i, x)$, as long as the time discretization step δ is small enough.

Secondly, we would like to show that we can choose a set of orthonormal functions which represents each $v_{i+1}(\tau_i, x)$, $(1 \leq i \leq N)$ well, and the number of the orthonormal functions can be bounded by a polynomial of the observation dimension m.

In fact, the solution to the auxiliary equation (10.114) can be represented by

$$v_{i+1}(x, \tau_i) = \left(\prod_{j=1}^{i} (b_j U_\delta)\right) \sigma_0, \ i = 1, \cdots, N. \quad (10.118)$$

with U_δ the operator in the semigroup generated by (10.94). Therefore, according to the definition of $b_i(x)$, for each $i = 1, \cdots, N$, the function $v_{i+1}(x, \tau_i)$ is a linear combination of the functions in the set S_T given by

$$\mathcal{S}_T = \left\{ \left(\prod_{j=1}^{i}(H_j U_\delta) \right) \sigma_0 : H_j = 1, \, h_{j_1}, \, h_{j_1} h_{j_2}, \, 1 \leq j_1, j_2 \leq m, \, i = 1, \cdots, N \right\}, \tag{10.119}$$

and the coefficients of the linear combinations will be the products of those numbers $y_{\tau_i}^j - y_{\tau_{i-1}}^j$, $1 \leq i \leq N$, $1 \leq j \leq m$, given by the observations.

In the meanwhile, because y_t is a standard Brownian motion under the reference probability $\tilde{P}$, "high-order" functions in $\mathcal{S}_T$, where h_j's exist at least three times, contribute little to the linear combinations after taking expectations. In fact, the coefficients of these terms are of order at least $\delta^{\frac{3}{2}}$.

Therefore, the solution of the auxiliary equations can be well represented by linear combinations of those "low-order" functions in $\mathcal{S}_T$, where h_j's exist at most two times. The "low-order" functions in $\mathcal{S}_T$ are summarized in Table 10.1 and the number of "low-order" functions can be bounded by a quadratic polynomial of the observation dimension m.

Let us define $\mathcal{S}_M$ to be the linear space spanned by those "low-order" functions in $\mathcal{S}_T$, and the dimension M of $\mathcal{S}_M$ can be bounded by a quadratic polynomial of m. Let us find an orthonormal basis of $\mathcal{S}_M$, $\{\phi_i\}_{i=1}^{M}$, which can be done by standard Gram-Schmidt orthogonalization procedure. Then, all the functions $v_i(\tau_{i-1}, x)$ can be represented by the form

$$v_i(\tau_{i-1}, x) = \sum_{j=1}^{M} \lambda_{i,j} \phi_j(x), \tag{10.120}$$

for all $i = 1, \cdots, N$, where we ignore higher order terms with respect to δ.

Furthermore, the evolution of the coefficients λ_i can be described by the recursive formula as (10.113). In fact, since

$$v_{i+1}(\tau_i, x) = b_i(x) v_i(\tau_i, x) = b_i(x) U_\delta v_i(\tau_{i-1}, x). \tag{10.121}$$

With the expression (10.120),

Table 10.1 A summary of "low-order" functions in the set $|\mathcal{S}_T|$

Time	Basis functions
$i = 1$	σ_0
$i = 2$	$\mathcal{U}\sigma_0, h_j \mathcal{U}\sigma_0, h_j h_k \mathcal{U}\sigma_0$
$i = 3$	$\mathcal{U}^2 \sigma_0, \mathcal{U} h_j \mathcal{U}\sigma_0, h_j \mathcal{U}^2 \sigma_0, h_j h_k \mathcal{U}^2 \sigma_0, h_j \mathcal{U} h_k \mathcal{U}\sigma_0, \mathcal{U} h_j h_k \mathcal{U}\sigma_0$
$i = 4$	$\mathcal{U}^3 \sigma_0, \mathcal{U}^2 h_j \mathcal{U}\sigma_0, \mathcal{U} h_j \mathcal{U}^2 \sigma_0, h_j \mathcal{U}^3 \sigma_0, h_j h_k \mathcal{U}^3 \sigma_0, \cdots$
$\cdots$	$\cdots$
$i = N_T$	$\mathcal{U}^{N_T-1} \sigma_0, \mathcal{U}^{N_T-2} h_j \mathcal{U}\sigma_0, \cdots, h_j \mathcal{U}^{N_T-1} \sigma_0, h_j h_k \mathcal{U}^{N_T-1} \sigma_0, \cdots$

10.6 The Application of Deep Learning in Nonlinear Filtering

$$\sum_{j=1}^{M} \lambda_{i+1,j}\phi_j(x) = \sum_{j=1}^{M} \lambda_{i,j} b_i(x) U_\delta \phi_j(x)$$

$$= \sum_{j=1}^{M} \lambda_{i,j} U_\delta \phi_j + \sum_{j=1}^{M} \sum_{l=1}^{m} \lambda_{i,j}(y^l_{\tau_i} - y^l_{\tau_{i-1}}) h_l U_\delta \phi_j \qquad (10.122)$$

$$+ \frac{1}{2} \sum_{j=1}^{M} \sum_{l=1}^{m} \sum_{k=1}^{m} \lambda_{i,j}(y^l_{\tau_i} - y^l_{\tau_{i-1}})(y^k_{\tau_i} - y^k_{\tau_{i-1}}) h_l h_k U_\delta \phi_j.$$

Since $\{\phi_j\}_{j=1}^{M}$ is an orthonormal basis of S_M, by taking inner products, we have

$$\lambda_{i+1,s} = \sum_{j=1}^{M} \lambda_{i,j} \langle \phi_j, \phi_s \rangle$$

$$= \sum_{j=1}^{M} \lambda_{i,j} \langle U_\delta \phi_j, \phi_s \rangle + \sum_{j=1}^{M} \sum_{l=1}^{m} \lambda_{i,j}(y^l_{\tau_i} - y^l_{\tau_{i-1}}) \langle h_l U_\delta \phi_j, \phi_s \rangle$$

$$+ \frac{1}{2} \sum_{j=1}^{M} \sum_{l=1}^{m} \sum_{k=1}^{m} \lambda_{i,j}(y^l_{\tau_i} - y^l_{\tau_{i-1}})(y^k_{\tau_i} - y^k_{\tau_{i-1}}) \langle h_l h_k U_\delta \phi_j, \phi_s \rangle,$$

$$(10.123)$$

for each $1 \leq s \leq M$, where $\langle \cdot, \cdot \rangle$ denotes the inner product in the Hilbert space $L^2(B_r)$. The right-hand side (10.123) is indeed a continuous function with respect to $\lambda_i = (\lambda_{i,1}, \cdots, \lambda_{i,M})^\top$ and $y_{\tau_i} - y_{\tau_{i-1}}$.

In this way, we have found a set of M orthonormal functions, $\{\phi_j\}_{j=1}^{M}$ with M bounded by a quadratic polynomial of m, and functions

$$\widetilde{v}_i(x) = \sum_{j=1}^{M} \lambda_{i+1,j} \phi_j(x), \qquad (10.124)$$

which provide good estimations to $\widetilde{u}_{i+1}(\tau_i, x)$. Also, the evolution of λ_i satisfies the open dynamical system

$$\lambda_{i+1} = \eta(\lambda_i, y_{\tau_i} - y_{\tau_{i-1}}). \qquad (10.125)$$

In this way, we have obtained the desired result of Theorem 10.8. □

10.6.4.3 Implementation of the Uniform Framework of Yau-Yau Algorithm by RNN

Despite the elegant theoretical results, in general, the functions in $\mathcal{S}_T$ in the proof of Theorem 10.8 cannot be written down in explicit form, and neither can those $\{\phi_j\}_{j=1}^M$. Fortunately, recurrent neural network has the capability to provide a good approximation to the open dynamical system (10.113), and according to the uniform framework of Yau-Yau algorithm summarized in Algorithm 7, it is the coefficients λ_i rather than those basis functions that matter in the computation of conditional expectations of a given test function.

Therefore, we can train and apply an RNN to track the propagation of the coefficients and fulfill the goal of filtering, which is sequentially computing the conditional expectations.

The implementation of the uniform framework of Yau-Yau algorithm based on RNN is similar to the discrete-time case. We can first generate samples by simulating the filtering system (10.93). For a test function φ, we then obtain the training data $\{(y_{\tau_i}(\omega_s), \varphi(x_{\tau_i}(\omega_s))) : 1 \leq s \leq S, 1 \leq i \leq N\}$. The RNN is trained with these data and we obtain

$$\hat{\varphi}_i = RNN(y_{\tau_i} - y_{\tau_{i-1}}, \theta), \tag{10.126}$$

which can be used as an estimation to the conditional expectation $E[\varphi(X_{\tau_i})|\mathcal{Y}_{\tau_i}]$ at time τ_i, for each $1 \leq i \leq N$.

10.6.5 Numerical Results

In the numerical experiment, we consider the following discrete-time nonlinear filtering system:

$$\begin{cases} x_{k+1} = (I_d + \alpha A_d)x_k + \alpha \cos(x_k) + v_k, \\ y_{k,i} = \alpha x_{k,i}^3 + w_{k,i}, \ i = 1, \ldots, d, \end{cases} \tag{10.127}$$

where $x_{k,i}$ denotes the i-th entry of the vector x_k, with $k = 1, \ldots, K$ where $K = 200$, $\alpha = 0.01$, and $d = 6$. The sequences $\{v_k\}_{k=1}^K$ and $\{w_k\}_{k=1}^K$ are mutually independent Gaussian random vectors with zero means and covariance matrices $\mathbb{E}[v_k v_k^\top] = 0.01 I_d$ and $\mathbb{E}[w_k w_k^\top] = 0.01\sigma^2 I_d$ where $\sigma^2 \in \{0.01, 0.1, 1\}$. The matrix $A_d = [a_{ij}]$ has elements defined as follows:

$$a_{ij} = \begin{cases} 0.5, & \text{if } i+1 = j, \\ -1, & \text{if } i = j, \\ 0, & \text{otherwise.} \end{cases}$$

Here, we would like to use the RNN-based filtering algorithm (RNNF) to solve this six-dimensional filtering problem, with the performance of extended Kalman filter (EKF) and particle filter (PF) introduced in Chap. 9 as a comparison.

The initial value of the true state is set to $x_0 = [1, 1, \ldots, 1]^\top$. To investigate the influence of the initial values of the filtering algorithms on the estimation results, the initial mean and covariance of the EKF and PF are set to the zero vector and the identity matrix, respectively. The initial hidden state of the RNNF is also initialized as the zero vector. Furthermore, to evaluate the performance of the algorithms under different levels of observation noise, we choose $\sigma^2 \in \{0.01, 0.1, 1\}$. As expected, the accuracy of the state estimation decreases with increasing observation noise, i.e., with larger σ^2. In the PF, we use 1000 particles, and all parameters used in the RNNF are listed in Table 10.2.

To evaluate the performance of these methods, we introduce two metrics: the mean squared error (MSE) and the mean absolute error (MAE), based on 20 realizations. These metrics are defined as follows:

$$\text{MSE} := \frac{1}{20} \sum_{l=1}^{20} \frac{1}{K+1} \sum_{k=0}^{K} \left| x_k^{(l)} - \hat{x}_k^{(l)} \right|^2,$$

$$\text{MAE}(k) := \frac{1}{20} \sum_{l=1}^{20} \left| x_k^{(l)} - \hat{x}_k^{(l)} \right|,$$

(10.128)

where $x_k^{(l)}$ represents the true state at time instant k in the l-th experiment and $\hat{x}_k^{(l)}$ is the corresponding estimate, with $0 \leq k \leq K$, and $K \in \mathbb{N}$ is the total number of time steps.

In the subsequent numerical evaluations, all experiments were performed using NVIDIA RTX2060 GPUs on a computational platform equipped with 16 Intel Core i7-10700 CPUs running at 2.90 GHz. The RNNF implementation was carried out using PyTorch, while the EKF and PF implementations utilized NumPy, a Python library for scientific computing.

The MAE of the three algorithms, based on 20 realizations under different levels of observation noise, is shown in Fig. 10.4. It can be observed that, in all cases, the RNNF performs best in terms of MAE and is less sensitive to the initial value. The MSE and running time of the three methods are summarized in Table 10.3. The results show that the RNNF achieves the lowest MSE and the shortest running time, even outperforming the EKF in speed.

10.7 Exercises

1. For a function J that maps a column vector $\mathbf{w} \in \mathbb{R}^n$ to $\mathbb{R}$, the gradient is defined as

Table 10.2 Parameters used in RNNF

Parameter	Value
Paths in training set	1500
Paths in test set	20
Activation function	ReLU
Optimizer	Adam
Total epochs	2000
Batch size	64
Hidden layer neurons	40
Learning rate	0.0004

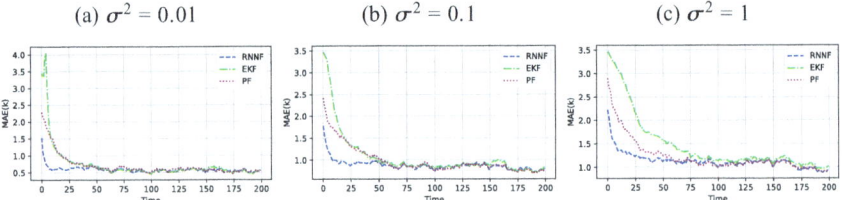

Fig. 10.4 Mean absolute error (MAE) of three filtering algorithms based on 20 experiments with different levels of observation noise. (**a**) $\sigma^2 = 0.01$. (**b**) $\sigma^2 = 0.1$. (**c**) $\sigma^2 = 1$

Table 10.3 Average performance of different methods based on 20 simulations for system (10.127)

Observation noise σ^2	Algorithm	MSE	Running time (s)
0.01	RNNF	**0.5922**	**0.0034**
	EKF	0.7495	0.0260
	PF	0.7124	0.5570
0.1	RNNF	**0.8810**	**0.0036**
	EKF	1.0752	0.0070
	PF	0.9924	0.5625
1	RNNF	**1.1416**	**0.0036**
	EKF	1.4829	0.0047
	PF	1.2363	0.5732

$$\nabla J(\mathbf{w}) = \begin{pmatrix} \frac{\partial J(\mathbf{w})}{\partial w_1} \\ \vdots \\ \frac{\partial J(\mathbf{w})}{\partial w_n} \end{pmatrix}, \quad (10.129)$$

where $\partial J(\mathbf{w})/\partial w_i$ are the partial derivatives of $J(\mathbf{w})$ with respect to the i-th element of the vector $\mathbf{w} = (w_1, \ldots, w_n)^\top$ (in the standard basis). Alternatively, it is defined to be the column vector $\nabla J(\mathbf{w})$ such that

$$J(\mathbf{w} + \epsilon \mathbf{h}) = J(\mathbf{w}) + \epsilon (\nabla J(\mathbf{w}))^\top \mathbf{h} + O\left(\epsilon^2\right) \quad (10.130)$$

for an arbitrary perturbation $\epsilon \mathbf{h}$. This phrases the derivative in terms of a first-order, or affine, approximation to the perturbed function $J(\mathbf{w} + \epsilon \mathbf{h})$. The derivative ∇J is a linear transformation that maps $\mathbf{h} \in \mathbb{R}^n$ to $\mathbb{R}$.

Use either definition to determine $\nabla J(\mathbf{w})$ for the following functions where $\mathbf{a} \in \mathbb{R}^n$, $\mathbf{A} \in \mathbb{R}^{n \times n}$ and $f : \mathbb{R} \to \mathbb{R}$ is a differentiable function.

1. $J(\mathbf{w}) = \mathbf{a}^\top \mathbf{w}$.
2. $J(\mathbf{w}) = \mathbf{w}^\top \mathbf{A} \mathbf{w}$.
3. $J(\mathbf{w}) = \mathbf{w}^\top \mathbf{w}$.
4. $J(\mathbf{w}) = \|\mathbf{w}\|_2$.
5. $J(\mathbf{w}) = f(\|\mathbf{w}\|_2)$.
6. $J(\mathbf{w}) = f(\mathbf{w}^\top \mathbf{a})$.

2. In this exercise, you are required to implement an RNNF using PyTorch and then compare its performance with that of the Kalman filter (KF). The system under consideration is a linear Gaussian system with independent noises, described by the following equations:

$$\begin{cases} x_k = (\alpha A_n + I_n) x_{k-1} + \sqrt{\alpha} w_{k-1} \\ y_k = \alpha x_k + \sqrt{\alpha} v_k \end{cases} \quad (10.131)$$

Here, $x_0 \sim \mathcal{N}(0, I_n)$, where I_n is the identity matrix in $\mathbb{R}^n$, with $n = 10$. The parameter $\alpha = 0.01$, and w_k and v_k are standard white noises. $A_n = [a_{ij}]$ is a matrix with elements as follows:

$$a_{ij} = \begin{cases} 0.1, & \text{if } i + 1 = j \\ -0.4, & \text{if } i = j \\ 0, & \text{otherwise} \end{cases} \quad (10.132)$$

You are given the freedom to choose $K_1 = 1000$ and $N = 2000$ for this exercise. Your task is to implement the RNNF using PyTorch and then compare its performance with that of the Kalman filter.

References

1. D. Bahdanau, K. Cho, and Y. Bengio. Neural machine translation by jointly learning to align and translate. *International Conference on Learning Representations*, 2015.
2. L. Bottou, F. E. Curtis, and J. Nocedal. Optimization methods for large-scale machine learning. *Siam Review*, 60(2):223–311, 2018.
3. W. Byeon, T. M. Breuel, F. Raue, and M. Liwicki. Scene labeling with lstm recurrent neural networks. *IEEE Conference on Computer Vision and Pattern Recognition (CVPR)*, 3547–3555, 2015.
4. X. Chen, Y. Tao, W. Xu, and S. S.-T. Yau. Recurrent neural networks are universal approximators with stochastic inputs. *IEEE Transactions on Neural Networks and Learning Systems*, 34(10):7992–8006, 2023.

5. X. Chen, Z. Sun, Y. Tao, and S. S.-T. Yau. A uniform framework of Yau-Yau algorithm based on deep learning with the capability of overcoming the curse of dimensionality. *IEEE Transactions on Automatic Control*, 2024, doi.org/10.1109/TAC.2024.3424628
6. J. Chung, C. Gulcehre, K. Cho, and Y. Bengio. Empirical evaluation of gated recurrent neural networks on sequence modeling. *arxiv:1412.3555*, 2014.
7. J. B. Conway. *A course in functional analysis*. Springer, 2019.
8. G. V. Cybenko. Approximation by superpositions of a sigmoidal function. *Mathematics of Control, Signals and Systems*, 2:303–314, 1989.
9. J. L. Elman. Finding structure in time. *Cognitive science*, 14:179–211, 1990.
10. A. Gelb. *Applied optimal estimation*. MIT Press, 1974.
11. I. Goodfellow, J. Pouget-Abadie, M. Mirza, B. Xu, D. Warde-Farley, S. Ozair, A. Courville, and Y. Bengio. Generative adversarial nets. *Advances in Neural Information Processing Systems*, 27, 2014.
12. A. Graves and N. Jaitly. Towards end-to-end speech recognition with recurrent neural networks. *International Conference on Machine Learning*, 1764–1772, 2014.
13. S. S. Haykin. *Kalman filtering and neural networks*. Wiley Online Library, 2001.
14. J. P. Hespanha. *Linear systems theory*. Princeton University Press, 2018.
15. S. Hochreiter and J. Schmidhuber. Long short-term memory. *Neural Computation*, 9(8):1735–1780, 1997.
16. K. Hornik, M. B. Stinchcombe, and H. L. White. Multilayer feedforward networks are universal approximators. *Neural Networks*, 2:359–366, 1989.
17. D. P. Kingma and J. Ba. Adam: A method for stochastic optimization. *arXiv:1412.6980*, 2014.
18. D. P. Kingma and M. Welling. Auto-encoding variational Bayes. *arXiv:1312.6114*, 2013.
19. A. N. Kolmogorov. On the representation of continuous functions of many variables by superposition of continuous functions of one variable and addition. *Doklady Akademii Nauk*, 114(5):953–956. Russian Academy of Sciences, 1957.
20. R. Krishnan, U. Shalit, and D. Sontag. Structured inference networks for nonlinear state space models. In *Proceedings of the AAAI Conference on Artificial Intelligence*, 31(1), 2017.
21. K. Law, A. Stuart, and K. Zygalakis. *Data assimilation*. Cham, Switzerland: Springer, 2015.
22. Y. LeCun, Y. Bengio, and G. Hinton. Deep learning. *Nature*, 521(7553):436–444, 2015.
23. Y. LeCun. A theoretical framework for back-propagation. *Proceedings of the 1988 Connectionist Models Summer School*, 1:21–28, 1988.
24. S. Li. Comparative analysis of backpropagation and extended Kalman filter in pattern and batch forms for training neural networks. *International Joint Conference on Neural Networks*, 1:144–149, 2001.
25. L. Ljung. System identification. *Signal Analysis and Prediction*, 163–173. Springer, 1998.
26. J. T.-H. Lo. Synthetic approach to optimal filtering. *IEEE Transactions on Neural Networks*, 5(5):803–811, 1994.
27. G. J. McLachlan and T. Krishnan. *The EM algorithm and extensions*. John Wiley & Sons, 2007.
28. T. Mikolov, M. Karafiát, L. Burget, J. Cernock, and S. Khudanpur. Recurrent neural network based language model. *Interspeech, Conference of the International Speech Communication Association*, 2015.
29. A. Paszke, S. Gross, F. Massa, et al. Pytorch: An imperative style, high-performance deep learning library. *Conference and Workshop on Neural Information Processing Systems*, 2019.
30. H. Qian, Q. Zhang, and G. Yin.
Filtering with degenerate observation noise: a stochastic approximation approach. *Automatica*, 142:110376, 2022.
31. H. Qian, G. Yin, and Q. Zhang. Deep filtering with adaptive learning rates. *IEEE Transactions on Automatic Control*, 68(6):3285–3299, 2023.
32. D. E. Rumelhart, G. E. Hinton, and R. J. Williams. Learning representations by back-propagating errors. *Nature*, 323(6088):533–536, 1986.
33. A. M. Schäfer and H. G. Zimmermann. Recurrent neural networks are universal approximators. *International Journal of Neural Systems*, 17(4):253–263, 2006.
34. B. Schottky and D. Saad. Statistical mechanics of EKF learning in neural networks. *Journal of Physics A: Mathematical and General*, 32(9):1605, 1999.

References

35. Y. Tao, J. Kang, and S. S.-T. Yau. Neural projection filter: learning unknown dynamics driven by noisy observations. *IEEE Transactions on Neural Networks and Learning Systems*, 35(7): 9508–9522, 2024.
36. E. A. Wan and R. Van Der Merwe. The unscented Kalman filter for nonlinear estimation. *Proceedings of the IEEE 2000 Adaptive Systems for Signal Processing, Communications, and Control Symposium*, 153–158, 2000.
37. L. Y. Wang, G. Yin, and Q. Zhang. Deep filtering. *Communications in Information and Systems*, 21(4):651–667, 2021.
38. S. H. Zak. *Systems and control*. Oxford University Press New York, 2003.
39. Z.-H. Zhou. *Machine learning*. Springer Nature, 2021.

Chapter 11
Solutions

Problems of Chap. 1

1. $\sigma(I) = \{\emptyset, \{1,2,3,4\}, \{1\}, \{3\}, \{1,3\}, \{2,4\}, \{1,2,4\}, \{2,3,4\}\}$
2. Let $\Omega_i = \{j\}_{j=1}^i$ and denote $\mathcal{I}_i$ be the σ algebra consisting of all subset in Ω_i. Assume that $\cup_{n=1}^{\infty} \mathcal{I}_n$ is also a σ algebra. Notice that $\{i\} \in \mathcal{I}_i$ which leads to result $\{i\} \in \cup_{n=1}^{\infty} \mathcal{I}_n$. By assumption, we get $\cup_{n=1}^{\infty}\{i\} \in \cup_{n=1}^{\infty} \mathcal{I}_n$. Hence there exists certain k such that $\cup_{n=1}^{\infty}\{i\} \in \mathcal{I}_k$. That will imply $\cup_{n=1}^{\infty}\{i\} \in \Omega_k$ which generates a contradiction.
3. Assume that the equation $E(f(\alpha X)) = \alpha^2 E(f(X))$ holds for arbitrary X and α. By applying condition of f, it is equivalent to the following:

$$(a_1 \alpha - \alpha^2 a_1) E(X) + (1 - \alpha^2) E(a_0) = 0$$

   Consider a random variable X with $E(X) \neq 0$. Let $\alpha = -1$ which will lead to $a_1 = 0$. Furthermore, put $\alpha = 0$ we shall obtain $a_0 = 0$. Therefore, we obtain the equation holds if $f(x) = a_2 x^2$. Conversely, it is easy to check if $f(x) = a_2 x^2$ is satisfied, the original equation holds.
4. By applying Cauchy-Schwarz inequality, we get

$$\begin{aligned}
|Corr(X,Y)| &= \left|\frac{Cov(X,Y)}{\sqrt{Var(X)Var(Y)}}\right| \\
&= \left|\frac{E((X-\mu_X)(Y-\mu_Y))}{\sqrt{E((X-\mu_X)^2)E((Y-\mu_Y)^2)}}\right| \\
&\leq \frac{E(|(X-\mu_X)(Y-\mu_Y)|)}{\sqrt{E((X-\mu_X)^2)E((Y-\mu_Y)^2)}} \\
&\leq 1
\end{aligned}$$

5. Consider following two discrete distribution with discrete probability density:

$$(p_1, p_2, p_3, p_4) = (\frac{8}{96}, \frac{54}{96}, \frac{12}{96}, \frac{22}{96})$$

and

$$(p_1, p_2, p_3, p_4) = (\frac{24}{96}, \frac{6}{96}, \frac{60}{96}, \frac{6}{96})$$

It can be directly examine that two distributions have the same expectation $E(X) = E(Y) = \frac{240}{96}$ and $E(X^2) = E(Y^2) = \frac{684}{96}$ but different third order moments $E(X^3) = \frac{2172}{96} \neq \frac{2076}{96} = E(Y^3)$.

6. Notice $P(X_1 = 0, X_2 = 0) = 0$ which will imply $P(X_1 = 0, X_2 = 1) = p(X_1 = 0) - P(X_1 = 0, X_2 = 0) = \frac{1}{2}$. Since $P(X_2 = 1) = \frac{1}{2}$, $P(X_1 = -1, X_2 = 1) = P(X_1 = 1, X_2 = 1) = 0$. Then $P(X_1 = -1, X_2 = 0) = P(X_1 = -1) - P(X_1 = -1, X_2 = 1) = \frac{1}{4}$ and similarly we get $P(X_1 = 1, X_2 = 0) = \frac{1}{4}$. Since there exists zeros in jointly distribution however marginal distributions are nonzero, two random variables are not independent.

7. Consider for arbitrary m-dimensional real vector t, characteristic function of η is defined as follows:

$$f_\eta(i) = Ee^{it^T\eta} = Ee^{it^T C\xi} = Ee^{i(C^Tt)^T\xi}$$

$$= \exp(i\mu^T(C^Tt) - \frac{1}{2}(C^Tt)^T\Sigma(C^Tt))$$

$$= \exp(i(C\mu)^Tt - \frac{1}{2}t^T(C\Sigma C^T)t)$$

By definition, η satisfies m-dimensional Gaussian distribution with parameters $N(C\mu, C\Sigma C^T)$.

8. Necessity property holds naturally and easy to present. In the following, we shall assume each two of random variables $\xi_1, \xi_2, \cdots, \xi_n$ are irrelevant, i.e.,

$$\rho_{jk} = \frac{E[(\xi_j - E\xi_j)(\xi_k - E\xi_k)]}{\sqrt{D\xi_j}\sqrt{D\xi_k}} = 0, \text{ for } j \neq k$$

Hence we deduce that $\sigma_{jk} = E[(\xi_j - E\xi_j)(\xi_k - E\xi_k)] = 0$ therefore

11 Solutions

$$f(t_1,\cdots,t_n) = \exp(i\sum_{k=1}^{n}\mu_k t_k - \frac{1}{2}\sum_{k=1}^{n}\sigma_{kk} t_k^2)$$

$$= \prod_{k=1}^{n}\exp(i\mu_k t_k - \frac{1}{2}\sigma_{kk} t_k^2)$$

$$= \prod_{k=1}^{n} f_{\xi_k}(t_k)$$

This implies the independence property of random variables.

9. By definition,

$$f(t) = \int e^{it^T x} p(x) dx$$

$$= \frac{1}{(2\pi)^{n/2}(det\,\Sigma)^{1/2}} \int e^{it^T x} \exp(-\frac{1}{2}(x-\mu)^T \Sigma (x-\mu)) dx$$

Taking the linear transformation $y = L^{-1}(x-\mu)$, we shall get

$$it^T x = it^T \mu + i(L^T t)^T y := it^T \mu + is^T y$$

Then

$$it^T x - \frac{1}{2}(x-\mu)^T \Sigma(x-\mu) = i\mu^T t - \frac{1}{2}t^T \Sigma t - \frac{1}{2}\sum_{k=1}^{n}(y_k - is_k)^2$$

and

$$f(t) = \frac{e^{i\mu^T t - \frac{1}{2}t^T \Sigma t}}{(2\pi)^{n/2}(det\,\Sigma)^{1/2}} \int \exp(-\frac{1}{2}\sum_{k=1}^{n}(y_k - is_k)^2)(det\,\Sigma)^{1/2} dy_1 \cdots dy_n$$

$$= e^{i\mu^T t - \frac{1}{2}t^T \Sigma t}$$

10.

$$P(\eta < a) = P(g(\xi) < a)$$

$$= \int_{-\infty}^{g^{-1}(a)} p(x) dx$$

$$= \int_{-\infty}^{a} p(g^{-1}(y))|g^{-1}(y)'| dy$$

which implies η admits density function $p(g^{-1}(y))|g^{-1}(y)'|$.

Problems of Chap. 2

1. Let $\{M_n\}_{n\geq 0}$ and $\{N_n\}_{n\geq 0}$ be two martingales with respect to the same filtration $\{\mathcal{F}_n\}_{n\geq 0}$. We need to show that $\{M_n + N_n\}_{n\geq 0}$ is also a martingale with respect to $\{\mathcal{F}_n\}_{n\geq 0}$.
   By definition, $\{M_n\}_{n\geq 0}$ and $\{N_n\}_{n\geq 0}$ satisfy
   $$E[M_{n+1} \mid \mathcal{F}_n] = M_n \quad \text{and} \quad E[N_{n+1} \mid \mathcal{F}_n] = N_n \quad \text{for all } n \geq 0.$$
   Consider $E[M_{n+1} + N_{n+1} \mid \mathcal{F}_n]$
   $$E[M_{n+1} + N_{n+1} \mid \mathcal{F}_n] = E[M_{n+1} \mid \mathcal{F}_n] + E[N_{n+1} \mid \mathcal{F}_n] = M_n + N_n.$$
   Therefore, $\{M_n + N_n\}_{n\geq 0}$ is a martingale.
   For the continuous case, let $\{M_t\}_{t\geq 0}$ and $\{N_t\}_{t\geq 0}$ be continuous-time martingales. By the same argument, we have
   $$E[M_{t+s} + N_{t+s} \mid \mathcal{F}_t] = E[M_{t+s} \mid \mathcal{F}_t] + E[N_{t+s} \mid \mathcal{F}_t] = M_t + N_t,$$
   which shows that $\{M_t + N_t\}_{t\geq 0}$ is also a martingale.

2. (1) Consider a simple symmetric random walk X_n defined as
   $$X_{n+1} = \begin{cases} X_n + 2 & \text{with probability } \frac{1}{2}, \\ X_n - 1 & \text{with probability } \frac{1}{2}. \end{cases}$$
   Clearly, this is a Markov process because the future state depends only on the current state and not on the past states. However, this process is not a martingale because the expected value of the next state is not equal to the current state:
   $$E[X_{n+1} | X_n] = X_n + 0.5$$

   (2) Not every martingale is Markovian. Here is a counterexample: Let $\{Z_n\}$ be a sequence of independent identically distributed random variables with mean 0 and variance 1. Define $S_0 = 0$, $S_n = \sum_{i=1}^n Z_i$. Let
   $$X_n = S_n - \frac{S_{n-1}}{n}, n \geq 1$$
   It can be verified that $\{X_n, \mathcal{F}_n\}$ is a martingale, where $\mathcal{F}_n = \sigma(Z_1, \cdots, Z_n)$. But $\{X_n\}$ is not a Markov chain.

3. (1) We need to ensure that
   $$E[S_{n+1}^2 + b_{n+1} S_{n+1} + c_{n+1} | \mathcal{F}_n] = S_n^2 + b_n S_n + c_n.$$

11 Solutions

Expanding the left-hand side

$$E[S_{n+1}^2 + b_{n+1}S_{n+1} + c_{n+1}|\mathcal{F}_n]$$
$$= E[(S_n + X_{n+1})^2 + b_{n+1}(S_n + X_{n+1}) + c_{n+1}|\mathcal{F}_n]$$
$$= E[S_n^2 + 2S_n X_{n+1} + X_{n+1}^2 + b_{n+1}S_n + b_{n+1}X_{n+1} + c_{n+1}|\mathcal{F}_n]$$
$$= S_n^2 + 2S_n m_{n+1} + \sigma_{n+1}^2 + m_{n+1}^2 + b_{n+1}S_n + b_{n+1}m_{n+1} + c_{n+1}.$$

Equating coefficients with the right-hand side

$$b_n = b_{n+1} + 2m_{n+1},$$
$$c_n = c_{n+1} + \sigma_{n+1}^2 + m_{n+1}^2 + b_{n+1}m_{n+1}.$$

The solution is

$$b_n = -2\sum_{i=0}^{n} m_i,$$

$$c_n = -\sum_{i=0}^{n}(\sigma_i^2 + m_i^2) - 2\sum_{k=0}^{n}\left(\sum_{i=0}^{k} m_i\right)m_k.$$

(2) For the sequence to be an $\mathcal{F}_n$-martingale, it must satisfy

$$E[\exp(\lambda S_{n+1} - a_{\lambda_{n+1}})|\mathcal{F}_n] = \exp(\lambda S_n - a_{\lambda_n})$$

Using the independence of X_i's and the properties of conditional expectation, we have

$$E[\exp(\lambda S_{n+1} - a_{\lambda_{n+1}})|\mathcal{F}_n] = E[\exp(\lambda(S_n + X_{n+1}) - a_{\lambda_{n+1}})|\mathcal{F}_n]$$
$$= \exp(\lambda S_n - a_{\lambda_{n+1}}) \cdot E[\exp(\lambda X_{n+1})|\mathcal{F}_n]$$
$$= \exp(\lambda S_n - a_{\lambda_{n+1}}) \cdot G_{n+1}(\lambda)$$

For the martingale property to hold, we need

$$\exp(\lambda S_n - a_{\lambda_{n+1}}) \cdot G_{n+1}(\lambda) = \exp(\lambda S_n - a_{\lambda_n})$$

Taking the logarithm on both sides, we get

$$\lambda S_n - a_{\lambda_{n+1}} + \log G_{n+1}(\lambda) = \lambda S_n - a_{\lambda_n}$$

Simplifying, we have

$$a_{\lambda_{n+1}} - a_{\lambda_n} = \log G_{n+1}(\lambda)$$

Therefore, we can choose the sequence $(a_{\lambda_n})_{n \geq 0}$ as

$$a_{\lambda_n} = \sum_{i=1}^{n} \log G_i(\lambda)$$

With this choice of $(a_{\lambda_n})_{n \geq 0}$, the sequence $\{\exp(\lambda S_n - a_{\lambda_n})\}_{n \geq 0}$ is an $\mathcal{F}_n$-martingale.

4. To prove that the process $N = \{N_t : t \in \{0, 1, 2, \ldots\}\}$ is a martingale, we need to verify the following two conditions:

   (a) $E[|N_t|] < \infty$ for all $t \in \{0, 1, 2, \ldots\}$.
   (b) $E[N_t|\mathcal{F}_{t-1}] = N_{t-1}$ for all $t \in \{1, 2, \ldots\}$.

Proof of Condition 1:
Using the definition of N_t and the triangle inequality, we have

$$E[|N_t|] = E\left[\left|N_0 + \sum_{k=1}^{t} \Theta_k(M_k - M_{k-1})\right|\right]$$

$$\leq E[|N_0|] + \sum_{k=1}^{t} E[|\Theta_k(M_k - M_{k-1})|]$$

Since N_0 is $\mathcal{F}_0$-measurable and square-integrable, $\mathbb{E}[|N_0|] < \infty$.
For each term in the sum, we apply the Cauchy-Schwarz inequality:

$$E[|\Theta_k(M_k - M_{k-1})|] \leq \sqrt{E[|\Theta_k|^2]}\sqrt{E[|M_k - M_{k-1}|^2]}$$
$$< \infty$$

The last inequality holds because Θ_k is square-integrable by assumption and M is a square-integrable martingale. Therefore, $E[|N_t|] < \infty$ for all $t \in \{0, 1, 2, \ldots\}$.

Proof of Condition 2:
For $t \in \{1, 2, \ldots\}$, we have

$$E[N_t|\mathcal{F}_{t-1}] = E[N_0 + \sum_{k=1}^{t} \Theta_k(M_k - M_{k-1})|\mathcal{F}_{t-1}]$$

$$= N_0 + \sum_{k=1}^{t-1} \Theta_k(M_k - M_{k-1}) + E[\Theta_t(M_t - M_{t-1})|\mathcal{F}_{t-1}]$$

$$= N_{t-1} + \Theta_t E[M_t - M_{t-1}|\mathcal{F}_{t-1}]$$

$$= N_{t-1}$$

The last equality holds because M is a martingale, so $E[M_t - M_{t-1}|\mathcal{F}_{t-1}] = 0$. Therefore, both conditions are satisfied, and the process $N = \{N_t : t \in \{0, 1, 2, \ldots\}\}$ is a martingale.

5. To prove that the process $M = \{M_t : t \in \{0, 1, 2, \ldots\}\}$ is a martingale, we need to verify the following two conditions:

   (a) $E[|M_t|] < \infty$ for all $t \in \{0, 1, 2, \ldots\}$.
   (b) $E[M_t|\mathcal{F}_s] = M_s$ for all $s, t \in \{0, 1, 2, \ldots\}$ with $s \leq t$.

   Proof of Condition 1:
   By the definition of conditional expectation and the monotonicity of the expectation operator, we have

   $$E[|M_t|] = E[|\mathbb{E}_P[X \mid \mathcal{F}_t]|]$$
   $$\leq E[\mathbb{E}_P[|X| \mid \mathcal{F}_t]]$$
   $$= E[|X|]$$
   $$< \infty$$

   The last inequality holds because X is square-integrable, which implies $E[|X|] < \infty$. Therefore, $E[|M_t|] < \infty$ for all $t \in \{0, 1, 2, \ldots\}$.

   Proof of Condition 2:
   For $s, t \in \{0, 1, 2, \ldots\}$ with $s \leq t$, we have

   $$E[M_t|\mathcal{F}_s] = E[E_P[X \mid \mathcal{F}_t]|\mathcal{F}_s]$$
   $$= E_P[X \mid \mathcal{F}_s]$$
   $$= M_s$$

   The second equality holds due to the tower property of conditional expectation, which states that for any sub-sigma-algebra $\mathcal{G} \subseteq \mathcal{F}$, we have $E[E[Y \mid \mathcal{F}] \mid \mathcal{G}] = E[Y \mid \mathcal{G}]$ for any integrable random variable Y.
   Therefore, both conditions are satisfied, and the process $M = \{M_t : t \in \{0, 1, 2, \ldots\}\}$ is a martingale.

6. Let $M = \{M_t : t \geq 0\}$ be a $\{\mathcal{F}_t\}$-martingale and let $N = \{N_t : t \geq 0\}$ be a $\{\mathcal{G}_t\}$-martingale.
   Since $\mathcal{F}_t \subseteq \mathcal{G}_t$ for all t, we have that for any $s \leq t$,

   $$E[M_t|\mathcal{G}_s] = E[E[M_t|\mathcal{F}_t]|\mathcal{G}_s]$$
   $$= E[M_t|\mathcal{F}_s]$$
   $$= M_s$$

   where the second equality holds due to the tower property of conditional expectation. Therefore, M is also a $\{\mathcal{G}_t\}$-martingale.

However, N is not necessarily a $\{\mathcal{F}_t\}$-martingale. Since $\mathcal{F}_t \subseteq \mathcal{G}_t$, the equality $E[N_t|\mathcal{F}_s] = N_s$ may not hold for all $s \leq t$.

Let τ be a $\{\mathcal{F}_t\}$-stopping time and σ be a $\{\mathcal{G}_t\}$-stopping time. Since $\mathcal{F}_t \subseteq \mathcal{G}_t$ for all t, we have that for any $t \geq 0$,

$$\{\sigma \leq t\} = \{\omega \in \Omega : \sigma(\omega) \leq t\} \in \mathcal{G}_t \subseteq \mathcal{F}_t$$

Therefore, σ is also a $\{\mathcal{F}_t\}$-stopping time.

However, τ is not necessarily a $\{\mathcal{G}_t\}$-stopping time. Since $\mathcal{F}_t \subseteq \mathcal{G}_t$, the set $\{\tau \leq t\}$ may not be in $\mathcal{G}_t$ for all $t \geq 0$.

7. To show that $\{M_n, n \geq 0\}$ converges almost surely (a.s.) and in L^1 toward a limiting M_∞:

Since $\{M_n\}$ is a UI martingale, we have:

- $E[|M_n|] < \infty$ for all $n \geq 0$, and
- $\lim_{n \to \infty} E[|M_n|\mathbf{1}_{\{|M_n|>k\}}] = 0$ for all $k > 0$.

By the martingale convergence theorem, a UI martingale converges a.s. and in L^1 to a limiting random variable M_∞. Therefore, $\{M_n, n \geq 0\}$ converges a.s. and in L^1 to M_∞.

To show that for any $n \in \mathbb{N}$, $M_n = E[M_\infty|\mathcal{F}_n]$:

Since $\{M_n\}$ is a martingale with respect to $\{\mathcal{F}_n\}$, we have

$$E[M_{n+1}|\mathcal{F}_n] = M_n$$

Taking the limit as $n \to \infty$ on both sides, and using the dominated convergence theorem (since $\{M_n\}$ is UI), we get

$$\lim_{n \to \infty} E[M_{n+1}|\mathcal{F}_n] = \lim_{n \to \infty} M_n$$

$$E[\lim_{n \to \infty} M_{n+1}|\mathcal{F}_n] = M_\infty$$

$$E[M_\infty|\mathcal{F}_n] = M_n$$

Therefore, for any $n \in \mathbb{N}$, $M_n = E[M_\infty|\mathcal{F}_n]$.

8. To show that $\{X_n, n \geq 0\}$ is a martingale, we need to verify that:

- $E[|X_n|] < \infty$ for all $n \geq 0$, and
- $E[X_{n+1}|X_0, \ldots, X_n] = X_n$ for all $n \geq 0$.

For the first condition, note that $|X_n| \leq n$ for all $n \geq 0$, so $E[|X_n|] \leq n < \infty$.

For the second condition, we can show it by induction. For $n = 0$, $E[X_1|X_0] = \frac{1}{2}(1) + \frac{1}{2}(-1) = 0 = X_0$. Assume the condition holds for $n = k$. For $n = k + 1$:

$$E[X_{k+2}|X_0, \ldots, X_{k+1}] = E[X_{k+2}|X_{k+1}]$$

$$= \begin{cases} \frac{1}{2^{k+1}}(1) + \frac{1}{2^{k+1}}(-1) + (1-\frac{1}{2^{k+1}})(0) = 0 = X_{k+1}, & \text{if } X_{k+1} = 0 \\ \frac{1}{k+1}((k+1)X_{k+1}) + (1-\frac{1}{k+1})(0) = X_{k+1}, & \text{if } X_{k+1} \neq 0 \end{cases}$$

Therefore, $\{X_n, n \geq 0\}$ is a martingale. X_n converges almost surely. Consider the event $A = \{X_n = 0 \text{ infinitely often}\}$. By the Borel-Cantelli lemma, since $\sum_{n=1}^{\infty} P(X_n = 0) \geq \sum_{n=1}^{\infty} \min(P(X_n = 0 \mid X_{n-1} \neq 0), P(X_n = 0 \mid X_{n-1} = 0)) = \sum_{n=1}^{\infty} \left(1 - \frac{1}{2^{n-1}}\right) = \infty$ we have $P(A) = 1$. This implies that X_n converge almost surely and converge in probability. We can show that $E[|X_n|] \geq \frac{1}{2}$ for all $n \geq 1$, which implies that X_n does not converge in L^1.

9. To show that $\{X_n\}$ is a $\{F_n\}$-martingale, we need to verify that:

- X_n is F_n-measurable for all $n \geq 0$: This is clear since X_n is a function of $V_1, \ldots, V_n$, which are all F_n-measurable.
- $E[|X_n|] < \infty$ for all $n \geq 0$: Since V_i are nonnegative, $|X_n| = X_n = \prod_{i=1}^{n} V_i$. By the assumption that $E[V_i] = 1$, we have $E[|X_n|] = E[X_n] = E[\prod_{i=1}^{n} V_i] = \prod_{i=1}^{n} E[V_i] = 1 < \infty$.
- $E[X_{n+1}|F_n] = X_n$ for all $n \geq 0$:

$$E[X_{n+1}|F_n] = E[\prod_{i=1}^{n+1} V_i | F_n]$$

$$= E[(\prod_{i=1}^{n} V_i) \cdot V_{n+1} | F_n]$$

$$= (\prod_{i=1}^{n} V_i) \cdot E[V_{n+1}|F_n]$$

$$= (\prod_{i=1}^{n} V_i) \cdot E[V_{n+1}]$$

$$= X_n \cdot 1 = X_n.$$

The third equality follows from the fact that $\prod_{i=1}^{n} V_i$ is F_n-measurable and the fourth equality follows from the independence of V_{n+1} and F_n.

Therefore, $\{X_n\}$ is a $\{F_n\}$-martingale.

To determine the convergence of $\{X_n\}$, we can apply the martingale convergence theorem. However, we need to check the conditions first.

- $\sup_n E[|X_n|] < \infty$: We have shown that $E[|X_n|] = 1$ for all n, so $\sup_n E[|X_n|] = 1 < \infty$.
- $X_n \to X_\infty$ almost surely for some random variable X_∞: Let's consider the event $A = \{\omega : V_i(\omega) = 1 \text{ for infinitely many } i\}$. By the Borel-Cantelli Lemma, since $\sum_{i=1}^{\infty} P(V_i = 1) = \infty$ (as $P(V_i = 1) = p > 0$ for all i),

we have $P(A) = 1$. On the event A, $X_n(\omega) = \prod_{i=1}^{n} V_i(\omega) = 0$ for infinitely many n, which means X_n does not converge almost surely.

Therefore, $\{X_n\}$ does not converge almost surely. It also does not converge in L^1, as $E[|X_n - X_m|] \geq |E[X_n] - E[X_m]| = |1 - 1| = 0$ does not converge to 0 as $n, m \to \infty$.

Problems of Chap. 3

1. Let $f(x) = \frac{1}{2}x^2$. According to Itô's formula

$$f(W_t) = f(W_0) + \int_0^t f'(W_s)dW_s + \frac{1}{2}\int_0^t f''(W_s)ds$$

$$= \int_0^t W_s dW_s + \frac{1}{2}\int_0^t 1 ds$$

$$= \int_0^t W_s dW_s + \frac{1}{2}t.$$

Hence,

$$\int_0^t W_s dW_s = \frac{1}{2}\left(W_t^2 - t\right).$$

2. Let us denote

$$U_t = \sum_{i=1}^{d} \int_0^t X_s^i dW_s^i - \frac{1}{2}\int_0^t \|X_s\|^2 ds,$$

then according to Itô's formula,

$$Z_t = e^{U_t} = e^{U_0} + \int_0^t e^{U_s} dU_s + \frac{1}{2}\int_0^t e^{U_s} d\langle U \rangle_s.$$

Since,

$$dU_t = \sum_{i=1}^{d} X_t^i dW_t^i - \frac{1}{2}\|X_s\|^2 ds,$$

and

$$d\langle U \rangle_t = \|X_t\|^2 dt,$$

we have

$$Z_t = 1 + \int_0^t Z_s \sum_{i=1}^d X_s^i dW_s^i - \frac{1}{2}\int_0^t Z_s \|X_s\|^2 ds + \frac{1}{2}\int_0^t Z_s \|X_s\|^2 ds$$

$$= 1 + \sum_{i=1}^d \int_0^t Z_s X_s^i dW_s^i.$$

3. Let us define $f(x, y) = xy$, then, according to the multi-dimensional Itô's formula,

$$X_t Y_t = X_0 Y_0 + \int_0^t \left[\frac{\partial f}{\partial x} dX_s + \frac{\partial f}{\partial y} dY_s\right]$$
$$+ \frac{1}{2}\int_0^t \left[\frac{\partial^2 f}{\partial x^2} d\langle X\rangle_s + 2\frac{\partial^2 f}{\partial x \partial y} d\langle X,Y\rangle_s + \frac{\partial^2 f}{\partial y^2} d\langle Y\rangle_s\right]$$
$$= X_0 Y_0 + \int_0^t X_s dY_s + \int_0^t Y_s dX_s + \langle X, Y\rangle_t.$$

which is the integration-by-part formula in Itô's sense:

$$\int_0^t X_s dY_s = X_t Y_t - X_0 Y_0 - \int_0^t Y_s dX_s - \langle X, Y\rangle_t.$$

4. According to the relationship between Itô's integral and Stratonovich integral:

$$\int_0^t Y_s \circ dX_s = \int_0^t Y_s dX_s + \frac{1}{2}\langle X, Y\rangle_s,$$

we obtain the integration-by-part formula in Stratonovich sense based on that in Itô's sense obtained in the previous exercise:

$$\int_0^t X_s \circ dY_s = X_t Y_t - X_0 Y_0 - \int_0^t Y_s \circ dX_s,$$

which coincides the integration-by-part formula for regular integrations.

5. According to Itô's lemma,

$$X_t = e^{W_t} = e^{W_0} + \int_0^t e^{W_s} dW_s + \frac{1}{2}\int_0^t e^{W_s} ds.$$

Writing in differential form,

$$dX_t = \frac{1}{2} X_t dt + X_t dW_t \tag{11.1}$$

6. Formally, if we write $dW_t = W'(t)dt$, then the solution of the linear 'ordinary differential equation'

$$\frac{dX_t}{dt} = AX_t + BW'(t)$$

can be written as

$$X_t = e^{tA}\left(x_0 + \int_0^t e^{-sA}BW'(s)ds\right).$$

Let us verify that

$$X_t = e^{tA}\left(x_0 + \int_0^t e^{-sA}BdW_s\right)$$

is a solution to the linear stochastic differential equation. In fact, applying Itô's formula to $f(x, y) = xy$, we have

$$X_t = x_0 + \int_0^t e^{sA}e^{-sA}BdW_s + A\int_0^t e^{sA}\left(x_0 + \int_0^t e^{-sA}BdW_s\right)ds$$

$$= x_0 + \int_0^t BdW_s + \int_0^t AX_s ds,$$

which is the stochastic differential equation we would like to solve.

Problems of Chap. 4

1. If $z \in \mathcal{X}$, the conclusion holds naturally. Otherwise $dist(z, \mathcal{X}) \leq dist(z, x)$ holds for any point $x \in \mathcal{X}$. Therefore there exists certain point sequence $\{x^k\} \subset \mathcal{X}$ satisfying $x^k \to \bar{x}$ and $\|z - \bar{x}\| = \min\{\|z - x\| | x \in \mathcal{X}\}$.
For any $x \in \mathcal{X}$, let $\hat{x} = \alpha x + (a - \alpha)\bar{x}$ and $0 < \alpha \leq 1$. Considering following

$$\|z - \bar{x}\|^2 \leq \|z - \hat{x}\|^2$$

which is equivalent to $2(z - \bar{x})^T(\bar{x} - x) + \alpha\|\bar{x} - x\|^2 \geq 0$. This implies $(z - \bar{x})^T(x - \bar{x}) \leq 0$ since α is arbitrary. Finally we discuss the uniqueness and assume there exists another x^* such that $\|z - x^*\| = \min\{\|z - x\| | x \in \mathcal{X}\}$ and $x^* \neq \bar{x}$.

$$(\bar{x} - x^*)^T(\bar{x} - x^*) = (z - \bar{x})^T(x^* - \bar{x}) + (z - x^*)^T(\bar{x} - x^*) \leq 0$$

A contradiction!

11 Solutions

2. Due to the condition $z \notin cl(X)$, $dist(z, cl(X)) > 0$. Assume $\|z-\bar{x}\| = \min\{\|z-x\| \mid x \in cl(X)\}$ then $(z-\bar{x})^T(z-\bar{x}) > 0$. By applying result of problem 1, we obtain $(z-\bar{x})^T(x-\bar{x}) \leq 0$. Integrating two inequality implies that

$$(\bar{x}-z)^T x \geq (\bar{x}-z)^T \bar{x} > (\bar{x}-z)^T z$$

holds for arbitrary $x \in X$. Finally let $a = \bar{x} - z$ and $b = (\bar{x}-z)^T \bar{x}$, the desired result is obtained.

3. (i) For arbitrary $y \in (\mathbb{R}_+^n)^*$, we shall take vectors in $\mathbb{R}_+^n$ as $(1, 0, \cdots, 0)^T$, $(0, 1, \cdots, 0)^T$,..., $(0, 0, \cdots, 1)^T$. By definition of dual set, we obtain $y \geq 0$ which implies $(\mathbb{R}_+^n)^* \subset \mathbb{R}_+^n$ and vice versa by the same technique.

   (ii) Given any $A \in \mathcal{S}_+^n$, it can be decomposed as $A = C^T C$. Similarly, for given $B \mathcal{S}_+^n$, we get decomposition $B = D^T D$. Then

$$B \circ A = tr(B^T A) = tr(D^T D C^T C) = tr((DC^T)^T (DC^T)) \geq 0$$

   which implies $\mathcal{S}_+^n \subset (\mathcal{S}_+^n)^*$. On the other hand, given any $B \in (\mathcal{S}_+^n)^*$ and any x, we shall take $A = xx^T \in \mathcal{S}_+^n$. Condition $A \circ B = tr(Bxx^T) = x^T Bx \geq 0$ holds for any x which implies B is positive semi-definite matrix. From this inclusion fact, we conclude $(\mathcal{S}_+^n)^* \subset \mathcal{S}_+^n$.

4. Lagrange function of linear programming is given by

$$L(x, \lambda) = \begin{cases} (c - A^T \lambda)^T x + \lambda^T b, & x \in \mathbb{R}_+^n \\ +\infty, & x \notin \mathbb{R}_+^n \end{cases}$$

Then we derive

$$\max_{\lambda \in \mathbb{R}_+^n} v(\lambda) = \max_{\lambda \in \mathbb{R}_+^n} \min_{x \in \mathbb{R}_+^n} ((c - A^T \lambda)^T x + \lambda^T b) = \max_{\{\lambda : A^T \lambda \leq c\}} b^T \lambda$$

5. Lagrange function can be calculated as follows:

$$L(x, \sigma) = \frac{1}{2} x^T (A + \sigma B) x - \sigma$$

Then Lagrange duality problem can be derived as

$$\max_{\sigma \geq 0} \min_{x \in \mathbb{R}^n} L(x, \sigma) = \max_{\{\sigma : A + \sigma B \in \mathcal{S}_+^n\}} -\sigma$$

6. It is direct to verify that the conjugate function

$$h(y) = \sup_{x \leq 0}\{xy + 2\sqrt{-x}\} = \begin{cases} \frac{1}{y}, & y \geq 0 \\ +\infty, & \text{otherwise} \end{cases}$$

7. By definition of conjugate function, given arbitrary $y \in \mathcal{Y}$, we have $h(y) = y^T x - f(x)$ hold for any x.
   If there exists x, y satisfying $x^T y = f(x) + h(y)$. It can be derived that for any $\hat{x}$,
   $$y^T \hat{x} - f(\hat{x}) \leq h(y) = y^T x - f(x)$$
   The equation can be equivalent to transform to
   $$f(\hat{x}) \geq y^T (\hat{x} - x) + f(x)$$
   which implies that $y \in \partial f(x)$. The necessity is also obtained by reversing this procedure.

8.
$$\min x_1^2$$
$$s.t. x_1 x_2 = 1, x_1, x_2 \in R$$

9. Notice that $A \in S^n$ is a symmetric matrix which will lead to the orthogonal decomposition $A = Q \Lambda Q^T$ where $\Lambda = diag(\lambda_1, \lambda_2, \cdots, \lambda_n)$ and Q is an orthogonal matrix. Then
$$\frac{x^T A x}{x^T x} = \frac{(Q^T x)^T \Lambda Q^T x}{(Qx)^T Qx} \leq \max\{\lambda_i\}$$
By the similar technique, it is derived that $\frac{x^T A X}{x^T x} \geq \min\{\lambda_i\}$.

Problems of Chap. 5

1. It can be obtained directly from the contents of this chapter.
2. It can be obtained directly from the contents of this chapter.
3. First, we express the local weak error as
$$E[f(Y_{\Delta t})] - E[f(\hat{Y}_1)] = E[f(Y_{\Delta t}) - f(\hat{Y}_1)].$$
Using Taylor's theorem, we can expand $f(Y_{\Delta t})$ around $\hat{Y}_1$ as
$$f(Y_{\Delta t}) = f(\hat{Y}_1) + f'(\hat{Y}_1)(Y_{\Delta t} - \hat{Y}_1) + \frac{1}{2} f''(\hat{Y}_1)(Y_{\Delta t} - \hat{Y}_1)^2 + R_{\Delta t}$$
$$= f(\hat{Y}_1) + f'(\hat{Y}_1)\left(Y_{\Delta t} - \hat{Y}_1 - \frac{1}{2}\sigma^2 \hat{Y}_1 \Delta t\right)$$
$$+ \frac{1}{2} f''(\hat{Y}_1)\left(Y_{\Delta t} - \hat{Y}_1 - \frac{1}{2}\sigma^2 \hat{Y}_1 \Delta t\right)^2 + R_{\Delta t},$$

11 Solutions

where the remainder term $R_{\Delta t}$ satisfies $\lim_{\Delta t \to 0} \frac{R_{\Delta t}}{\Delta t^2} = 0$.
Taking expectations and using the properties of the Euler approximation, we obtain

$$E[f(Y_{\Delta t}) - f(\hat{Y}_1)] = \frac{1}{2}\sigma^2 \hat{Y}_1 \Delta t \, E[f''(\hat{Y}_1)] + E[R_{\Delta t}].$$

Dividing both sides by Δt^2 and taking the limit superior as $\Delta t \to 0$, we get

$$\limsup_{\Delta t \to 0} \left(\frac{E[f(Y_{\Delta t})] - E[f(\hat{Y}_1)]}{\Delta t^2} \right) = \frac{1}{2}\sigma^2 \, E[f''(1)] < \infty,$$

since f and its derivatives are bounded. Therefore, the local weak error is indeed of second order.

4. To derive the equations for $\hat{X}_t$ and $\hat{\Sigma}_t$, we use the Duncan-Mortensen-Zakai (DMZ) equation, which describes the evolution of the unnormalized conditional density $p(t, x)$ given the observations Y_t:

$$dp(t, x) = \mathcal{L}_t p(t, x) \, dt + p(t, x) \frac{L_1 x_t + L_0}{B^2} (dy_t - (L_1 x_t + L_0) \, dt)$$

Here, $\mathcal{L}$ is given by

$$\mathcal{L}_t \phi(x) = -(a_0 + a_1 x) \frac{\partial \phi(x)}{\partial x} + \frac{1}{2} b^2 \frac{\partial^2 \phi(x)}{\partial x^2}$$

We assume that the conditional density $p(x_t \mid y_t)$ is Gaussian. Substituting this form into the Kushner equation and using the properties of Gaussian densities, we can derive the equations for $\hat{X}_t$ and $\hat{\Sigma}_t$.

5. To show that the unnormalized filter is given by

$$r_t(A) = \int_A \exp\left(\frac{s^2}{2} x Y_t - \frac{1}{2} s^2 x^2 t \right) \mu(dx),$$

we can use the Girsanov theorem to define an equivalent probability measure $\tilde{P}$ on $\mathcal{F}_T$ such that Y becomes a martingale and independent of X.
Let $\tilde{W}_t = W_t - \frac{s}{2} \int_0^t X \, ds$ be a Brownian motion under $\tilde{P}$, and define

$$\frac{d\tilde{P}}{dP} = \exp\left(-\frac{s^2}{2} X^2 T + s X W_T \right).$$

Then, under $\tilde{P}$, we have

$$Y_t = tX + \sqrt{s} \tilde{W}_t,$$

which is a martingale and independent of X.

The unnormalized filter $r_t(A)$ is given by

$$r_t(A) = \tilde{E}\left[1\!\!1_A(X)\frac{dP}{d\tilde{P}} \mid \mathcal{F}_t^Y\right],$$

where $\mathcal{F}_t^Y$ is the filtration generated by Y.

Substituting the expression for $\frac{dP}{d\tilde{P}}$ and using properties of the Brownian motion, we obtain

$$r_t(A) = \int_A \exp\left(\frac{s^2}{2}xY_t - \frac{1}{2}s^2x^2 t\right)\mu(dx),$$

as desired.

6.

$$\hat{X}_t = E[X|\mathcal{F}_t^Y]$$
$$= E[X] + \frac{\text{Cov}(X, Y_t)}{\text{Var}(Y_t)}(Y_t - E[Y_t])$$
$$= E[X] + \frac{t\text{Var}(X)}{t^2\text{Var}(X) + s}(Y_t - tE[X]).$$

Covariance:

$$\hat{\Sigma}_t = \text{Var}(X|\mathcal{F}_t^Y) = \text{Var}(X) - \frac{(\text{Cov}(X, Y_t))^2}{\text{Var}(Y_t)} = \text{Var}(X) - \frac{t^2(\text{Var}(X))^2}{t^2\text{Var}(X) + s}$$

7. Consider the following model for population growth with noisy observations:

$$dX_t = rX_t\, dt,$$
$$dY_t = X_t\, dt + m\, dW_t,$$

with $X_0 \sim \mathcal{N}(b, a^2)$ and $Y_0 = 0$ for some constants $r, m, b, a > 0$.

We can apply the Kalman-Bucy filter to estimate the state X_t given the observations Y_t. The asymptotic covariance matrix $\hat{\Sigma}_t$ as $t \to \infty$ satisfies the algebraic Riccati equation:

$$r\hat{\Sigma}_t + \hat{\Sigma}_t r - \frac{\hat{\Sigma}_t^2}{m^2} + 1 = 0.$$

Solving this equation, we get

$$\lim_{t \to \infty} \hat{\Sigma}_t = \frac{m^2}{2}\left(r + \sqrt{r^2 + \frac{4}{m^2}}\right).$$

The asymptotic precision of the filter, which is inversely proportional to $\hat{\Sigma}_t$, decreases as the growth rate r increases. This means that for rapidly growing populations, the uncertainty in the state estimate increases.

8. The Python 3 Code is provided as follows:

```
import numpy as np
import matplotlib.pyplot as plt

# Set parameters
r = 0.5
m = 1
b = 1
a = 0.5
T = 1  # Total time
N = 1000  # Number of time steps
dt = T / N  # Time step size

# Initialize
t = np.linspace(0, T, N+1)
X = np.zeros(N+1)
Y = np.zeros(N+1)
X[0] = np.random.normal(b, a)
Y[0] = 0

# Simulate path using Euler-Maruyama scheme
for i in range(N):
dW = np.random.normal(0, np.sqrt(dt))
X[i+1] = X[i] + r * X[i] * dt
Y[i+1] = Y[i] + X[i] * dt + m * dW

# Kalman-Bucy filter
mu = np.zeros(N+1)
P = np.zeros(N+1)
mu[0] = b
P[0] = a**2

for i in range(N):
# Prediction step
mu_pred = mu[i] + r * mu[i] * dt
P_pred = P[i]
```

```
# Update step
K = P_pred
mu[i+1] = mu_pred + K * (Y[i+1] - Y[i] - mu_pred*dt)
P[i+1] = P_pred + (2*r*P[i] -P[i]**2) * dt

# Plot results
plt.figure(figsize=(12, 6))
plt.plot(t, X, label='True X')
plt.plot(t, Y, label='Observed Y')
plt.plot(t, mu, label='Estimated X')
plt.fill_between(t, mu - 1.96*np.sqrt(P),
mu + 1.96*np.sqrt(P), alpha=0.2, label='95% CI')
plt.legend()
plt.xlabel('Time')
plt.ylabel('Value')
plt.title('Kalman Filter for Population Growth Model')
plt.show()
```

When running this code, you will observe:

(a) The true X_t shows an exponential growth trend, which is consistent with the population growth model.
(b) The observed values Y_t fluctuate around the true X_t, demonstrating the effect of observation noise.
(c) The X_t estimated by the Kalman-Bucy filter tracks the true X_t well, although there is some lag.
(d) Over time, the estimated uncertainty (represented by the confidence interval) gradually decreases, indicating that the filter improves its estimation as time progresses.

9. (a) To show that Y_t is a solution to the given SDE, we can use Ito's formula. Let $f(t, x) = \exp(x - X_0 - \frac{1}{2}\langle X, X \rangle_t)$. Then

$$dY_t = \frac{\partial f}{\partial x} dX_t + \frac{\partial f}{\partial t} dt + \frac{1}{2}\frac{\partial^2 f}{\partial x^2} d\langle X, X \rangle_t$$

$$= Y_t dX_t - \frac{1}{2} Y_t d\langle X, X \rangle_t + \frac{1}{2} Y_t d\langle X, X \rangle_t$$

$$= Y_t dX_t$$

Also, at $t = 0$, $Y_0 = \exp(X_0 - X_0 - 0) = 1$. Thus, Y_t satisfies the given SDE.

(b) If $\mu = 0$, then $dX_t = s_t dW_t$. Substituting this into the SDE for Y

$$dY_t = Y_t s_t dW_t$$

11 Solutions

This is the SDE for a local martingale. Therefore, Y is a local martingale when $\mu = 0$.

(c) To show that Y is a martingale when $\mu = 0$ and s is bounded, we can use Novikov's condition. When $\mu = 0$, we have

$$Y_t = \exp\left(\int_0^t s_u dW_u - \frac{1}{2}\int_0^t s_u^2 du\right)$$

If s is bounded, say $|s_t| \leq K$ for all t, then

$$\mathbb{E}\left[\exp\left(\frac{1}{2}\int_0^T s_u^2 du\right)\right] \leq \exp\left(\frac{1}{2}K^2 T\right) < \infty$$

This satisfies Novikov's condition, which implies that Y is a true martingale, not just a local martingale.

Problems of Chap. 6

1. For arbitrary test function ϕ,

$$\frac{1}{2}\left(\sum_{i=1}^n D_i^2 - \eta\right)\phi = \frac{1}{2}[\sum_{i=1}^n D_i(D_i\phi) - \eta\phi]$$

$$= \frac{1}{2}[\sum_{i=1}^n (\frac{\partial^2 \phi}{\partial x_i^2} - \frac{\partial f_i}{\partial x_i}\phi - 2f_i\frac{\partial \phi}{\partial x_i} + f_i^2\phi) - \eta\phi]$$

$$= \frac{1}{2}\sum_{i=1}^n \frac{\partial^2 \phi}{\partial x_i^2} - \sum_{i=1}^n \frac{\partial f_i}{\partial x_i}\phi - \sum_{i=1}^n f_i\frac{\partial \phi}{\partial x_i} - \frac{1}{2}\sum_{i=1}^n h_i^2 \phi$$

2. (1)

$$= \frac{1}{2}[\sum_i (D_i^2 - \eta), x_i]$$

$$= \frac{1}{2}[D_i^2, x_i] = D_i$$

(2)

$$[[L_0, \phi], \phi] = [[\sum_i \phi_{x_i} D_i + \frac{1}{2}\phi_{x_i x_i}, \phi] = |\nabla \phi|^2$$

(3)
$$=[\frac{1}{2}\left(\sum_{i=1}^{n} D_i^2 - \eta\right), D_j]$$

$$= \sum_{i=1}^{n} \omega_{ji} D_i + \frac{1}{2}\frac{\partial \eta}{\partial x_j} + \frac{1}{2}\sum_{i=1}^{n}\frac{\partial \omega_{ji}}{\partial x_i}$$

(4)
$$[L_0, x_j^2] = \frac{1}{2}[\sum_i D_i^2, x_j^2] = 2x_j D_j + 1$$

3. It can be easily calculated by employing rules of Lie bracket.
4. Let $\phi = x_1^m \zeta$. Then

$$E_l(\phi) = \sum_{i=1}^{l} x_i \frac{\partial}{\partial x_i}(x_1^m \zeta) = x_1^m(m\zeta + E_l(\zeta)) = 0$$

$$\phi(x_1, \cdots, x_l, \cdots, x_n) - \phi(\epsilon x_1, \cdots, \epsilon x_l, \cdots, x_n)$$

$$= \int_\epsilon^1 \frac{d\phi}{dt}(tx_1, \cdots, tx_l, x_{l+1}, \cdots, x_n) dt$$

$$= \int_\epsilon^1 \frac{1}{t}(E_l(\phi))(tx_1, \cdots, tx_l, x_{l+1}, \cdots, x_n) dt = 0$$

Hence $\phi(x_1, \cdots, x_l, \cdots, x_n) = \phi(\epsilon x_1, \cdots, \epsilon x_l, \cdots, x_n)$. If we take $\epsilon \to 0$, it can be obtained that $\phi = 0$ thus $\zeta = 0$.

5. For $1 \leq j \leq l$,

$$\frac{\partial}{\partial x_j} E_l(\zeta) = \frac{\partial \zeta}{\partial x_j} + E_l(\frac{\partial \zeta}{\partial x_j}) = 0$$

By applying the conclusion of problem 4, we derive that $\frac{\partial \zeta}{\partial x_j} = 0$.

6. First

$$D_j D_i = \frac{\partial^2}{\partial x_i \partial x_j} - f_j \frac{\partial}{\partial x_i} - \frac{\partial f_i}{\partial x_j} - f_i \frac{\partial}{\partial x_j} + f_j f_i$$

Similarly

$$D_i D_j = \frac{\partial^2}{\partial x_i \partial x_j} - f_j \frac{\partial}{\partial x_i} - \frac{\partial f_j}{\partial x_i} - f_i \frac{\partial}{\partial x_j} + f_j f_i$$

Then $D_i D_j = D_j D_i + \omega_{ji} = D_j D_i$, mod U_0. Without loss of generality, let $k_1 \neq 1$ and by induction we get

$$gD_1^{i_1} \cdots D_n^{i_n} = gD_1^{i_1} \cdots D_{k_1-1}^{i_{k_1}-1} D_{k_1}^{i_{k_1}} \cdots D_n^{i_n}$$
$$= gD_1^{i_1} \cdots (D_{k_1}^{i_{k_1}} D_{k_1-1}^{i_{k_1}-1}, \text{ mod } U_{i_{k_1}+i_{k_1-1}-2}) \cdots D_n^{i_n}$$
$$= \cdots$$
$$= gD_{k_1}^{i_{k_1}} \cdots D_{k_n}^{i_{k_n}}, \text{ mod } U_{|i|-2}$$

7. Assume $\frac{\partial p^{(2)}}{\partial x_j}(x) \neq 0$ for some $k_1 + 1 \leq j \leq n - k_2$. Next we can derive a contradiction.

First maximal rank quadratic polynomial is

$$p_0 = x^T \begin{pmatrix} I_{k_1 \times k_1} & 0 & 0 \\ 0 & 0 & 0 \\ 0 & 0 & I_{k_2 \times k_2} \end{pmatrix} x,$$

and homogeneous quadratic part of $p \in E$ is

$$p^{(2)} = x^T \begin{pmatrix} A_{11} & A_{12} & A_{13} \\ A_{12}^T & A_{22} & A_{23} \\ A_{13}^T & A_{23}^T & A_{33} \end{pmatrix} x,$$

where $x = (x_1, x_2, \cdots, x_n)^T$, $A_{11} \in \mathbb{R}^{k_1 \times k_1}$, $A_{22} \in \mathbb{R}^{(n-k) \times (n-k)}$, $A_{33} \in \mathbb{R}^{k_2 \times k_2}$ are symmetric matrices and $A_{12} \in \mathbb{R}^{k_1 \times (n-k)}$, $A_{13} \in \mathbb{R}^{k_1 \times k_2}$, $A_{23} \in \mathbb{R}^{(n-k) \times k_2}$. Next we consider quadratic polynomial $tp_0 + p$. We will prove if t is large enough, $rank(tp_0 + p)$ will be larger than p_0 that derives a contradiction. Since $p_0.p \in E$, naturally we have $tp_0 + p \in E$.

$$(tp_0 + p)^{(2)} = tp_0 + p^{(2)}$$
$$= x^T \begin{pmatrix} tI + A_{11} & A_{12} & A_{13} \\ A_{12}^T & A_{22} & A_{23} \\ A_{13}^T & A_{23}^T & tI + A_{33} \end{pmatrix} x,$$

Then we have

$$rank(tp_0 + p) := rank \begin{pmatrix} tI + A_{11} & A_{12} & A_{13} \\ A_{12}^T & A_{22} & A_{23} \\ A_{13}^T & A_{23}^T & tI + A_{33} \end{pmatrix}$$

$$= rank \begin{pmatrix} tI + A_{11} & A_{13} & A_{12} \\ A_{13}^T & tI + A_{33} & A_{23}^T \\ A_{12}^T & A_{23} & A_{22} \end{pmatrix}$$

$$= rank \left(\begin{array}{c|c} tI + \begin{pmatrix} A_{11} & A_{13} \\ A_{13}^T & A_{33} \end{pmatrix} & \begin{array}{c} A_{12} \\ A_{23}^T \end{array} \\ \hline A_{12}^T \ \ A_{23} & A_{22} \end{array} \right)$$

$$:= rank(A),$$

where in the second equality we used elementary transformation which do not change rank of matrix. Here we define

$$A := \left(\begin{array}{c|c} tI + \begin{pmatrix} A_{11} & A_{13} \\ A_{13}^T & A_{33} \end{pmatrix} & \begin{array}{c} A_{12} \\ A_{23}^T \end{array} \\ \hline A_{12}^T \ \ A_{23} & A_{22} \end{array} \right).$$

Since $\frac{\partial p^{(2)}}{\partial x_j}(x) \neq 0$ for some $k_1 + 1 \leq j \leq n - k_2$, then A_{12}, A_{23}, A_{22} are not all zero. By an elementary matrix transformation, we always can put a nonzero element in submatrices A_{12}, A_{23}, A_{22} into position (i, j) of A, where $i = k + 1, 1 \leq j \leq k + 1$ or $j = k + 1, 1 \leq i \leq k + 1$. Then we can derive

$$rank(A) \geq rank \left(\begin{array}{c|c} tI + \begin{pmatrix} A_{11} & A_{13} \\ A_{13}^T & A_{33} \end{pmatrix} & c \\ \hline c^T & b \end{array} \right) := rank \begin{pmatrix} \tilde{A} & c \\ c^T & b \end{pmatrix},$$

where we define

$$\tilde{A} := tI + \begin{pmatrix} A_{11} & A_{13} \\ A_{13}^T & A_{33} \end{pmatrix},$$

and $c \in R^k, b \in R$ and b, c are not both zero. Since $\begin{pmatrix} A_{11} & A_{13} \\ A_{13}^T & A_{33} \end{pmatrix}$ is a real symmetric matrix, it has orthogonal diagonalized decomposition

$$\begin{pmatrix} A_{11} & A_{13} \\ A_{13}^T & A_{33} \end{pmatrix} = U \begin{pmatrix} \lambda_1 & \cdots & 0 \\ \vdots & \ddots & \vdots \\ 0 & \cdots & \lambda_k \end{pmatrix} U^T := U \Lambda U^T$$

where $\Lambda := diag(\lambda_1, \cdots, \lambda_k)$ and $U \in R^{k \times k}$ is an orthogonal matrix and $\lambda_1, \cdots, \lambda_k$ are eigenvalues. Then we can take orthogonal transformation for matrix $\begin{pmatrix} \tilde{A} & c \\ c^T & b \end{pmatrix}$ and get

$$\begin{pmatrix} U^T & 0 \\ 0 & 1 \end{pmatrix} \begin{pmatrix} \tilde{A} & c \\ c^T & b \end{pmatrix} \begin{pmatrix} U & 0 \\ 0 & 1 \end{pmatrix} = \begin{pmatrix} tI_k + \Lambda & U^T c \\ c^T U & b \end{pmatrix},$$

where we denote $\tilde{c} = (\tilde{c}_1, \cdots, \tilde{c}_k)^T := U^T c$. Since b, c are not all zero, then $b, \tilde{c}_1, \cdots, \tilde{c}_k$ are not all zero. When we take t enough large, $t + \lambda_i > 0$ for $1 \le i \le k$ and $\tilde{A}$ is nonsingular. Then by rank formula of block matrix, we have

$$rank \begin{pmatrix} tI + \Lambda & \tilde{c} \\ \tilde{c}^T & b \end{pmatrix} = k + rank \left(b - \sum_{i=1}^{k} \frac{\tilde{c}_i^2}{t + \lambda_i} \right).$$

(i) if $b = 0, \tilde{c}_1, \cdots, \tilde{c}_k$ are not all zero, then

$$\sum_{i=1}^{k} \frac{\tilde{c}_i^2}{t + \lambda_i} \ne 0.$$

Then $rank \begin{pmatrix} tI + \Lambda & U^T c \\ c^T U & b \end{pmatrix} = k + 1$.

(ii) if $b \ne 0$, when t is enough large

$$b > \sum_{i=1}^{k} \frac{\tilde{c}_i^2}{t + \lambda_i} \implies rank \begin{pmatrix} tI + \Lambda & U^T c \\ c^T U & b \end{pmatrix} = k + 1.$$

Combining (i) and (ii), for enough large t,

$$rank(tp_0 + p) \ge rank \begin{pmatrix} \tilde{A} & c \\ c^T & b \end{pmatrix}$$

$$= rank \begin{pmatrix} tI + \Lambda & U^T c \\ c^T U & b \end{pmatrix}$$

$$= k + 1.$$

This is contradictory to that p_0 has greatest quadratic rank k in E. Then $\frac{\partial p^{(2)}}{\partial x_j} = 0$ for $j = k_1 + 1, \cdots, n - k_2$ holds.

8.

$$\frac{\partial \omega_{ij}}{\partial x_k} + \frac{\partial \omega_{jk}}{\partial x_i} + \frac{\partial \omega_{ki}}{\partial x_j}$$
$$= \frac{\partial}{\partial x_k}(\frac{\partial f_j}{\partial x_i} - \frac{\partial f_i}{\partial x_j}) + \frac{\partial}{\partial x_i}(\frac{\partial f_k}{\partial x_j} - \frac{\partial f_j}{\partial x_k}) + \frac{\partial}{\partial x_j}(\frac{\partial f_i}{\partial x_k} - \frac{\partial f_k}{\partial x_i}) = 0$$

9. **Sufficiency**. First we will prove the sufficiency of statement. Drift function is assumed to be quadratic polynomials plus a gradient form.

$$(f_1, \cdots, f_n) = (l_1, \cdots, l_n) + \left(\frac{\partial \psi}{\partial x_1}, \cdots, \frac{\partial \psi}{\partial x_n} \right).$$

Then

$$\frac{\partial f_j}{\partial x_i} - \frac{\partial f_i}{\partial x_j} = \frac{\partial}{\partial x_i}\left(l_j + \frac{\partial \psi}{\partial x_j}\right) - \frac{\partial}{\partial x_j}\left(l_i + \frac{\partial \psi}{\partial x_i}\right)$$
$$= \frac{\partial l_j}{\partial x_i} - \frac{\partial l_i}{\partial x_j} = P_1(x),$$

for all $1 \leq i, j \leq n$.

Necessity. Conversely, we assume $\frac{\partial f_j}{\partial f_i} - \frac{\partial f_i}{\partial f_j} = c_{ij} + D_{ij}^T x$ are degree at most 1 polynomials for $1 \leq i, j \leq n$. We observe that $c_{ij} = -c_{ji}$, $D_{ij} = -D_{ji}$. Let $B_{ij} = -\frac{1}{2}c_{ij}$, $A_{ij} = -\frac{1}{2}D_{ij}^T$. Then we have

$$B_{ji} - B_{ij} = -\frac{1}{2}c_{ji} - \left(-\frac{1}{2}c_{ij}\right) = c_{ij},$$

$$A_{ji} - A_{ij} = -\frac{1}{2}D_{ji}^T - \left(-\frac{1}{2}D_{ij}^T\right) = D_{ij}^T.$$

Next we assume $l_i(x) = \frac{1}{2}x^T A_i x + B_i^T x + C_i$ for $1 \leq i \leq n$, where $A_i \in \mathbb{R}^{n \times n}$, $B_i \in \mathbb{R}^{n \times 1}$, $C_i \in \mathbb{R}$. Then we assume A_i and B_i has the block form:

$$A_i = \begin{pmatrix} A_{i1} \\ A_{i2} \\ \vdots \\ A_{in} \end{pmatrix}, B_i = \begin{pmatrix} B_{i1} \\ B_{i2} \\ \vdots \\ B_{in} \end{pmatrix}.$$

Clearly exterior derivative of the differential forms $f_1 dx_1 + f_2 dx_2 + \cdots + f_n dx_n$ and $l_1 dx_1 + l_2 dx_2 + \cdots + l_n dx_n$ are given as follows:

$$d\left(\sum_{i=1}^{n} f_i dx_i\right) = \sum_{i=1}^{n} df_i dx_i$$

$$= \sum_{i=1}^{n} \left(\sum_{j=1}^{n} \frac{\partial f_i}{\partial x_j} dx_j\right) dx_i$$

$$= \sum_{i,j=1}^{n} \frac{\partial f_i}{\partial x_j} dx_j \wedge dx_i$$

$$= \sum_{i<j} \left(\frac{\partial f_j}{\partial x_i} - \frac{\partial f_i}{\partial x_j}\right) dx_i \wedge dx_j$$

$$= \sum_{i<j} (c_{ij} + D_{ij}^T x) dx_i \wedge dx_j,$$

and

$$d\left(\sum_{i=1}^{n} l_i dx_i\right) = \sum_{i<j} \left(\frac{\partial l_j}{\partial x_i} - \frac{\partial l_i}{\partial x_j}\right) dx_i \wedge dx_j$$

$$= \sum_{i<j} \left[\frac{\partial}{\partial x_i}\left(\frac{1}{2} x^T A_j x + B_j^T x + C_j\right)\right.$$

$$\left. - \frac{\partial}{\partial x_j}\left(\frac{1}{2} x^T A_i x + B_i^T x + C_i\right)\right] dx_i \wedge dx_j$$

$$= \sum_{i<j} (A_{ji} x + B_{ji} - A_{ij} x - B_{ij}) dx_i \wedge dx_j$$

$$= \sum_{i<j} [(A_{ji} - A_{ij}) x + (B_{ji} - B_{ij})] dx_i \wedge dx_j$$

$$= \sum_{i<j} (D_{ij}^T x + c_{ij}) dx_i \wedge dx_j.$$

Then we obtain

$$d\left(\sum_{i=1}^{n} f_i dx_i\right) = d\left(\sum_{i=1}^{n} l_i dx_i\right),$$

i.e.,

$$d\left(\sum_{i=1}^{n} f_i dx_i - \sum_{i=1}^{n} l_i dx_i\right) = 0.$$

By Poincare lemma, every closed differential form on $\mathbb{R}^n$ is exact. Therefore there exists a smooth function ψ such that

$$\sum_{i=1}^{n} f_i dx_i - \sum_{i=1}^{n} l_i dx_i = d\psi$$

$$= \sum_{i=1}^{n} \frac{\partial \psi}{\partial x_i} dx_i.$$

It implies

$$(f_1, \cdots, f_n) = (l_1, \cdots, l_n) + \left(\frac{\partial \psi}{\partial x_1}, \cdots, \frac{\partial \psi}{\partial x_n} \right).$$

Problems of Chap. 7

1. The DMZ equation satisfied by this one-dimensional system is given by

$$d\sigma(t,x) = \frac{1}{2} \frac{\partial^2}{\partial x^2} \sigma(t,x) dt + x^3 \sigma(t,x) dy_t.$$

2. The robust DMZ equation satisfied by $u(t,x)$ is given by

$$\frac{\partial u}{\partial t} = \frac{1}{2} \frac{\partial^2 u}{\partial x^2} + 3y_t x^2 \frac{\partial u}{\partial x} + \left(\frac{1}{2} x^6 + 3y_t x + \frac{9}{2} y_t^2 x^4 \right) u(t,x).$$

3. The equation satisfied by $u_i(t,x)$ is

$$\frac{\partial u_i}{\partial t} = \frac{1}{2} \frac{\partial^2 u}{\partial x^2} + 3y_{\tau_{i-1}} x^2 \frac{\partial u}{\partial x} + \left(\frac{1}{2} x^6 + 3y_{\tau_{i-1}} x + \frac{9}{2} y_{\tau_{i-1}}^2 x^4 \right) u(t,x),$$

and the equation satisfied by $\tilde{u}_i(t,x)$ is

$$\frac{\partial \tilde{u}}{\partial t} = \frac{1}{2} \frac{\partial^2 \tilde{u}}{\partial x^2} - \frac{1}{2} x^6 \tilde{u}(t,x), \quad t \in [\tau_{i-1}, \tau_i],$$

with initial value

$$\tilde{u}(\tau_{i-1}, x) = \exp\left(x^3 y_{\tau_{i-1}} \right) u(t,x).$$

11 Solutions

4. The variational form of the parabolic equation satisfied by $\tilde{u}_i(t,x)$ is given by

$$\frac{\partial}{\partial t}\langle \tilde{u}_i, \varphi \rangle = -\frac{1}{2}\left\langle \frac{\partial \tilde{u}_i}{\partial x}, \frac{\partial \varphi}{\partial x} \right\rangle - \frac{1}{2}\langle x^6 \tilde{u}_i, \varphi \rangle,$$

for all $\varphi \in L^2(\mathbb{R})$. Since

$$\tilde{u}_i^N(t,x) = \sum_{j=1}^{N} a_{i,j}^{(N)}(t) H_j(x),$$

and according to the Galerkin approximation, we have

$$\frac{\partial}{\partial t}\langle \tilde{u}_i^N, H_j \rangle = -\frac{1}{2}\left\langle \frac{\partial \tilde{u}_i^N}{\partial x}, H_j' \right\rangle - \frac{1}{2}\langle x^6 \tilde{u}, H_j \rangle, \quad j = 1, \cdots, N.$$

According to the orthogonality of H_j, we have

$$\frac{d}{dt}a_j^{(N)}(t) = -\frac{1}{2}\sum_{k=1}^{N} a_k^{(N)}(t)\langle H_k', H_j' \rangle - \frac{1}{2}\sum_{k=1}^{N} a_k^{(N)}(t)\langle x^6 H_k, H_j \rangle.$$

for all $j = 1, \cdots, N$.

5. An approximation of $E[X_t|\mathcal{Y}_t]$ at time $t = \tau_i$ can be given by

$$E[X_{\tau_i}|\mathcal{Y}_{\tau_i}] \approx \frac{\int x \tilde{u}_{i+1}^{(N)}(\tau_i, x)dx}{\int \tilde{u}_{i+1}(\tau_i, x)dx} = \frac{\sum_{j=1}^{N} a_{i+1,j}^{N}(\tau_i) \int x H_j(x)dx}{\sum_{j=1}^{N} a_{i+1,j}^{(N)}(\tau_i) \int H_j(x)dx}.$$

6. Since the identity

$$\mathcal{H}_{n+1}(x) = 2x\mathcal{H}_n(x) - 2n\mathcal{H}_{n-1}(x)$$

holds for Hermite polynomials, and

$$H_n(x) = \frac{1}{\sqrt{2^n n!}}\mathcal{H}_n(\alpha(x-\beta))e^{-\frac{1}{2}\alpha^2(x-\beta)},$$

we have

$$\mathcal{H}_{n+1}(\alpha(x-\beta)) = 2\alpha(x-\beta)\mathcal{H}_n(\alpha(x-\beta)) - 2n\mathcal{H}_{n-1}(\alpha(x-\beta))$$

and

$$\sqrt{2^{n+1}(n+1)!}H_{n+1}(x) = 2\alpha(x-\beta)\sqrt{2^n n!}H_n(x)$$
$$- 2n\sqrt{2^{n-1}(n-1)!}H_{n-1}(x)$$

Therefore,
$$2\alpha(x-\beta)H_n(x) = \sqrt{2n}H_{n-1}(x) + \sqrt{2(n+1)}H_{n+1}(x).$$

Next, because
$$H_n(x) = (-1)^n e^{x^2}\frac{d^n}{dx^n}e^{-x^2},$$

for $\alpha = 1, \beta = 0$, we have
$$H_n^{1,0}(x) = \frac{1}{\sqrt{2^n n!}}H_n(x)e^{-\frac{1}{2}x^2} = (-1)^n \frac{1}{\sqrt{2^n n!}}e^{\frac{1}{2}x^2}\frac{d^n}{dx^n}e^{-x^2}$$

Thus,
$$\left(H_n^{1,0}\right)'(x) = (-1)^n \frac{1}{\sqrt{2^n n!}}e^{\frac{1}{2}x^2}\left(\frac{d^{n+1}}{dx^{n+1}}e^{-x^2} + x\frac{d^n}{dx^n}e^{-x^2}\right)$$
$$= -\sqrt{2(n+1)}H_{n+1}^{1,0}(x) + xH_n^{1,0}(x)$$
$$= -\sqrt{2(n+1)}H_{n+1}(x) + \frac{1}{2}\sqrt{2n}H_{n-1}^{1,0}(x) + \frac{1}{2}\sqrt{2(n+1)}H_{n+1}^{1,0}(x).$$

That is,
$$\left(H_n^{1,0}\right)'(x) = \frac{1}{2}\sqrt{2n}H_{n-1}^{1,0}(x) - \frac{1}{2}\sqrt{2(n+1)}H_{n+1}^{1,0}(x)$$

For general $\alpha > 0, \beta \in \mathbb{R}$, we have
$$H_n^{\alpha,\beta}(x) = H_n^{1,0}(\alpha(x-\beta)).$$

and
$$\left(H_n^{\alpha,\beta}\right)'(x) = \alpha\left(H_n^{1,0}\right)'(\alpha(x-\beta)) = \frac{1}{2}\sqrt{\lambda_n}H_{n-1}^{\alpha,\beta}(x) - \frac{1}{2}\sqrt{\lambda_{n+1}}H_{n+1}^{\alpha,\beta}(x)$$

The quasi-orthogonality of $\{\partial_x H_n(x)\}_{n=0}^\infty$ is obtained by the above two formulae for H_n and H_n', as well as the orthogonality of $\{H_n\}_{n=0}^\infty$.

In fact,

11 Solutions

$$\int \partial_x H_n(x) \partial_x H_m(x) dx$$

$$= \int \left(\frac{1}{2}\sqrt{\lambda_n} H_{n-1} - \frac{1}{2}\sqrt{\lambda_{n+1}} H_{n+1} \right) \left(\frac{1}{2}\sqrt{\lambda_m} H_{m-1} - \frac{1}{2}\sqrt{\lambda_{m+1}} H_{m+1} \right) dx$$

$$= \begin{cases} \dfrac{\sqrt{\pi}}{4\alpha}(\lambda_n + \lambda_{n+1}), & m = n \\ -\dfrac{\sqrt{\pi}}{4\alpha}\sqrt{\lambda_{m+1}\lambda_{m+2}}, & n - m = 2 \\ -\dfrac{\sqrt{\pi}}{4\alpha}\sqrt{\lambda_{n+1}\lambda_{n+2}}, & m - n = 2 \\ 0 & otherwise. \end{cases}$$

7. The assumptions in the convergence analysis are

$$-\frac{1}{2}x^6 - 3y_t x + \frac{9}{2}y_t^2 x^4 + 3|y_t|x^2 \le c_1,$$

$$-\frac{1}{2}x^6 - 3y_t x + \frac{9}{2}y_t^2 x^4 + 12 + 2n + 12|y_t|x^2 \le c_2,$$

$$e^{-\sqrt{1+x^2}}(12 + 2n + 12|y_t|x^2) \le c_3.$$

Generally speaking, a sufficient condition for these assumptions to hold in this model is that the observations y_t are bounded for all $t \in [0, T]$.

Problems of Chap. 8

1. (1)

$$\nabla \cdot (fF) = \frac{\partial}{\partial x_i}(fF_i)$$
$$= f_{x_i} F_i + F_{i,x_i}$$
$$= \nabla f \cdot F + f \nabla \cdot F$$

(2)

$$\nabla \cdot (g\nabla f + f\nabla g) = (\nabla g \cdot \nabla f + g\nabla \cdot \nabla f) + (\nabla f \cdot \nabla g + f\nabla \cdot (\nabla g))$$
$$= g\nabla \cdot \nabla f + 2\nabla f \cdot \nabla g + f\nabla \cdot \nabla g$$

2.
$$\frac{1}{2}\sum_{i=1}^n H_i^2(x) - \frac{1}{2}\sum_{i=1}^n \frac{\partial H_i}{\partial x_i} - \frac{1}{2}\eta(x) + P(x)$$
$$= \frac{1}{2}\sum_{i=1}^n l_i^2 - \frac{1}{2}\sum_{i=1}^n l_{i,x_i} - \frac{1}{2}\eta + \frac{1}{2}\eta - \frac{1}{2}\sum_{i=1}^n l_i^2 + \frac{1}{2}\sum_{i=1}^n l_{i,x_i} = 0$$

3. It can be verified by direct computations.

4.
$$\frac{1}{2}\sum_{i=1}^n H_i^2(x) - \frac{1}{2}\sum_{i=1}^n \frac{\partial H_i}{\partial x_i} - \frac{1}{2}\eta(x) + P(x)$$
$$= \frac{1}{2}\sum_{i=1}^n f_i^2 - \frac{1}{2}\sum_{i=1}^n f_{i,x_i} - \frac{1}{2}\eta + \frac{1}{2}\eta - \frac{1}{2}\sum_{i=1}^n f_i^2 + \frac{1}{2}\sum_{i=1}^n f_{i,x_i} = 0$$

5. Sufficiency is easy to obtain by direct computation.

Necessity. Conversely, we assume $\frac{\partial f_j}{\partial f_i} - \frac{\partial f_i}{\partial f_j} = c_{ij}$ are degree at most 1 polynomials for $1 \leq i, j \leq n$. We observe that $c_{ij} = -c_{ji}$. Let $B_{ij} = -\frac{1}{2}c_{ij}$. Then we have

$$B_{ji} - B_{ij} = -\frac{1}{2}c_{ji} - \left(-\frac{1}{2}c_{ij}\right) = c_{ij}.$$

Next we assume $l_i(x) = B_i^T x + C_i$ for $1 \leq i \leq n$, where $B_i \in \mathbb{R}^{n \times 1}, C_i \in \mathbb{R}$. Then we assume B_i has the block form

$$B_i = \begin{pmatrix} B_{i1} \\ B_{i2} \\ \vdots \\ B_{in} \end{pmatrix}.$$

Clearly exterior derivative of the differential forms $f_1 dx_1 + f_2 dx_2 + \cdots + f_n dx_n$ and $l_1 dx_1 + l_2 dx_2 + \cdots + l_n dx_n$ are given as follows:

11 Solutions

$$d\left(\sum_{i=1}^{n} f_i dx_i\right) = \sum_{i=1}^{n} df_i dx_i$$

$$= \sum_{i=1}^{n}\left(\sum_{j=1}^{n} \frac{\partial f_i}{\partial x_j} dx_j\right) dx_i$$

$$= \sum_{i,j=1}^{n} \frac{\partial f_i}{\partial x_j} dx_j \wedge dx_i$$

$$= \sum_{i<j}\left(\frac{\partial f_j}{\partial x_i} - \frac{\partial f_i}{\partial x_j}\right) dx_i \wedge dx_j$$

$$= \sum_{i<j} c_{ij} dx_i \wedge dx_j,$$

and

$$d\left(\sum_{i=1}^{n} l_i dx_i\right) = \sum_{i<j}\left(\frac{\partial l_j}{\partial x_i} - \frac{\partial l_i}{\partial x_j}\right) dx_i \wedge dx_j$$

$$= \sum_{i<j} c_{ij} dx_i \wedge dx_j.$$

Then we obtain

$$d\left(\sum_{i=1}^{n} f_i dx_i\right) = d\left(\sum_{i=1}^{n} l_i dx_i\right),$$

i.e.,

$$d\left(\sum_{i=1}^{n} f_i dx_i - \sum_{i=1}^{n} l_i dx_i\right) = 0.$$

By Poincare lemma, every closed differential form on $\mathbb{R}^n$ is exact. Therefore there exists a smooth function ψ such that

$$\sum_{i=1}^{n} f_i dx_i - \sum_{i=1}^{n} l_i dx_i = d\psi$$

$$= \sum_{i=1}^{n} \frac{\partial \psi}{\partial x_i} dx_i.$$

It implies

$$(f_1, \cdots, f_n) = (l_1, \cdots, l_n) + \left(\frac{\partial \psi}{\partial x_1}, \cdots, \frac{\partial \psi}{\partial x_n}\right).$$

6. This numerical achievement will be left to reader himself.
7. Let $f = \tanh x$ and $h_1(x) = x \sin x, h_2(x) = x \cos x$.
8. Direct computations show that

$$\eta = |f|^2 + \nabla \cdot f + |h|^2$$
$$= |Lx + l + \nabla \phi|^2 + \nabla \cdot (Lx + l + \nabla \phi) + |h|^2$$
$$= |Lx + l|^2 + q(x)$$

Therefore, it is obvious that η is quadratic polynomial is equivalent to that $q(x)$ is also quadratic polynomial.

9. First we shall calculate the following derivatives in terms of t, x_i:

$$\frac{\partial u}{\partial t} = e^\phi \frac{\partial v}{\partial t} + e^\phi \frac{\partial \phi}{\partial t} v$$

$$\frac{\partial u}{\partial x_i} = e^\phi \frac{\partial v}{\partial x_i} + e^\phi \frac{\partial \phi}{\partial x_i} v$$

$$\frac{\partial^2 u}{\partial x_i^2} = e^\phi \frac{\partial^2 v}{\partial x_i^2} + 2e^\phi \frac{\partial \phi}{\partial x_i} \frac{\partial v}{\partial x_i} + e^\phi (\frac{\partial \phi}{\partial x_i})^2 v + e^\phi \frac{\partial^2 \phi}{\partial x_i^2} v$$

Substituting these equations to forward Kolmogorov equation satisfied by u, the desired result is obtained.

Problems of Chap. 9

1. We need to prove the following equality:

$$(A + BCD)(A^{-1} - A^{-1}B(C^{-1} + DA^{-1}B)^{-1}DA^{-1}) = I$$

The left-hand side equals:

$$(A + BCD)(A^{-1} - A^{-1}B(C^{-1} + DA^{-1}B)^{-1}DA^{-1})$$
$$= AA^{-1} - AA^{-1}B(C^{-1} + DA^{-1}B)^{-1}DA^{-1}$$
$$+ BCD(A^{-1} - A^{-1}B(C^{-1} + DA^{-1}B)^{-1}DA^{-1})$$

11 Solutions

$$= I - B(C^{-1} + DA^{-1}B)^{-1}DA^{-1}$$
$$+ BCD(A^{-1} - A^{-1}B(C^{-1} + DA^{-1}B)^{-1}DA^{-1})$$
$$= I - B(C^{-1} + DA^{-1}B)^{-1}(DA^{-1} - CDA^{-1}$$
$$+ CDA^{-1}B(C^{-1} + DA^{-1}B)^{-1}DA^{-1})$$
$$= I - B(C^{-1} + DA^{-1}B)^{-1}(C^{-1} + DA^{-1}B - C^{-1} - DA^{-1}B)$$
$$= I$$

Therefore, the equality holds.
2. You can refer to the exercises in Chaps. 3 and 5.
3. $K = \begin{pmatrix} 0.7824 \\ 0.7518 \end{pmatrix}$.
4. The Python 3 code is given as follows:

```python
import numpy as np

# Define system matrices
A = np.array([[1, 1], [0, 1]])
H = np.array([[1, 0]])
Q = np.array([[0.001, 0], [0, 0.001]])
R = np.array([[0.01]])

# Initial state and covariance
x = np.array([[0], [0]])
P = np.eye(2)

# Measurements
z = np.array([1.0, 1.3, 1.6, 1.9])

# Lists to store results
estimated_states = []
estimated_covariances = []

# Kalman Filter
for measurement in z:
# Predict
x_pred = A @ x
P_pred = A @ P @ A.T + Q

# Update
y = measurement - H @ x_pred
S = H @ P_pred @ H.T + R
K = P_pred @ H.T @ np.linalg.inv(S)
```

```
            x = x_pred + K @ y
            P = (np.eye(2) - K @ H) @ P_pred

            # Store results
            estimated_states.append(x)
            estimated_covariances.append(P)

            # Print results
            print("Estimated States:")
            for i, state in enumerate(estimated_states):
                print(f"Step {i+1}: {state.T[0]}")

            print("\nEstimated Covariances:")
            for i, cov in enumerate(estimated_covariances):
                print(f"Step {i+1}:")
                print(cov)
```

5. Resampling is a critical component of particle filters (PF) that addresses the issue of particle degeneracy. The process involves selecting and replicating particles based on their weights, effectively discarding particles with low weights and multiplying those with high weights. This is typically achieved by computing normalized weights for all particles, creating a cumulative sum of these weights, and then using uniform random numbers to select particles for the next iteration. The necessity of resampling stems from the tendency of particle filters to suffer from degeneracy, where after several iterations, most particles have negligible weights and only a few dominate. This phenomenon is quantified by the effective number of particles, $N_{eff} = 1/\sum_{i=1}^{N}(w_i^*)^2$, which becomes much smaller than the total number of particles N in degenerate cases. Resampling allows the filter to concentrate computational effort on the most promising regions of the state space and maintain a diverse set of particles that better represent the posterior distribution. However, it's important to note that resampling introduces a trade-off between particle degeneracy and sample impoverishment. While addressing degeneracy, frequent resampling can lead to a loss of diversity in the particle set. Therefore, finding the right balance, often through adaptive resampling techniques, is crucial for the effective performance of particle filters in various applications, such as robot localization, target tracking, and financial modeling.

6. The Python 3 code is given in follows:

```
    import numpy as np
    import matplotlib.pyplot as plt

    # Set random seed for reproducibility
    np.random.seed(42)

    # Model parameters
    num_particles = 1000
```

```python
num_steps = 5

# Initialize particles and weights
particles = np.zeros((num_particles, num_steps))
weights = np.ones(num_particles) / num_particles

# True state and measurements
true_state = np.zeros(num_steps)
measurements = np.zeros(num_steps)

# Generate true state and measurements
for t in range(1, num_steps):
    true_state[t]=true_state[t-1]+np.random.normal(0, 1)
    measurements[t]=true_state[t]+np.random.normal(0, 1)

# Particle filter
for t in range(1, num_steps):
    # Predict
    particles[:, t] = particles[:, t-1]
        + np.random.normal(0, 1, num_particles)

    # Update weights
    weights *= np.exp(-0.5 * (measurements[t]
        - particles[:, t])**2)
    weights /= np.sum(weights)

    # Resample
    if 1. / np.sum(weights**2) < num_particles / 2:
        indices = np.random.choice(num_particles,
            num_particles, p=weights)
        particles[:, t] = particles[indices, t]
        weights = np.ones(num_particles) / num_particles

    # Estimate
    estimate = np.sum(particles[:, t] * weights)

    print(f"Step {t}:")
    print(f"  True state: {true_state[t]:.4f}")
    print(f"  Measurement: {measurements[t]:.4f}")
    print(f"  Estimate: {estimate:.4f}")
    print()

# Plot results
plt.figure(figsize=(10, 6))
plt.plot(range(num_steps), true_state,
```

```
        'r-', label='True State')
plt.plot(range(num_steps), measurements,
        'g*', label='Measurements')
plt.plot(range(num_steps), np.mean(particles, axis=0),
        'b--', label='PF Estimate')
plt.legend()
plt.xlabel('Time Step')
plt.ylabel('State')
plt.title('Particle Filter Estimation')
plt.show()
```

7. (a) Discrete-time state equations:
   Let $\Delta t = 0.1$ seconds be the sampling time, then the discrete-time state equations can be written as

$$\begin{bmatrix} x_1(k+1) \\ x_2(k+1) \\ x_3(k+1) \end{bmatrix} = \begin{bmatrix} x_1(k) \\ x_2(k) \\ x_3(k) \end{bmatrix}$$

$$+ \begin{bmatrix} \Delta t \cdot x_2(k) \\ \Delta t \cdot \left(x_2^2(k) - x_1(k) - 0.1x_3(k) + 0.5\cos(1.2k\Delta t)\right) \\ 0 \end{bmatrix}$$

(b) Calculate the Jacobians required for EKF:
    Jacobian of the state transition model:

$$\frac{\partial f}{\partial x} = \begin{bmatrix} 1 & \Delta t & 0 \\ -\Delta t & 1 + 2x_2(k)\Delta t & -0.1\Delta t \\ 0 & 0 & 1 \end{bmatrix}$$

where f is the state transition function

Jacobian of the measurement model:

$$\frac{\partial h}{\partial x} = \begin{bmatrix} 1 & 0 & 0 \end{bmatrix}$$

where h is the measurement function $z = x_1 + v$.

8. The Python 3 code is given as follows:

```
import numpy as np
from scipy.linalg import cholesky

def state_transition(x):
return np.sin(x)
```

11 Solutions

```
def measurement_model(x):
return x**2

def ukf(z, x0, P0, Q, R, n_steps):
n = 1  # state dimension
m = 1  # measurement dimension
alpha = 1e-3
beta = 2
kappa = 0
lambda_ = alpha**2 * (n + kappa) - n

# Weights for mean and covariance
Wm = np.full(2*n+1, 1/(2*(n+lambda_)))
Wm[0] = lambda_ / (n + lambda_)
Wc = Wm.copy()
Wc[0] += (1 - alpha**2 + beta)

x = x0
P = P0

x_history = [x]

for t in range(n_steps):
# Generate sigma points
L = cholesky((n + lambda_) * P)
X = np.zeros((2*n+1, n))
X[0] = x
for i in range(n):
X[i+1] = x + L[i]
X[n+i+1] = x - L[i]

# Prediction step
X_pred = np.array([state_transition(Xi) for Xi in X])
x_pred = np.sum(Wm[:, np.newaxis] * X_pred, axis=0)
P_pred = np.zeros((n, n))
for i in range(2*n+1):
diff = X_pred[i] - x_pred
P_pred += Wc[i] * np.outer(diff, diff)
P_pred += Q

# Update step
Z = np.array([measurement_model(Xi) for Xi in X_pred])
z_pred = np.sum(Wm[:, np.newaxis] * Z, axis=0)
Pzz = np.zeros((m, m))
Pxz = np.zeros((n, m))
```

```
        for i in range(2*n+1):
            diff_z = Z[i] - z_pred
            diff_x = X_pred[i] - x_pred
            Pzz += Wc[i] * np.outer(diff_z, diff_z)
            Pxz += Wc[i] * np.outer(diff_x, diff_z)
        Pzz += R

        K = Pxz @ np.linalg.inv(Pzz)
        x = x_pred + K @ (z[t] - z_pred)
        P = P_pred - K @ Pzz @ K.T

        x_history.append(x)

    return np.array(x_history)

# Set up the system
np.random.seed(42)
n_steps = 10
true_x = np.zeros(n_steps + 1)
true_x[0] = 1
z = np.zeros(n_steps)

for t in range(n_steps):
    true_x[t+1] = np.sin(true_x[t]) + 0.1 * np.random.randn()
    z[t] = true_x[t+1]**2 + np.random.randn()

# UKF parameters
x0 = np.array([1.0])
P0 = np.array([[1.0]])
Q = np.array([[0.01]])   # Process noise covariance
R = np.array([[1.0]])    # Measurement noise covariance

# Run UKF
estimated_x = ukf(z, x0, P0, Q, R, n_steps)

# Print results
print("True states:")
print(true_x)
print("\nEstimated states:")
print(estimated_x)
```

9. The Python 3 code is given as follows:

```
import numpy as np
import matplotlib.pyplot as plt
```

11 Solutions

```python
# System parameters
A = np.array([[-0.5, 1], [-1, -0.5]])
H = np.array([[1, 0]])
Q = np.eye(2) * 0.1
R = np.array([[0.1]])

# Simulation parameters
T = 10.0  # Total simulation time
dt = 0.01  # Time step
N = int(T / dt)  # Number of time steps
t = np.linspace(0, T, N)

# True system simulation
def true_system(x, dt):
dw = np.sqrt(dt) * np.random.randn(2)
return x + A @ x * dt + dw

x0 = np.array([1.0, 1.0])
x_true = np.zeros((N, 2))
x_true[0] = x0

for i in range(1, N):
x_true[i] = true_system(x_true[i-1], dt)

# Generate measurements
y = np.zeros(N)
for i in range(N):
y[i] = H @ x_true[i]
+ np.sqrt(R[0, 0]) * np.random.randn()

# Feedback Particle Filter
def fpf_update(x, y, K, dt):
dw = np.sqrt(dt) * np.random.randn(2)
dy = y - H @ x
return x + (A @ x + K @ dy) * dt + dw

# Continuous-time Kalman Filter
def kalman_filter_update(x, y, P, dt):
dx = (A @ x + P @ H.T @ np.linalg.inv(R)
@ (y - H @ x)) * dt
return x + dx

def riccati_update(P, dt):
dP = (A @ P + P @ A.T + Q
```

```python
        - P @ H.T @ np.linalg.inv(R) @ H @ P) * dt
    return P + dP

# Initialize filters
M = 1000  # Number of particles for FPF
x_fpf = np.random.randn(M, 2) + x0
x_kf = x0.copy()
P = np.eye(2)

# Simulate filters
x_fpf_mean = np.zeros((N, 2))
x_kf_est = np.zeros((N, 2))

for i in range(N):
    # FPF update
    K = np.cov(x_fpf.T) @ H.T @ np.linalg.inv(R)
    x_fpf = np.array([fpf_update(x, y[i], K, dt)
                      for x in x_fpf])
    x_fpf_mean[i] = np.mean(x_fpf, axis=0)

    # Kalman Filter update
    P = riccati_update(P, dt)
    x_kf = kalman_filter_update(x_kf, y[i], P, dt)
    x_kf_est[i] = x_kf

# Calculate MSE
mse_fpf = np.mean((x_fpf_mean - x_true)**2)
mse_kf = np.mean((x_kf_est - x_true)**2)

# Plot results
plt.figure(figsize=(12, 8))
plt.subplot(2, 1, 1)
plt.plot(t, x_true[:, 0], 'k-', label='True')
plt.plot(t, x_fpf_mean[:, 0], 'r--', label='FPF')
plt.plot(t, x_kf_est[:, 0], 'b--', label='KF')
plt.ylabel('State 1')
plt.legend()
plt.title('State Estimation Comparison')

plt.subplot(2, 1, 2)
plt.plot(t, x_true[:, 1], 'k-', label='True')
plt.plot(t, x_fpf_mean[:, 1], 'r--', label='FPF')
plt.plot(t, x_kf_est[:, 1], 'b--', label='KF')
plt.xlabel('Time')
plt.ylabel('State 2')
```

11 Solutions

```
plt.legend()

plt.tight_layout()
plt.show()

print(f"MSE for FPF: {mse_fpf}")
print(f"MSE for KF: {mse_kf}")

# Analyze the performance
fpf_error = np.abs(x_fpf_mean - x_true)
kf_error = np.abs(x_kf_est - x_true)

plt.figure(figsize=(12, 6))
plt.subplot(2, 1, 1)
plt.plot(t, fpf_error[:, 0], 'r-', label='FPF')
plt.plot(t, kf_error[:, 0], 'b-', label='KF')
plt.ylabel('Absolute Error (State 1)')
plt.legend()
plt.title('Absolute Error Comparison')

plt.subplot(2, 1, 2)
plt.plot(t, fpf_error[:, 1], 'r-', label='FPF')
plt.plot(t, kf_error[:, 1], 'b-', label='KF')
plt.xlabel('Time')
plt.ylabel('Absolute Error (State 2)')
plt.legend()

plt.tight_layout()
plt.show()
```

10. (a) The FPF formulation for this continuous multi-sensor setup is $dx_t^i = v_t \, dt + K_{t,1}(dz_{t,1} - x_t^i \, dt) + K_{t,2}(dz_{t,2} - x_t^i \, dt) + \sqrt{Q} \, dB_t^i$, where $K_{t,1} = \frac{P_t}{R_1}$ and $K_{t,2} = \frac{P_t}{R_2}$ are feedback gains for sensors 1 and 2, and P_t is the particle distribution variance at time t.

(b) Feedback gains $K_{t,1}$ and $K_{t,2}$ are inversely proportional to noise intensities R_1 and R_2. Higher noise intensity decreases gain, reducing sensor influence on state estimation. If $R_1 < R_2$, then $K_{t,1} > K_{t,2}$, giving sensor 1 more influence; if $R_1 > R_2$, then $K_{t,1} < K_{t,2}$, favoring sensor 2; if $R_1 = R_2$, then $K_{t,1} = K_{t,2}$, indicating equal influence.

(c) To implement and analyze the two-sensor FPF: initialize particles, propagate them using the process model, update with both sensors' measurements, and estimate state as particle distribution mean. Compare with single-sensor FPF by calculating mean squared error (MSE) and particle distribution variance. Benefits of two sensors include reduced estimation error, increased robustness, faster convergence, and improved handling of nonlinearities and non-Gaussian noise. Quantify these benefits by comparing MSE, convergence time, and particle distribution characteristics between single and two-sensor setups in various scenarios.

Problems of Chap. 10

1. The gradients of the functions are listed as follows:

    (1) $\nabla J(w) = a$;
    (2) $\nabla J(w) = (A + A^\top)w$;
    (3) $\nabla J(w) = 2w$;
    (4) $\nabla J(w) = \frac{w}{\|w\|_2}$;
    (5) $\nabla J(w) = f'(\|w\|_2)\frac{w}{\|w\|_2}$;
    (6) $\nabla J(w) = f'(w^\top a)a$.

2. The simulation result is omitted here. Readers can refer to [1] in Chap. 10 to find typical simulation results of this kind of filtering problems and compare their results with those in the paper.

Reference

1. X. Chen, Y. Tao, W. Xu, and S. S.-T. Yau. Recurrent neural networks are universal approximators with stochastic inputs. *IEEE Transactions on Neural Networks and Learning Systems*, 34(10):7992–8006, 2023.

Correction to: Principles of Nonlinear Filtering Theory

Stephen S.-T. Yau, Xiuqiong Chen, Xiaopei Jiao, Jiayi Kang, Zeju Sun, and Yangtianze Tao

Correction to:
S. S.-T. Yau et al., *Principles of Nonlinear Filtering Theory*, Algorithms and Computation in Mathematics 33, https://doi.org/10.1007/978-3-031-77684-7

After initial publication of the book, various errors were identified and they have been corrected and listed below.

FM: The below mentioned acknowledgements have been included in the preface.

Acknowledgments. We gratefully acknowledge the financial support from the National Natural Science Foundation of China (Grants Nos. 42450242, 12201631, 123B2020, 11471184, 11961141005) and the Tsinghua University Education Foundation. We extend our sincere gratitude to the anonymous reviewers for their insightful comments and constructive suggestions, which significantly improved the quality of this work. Special thanks to our colleagues, collaborators, and family members for their support and encouragement throughout this project.

1 Chapter 1. Probability theory

Revision 1.1 (Page 19).

- In equation (1.82), the left hand side should be corrected from

$$\varphi_{X_1+\cdots+X_n}(u_1,\cdots,u_n)$$

to

$$\varphi_{X_1,\cdots,X_n}(u_1,\cdots,u_n)$$

The updated version of this book can be found at
https://doi.org/10.1007/978-3-031-77684-7

- After Eq. (1.83), add the sentence "where $\varphi^{(k)}(0)$ denote the k-th order derivative of φ, $\varphi^{(k)} = \frac{d^k \varphi}{dx^k}$".
- In the right hand side of Eq. (1.84), phrase "for all λ" should be "for all u".

Revision 1.2 (Page 20).

- The right hand side of Eq. (1.86) should be updated as

$$\int_{-\infty}^{\infty} e^{iux} p_X(x) dx$$

- In Eq. (1.87), the right hand side should be

$$\int_{-\infty}^{\infty} e^{-iux} \varphi_X(u) du$$

- In Eq. (1.88), the left hand side should be updated from

$$\varphi_{X_1 + \cdots + X_n}(u_1, \cdots, u_n)$$

to

$$\varphi_{X_1, \cdots, X_n}(u_1, \cdots, u_n)$$

Revision 1.3 (Page 20). Supplement of proof of (iii). That means replacing original sentence "See [1] for the proof of (iii)." by the following
"

1. **Fourier Inversion:** For any continuity points $a < b$ of F_X:

$$F_X(b) - F_X(a) = \lim_{T \to \infty} \frac{1}{2\pi} \int_{-T}^{T} \frac{e^{-i\lambda a} - e^{-i\lambda b}}{i\lambda} \phi_X(\lambda) d\lambda$$

   The identical formula holds for F_Y using ϕ_Y.
2. **Equality on Continuity Points:** Since $\phi_X = \phi_Y$, we have:

$$F_X(b) - F_X(a) = F_Y(b) - F_Y(a)$$

   for all common continuity points $a < b$.
3. **Extension to All Points:**
   - Let $\mathcal{D}$ be the set of common continuity points of F_X and F_Y, which is co-countable.
   - Fix $x \in \mathbb{R}$. Take any sequence $\{d_n\} \subset \mathcal{D}$ with $d_n \to -\infty$. From Step 2:

$$F_X(x) - F_X(d_n) = F_Y(x) - F_Y(d_n)$$

   for any $x \in \mathcal{D}$.

- Taking $n \to \infty$ and using $F_X(-\infty) = F_Y(-\infty) = 0$:

$$F_X(x) = F_Y(x) \quad \text{for all } x \in \mathcal{D}$$

- For $x \notin \mathcal{D}$ (a discontinuity point), take $y_n \in \mathcal{D}$ with $y_n \downarrow x$:

$$F_X(x) = \lim_{n \to \infty} F_X(y_n) = \lim_{n \to \infty} F_Y(y_n) = F_Y(x)$$

by right-continuity and the equality on $\mathcal{D}$.

Thus, $F_X(x) = F_Y(x)$ for all $x \in \mathbb{R}$."

Revision 1.4 (Page 30). At the beginning of section 1.5.1, add the following part about Kolmogorov's Maximal Inequality and its proof.

"In order to make some preparations, we provide the following important ingredient about Kolmogorov inequality.

Theorem 1.1 (Kolmogorov's Maximal Inequality). *Let $X_1, \ldots, X_n$ be independent random variables with $\mathbb{E}[X_k] = 0$ and $\text{Var}(X_k) < \infty$ for each k. Then for any $a > 0$,*

$$P\left(\max_{1 \le k \le n} |S_k| \ge a\right) \le \frac{1}{a^2} \sum_{k=1}^{n} \text{Var}(X_k),$$

where $S_k = \sum_{i=1}^{k} X_i$.

Proof. We proceed with a stopping time argument:

1. **Define the stopping time**: Let $\tau = \inf\{k \le n : |S_k| \ge a\}$, with $\tau = \infty$ if the maximum is never exceeded.

2. **Decompose the variance**: Since $\mathbb{E}[S_n^2] \ge \mathbb{E}[S_\tau^2 \mathbf{1}_{\tau \le n}]$, we have:

$$\sum_{k=1}^{n} \text{Var}(X_k) = \mathbb{E}[S_n^2] \ge \mathbb{E}[S_\tau^2 \mathbf{1}_{\tau \le n}].$$

3. **Analyze the stopped process**: On $\{\tau \le n\}$, we have $|S_\tau| \ge a$ by definition. Thus:

$$\mathbb{E}[S_\tau^2 \mathbf{1}_{\tau \le n}] \ge a^2 P\left(\max_{1 \le k \le n} |S_k| \ge a\right).$$

4. **Combine the inequalities**:

$$\sum_{k=1}^{n} \text{Var}(X_k) \ge a^2 P\left(\max_{1 \le k \le n} |S_k| \ge a\right).$$

Rearranging gives the desired result. $\square$

"

2 Chapter 2. Stochastic process

Revision 2.1 (Page 41).

- The statement of Lemma 2.1 should be updated as:
  "**Lemma 2.1** If T and S are stopping times, then so are $T \wedge S$, $T \vee S$ and $T + S$."
- In the statement of Proposition 2.5, the second equation should be

$$\mathcal{G}_T = \{A \in \mathcal{F}_\infty | t > 0, A \cap \{T < t\} \in \mathcal{F}_t\}.$$

Revision 2.2 (Page 42). In the last line of this page, the euqation should be

$$\mathcal{F}_{S^+} = \bigcap_{1 \leq n} \mathcal{F}_{S_n^+}$$

Revision 2.3 (Page 43).

- At the top of this page, in statement (h) of Proposition 2.6, sentence "the restriction of Y to the set $\{T \leq t\}$ is $\mathcal{F}_T$-measurable"
- The sentence "And conversely, if $A \cap_n \mathcal{F}_{S_n^+}$, ..." should be corrected as "And conversely, if $A \in \cap_n \mathcal{F}_{S_n^+}$,".

Revision 2.4 (Page 57). In the Proposition 2.7, the original proof can be replaced by:

"**Proposition 2.7.** An increasing sequence $\{A_n, \mathcal{F}_n\}$ is predictable iff it is natural.
Proof of Proposition 2.7. Predictable $\implies$ Natural: A_n is $\mathcal{F}_{n-1}$ measurable, then

$$E[M_n A_n] = \sum_{i=1}^n E\left[E\left[M_n(A_i - A_{i-1})|\mathcal{F}_{i-1}\right]\right]$$

$$= \sum_{i=1}^n E\left[(A_i - A_{i-1})E[M_n|\mathcal{F}_{i-1}]\right]$$

$$= E\left[\sum_{i=1}^n M_{i-1}(A_i - A_{i-1})\right]$$

Natural $\implies$ Predictable: First we show by induction:

$$E\left[\sum_{i=1}^n A_i(M_i - M_{i-1})\right] = 0$$

When $n = 1$ this comes directly from the definition of naturality.
Suppose it holds for $n = k$, then for $n = k + 1$:

$$E\left[\sum_{i=1}^{k+1} A_i(M_i - M_{i-1})\right] - E\left[\sum_{i=1}^{k} A_i(M_i - M_{i-1})\right] = E[M_{k+1}A_{k+1} - M_k A_{k+1}]$$

$$= -E\left[\sum_{i=1}^{k} A_i(M_i - M_{i-1})\right] = 0$$

Then we have among the way

$$E[A_n M_n] = E[A_n M_{n-1}] = E[E[A_n|\mathcal{F}_{n-1}]M_{n-1}]$$

Hence for any bounded martingale M_n we have

$$E[(A_n - E[A_n|\mathcal{F}_{n-1}])M_{n-1}] = 0$$

Now we fix n and let $M_k = (A_n - E[A_n|\mathcal{F}_{n-1}])I_{k \geq n-1}$, we deduce

$$0 = E[(A_n - E[A_n|\mathcal{F}_{n-1}])M_{n-1}] = E[(A_n - E[A_n|\mathcal{F}_{n-1}])]^2$$

which implies A_n is $\mathcal{F}_{n-1}$ measurable. "

Revision 2.5 (Page 58). In the Remark 2.2 (i), the equation should be corrected from

$$I_t^{\pm}(\omega) := \int_{(0..t]} X_s^{\pm}(\omega) dA_s(\omega)$$

to

$$I_t^{\pm}(\omega) := \int_{(0,t]} X_s^{\pm}(\omega) dA_s(\omega)$$

Revision 2.6 (Page 61). In the Remark 2.3 (i), the sentence "but is not a martingale if $E[|Z|] = 1$" should be replaced by "but is not a martingale if $E[|Z|] = \infty$".

Revision 2.7 (Page 65).

- In the statement of Theorem 2.19, "Let X be in $\mathcal{F}_2$" should be "Let X be in $\mathcal{M}_2$".
- In the first line of proof of Theorem 2.19, "continuous process in $\mathcal{F}_2$" should be "continuous process in $\mathcal{M}_2$".

Revision 2.8 (Page 67). In the proof of Proposition 2.13, "$M_{t \wedge T}^2 - (\langle M, M \rangle_{t \wedge T})$" should be corrected as "$M_{t \wedge T}^2 - \langle M, M \rangle_{t \wedge T}$".

Revision 2.9 (Page 69).

- The second equation should be corrected as

$$E[M_{t \wedge T_n}^2] = E[\langle M, M \rangle_{t \wedge T_n}] < \infty$$

- The sentence "for every t_0" should be "for every $t \leq 0$".

3 Chapter 3. Stochastic differential equation

Revision 3.1 (Page 76). At the bottom of this page, the last equality should be

$$E\left[(I_t^M(X) - I_s^M(X))\Big|\mathcal{F}_s\right] = \sum_{i=0}^{\infty} \xi_i E\left[M_{t\wedge t_{i+1}} - M_{s\wedge t_{i+1}} - M_{t\wedge t_i} + M_{s\wedge t_i}\Big|\mathcal{F}_s\right].$$

Revision 3.2 (Page 82). Equation (3.19) should become:

$$\widetilde{E}_T[Y|\mathcal{F}_s] = \frac{1}{Z_s} E[YZ_t|\mathcal{F}_s], \text{ a.s. } P \text{ and } \widetilde{P}_T. \qquad (1)$$

Revision 3.3 (Page 83). In proof of Lemma 3.2, the second line of formula, towel rule is used however the symbol of expectation is lost. The correct one should be like

$$\widetilde{E}_T[1_A \frac{1}{Z_s} E[YZ_t|\mathcal{F}_s]] = E[1_A \frac{Z_T}{Z_s} E[YZ_s|\mathcal{F}_s]]$$
$$= E[E[1_A \frac{Z_T}{Z_s} E[YZ_s|\mathcal{F}_s]|\mathcal{F}_s]] \qquad (2)$$
$$= \cdots$$

Revision 3.4 (Page 97).

- Above the equation (3.26), the equation after "Then ..." should be replaced by

$$Z_t[\widetilde{W}_t^{(k)}\widetilde{W}_t^{(j)} - \langle W_t^{(k)}, W_t^{(j)}\rangle_t] = \int_0^t Z_s \widetilde{W}_s^{(k)} dW_s^{(j)} + \int_0^t Z_s \widetilde{W}_s^{(j)} dW_s^{(k)} + \cdots \qquad (3)$$

- After the sentence "To this end, we first notice that ...", where the definition of B_t should be revised as

$$B_t = \int_0^t (b(s, X_s^{(k)}) - b(s, X_s^{(k-1)})) ds \qquad (4)$$

Revision 3.5 (Page 98).

- On the top of the page, first equation should be replaced by

$$E\|B_t\|^2 \le t \cdot K^2 E \int_0^t \|X_s^{(k+1)} - X_s^{(k)}\|^2 ds.$$

- The equation above the Eq. (3.67) should be corrected as

$$E\left[\max_{0\le s\le t} \|X_s^{(k+1)} - X_s^{(k)}\|^2\right] \le L \int_0^t E[\|X_s^{(k)} - X_s^{(k-1)}\|^2] ds.$$

Revision 3.6 (Page 102). Eq. 3.69

$$\mathcal{A}_t f \triangleq \frac{1}{2} \sum_{i,j=1}^{d} a_{ij}(t,x) \frac{\partial^2}{\partial x_i \partial x_j} f + \sum_{i=1}^{d} b(t,x) \frac{\partial}{\partial x_i} f; \quad f \in C^2(R^d). \tag{5}$$

4 Chapter 4. Optimization

Revision 4.1 (Page 120). Equation (4.13) should be replaced by

$$\mathcal{I}(x) := \{i \mid g_i(x) = 0\} \tag{6}$$

Revision 4.2 (Page 122).

- Below Eq. (4.21), the sentence "Letting $d^k \to 0, \cdots$" should be corrected as "Letting $\delta^k \to 0, \cdots$".
- In the statement of Theorem 4.6 and its corresponding proof, two indices "(4.26)" should be replaced by "(4.25)".

Revision 4.3 (Page 134). In Eq. (4.57), "$A(X_k)$" should be corrected as "$A(x_k)$", i.e, Eq. (4.57) should formulate as

$$\begin{pmatrix} B(x_k, \mu_k) & -A(x_k)^T \\ A(x_k) & 0 \end{pmatrix} \begin{pmatrix} d_k \\ \bar{\mu}_k \end{pmatrix} = - \begin{pmatrix} \nabla f(x_k) \\ h(x_k) \end{pmatrix}. \tag{7}$$

Revision 4.4 (Page 135).

- In the quation (4.60), "$i \in I = \{1, 2, \cdots, m\}$" should be replaced by "$i \in I_x = \{1, 2, \cdots, m\}$".
- The original Lemma 4.2 has been rewritten as:
  Lemma 4.2. Let $B(x, \mu) > 0$ be positive definite. Thus d^* satisfies Eq. (4.57) if and only if that d^* is global minimum point of following strictly convex quadratic programming for any given pair (x, μ):

$$\begin{cases} \min_d & q(d) := \frac{1}{2} d^T B(x, \mu) d + \nabla f(x)^T d, \\ s.t. & h(x) + A(x) d = 0. \end{cases} \tag{8}$$

Proof. We use equivalence form to prove this lemma. Here first we claim that "d_k is global minimum point" is equivalent to "there exists pair (d^*, μ^*) satisfying Karush-Kuhn-Tucker (KKT) condition". the necessity is obvious because any local minimizer will satisfy Karush-Kuhn-Tucker condition. On the converse, since matrix $B(x, \mu) > 0$ is positive definite, there exists a unique global minimizer satisfying Karush-Kuhn-Tucker condition. Therefore, we shall demonstrate that Eq. (4.57) is equivalent to Karush-Kuhn-Tucker condition.

First of all, we notice Lagrange function

$$L(d, \lambda) = q_k(d) - \lambda^T (h(x_k) + A(x_k)d) \qquad (9)$$

with its gradient can be obtained.

$$\nabla L(d_k, \lambda_k) = \begin{pmatrix} \nabla_d L(d_k, \lambda_k) \\ \nabla_\lambda L(d_k, \lambda_k) \end{pmatrix} = 0 \qquad (10)$$

More precisely,

$$\begin{cases} B(x, \mu)d^* + \nabla f(x) - A(x)^T \mu^* = 0 \\ h(x) + A(x)d^* = 0 \end{cases} \qquad (11)$$

where (d^*, μ^*) denotes the Karush-Kuhn-Tucker points and corresponding Lagrange multiplier. This is directly noted the same as Eq. (4.57) and the desired result is obtained. □

- In Eq. (4.60) third line, notation of active set $\mathcal{I}$ should be $\mathcal{I}(x_k)$. The revised formula is updated as

$$\begin{aligned} \min_{d} \quad & \tfrac{1}{2} d^T B_k d + \nabla f(x_k)^T d \\ s.t. \quad & h_i(x_k) + \nabla h_i(x_k)^T d = 0, i \in E = \{1, 2, \cdots, l\} \\ & g_i(x_k) + \nabla g_i(x_k)^T d \geq 0, i \in \mathcal{I}(x_k) = \{1, 2, \cdots, m\}, \end{aligned} \qquad (12)$$

5 Chapter 5. Filtering equation

Revision 5.1 (Page 144). Below Eq.(5.10), sentence "where the result $\langle T, Y^n \rangle_t = T_t \langle X, M \rangle_t$ holds due to the Kunita-Watanabe identity" is possibly changed to "where the result $d\langle T, Y^n \rangle_t = T_t d\langle X, M \rangle_t$ holds due to the Kunita-Watanabe identity"

Revision 5.2 (Page 146).

- In the proof of Lemma 5.1, the equation below Eq. (5.17)

$$\sum_{i=1}^{m} \int_0^t (h(X_t)^i)^2 ds.$$

should be

Correction to: Principles of Nonlinear Filtering Theory

$$\sum_{i=1}^{m} \int_0^t (h(X_s)^i)^2 ds.$$

- In the phrase before Eq. (5.19), "...by the condition (5.15)." should be corrected by "...by the condition (5.16)."
- There is possible mistake about sign in Eq. (5.19), i.e., first plus on the right hand side should be minus. Corrected formula should be

$$\frac{T_t}{1+\epsilon T_t} = \frac{1}{1+\epsilon} - \sum_{i=1}^{m} \int_0^t \frac{T_t}{(1+\epsilon T_t)^2} h^i(X_s) dW_s^i$$

$$- \sum_{i=1}^{m} \int_0^t \frac{\epsilon T_t^2}{(1+\epsilon T_t)^3} h^i(X_t)^2 ds. \qquad (13)$$

Revision 5.3 (Page 147). Eq. (5.22) should be replaced from

$$d(T_t^{-1}) = -(T_t^{-1})^2 dT_t = T_t^{-1} \left(\sum_{i=1}^{m} h^i(X_t) dW_t^i \right) \text{ with } T_0^{-1} = 1, \qquad (14)$$

to

$$d(T_t^{-1}) = T_t^{-1} (\sum_{i=1}^{m} h^i(X_t) dW_t^i) + T_t^{-1} \sum_{i=1}^{m} (h^i)^2(X_t) dt, \text{ with } T_0^{-1} = 1, \qquad (15)$$

Revision 5.4 (Page 149).

- In the statement of Definition 5.1, "bound and measure" should be replaced by "bound and measurable".
- In statement of Lemma 5.2,

$$\rho_t(\varphi) = \tilde{E}[\tilde{T}_t \varphi(X)_t | \mathcal{Y}_t] \quad \tilde{P} - a.s.. \qquad (16)$$

should be corrected as

$$\rho_t(\varphi) = \tilde{E}[\tilde{T}_t \varphi(X_t) | \mathcal{Y}_t] \quad \tilde{P} - a.s.. \qquad (17)$$

- In the first line of proof of Lemma 5.2, "By construction $\{\zeta, t \geq 0\}$ is also RCLL." should be updated as "By construction $\{\zeta_t, t \geq 0\}$ is also RCLL."
- In the statement of Corollary 5.1, the formula

$$\pi_t(\varphi) = \frac{\rho_t(\varphi)}{\rho(1)} \quad \forall t \in [0, \infty). \qquad (18)$$

should read as

$$\pi_t(\varphi) = \frac{\rho_t(\varphi)}{\rho_t(1)} \quad \forall t \in [0, \infty). \tag{19}$$

- The proof of Eq. (5.34) "The result is a direct consequence of Definition 5.1." should be corrected as "The result is a direct consequence of Proposition 5.1."

Revision 5.5 (Page 150). Eq. (5.39) should be replaced from

$$\tilde{E}\left[\int_0^t \tilde{T}_t dM_s^\varphi | \mathcal{Y}_t\right] = 0. \tag{20}$$

to

$$\tilde{E}\left[\int_0^t \tilde{T}_s dM_s^\varphi | \mathcal{Y}_t\right] = 0. \tag{21}$$

Revision 5.6 (Page 153). Eq. (5.48)

$$E[\pi_r(h)|\mathcal{Y}_s] = E[E[h_r(X_r)|\mathcal{Y}_r]|\mathcal{Y}_s] = E[h(X_r)|\mathcal{Y}_s], \tag{22}$$

should be corrected as

$$E[\pi_r(h)|\mathcal{Y}_s] = E[E[h(X_r)|\mathcal{Y}_r]|\mathcal{Y}_s] = E[h(X_r)|\mathcal{Y}_s], \tag{23}$$

Revision 5.7 (Page 161). Eq. (5.88) should be updated as

$$\tilde{T}_t = \exp\left(\int_0^t h(X_s)^\top dY_s - \frac{1}{2}\int_0^t \|h(X_s)\|^2 ds\right). \tag{24}$$

after changing the subscripts of the integrated variable.

6 Chapter 6. Estimation algebra

Revision 6.1 (Page 175). In the proof of Lemma 6.1, "$\frac{\partial^s a_{j_1,\cdots,j_n}}{x_k^s} \neq 0, \quad \forall s \geq 0$" should be "$\frac{\partial^s a_{j_1,\cdots,j_n}}{\partial x_k^s} \neq 0, \quad \forall s \geq 0$". "$\frac{\partial a_{j_1,\cdots,j_n}}{\partial x_1} + \frac{\partial a_{j_1+1,j_2-1,\cdots,j_n}}{\partial x_2} + \frac{\partial a_{j_1+1,j_2,\cdots,j_n-1}}{\partial x_n}$" should be replaced by "$\frac{\partial a_{j_1,\cdots,j_n}}{\partial x_1} + \frac{\partial a_{j_1+1,j_2-1,\cdots,j_n}}{\partial x_2} + \cdots + \frac{\partial a_{j_1+1,j_2,\cdots,j_n-1}}{\partial x_n}$"

Revision 6.2 (Page 176). In printed book, left hand side of Eq. (6.15) (6.16) and (6.17) are missing which should be $[gD_i, hD_j]$, $[gD_i^2, h]$ and $[D_i^2, hD_j]$.

Revision 6.3 (Page 177). Below Eq. (6.18), notation "... for $1 \leq x_{i_1} \cdots \partial x_{i_k} \leq n$, ..." should be replaced by "... for $1 \leq i_1, \cdots, i_k \leq n$, ..."

Correction to: Principles of Nonlinear Filtering Theory

Revision 6.4 (Page 180). In Eq. (6.35) the summation index in the first summation should be better to use i instead of k. Therefore, Eq. (6.35) should be replaced from "$p_i = \sum_{i=k_{i-1}+1}^{k_i} x_i^2$" to "$p_i = \sum_{j=k_{i-1}+1}^{k_i} x_j^2$".

Revision 6.5 (Page 186). In the first paragraph of the proof of Lemma 6.9, "$\alpha_i = \frac{1}{2}[[L_0, D_j], p_0] \in E$" should read as "$\alpha_j = \frac{1}{2}[[L_0, D_j], p_0] \in E$".

Revision 6.6 (Page 187). Eq. (6.65) should be updated from

$$\alpha_i(0, \bar{X}_1) = 0 \quad \forall i \in S_i. \tag{25}$$

to

$$\alpha_i(0, \bar{X}_1) = 0 \quad \forall i \in S_1. \tag{26}$$

Revision 6.7 (Page 191). In the proof of Lemma 6.12, "from (vi) of lemma 6.3" should be corrected as "from Lemma 6.3 (6)".

Revision 6.8 (Page 195). In the statement of Theorem 6.13, "for some $p < p$" should be "for some $p < q$".

Revision 6.9 (Page 210).

- In Eq. (6.147) right hand side, "mod $U_k - 1$" should be replaced by "mod U_{k-1}".
- Eq. (6.148) should be updated as

$$Z_r = \sum_{j=0}^{l-r} c_{rj} \sum_{i=0}^{k} b_i^{(j)} D_1^{k-i} D_2^i \in E, \text{ mod } U_{k-1}, \tag{27}$$

and sentence "(1) $Z_0 = y$, i.e., $c_{0j} = 1, 1 \leq j \leq l$;" should be "(1) $Z_0 = Y$, i.e., $c_{0j} = 1, 0 \leq j \leq l$;"

Revision 6.10 (Page 214). In the proof of Theorem 6.29, sentence "If $c_1 = 0$ and $c2 \neq 0$, from $2ac_2x_2 + 2ac_0x_2$ and $ax_1^2 + ax_2^2 + d$" should be "If $c_1 = 0$ and $c_2 \neq 0$, from $2ac_2x_2^2 + 2ac_0x_2$ and $ax_1^2 + ax_2^2 + d$".

Revision 6.11 (Page 215). In the proof of Lemma 6.23, "...η is a polynomial in x1..." should be "...η is a polynomial in x_1...".

Revision 6.12 (Page 224). After Eq. (6.197),

$$=(?ax_1 + dx_2)D_1 + (2bx_2 + dx_!)D_2 + (2cx_3 + g)D_3 + a + b + c \in E \tag{28}$$

should be

$$=(2ax_1 + dx_2)D_1 + (2bx_2 + dx_1)D_2 + (2cx_3 + g)D_3 + a + b + c \in E \tag{29}$$

Revision 6.13 (Page 225). In Eq. (6.209), last line,

$$N_n = [M_n, N_0] = 2^{2n}(x_3 + g)^2 D_2^{n+1}, \text{mod } U_1 \in E.$$

should be

$$N_n = [M_n, N_0] = 2^{2n}(x_3 + g)^2 D_2^{n+1}, \text{mod } U_n \in E.$$

Revision 6.14 (Page 227). After Eq. (6.217), "...lower order in $D3$." should be "lower order in D_3.".

Revision 6.15 (Page 228). Eq. (6.220) "$A := \ldots$" should be "$A_1 := \ldots$". "That is, $b_i = b_{i+1} = b_{i+1} = 0, 0 \leq i \leq k-2$," should be "That is, $b_i = b_{i+1} = b_{i+2} = 0, 0 \leq i \leq k-2$,".

Revision 6.16 (Page 229). After Eq. (6.227), "If $c \neq 0$, then $[D_1, a_l D_3^{l+1}] = c_1 D_3^{l+1}$, mod $U_l \in E$." should be "If $c_1 \neq 0$, then $[D_1, a_l D_3^{l+1}] = c_1 D_3^{l+1}$, mod $U_l \in E$.".

Revision 6.17 (Page 231). The sentence "... and $K, B.C$ satisfy ..." should be "...and K, B, C satisfy ...".

Revision 6.18 (Page 232).

- After Eq. (6.236), "assume that $c1 \neq 0$" should be "assume that $c_1 \neq 0$".
- Below Eq. (6.239), "infinite sequence $\{Bn\}$ in E," should be "infinite sequence $\{B_n\}$ in E,".

Revision 6.19 (Page 233). In the statement of Lemma 6.35, "Case (A): ... $\psi^{(2)} = cst \cdot x_3^2$." should be "Case (A): ...$\phi^{(2)} = cst \cdot x_3^2$.".

Revision 6.20 (Page 235).

- In Eq. (6.248), third line, "$\ldots \Longrightarrow D_3^2$, mod $U_0 \in E$" should be "$\ldots \Longrightarrow K = \frac{1}{2} D_3^2$, mod $U_0 \in E$"
- In Eq. (6.249), last equality, "$[D_2, [M, D_2]]$" should be "$[D_2, [M, D_1]]$".

Revision 6.21 (Page 237).

- In Eq.(6.256), first equality, "$ex_1^2 + fx_3^2 + gx_3^2 \in E$" should be "$ex_1^2 + fx_3^2 + gx_3 \in E$"
- In Eq. (6.257) last equality, "$\alpha_0 x_3 \in E \Longrightarrow \alpha_0$." should be "$\alpha_0 x_3 \in E \Longrightarrow \alpha_0 = 0$."

Revision 6.22 (Page 239).

- After Eq. (6.265), "Then we assume $\eta = a_3 x_1^3 + a_2 x_1^2 + a_2 x_1^2 + a_0$," should be "Then we assume $\eta = a_3 x_1^3 + a_2 x_1^2 + a_1 x_1 + a_0$,".
- In Eq. (6.266), second line, "$-\omega_{23}^2 - \frac{\beta_2}{2} \left(\frac{\partial a_0}{\partial x_2} + \frac{\partial a_1}{\partial x_2} x_1 \right)$" should be "$-\omega_{23}^3 - \frac{\beta_2}{2} \left(\frac{\partial a_0}{\partial x_2} + \frac{\partial a_1}{\partial x_2} x_1 \right)$".

Revision 6.23 (Page 243). Sentence "(i) $f_i = a_i x_1 + \phi(x_2, x_3)$, $1 \leq i \leq 3$," should be "(i) $f_i = a_i x_1 + \phi_i(x_2, x_3)$, $1 \leq i \leq 3$,".

Revision 6.24 (Page 249). Eq. (6.304) should be replaced by

$$\begin{cases} dy_1 = H_1^T x dt + dv_1 = (x_2 - x_3) dt + dv_1, \\ dy_2 = H_2^T x dt + dv_2 = (x_2 - x_4) dt + dv_2, \\ dy_3 = H_3^T x dt + dv_3 = (x_2 - x_5) dt + dv_3, \\ \cdots \\ dy_{n-2} = H_{n-2}^T x dt + dv_{n-2} = (x_2 - x_n) dt + dv_{n-2}, \end{cases} \quad (30)$$

Revision 6.25 (Page 253). Before Eq. (6.323), "$\Gamma = a_1 \Omega + a_2 J_\eta a_3 I, a_i \in \mathbb{R}$" should be "$\Gamma = a_1 \Omega + a_2 J_\eta + a_3 I, a_i \in \mathbb{R}$".

7 Chapter 7. Yau-Yau algorithm

Revision 7.1 (Page 276). Eq. (7.42) should be corrected as

$$\left| \frac{m^2}{2} |x|^{2m-2} - \frac{m(m+n-2)}{2} |x|^{m-2} - m|x|^{m-2} (f - \nabla K) \cdot x \right| \\ \leq \frac{m(m+1)}{2} |x|^{2m-2} + c'', \quad (31)$$

Revision 7.2 (Page 277). In equation after Eq. (7.45), the last line should be replaced by

$$+ \int_{B_{r_0}} \left(-\frac{\Delta K}{2} - \frac{1}{2} |h|^2 - f \cdot \nabla K + \frac{1}{2} |\nabla K|^2 \right) \phi u_r$$

Revision 7.3 (Page 278). The equation in the last line of Page 278 should be corrected:

$$\int_{B_r} u_r(t, x) \leq e^{c_1 t} \int_{B_r} u_r(0, x), \ 0 \leq t \leq T$$

Revision 7.4 (Page 284). Before Eq. (7.65), "$H_1(x) = 2x$" should be "$\mathcal{H}_1(x) = 2x$".

Revision 7.5 (Page 286).

- Eq. (7.71) should be

$$\|x^{r_1}\partial_x^{r_2}u\|^2 \leq C\alpha^{-2r_1-1}\max\{(\alpha\beta)^{2r_1},1\}\|u\|_{r_1+r_2}^2 \tag{32}$$

- Eq. (7.72) should be "$\|x^{r_1}\partial_x^{r_2}u\|^2 = ...$".

Revision 7.6 (Page 288). Right hand side of Eq. (7.82) should be

$$|P_Nu_x-(P_Nu)_x|_{\mu-1}^2 \leq C\alpha^{-2r-1}N^{-r}\lambda_{N+1}^\mu\|u\|_r^2 \leq C\alpha^{2\mu-2r-1}N^{\mu-r}\|u\|_r^2. \tag{33}$$

Revision 7.7 (Page 290). Sentences "From Lemma 7.3" should be "From Theorem 7.8".

Revision 7.8 (Page 292). In algorithm 1, step 3, "the nuw observation" should be updated as "the new observation".

8 Chapter 8. Direct method.

Revision 8.1 (Page 304). In the second line of Theorem 8.4, "$u(t,x)$ is a solution of (3.1)" should be "$u(t,x)$ is a solution of (8.17)".

Revision 8.2 (Page 305). Below Eq. (8.30), "we can calculate (3.4) and (3.5)" should be "we can calculate (8.22)".

Revision 8.3 (Page 308). Eq. (8.52) should be updated as:

$$\begin{cases} c'(t) = -\frac{1}{2}\sum_{i=1}^m (b_i'(t))^2 + \sum_{i=1}^n a_i(t)b_i'(t) - \sum_{i=1}^n d_ib_i'(t) \\ +\frac{1}{2}\sum_{i,j=1}^n \eta_{ij}b_i(t)b_j(t) + \frac{1}{2}\sum_{i=1}^n \eta_i b_i \\ c(0) = 0 \end{cases} \tag{34}$$

Revision 8.4 (Page 309). In the first line of Eq. (8.56), there is an extra minus sign "$c'(t) - -\frac{1}{2}\sum_{i=1}^n(b_i')^2...$" which should be corrected as "$c'(t) - \frac{1}{2}\sum_{i=1}^n(b_i')^2$".

Revision 8.5 (Page 311).

- In the last equation of Eq. (8.63), parentheses are missing and correct one should be

$$\frac{\partial^2 \hat{u}_k}{\partial x_i^2} = e^{\Lambda(x)}\left\{\left(\frac{\partial^2\Lambda(x)}{\partial x_i^2}+\left(\frac{\partial\Lambda(x)}{\partial x_i}\right)^2\right)\tilde{u}_k + 2\frac{\partial\Lambda(x)}{\partial x_i}\frac{\partial\tilde{u}_k}{\partial x_i}+\frac{\partial^2\tilde{u}_k}{\partial x_i^2}\right\} \tag{35}$$

- Eq. (8.60) should be updated as:

$$\theta(x) = \frac{1}{2}(\sum_{i=1}^{n}\theta_i^2(x) + \sum_{i=1}^{n}\frac{\partial\theta_i}{\partial x_i} - \eta(x))$$

$$= \frac{1}{2}\sum_{i=1}^{n}f_i^2(x) - \frac{1}{2}\eta + \sum_{i=1}^{n}\frac{\partial\theta_i}{\partial x_i} + \frac{1}{2}\sum_{i=1}^{n}\frac{\partial f_i}{\partial x_i} \quad (36)$$

$$= -(\sum_{i=1}^{n}\frac{\partial f_i}{\partial x_i} + \frac{1}{2}\sum_{i=1}^{m}h_i^2).$$

- Eq. (8.61) should be replaced by

$$\begin{cases} \dfrac{\partial \hat{u}_k}{\partial t}(t,x) = \dfrac{1}{2}\Delta\hat{u}_k(t,x) - \sum_{i=1}^{n}H_i(x)\dfrac{\partial \hat{u}_k}{\partial x_i} - P(x)\hat{u}_k(t,x) \\ \hat{u}_k(\tau_{k-1},x) = e^{\Lambda(x)}\tilde{u}_k(\tau_{k-1},x), \end{cases} \quad (37)$$

Revision 8.6 (Page 312).

- After Eq. (8.67), "coefficients of $\hat{u}_k(t,x)$, $\frac{\partial \hat{u}_k}{\partial x_i}$ are degree one polynomial and degree two polynomial" should be "coefficients of $\frac{\partial \hat{u}_k}{\partial x_i}$, $\hat{u}_k(t,x)$ are degree one polynomial and degree two polynomial".
- The first equation in Eq. (8.64) should be replaced by "$\theta_i(x) = \frac{\partial \Lambda}{\partial x_i} - H_i(x)$".

Revision 8.7 (Page 317). Eq. (8.89) should be replaced by

$$q(B^{-1}(t)\tilde{x} - B^{-1}(t)b(t)) = \tilde{x}^T B(t)QB(t)^T \tilde{x} - [2b(t)^T B(t)QB(t)^T - P^T B(t)^T]\tilde{x}$$
$$+ b(t)^T B(t)QB(t)^T b(t) - P^T B(t)^T b(t) + r. \quad (38)$$

Revision 8.8 (Page 318).

- After (8.90), "$\tilde{b}(t) = (\tilde{b}_{ij}(t))$" should be "$\tilde{B}(t) = (\tilde{b}_{ij}(t))$"
- Eq. (8.92) should be

$$\frac{d\tilde{B}(t)}{dt} = 2\tilde{A}(t)\tilde{B}(t) \quad (39)$$

- Eq. (8.94) should be updated as

$$\frac{d\tilde{D}(t)}{dt} = 2\tilde{A}(t)\tilde{D}(t) + \tilde{B}(t)Q\tilde{B}(t)^T b(t) - \frac{1}{2}\tilde{B}(t)P \quad (40)$$

- Eq. (8.97) should be

$$\frac{\partial K}{\partial t}(t,\tilde{x},\tilde{y}) = [\tilde{x}^T \frac{d\tilde{A}}{dt}\tilde{x} + \tilde{x}^T \frac{d\tilde{B}}{dt}\tilde{y} + \tilde{y}^T \frac{d\tilde{C}}{dt}\tilde{y} + \frac{d\tilde{D}^T}{dt}\tilde{x} + \frac{d\tilde{E}^T}{dt}\tilde{y}$$
$$+ \frac{ds}{dt} - \frac{n}{2t}]K(t,\tilde{x},\tilde{y}) \quad (41)$$

- Eq. (8.99) should be

$$\frac{1}{2}\Delta_{\tilde{x}}K(t,\tilde{x},\tilde{y}) = [\frac{1}{2}\tilde{x}^T(\tilde{A}+\tilde{A}^T)^2\tilde{x} + \tilde{x}^T(\tilde{A}+\tilde{A}^T)\tilde{B}\tilde{y}$$
$$+ \frac{1}{2}\tilde{y}\tilde{B}^T\tilde{B}\tilde{y} + \tilde{x}^T(\tilde{A}+\tilde{A}^T)\tilde{D} + \tilde{y}^T\tilde{B}^T\tilde{D} + \frac{1}{2}\tilde{D}^T\tilde{D}$$
$$+ tr(\tilde{A})]K(t,\tilde{x},\tilde{y}). \quad (42)$$

Revision 8.9 (Page 320).

- The first equation in Eq. (8.105) should be

$$-\tilde{E}_{-1} = \tilde{B}_{-1}^T \tilde{D}_{-1}$$

- The third equation of Eq. (8.106) should be

$$s_1 = \frac{1}{2}(\tilde{D}_{-1}^T \tilde{D}_1 + \tilde{D}_0^T \tilde{D}_0 + \tilde{D}_1^T \tilde{D}_{-1}) + \text{tr}\tilde{A}_0$$
$$- \frac{1}{2}(b^{0^T} B_0 Q B_0^T b^0 - P^T B_0^T b^0 + r) \quad (43)$$

Revision 8.10 (Page 321).

- Eq. (8.108) should be corrected as:

$$q(B^{-1}(t)\tilde{x} - B^{-1}(t)b(t)) = \tilde{x}^T B(t) Q B(t)^T \tilde{x} - [2b(t)^T B(t) Q B(t)^T - P^T B(t)^T]\tilde{x}$$
$$+ b(t)^T B(t) Q B(t)^T b(t) - P^T B(t)^T b(t) + r \quad (44)$$

- The first equation of Eq. (8.110) should be

$$\tilde{A}_1 = -\frac{1}{6}B_0 Q_0 B_0^T$$

Revision 8.11 (Page 324). Eq. (8.121) should be

$$q(B^{-1}(t)\tilde{x} - B^{-1}(t)b(t)) = \tilde{x}^T B(t) Q B(t)^T \tilde{x} - [2b(t)^T B(t) Q B(t)^T - P^T B(t)^T]\tilde{x}$$
$$+ b(t)^T B(t) Q B(t)^T b(t) - P^T B(t)^T b(t) + r \quad (45)$$

Correction to: Principles of Nonlinear Filtering Theory

Revision 8.12 (Page 326). Eq. (8.128) should be replaced by

$$\frac{\partial u}{\partial t}(t,x) = \frac{1}{2}\sum_{i,j=1}^{n} G_{ij}(t)\frac{\partial^2 u}{\partial x_i \partial x_j}(t,x) - \sum_{i=1}^{n} f_i \frac{\partial u}{\partial x_i}(t,x)$$
$$+ \left(-\frac{\partial}{\partial t}\left(h^T S^{-1}\right)^T y_t\right)$$
$$+ \frac{1}{2}\sum_{i,j=1}^{n} G_{ij}(t)\left[\frac{\partial^2 \tilde{K}}{\partial x_i \partial x_j} + \frac{\partial \tilde{K}}{\partial x_i}\frac{\partial \tilde{K}}{\partial x_j}\right]$$
$$- \sum_{i=1}^{n} f_i \frac{\partial \tilde{K}}{\partial x_i}(t,x) - \sum_{i=1}^{n} \frac{\partial f_i}{\partial x_i}(t,x)$$
$$- \frac{1}{2}\left(h^T S^{-1} h\right) u(t,x)$$
$$u(0,x) = \sigma_0(x) \qquad (46)$$

Revision 8.13 (Page 330). In Eq. (8.147), the fourth equation should be corrected as:

$$\frac{d\tilde{D}}{dt}(t) = 2\tilde{A}\tilde{D} + F^T(t)Q(t)F(t)b(t) - \frac{1}{2}F^T(t)p(t)$$

Revision 8.14 (Page 331). The last equation in this page should be corrected as:

$$\tilde{D}_{n+1} = \frac{2}{n+2}\sum_{i=0}^{n+1}\tilde{A}_i\tilde{D}_{n-i} - \frac{1}{2(n+2)}\sum_{i=0}^{n} F_i^T p_{n-i} + \frac{1}{n+2}\sum_{j=0}^{n}\sum_{i=0}^{j}\sum_{l=0}^{i} F_l^T Q_{i-l} F_{j-l} b_{n-j}$$

9 Chapter 9. Classical filtering methods

Revision 9.1 (Page 351). At the beginning of section 9.2.5.2, "But the posterior distribution $pp(X_k \mid Y_{1:k})$ is very complex." should be updated as "But the posterior distribution $p(X_k \mid Y_{1:k})$ is very complex."

10 Chapter 10. Estimation algorithm based on deep learning

Revision 10.1 (Page 387). After Eq. (10.3), The phrase "... nonlinear ovservation on W_k" should read as "... nonlinear ovservation on $\mathbf{w}_k$".

Revision 10.2 (Page 403). In the statement of Lemma 10.3, "$\mathcal{T}_K$ is defined in (2)." should be replaced by "$\mathcal{T}_K$ is defined in (10.47)."

Revision 10.3 (Page 405). After equation (10.62), "are the state and the output of the RNN system (24)" should be "are the state and the output of the RNN system (10.58)".

Revision 10.4 (Page 408). In the first point of the Remark 10.3, "$\sup_{k\geq 1} \mathbb{E}[|a_k|^{p_2}] < \infty$" should be "$\sup_{k\geq 1} \mathbb{E}[|\alpha_k|^{p_2}] < \infty$".

Revision 10.5 (Page 415). Eq. (10.105) should become

$$\tilde{u}_{i+1}(\tau_i, x) = \exp\left(\sum_{k=1}^{m}(y_{\tau_i}^k - y_{\tau_{i-1}}^k)h_k(x)\right)\tilde{u}_i(\tau_i, x)$$

$$\approx \sum_{l=1}^{M}\sum_{j=1}^{M} a_{i,l}(\tau_{i-1})d_{l,j}\exp\left(\sum_{k=1}^{m}(y_{\tau_i}^k - y_{\tau_{i-1}}^k)h_k(x)\right)\phi_j(x) \quad (47)$$

$$\approx \sum_{l=1}^{M}\sum_{j=1}^{M}\sum_{j_1=1}^{M} a_{i,l}(\tau_{i-1})d_{l,j}r_{j,j_1}(y_{\tau_i} - y_{\tau_{i-1}})\phi_{j_1}(x),$$

Revision 10.6 (Page 416).

- Eq. (10.106) should be

$$r_{j,j_1}(y_{\tau_i} - y_{\tau_{i-1}}) = \left\langle \exp\left(\sum_{k=1}^{m}(y_{\tau_i}^k - y_{\tau_{i-1}}^k)h_k(x)\right)\phi_j, \phi_{j_1}\right\rangle, \quad 1 \leq j, j_1 \leq M.$$
(48)

- The second equation in Eq. (10.110) should be

$$\beta_{1,j} = \int \phi_j(x)\,dx, \quad 1 \leq j \leq M.$$

Revision 10.7 (Page 418). The sentence "... which depends on T, R ..." should be "... which depends on T, r ...".

Revision 10.8 (Page 419). Eq. (10.115) should become

$$b_i(x) = 1 + \sum_{k=1}^{m} h_k(x)(y_{\tau_i}^k - y_{\tau_{i-1}}^k) + \frac{1}{2}\sum_{j=1}^{m}\sum_{k=1}^{m} h_j(x)h_k(x)(y_{\tau_i}^k - y_{\tau_{i-1}}^k)(y_{\tau_i}^j - y_{\tau_{i-1}}^j).$$

Correction to: Principles of Nonlinear Filtering Theory

Revision 10.9 (Page 420). Eq. (10.119) should become

$$S_T = \left\{ \left(\prod_{j=1}^{i-1} (H_j U_\delta) \right) \sigma_0 : H_j = 1,\ h_{j_1},\ h_{j_1} h_{j_2},\ 1 \leq j_1, j_2 \leq m,\ i = 1, \ldots, N \right\}$$

Revision 10.10 (Page 421). The right-hand side of the first equality of Eq. (10.123) should become

$$\lambda_{i+1,s} = \sum_{j=1}^{M} \lambda_{i+1,j} \langle \phi_j, \phi_s \rangle$$

The manufacturer's authorised representative in the EU is Springer Nature Customer Service Centre GmbH, Europaplatz 3, 69115 Heidelberg, Germany. If you have any concerns regarding our products, please contact ProductSafety@springernature.com

Printed and bound by CPI Group (UK) Ltd, Croydon, CR0 4YY

26/03/2026

02078939-0012